Beaches and Coasts

Richard A. Davis Jr and *Duncan M. FitzGerald*

BLACKWELL PUBLISHING
350 Main Street, Malden, MA 02148-5020, USA
9600 Garsington Road, Oxford OX4 2DQ, UK
550 Swanston Street, Carlton, Victoria 3053, Australia

First published 2004 by Blackwell Science Ltd

2 2005

Library of Congress Cataloging-in-Publication Data

Davis, Richard.
Beaches and coasts / by Richard Davis Jr. and Duncan FitzGerald.
p. cm.
Includes bibliographical references (p.).
ISBN 0-632-04308-3 (alk. paper)
1. Beaches. 2. Coasts. 3. Coast changes. I. FitzGerald, Duncan. II. Title.

GB451.2 .D385 2003
551.45′7—dc21

2002151119

ISBN-13: 978-0-632-04308-8

A catalogue record for this title is available from the British Library.

Set in 10/12 pt Caslon
by Graphicraft Ltd, Hong Kong
Printed and bound in the United Kingdom
by The Bath Press

The publisher's policy is to use permanent paper from mills that operate a sustainable forestry policy, and which has been manufactured from pulp processed using acid-free and elementary chlorine-free practices. Furthermore, the publisher ensures that the text paper and cover board used have met acceptable environmental accreditation standards.

For further information on
Blackwell Publishing, visit our website:
www.blackwellpublishing.com

Brief contents

Contents

Acknowledgments

The authors and publisher gratefully acknowledge the permission granted to reproduce the copyright material in this book.

Fig. 2.21: from G. Plafker et al. 1969. Effects of the earthquake of March 27, 1964, on various communities. USGS Professional Paper 542-G.

Figs 3.8, 4.19, 8.4, 9.38, and 10.7: from R. A. Davis, *Geology of Holocene Barrier Island Systems*. © 1994 Springer-Verlag.

Figs 4.2 and 4.3: from P. D. Komar & D. B. Enfield, Short-term sea-level changes and coastal erosion. In D. Nummedal, O. H. Pilkey & J. D. Howard (eds), *Sea-level Fluctuation and Coastal Evolution*. SEPM Spec. Publ. No. 41, 1987, p. 18. © SEPM.

Fig. 4.7: from B. J. Skinner & S. C. Porter, *Dynamic Earth*. © 1989 Wiley. This material is used by permission of John Wiley and Sons, Inc.

Fig. 4.13: modified from V. Gornitz & S. J. Lebedeff, Global sea-level changes during the past century. In D. Nummedal, O. H. Pilkey & J. D. Howard (eds), *Sea-level Fluctuation and Coastal Evolution*. SEPM Spec. Publ. No. 41, 1987, p. 10. © SEPM.

Fig. 7.1: from R. A. Davis, *Coastal Sedimentary Environments*, p. 386, Fig. 6.5. © 1985 Springer-Verlag.

Fig. 7.6: from H. E. Clifton, R. E. Hunter & R. L. Phillips, Depositional structures and processes in the non-barred high-energy nearshore. *J. Sed. Petrology*, 1971, vol. 41, p. 722. © SEPM.

Fig. 7.15: from L. D. Wright, J. Chappell, B. G. Thom, M. P. Bradshaw & P. J. Cowell, Morphodynamics of reflective and dissipative beach and inshore systems, Southeastern Australia. *Mar. Geol.*, 1979, vol. 32, pp. 105–40, Fig. 6. With permission from Elsevier Science.

Fig. 7.25: modified from R. A. Davis, W. T. Fox, M. O. Hayes & J. C. Boothroyd, Comparison of ridge and runnel systems in tidal and non-tidal environments. *J. Sed. Petrology*, 1972, vol. 42, p. 426. © SEPM.

Fig. 8.3: reprinted from M. O. Hayes, 1979, Barrier island morphology as a function of tidal and wave regime. In S. P. Leatherman (ed.), *Barrier Islands*. New York: Academic Press, pp. 1–28. Copyright © 1979, with permission from Elsevier Science.

Fig. 8.27: from D. M. FitzGerald, Geomorphic variability and morphologic and sedimentologic controls on tidal inlets. *Journal of Coastal Research*, vol. SI23, pp. 47–71, Fig. 4. © Coastal Education & Research Foundation. Reproduced with permission.

Fig. 8.28: from R. Boyd, A. J. Bowen & R. K. Hall, 1987, An evolutionary model for transgressive sedimentation on the eastern shore of Nova Scotia. In D. M. FitzGerald & P. S. Rosen (eds), *Glaciated Coasts*. New York: Academic Press, pp. 87–114, Fig. 13. Copyright © 1987. Reprinted with permission from Elsevier Science.

Fig. 12.32: from P. L. Barnard & R. A. Davis, Jr, Anthropogenic vs. natural influences on inlet evolution: west-central Florida. In N. C. Kraus & W. G. McDougal (eds), *Coastal Sediments '99, vol. 2.* Proceedings of the 4th International Symposium on Coasting Engineering and Science of Coastal Sediment Processes, Hauppage, New York, June 21–23, 1999. Reston, VA: ASCE, p. 1495. Reproduced by permission of ASCE.

Figs 13.4, 13.13(a), 14.11, and 15.8: from R. A. Davis, *Coastal Sedimentary Environments*, Figs 3.2, 3.14, 2.6 and 2.48. © 1978 Springer-Verlag.

Fig. 13.23: from H. Reineck & I. B. Singh, *Depositional Sedimentary Environments with Reference to Terrigenous Clastics*, 2 edn, p. 432, Fig. 591. © 1980 Springer-Verlag.

Figs 14.7, 14.9, 14.10, and 14.21: from C. J. Dawes, *Marine Botany*. © 1981 Wiley. This material is used by permission of John Wiley and Sons, Inc.

Fig. 14.18: from J. M. Edwards & R. W. Frey, Substrate characteristics within a Holocene salt marsh, Sapelo Island, Georgia. *Senckenbergiana Maritima*, 1977, vol. 9, pp. 215–59. © Senckenberg Institute.

Fig. 15.9: from M. Nichols, Effect of increasing depth on salinity in the James River Estuary. In B. W. Nelson (ed.), *Environmental Framework of Coastal Plain Estuaries*. Geol. Soc. Amer. Memoir, 1972, 133, p. 209. © GSA.

Fig. 16.2: from J. R. L. Allen, Sediments of the modern Niger Delta: a summary and review. In J. P. Morgan (ed.), *Deltaic Sedimentation*. SEPM Spec. Publ. 15, 1970, p. 141. © SEPM.

Fig. 16.8: from J. M. Coleman & S. W. Gagliano, Cyclic sedimentation in the Mississippi River Deltaic Plain. *Trans. Gulf Coast Assoc. Geol. Soc.*, 1964, vol. 14, p. 69. © Gulf Coast Association of Geological Societies.

Fig. 16.20: from L. D. Wright & J. M. Coleman, Variations in morphology of major river deltas as a function of ocean wave and river discharge regimes. *Amer. Assoc. Petroleum Geol. Bull.*, 1973, vol. 57, p. 377. AAPG © 1973. Reprinted by permission of the AAPG, whose permission is required for further use.

Fig. 16.21: modified from J. R. L. Allen, Late Quaternary Niger Delta and adjacent areas: sedimentary environments and lithofacies. *Amer. Assoc. Petroleum Geol. Bull.*, 1965, vol. 49, p. 558. AAPG © 1965. Reprinted by permission of the AAPG, whose permission is required for further use.

Fig. 17.8: from F. K. Lutgens & E. J. Tarbuck, 1999, *Earth: An Introduction to Physical Geology*, 6th edn. © Prentice Hall. Reproduced with permission.

Fig. 17.23: from E. Uchupi, G. S. Giese, D. G. Aubrey & D. J. Kim, The Late Quaternary construction of Cape Cod, Massachusetts: a reconsideration of the W. M. Davis model. Boulder, CO: Geological Society of America, Special Paper 309, 1996, Figs 19, 20, 21, 22, 23, and 24. Reprinted with the permission of GSA.

Fig. 17.29: from H. E. Wright & D. Frey, *The Quaternary of the United States*. Copyright © 1965 by Princeton University Press; renewed 1993. Reprinted by permission of Princeton University Press.

Fig. 19.16: from W. G. H. Maxwell, 1968, *Atlas of the Great Barrier Reef*. Elsevier, New York. With permission from Elsevier Science.

Fig. 20.11: modified from F. P. Shepard, *Submarine Geology*, 2nd edn. 1973, p. 140. Reprinted by permission of Pearson Education, Inc.

Fig. 21.30: modified from M. A. Lynch-Blosse & R. A. Davis, Stabililty of Dunedin and Hurricane Passes, Florida. *Coastal Sediments '77*. © 1977 ASCE. Reproduced by permission of the ASCE.

Every effort has been made to trace copyright holders and to obtain their permission for the use of copyright material. The publisher apologizes for any errors or omissions and would be grateful if notified of any corrections that should be incorporated in future reprints or editions of this book.

1 Coastline variability and functions in the global environment

1.1 Introduction

The surface of the Earth is covered by two contrasting media: land and sea. They meet at the coast. There are, of course, glaciers that span parts of both the land and sea, such as in Greenland, in parts of the Canadian Arctic, and on Antarctica. Each of the two surfaces may cover millions of square kilometers over continents and oceans or much less in the case of small oceanic islands or some lakes within continental masses. Nevertheless, a narrow coastal zone separates these two major parts of the Earth's surface.

The world's coastline extends for about 440,000 km, but the coastal zone comprises less than 0.05% of the area of the landmasses combined. Because nearly half of the global population lives within less than 100 km of the coastline, the coastal zone has become arguably the most critical part of the Earth's surface in terms of global economy, strategies, and management needs.

1.2 Coastal settings

What do we actually mean by the coast? The coastline or shoreline is simply the contact between the land and the sea; an easy definition. The **coastal zone**, however, is a bit more difficult to delimit. For practical purposes it is any part of the land that is influenced by some marine conditions, such as tides, winds, biota, or salinity. The coast is global in its distribution but limited in width. We cannot give an average width, an average character, or any other average category that adequately typifies the coast. It is much too varied and complicated in its characteristics. In some places the coastal zone might be only a few hundred meters wide, whereas in others it might be more than 100 km wide. Some coastal zones include a wide range of environments that separate the true ocean from the terrestrial environment. In other situations, a single coastal environment may define the land–sea boundary.

In this book we consider the controlling factors that determine what type of coast develops. The processes that develop and maintain coastal environments, as well as those that destroy the coast, are discussed in order to convey the dynamic nature of all coastal environments. Each of the major environments is considered in light of these controlling factors and processes. The impact of human activity along the coast has been enormous, especially over the past century. Many examples of this impact appear throughout the book but a special chapter devoted to the topic is also included. Most of the emphasis in the book is directed toward geologic and physical attributes of the coast, although organisms are not overlooked.

Fig. 1.1 Photograph of an erosional coast in Oregon. The bluffs here are composed of a friable Miocene sandstone. Houses on top are in serious jeopardy.

Open coasts can be divided into two general categories: those that are dominantly erosional (Fig. 1.1) and those that are primarily depositional (Fig. 1.2) over long periods of time, i.e. thousands of years or more. Erosional coasts are extensive and have considerable variety, although they tend to be narrow. They typically are the high relief rocky coasts but also include some bluffs of unconsolidated sediments, beaches, and other local depositional features. Although erosional coasts are among the most beautiful and spectacular of all coastal types, there is less variation in this generally rocky type of coast than in those that are characterized by deposition.

Depositional coasts include a wide spectrum of systems, such as river deltas, barrier island systems, strandplain coasts, reef coasts, and glaciated coasts.

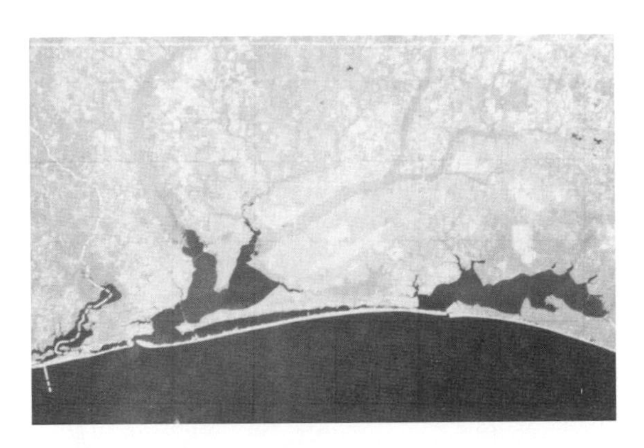

Fig. 1.2 Photograph of a depositional coast on the Atlantic coastal plain of the United States. This shows a barrier island coast with two large estuaries.

Each of these may contain numerous distinct environments. The variety of morphologic features and the complex interaction of depositional coasts deserves extensive attention and is emphasized in this book.

Climatic differences cause a wide variety of coastal types, in that temperature and rainfall exert a major influence on coastal development. Extreme climates such as those in the very high latitudes cause coastal areas to be covered with ice; all the time in some places, and for only a few months in others. Parts of Greenland and the Antarctic coast are covered with ice continually, whereas some of the coasts of Alaska, Canada, the Scandinavian countries, and Russia have ice cover for at least a few months each year. Desert conditions can directly influence coastal environments as well. Few significant rivers and therefore few river deltas are produced from desert areas. Some coastal deserts are dominated by huge sand dunes, such as along Namibia (Fig. 1.3) on the southwest coast of Africa. Along the Persian Gulf and in north Africa the arid, low-latitude environment produces extensive coastal environments called sabkhas that are nearly at sea level and have an almost horizontal surface (Fig. 1.4) dominated by chemically precipitated salts and other minerals.

The tectonics of the Earth's crust also exert a major influence on the coastal zone. Coasts that coincide with or that are near plate boundaries tend to have more relief and are narrow compared to those that are away from plate boundaries. Colliding plates provide for a particularly rugged coast, such as we see along the Pacific Northwest of the United States. The relationships between plate tectonics and coastal development are treated in detail in Chapter 2.

Fig. 1.3 Huge dunes that extend to the shoreline along the coast of Namibia in southwest Africa. These dunes may be up to 100 m high and the absence of vegetation makes them quite mobile. (Courtesy of N. Lancaster.)

Fig. 1.4 Sabkha along the northern coast of Libya. These environments are quite flat surfaces that are essentially at mean sea level and are located where tidal range is very small. (Courtesy of the US Geological Survey.)

1.3 Population and the coast

The coast is many things to many people. Depending on where and how we live, work, and recreate,

Fig. 1.5 Photograph of the coast at Alexandria, Egypt, an ancient city in the eastern Mediterranean. This development has been here for many centuries but this coast has only been erosional since the construction of the Aswan Dam in the twentieth century. (Courtesy of D. Stanley.)

our perception of it varies greatly. Large populations live on or near the coast because it is typically very beautiful and interesting. Many more visit the coast for the same reasons. A large number of people gain their livelihood directly or indirectly from the coast, and some have the task of protecting it from intruders or enemies.

1.3.1 History of coastal occupation

The ancient civilizations of the eastern Mediterranean Sea were largely associated with the coast, including the famous Greek, Roman, and Phoenician settlements and fortifications of biblical times and before. Many of the great cities of the time were located on the natural harbors afforded by the geologic and physiographic conditions along the coast (Fig. 1.5). These cities provided a setting that was conducive to trade and that could be defended against enemies.

Far to the north, Viking settlements in the Scandanavian countries of Norway, Sweden, and Denmark were typically located along the coast as well. Here the great fjords provided shelter, fortification, and ready access to the sea, which was a primary food source and the main avenue of transportation, and were the sites of many battles. At about the same time, the northern coast of what is now Germany and the Netherlands was also occupied for similar reasons but in a very different coastal setting: one of lowlands and barrier islands.

Many centuries later, cities in the New World such as Boston, New York, Baltimore, and San Francisco owe their location to the presence of a protected harbor. In their early stages of development many of the major civilizations of the world were directly on the coast or had important interaction with it.

In the early civilizations, reasons for this extensive occupation of coastal areas were strictly pragmatic. Coasts were essential for harboring ships, a primary means of transporting goods, one of the major activities of the time. The adjacent sea was also a primary source of food. Similar reasons were the cause for the settlement of many of the great cities of Europe, such as London, Amsterdam, Venice, Copenhagen, and others. All were settled on the water because their location fostered commerce that depended on transportation over water.

This pattern of coastal occupation and utilization continued until the latter part of the nineteenth century. By that time the interior areas of the United States had been settled and large cities were scattered all over the country. Many of these cities, however, are near water as well. In North America they are either on the Great Lakes (Chicago, Detroit, Cleveland, and Toronto) or on the banks of large rivers (St Louis, Cincinnati, Pittsburgh, and Montreal).

Since ancient times, the coast has been a strategic setting for military activity. At first the cities housed military installations. Later it became important as a staging ground for large-scale invasions; examples include the British conquest of France in the fourteenth century and Allied troop landings on the beaches of Iowa Jima and other islands of the Pacific and on the Normandy coast of France during the Second World War in the twentieth century.

It was only in the latter part of the nineteenth century that coastal activities expanded into broad-based recreational use, with related support industries. As a result of the Industrial Revolution and overall prosperity, both in North America and, to a lesser extent, in Europe, people began to look to the coast as a place to take family holidays.

Box 1.1 Venice, Italy: a city waiting to die

A unique city on the Adriatic Sea coast of northern Italy, Venice is a major historic treasure, and a very popular tourist stop for people from throughout the world. The city is commonly known for its canals, gondolas, and the absence of automobiles, as well as for its excellent cuisine. It is located within a large, shallow, backbarrier lagoon and is offshore from the mainland. Venice began in the eighth century as a city-state, with trading and fishing as its major industries. It became very famous and its merchants became very wealthy as the city developed into a major economic center of the eastern Mediterranean area.

The general setting is a few kilometers offshore, where there are several low islands that have become developed as residential and commercial centers. These islands have essentially no natural relief and they rest on thick muddy sediments deposited by rivers that used to flow into the lagoon. The main island, Venice, is almost entirely covered with buildings that are mostly several hundred years old and that have great historic significance to Italy. Included are numerous churches, governmental buildings, plazas, old military installations, etc. The entire city was built within a meter or two of sea level, but with the protection of the barrier islands a few kilometers seaward of them.

As time passed, three things occurred to jeopardize the future of the city: (i) the underlying mud began to compact; (ii) the rate of sea-level rise increased, (iii) the combination of these phenomena has resulted in a significant relative rise in sea level, essentially flooding the city. This has been exacerbated by the unfortunate decision to relocate the mouths of two rivers, the Sile River and the Brenta River, which naturally emptied into the Venice lagoon. It was feared that the discharge of sediment would fill up the lagoon. In order to prevent this from happening, the lower course was changed so that both rivers now empty directly into the Adriatic Sea, one north of the Venice lagoon and one to the south. This has resulted in a lack of sediment to nourish the marshes, which are vital to both the ecology of the lagoon and the protection of the developed islands from wave attack and erosion.

At the present time the city remains vital but the population, which peaked at more than 250,000 a couple of centuries ago, is now down to about 50,000. Much of this decrease is the result of the tremendous cost of renovating properties for residential or commercial use. Virtually all buildings fall within the historic preservation regulations and therefore must be restored in order to be occupied, and such expense is more than most people can bear. Another deterrent to Venetian residence is the lack of jobs.

The current rate of sea level rise is 2.5 mm yr^{-1}. With a city that is nearly at the level of spring high tide, the future is limited. The most popular location in Venice is Piazza San Marco, which is also one of the lowest elevations in the city. A hundred years ago this area was flooded only a few times per year, but with an increase in sea level of about 30 cm over that time, it is now flooded about 50 times each year, a couple of days during each spring tide situation plus times when winds blow for a sustained period. As sea-level rise increases and continues, the city will literally drown.

Numerous engineering approaches have been tried and others have been suggested to help to sustain the city. First, the barrier islands have been stabilized and nourished at great expense. This will: (i) protect the numerous communities that inhabit them; (ii) keep the barriers from migrating landward; and (iii) maintain the size and geometry of the lagoon, and therefore stabilize the tidal flux in and out of the three large inlets that serve the lagoon. The inlets have been stabilized to keep their channels fixed and to allow the large volume of ship traffic to the port of Venice to continue.

At the present time a plan has been developed, but not yet funded, to construct large floating gates that will prevent wind from blowing large amounts of water into the lagoon and raising its level beyond that which the city of Venice can withstand. These gates rest on the inlet channel floor during normal times, then when strong winds begin to raise the water level in the lagoon, water is evacuated from them, causing one end to float up above the surface and to act as a dam to water being blown into the lagoon.

In the long term, there seems to be little that can be done to save this beautiful and historically significant city other than encasing it in a large dike. It will be a very expensive and time-consuming project, but it might be the only way.

1.4 General coastal conditions

Varied geologic conditions provide different settings for the coast, and give variety and beauty to that part of the Earth's surface. As a consequence, some coasts are quite rugged, with bedrock cliffs and irregular shorelines, whereas others are low-lying, almost featureless areas with long, smooth shorelines. To be sure, with time, any coast can change extensively, but some important relationships continue through geologically significant periods of time, up to many millions of years.

Changes at a given part of the coast are typically slow and continuous, but they may be sporadic and rapid. Rocky cliffs tend to erode slowly but hurricanes can change beaches or reefs very quickly. Overprinted on this combination of slow and rapid processes of change is the very slow fluctuation in sea level over time, about 1–2 mm yr^{-1}. In the geologic past this rate was both much faster and even a bit slower. The point is that as coastal processes work to shape the substrate and the adjacent land, the position of the shoreline changes as well. This translates the processes and their effects across the shallow continental shelf and the adjacent coastal zone, producing long and slow, but relatively steady, coastal change.

Each specific coastal setting, regardless of scale, is unique yet is quite similar to other coastal settings of the same type. Although each delta is different, a common set of features characterizes all deltas. The approach of this book is to consider the general attributes of each of the various types of coastal environments. Numerous examples of each environmental type provide some idea of the range available. Finally, the overprint of time demonstrates the dynamic nature of all of these coastal elements.

1.5 Coastal environments

The variety of coastal environments is wide. This section briefly introduces each of the major environments to demonstrate this variety. All of these and more are discussed in detail in the following chapters.

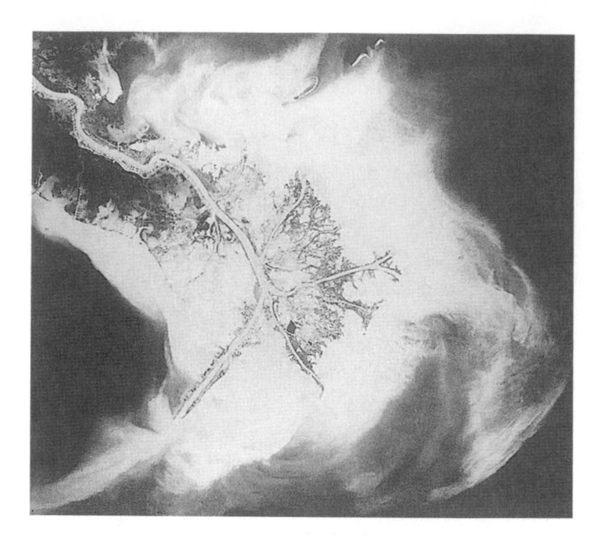

Fig. 1.6 Satellite photo of the Mississippi delta showing considerable suspended sediment being discharged through numerous distributaries. Although it appears that the sediment load is very large, it has been reduced greatly during the past century as the result of dams on the river. (Image from EROS Data Center.)

Rivers carry tremendous quantities of sediment to their mouths, where they deposit it. Much of the sediment is then entrained by waves and currents, but commonly there is a net accumulation of sediment at the river mouth: a delta (Fig. 1.6). In fact, most of the sediment along all types of depositional coasts owes its presence, at least indirectly, to a river. Deltas range widely in size and shape. Most are dominated by mud and sand but some have abundant gravel. The primary conditions for delta formation are a supply of sediment, a place for it to accumulate, and the inability of the open water processes to rework and remove all of the sediment from the river mouth.

Sea level has risen considerably over the past several thousand years as the result of glaciers melting and a combination of other factors. This increase in sea level flooded many parts of the land and developed extensive and numerous coastal bays. Streams feed most of these bays. These bays are called estuaries (Fig. 1.7) and are commonly surrounded by some combination of wetlands (usually either salt marshes or mangrove swamps) and tidal flats.

Fig. 1.7 Headland on the Oregon coast with an associated barrier spit protecting the small estuary. Even though this coast has a high tidal range, it is eroded at the headlands, with the sediment carried along the coast to form these barrier spits. (Photograph courtesy of W. T. Fox.)

Another common type of coastal bay is one that tends to parallel the coast and is protected from the open ocean by a barrier island. These elongate water bodies have little influx of fresh water or tidal exchange. They are lagoons. Tidal flats and marshes are uncommon along this type of bay. Other coastal embayments that cannot be considered as either an estuary or a lagoon are simply termed coastal bays.

Barrier islands are another important part of the scheme of coastal complexes. These islands are a protection in front of the mainland, typically fronting lagoons and/or estuaries. These barriers include beaches, adjacent dunes, and other environments. Wetlands, especially salt marshes, are widespread on the landward side of barrier islands. Tidal inlets dissect barrier islands and are among the most dynamic of all coastal environments. They not only separate adjacent barrier islands, but also provide for the exchange of water and nutrients between the open ocean and estuarine systems.

Strandplain coasts are low-relief coastal areas of a mainland that have many characteristics of the seaward side of a barrier island. They contain beaches and dunes but lack the coastal bay (Fig. 1.8). Examples include Myrtle Beach, South Carolina, and the Nayarit Coast of western Mexico.

Rocky or headland coasts can be present as short isolated sections within extensive sandy depositional

Fig. 1.8 Aerial photograph showing a strandplain coast where there is no significant estuary and no barrier island. Such a coast typically develops where there are no coastal bays and the gradient offshore is relatively steep, thus letting large waves reach the surf zone.

coasts, such as along parts of the east coast of Australia or the Pacific Northwest coast. Other geomorphically similar coasts may have their origin in glacial deposits, with New England being a good example. In both cases, the coast is characterized as erosional and they may provide sediments for nearby depositional beaches.

Reef coasts (Fig. 1.9) owe their origin to the construction of a framework by organisms; both plant and animal. Not only are these coastal environments very beautiful, but they also protect the adjacent mainland or island from erosion and wave attack from severe storms.

1.6 Historical trends in coastal research

Scientists have only recently undertaken comprehensive investigations of the coast. Some nineteenth-century publications considered the origin of barrier islands, scientists speculated on the development

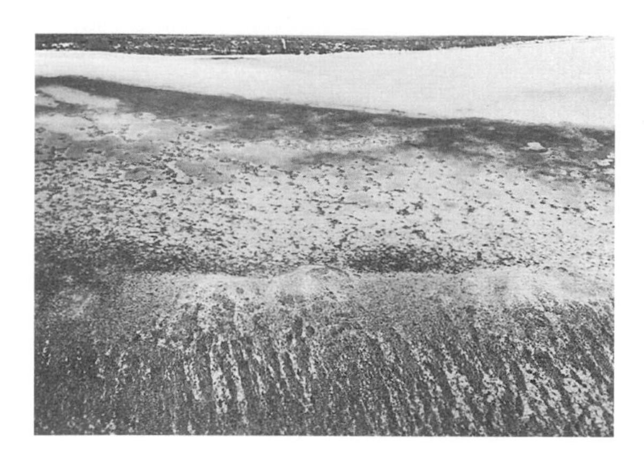

Fig. 1.9 Oblique aerial photograph showing a well developed reef along the Florida Keys coast. The portion of the reef that has the linear features is the seaward and deeper area, while the remainder of the reef is the flat, shallow, upper portion.

Fig. 1.10 Photograph of a seawall designed to protect the coast from erosion. These features are designed to stop erosion along the shoreline and to protect buildings or infrastructure that is close to the shoreline

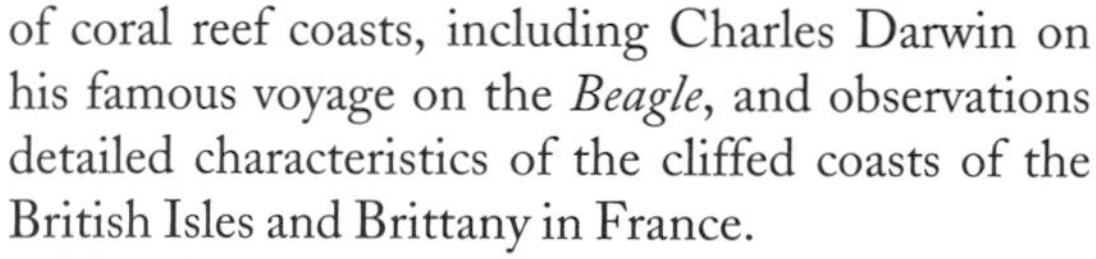

of coral reef coasts, including Charles Darwin on his famous voyage on the *Beagle*, and observations detailed characteristics of the cliffed coasts of the British Isles and Brittany in France.

The first systematic efforts at studying the coast in the early twentieth century were made by **geomorphologists**, those who study the morphology or landforms of the Earth. Geomorphologists also investigate mountains, deserts, rivers, and other Earth features. Their studies produced various classifications, maps, and reports on coastal landforms. Some scientists focused on the evolution of coasts and the processes responsible for molding them. For example, Douglas W. Johnson, a professor at Columbia University, wrote a classic and pioneering book in 1919 entitled *Shore Processes and Shoreline Development*, a monograph that is still commonly referenced.

Engineers have also given special attention to the coast over several centuries. Their interest was directed toward construction of dikes, harbors, docks, and bridges on the one hand, and stabilization of the open coast on the other. Although the two groups of professionals directed their efforts toward different aspects of the coast, their interests overlapped in many circumstances.

Ancient people recognized that the coast is potentially dangerous during storms and is continually changing due to processes associated with wind waves and storms. They understood that the shoreline is one of the most dynamic areas on the Earth. Erosion was a particularly important problem (Fig. 1.10) and settlements were lost or threatened as the shoreline retreated. For centuries dikes have been constructed along the North Sea coast of the Netherlands and Germany, both for protection (Fig. 1.11) and for land reclamation. In many other areas, however,

Fig. 1.11 Photograph of a dike along the Netherlands coast on the North Sea. Such structures protect the adjacent land, which is below sea level, from flooding. Like many of the dikes, this one supports a roadway. Some are used to graze sheep and cattle.

construction on the open coast was designed to slow or prevent erosion. As a result, various types of structures were emplaced at critical locations along densely inhabited areas of the coast in attempts to stabilize the beach and prevent erosion.

For decades these activities represented the major efforts of science and technology to understand and, in some respects, to control the response of the coast to natural processes. The Second World War was also an important period in furthering our understanding of the coast. Major war efforts took place along the coast, particularly the landing of troops, supplies, and equipment, whether on the European mainland or on Pacific islands. All branches of the military were involved in studying coastal geomorphology, coastal processes including waves, tides, and currents, and the analysis of weather patterns along the coast. Much of the world's coast was mapped in detail during this period. The Beach Erosion Board, a research branch of the US Army affiliated with the Corps of Engineers, made many very important contributions to our knowledge of coasts. This group conducted extensive research on beaches, waves, erosion, and other important aspects of the coast, using both their own staff and academic researchers from many of the best universities. Francis P. Shepard and Douglas L. Inman of the Scripps Institution of Oceanography (Fig. 1.12) were prominent contributors to the research programs of this group and later became among the most prominent coastal researchers in the world.

Fig. 1.12 The old Scripps Pier at Scripps Institution of Oceanography at La Jolla, California. Considerable data on nearshore processes have been collected from this structure over several decades. It has now been replaced by a new version. (Photograph courtesy of Scripps Institution of Oceanography.)

This coastal research effort continued after the Second World War, but with a distinctly engineering emphasis. The name of the original research organization was changed to Coastal Engineering Research Center, originally housed at Fort Belvoir, Virginia, and now located in Vicksburg, Mississippi. At about this time the Office of Naval Research (ONR) became heavily involved in basic research on the coast. Its first major efforts in this endeavor were through the Coastal Studies Institute of Louisiana State University. Although this organization conducted a variety of coastal research projects, its major effort was a global study of river deltas, beginning near home with the Mississippi delta. As time passed, the ONR expanded its coastal research support to emphasize beaches, inlets, and deltas – places where military activity could potentially take place. During the 1960s and 1970s this agency, through the leadership of Dr Evelyn Pruitt, supported most of the research on modern open coastal environments. This period began the modern era of coastal research, which emphasizes process-response systems. In other words, it is no longer enough to observe, describe, and classify coastal features and environments. The focus is now directed at determining the origin and development of these features, which necessitates the study of the physical and biological processes that operate on the coast and then the integration of these data with the resulting landforms. Thus began the process-response approach to coastal research. Coastal research of this type is only a few decades old.

Box 1.2 Beach Erosion Board of the US Army, Corps of Engineers

The Corps of Engineers is the part of the US Army that typically deals with construction: generally structures like dams, jetties at inlets, and seawalls to protect from erosion. It is also responsible for maintaining waterways, commonly through dredging the channels to permit shipping traffic, and for harbors for berthing the ships. During the Second World War there were some special needs associated with the deployment of ships and personnel in coastal waters. Landing troops on the beaches of Europe and in the Pacific were important military activities, and had to be done with careful planning and understanding of the coast and its various conditions.

As a consequence, the military, through the US Army, Corps of Engineers, enlisted a peacetime organization, the Beach Erosion Board, to assist in the task of learning as much as possible about coastal dynamics and geomorphology in order to aid the military effort in both the Eurpoean and Pacific theaters of war. The Beach Erosion Board, which had been a group of experts that advised the Corps on matters relating to ports, harbors, beach erosion, and navigation, now became an important aspect of the war effort. In order to carry out these activities several projects to be conducted at universities were developed and funded by the Army.

Much of the work was done in California at Scripps Institution of Oceanography and at the University of California at Berkeley. The Scripps researchers had long been noted for their marine research in physical oceanography and marine geology. The pier there was one of the first research piers equipped with instruments to record wave data, meteorology, and other coastal environmental phenomena. Such prominent scientists as Francis Shepard and Douglas Inman were in the group at Scripps that conducted this research. The coastal engineering group at Berkeley was also a leader in the field and was composed of experts at designing temporary facilities for military operations, including harbors, breakwaters, and bridges. Among them were Murrough P. O'Brien and Robert Weigel, two of the pioneers in coastal engineering.

Several prominent scientists and engineers were employed full-time as researchers for the military and some joined the military as scientific advisors. Although it is impossible to tell precisely, the efforts of both the academics and the military in aiding these war efforts saved many lives and countless dollars. Most of their work was unnoticed by the general public but is well known to the coastal research community.

Eventually, the Beach Erosion Board became the Coastal Engineering Research Center, the research branch of the US Army, Corps of Engineers. It has aided and continues to aid both civilian and military coastal and navigational operations. Originally located in Arlington, Virginia, near Washington DC, it is now combined with the Waterways Experiment Station of the Corps in Vicksburg, Mississippi. It is the premier coastal engineering research unit of the federal government of the United States.

Suggested reading

Barnes, R. S. K. (ed.) (1977) *The Coastline*. London: John Wiley & Sons.

Bird, E. C. F. & Schwartz, M. L. (eds) (1985) *The World's Coastline*. New York: Van Nostrand Reinhold.

Carter, R. W. F. (1988) *Coastal Environments*. New York: Academic Press.

Davies, J. L. (1980) *Geographic Variation in Coastal Development*. New York: Longman.

Fox, W. T. (1983) *At the Sea's Edge*. Englewood Cliffs, NJ: Prentice Hall.

Johnson, D. W. (1919) *Shorelines and Shoreline Development*. New York: John Wiley & Sons.

Shepard, F. P. & Wanless, H. R. (1971) *Our Changing Coastlines*. New York: McGraw-Hill.

2 The Earth's mobile crust

(a)

(b)

Fig. 2.1 Dissimilar tectonic settings produce very different types of coastlines. (a) The coastal plain setting of South Carolina fronted by barriers and tidal inlets (Murrells Inlet, 1977) is in sharp contrast to (b) the mountainous fjord coast of Kenai, Alaska.

2.1 Introduction

Coastlines of the world exhibit a wide range of morphologies and compositions in a variety of physical settings (Fig. 2.1). There are sandy barrier island coasts such as those of the East and Gulf coasts of the United States, deltaic coasts built by major rivers including those at the mouths of the Nile and Niger rivers, glacial alluvial fan coasts of the Copper River in Alaska and the Skiedarsar Sandar coast of southeast Iceland, coastlines fronted by expansive tidal flats such as the southeast corner of the North Sea in Germany, volcanic coasts of the Hawaiian Islands, carbonate coasts of the Bahamas and the South Pacific atolls, mangrove coasts of Malaysia and southwestern Florida, bedrock cliff and wave-cut platform coasts of the Alaskan peninsula and southwest Victoria in Australia, and many other types of coastlines. The diversity of the world's coastlines is largely a product of the Earth's mobile crust. The eruptions of Mount Saint Helens in Washington state (1980) and Mount Pinatubo in the Philippines (1991), and the devastating earthquakes in Mexico City (1985) and in Kobe, Japan (1995), are dramatic expressions of this mobility. The formation of pillow basalts and new oceanic crust at mid-ocean ridges and the presence of hydrothermal vents at these sites are also a manifestation of crustal movement. The theory that explains the mobility of the Earth's crust and the large-scale features of continents and ocean basins, including the overall geological character of coastlines, is known as plate tectonics.

Plate tectonics theory has done for geology what the theory of evolution did for biology, the big bang theory for astronomy, the theory of relativity for physics, and the establishment of the periodic table for chemistry. Each of these advancements revolutionized its respective field, explaining seemingly unrelated features and processes. For example, in geology the cause and distribution of earthquakes, the construction of mountain systems, the existence of deep ocean trenches, and the formation of ocean basins are all consequences of the unifying theory of plate tectonics. The germination of this theory began many centuries ago with scientists' and world explorers' interest in the distribution of continents and ocean basins. As early as the 1620s, Sir Francis Bacon recognized the jigsaw puzzle fit of the eastern outline of South America and the western outline of Africa. By 1858, Antonio Snider had published two maps illustrating how North and South America were joined with Africa and Europe during Carboniferous time (~300 million years ago) and how the continents had split apart to form the Atlantic Ocean. He reconstructed the positions of continents 300 million years ago to show why plant remains preserved in coal deposits of Europe are identical to those found in coal seams of eastern North America. Snider's maps were an important step in promoting a theory that later became known as continental drift (the theory that envisions continents moving slowly across the surface of the Earth).

In the early twentieth century, the idea of continental drift was popularized by Alfred Wegener, a German scientist at the University of Marburg (Fig. 2.2). Wegener was a meteorologist, astronomer, geologist, and polar explorer, and led several expeditions to Greenland. He was the first to present a sophisticated and well researched theory of continental drift, which he did in a series of lectures to European scientific societies in 1912. Three years later he published his ideas in a book entitled *Die Entstehung der Kontinente und Ozeane* (*The Origin of Continents and Oceans*). Wegener believed that all the continents were once joined together in a super continent he called Pangaea (Greek for "all Earth") which was surrounded by a single world ocean he named Panthalassa (all ocean) (Fig. 2.3). The northern portion of this supercontinent, encompassing North America and Eurasia, was called Laurasia, and the southern portion, consisting of all the other continents, was Gondwanaland. Partially separating these two landmasses was the Tethys Sea.

Fig. 2.2 This picture of Alfred Wegener was taken when he was 30 years old during one of his expeditions to Greenland. It was during this period of his life that he proposed his theory of "continental drift."

Wegener used a variety of supporting evidence to bolster his theory of a single continent, including the continuity of ancient mountain fold belts and other geological structures that extend across continents now separated by wide oceans. He noted that the coal deposits in frigid Antarctica and glacial

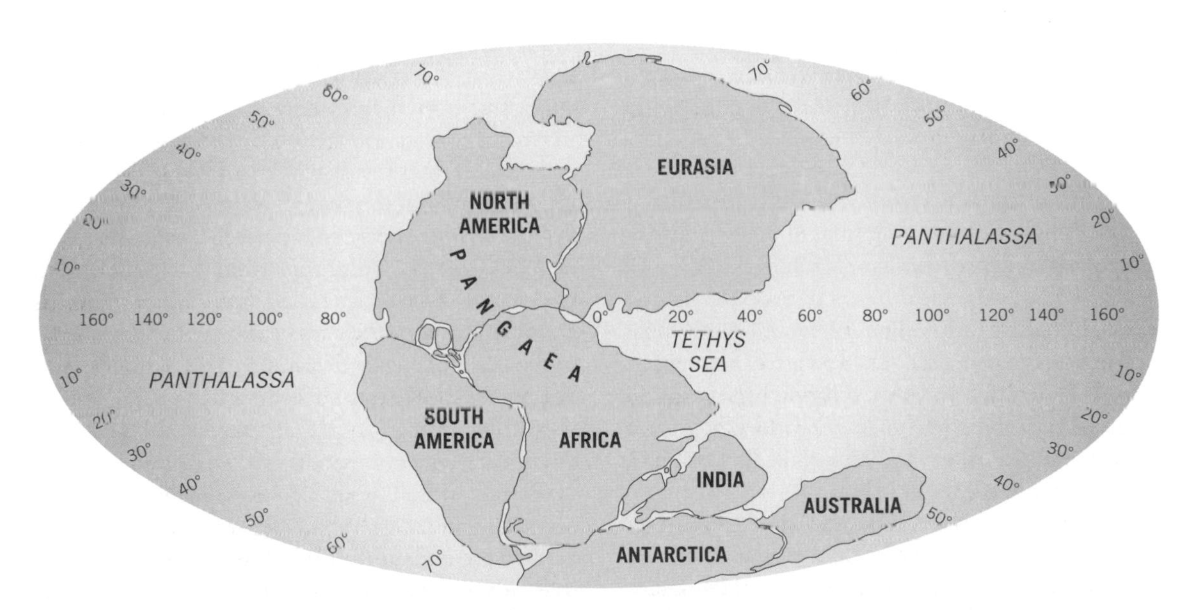

Fig. 2.3 One of the compelling pieces of evidence that Wegener used to argue for his continental drift theory was the geometric fit of South America and Africa. In fact, he believed that about 200 million years ago all the continents were joined together in a supercontinent he called Pangaea.

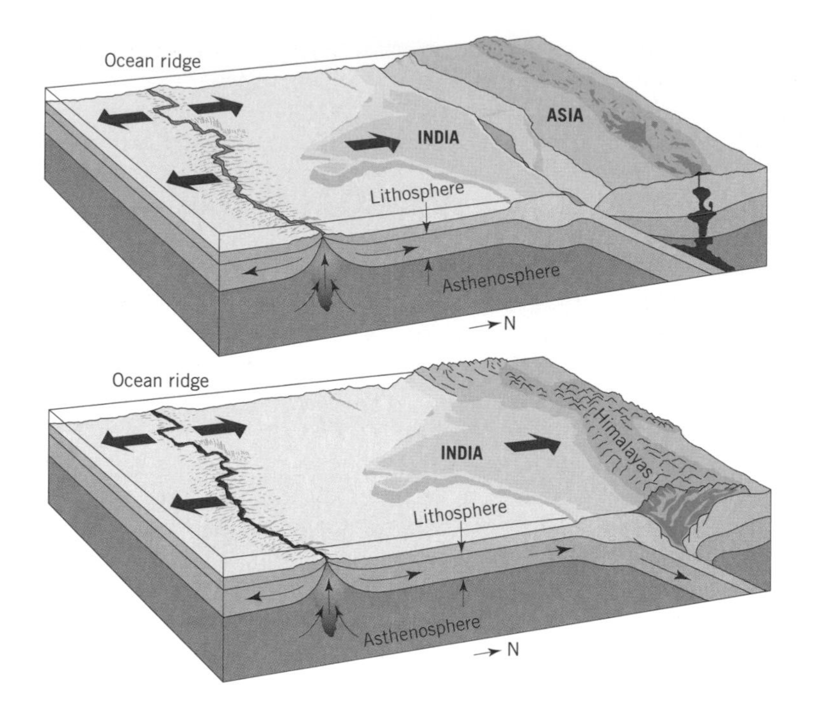

Fig. 2.4 According to Wegener the Indian subcontinent traveled northward following the breakup of Gondwanaland and eventually collided with Asia, forming the Himalayan Mountains.

sediments in what are now the tropical regions of South Africa, India, and Australia could only be logically explained by moving the continents to different latitudinal settings. He also demonstrated that identical fossils and rocks of similar age could be found on widely separated continents but actually plot side by side when Pangaea is reconstructed. Using additional fossil evidence, he theorized that Pangaea separated into a number of pieces approximately 200 million years ago, forming the Atlantic Ocean among other features. He reasoned that if continents could move, then the presence of glacial deposits in India meant that India had once been close to Antarctica and that following the breakup of Gondwanaland, its northward movement and ultimate collision with Asia caused the formation of the Himalayan Mountains (Fig. 2.4).

Although Wegener's book and ideas were initially very popular, the geological community of that time ultimately scorned his new theory. His failure to convince the scientific community stemmed from his weak argument that the continents plowed their way through or slid over the oceanic crust. He believed that this movement was caused by the gravitational pull of the sun and moon (tidal force) acting with a differential force on the surface crust relative to the underlying mantle. Contemporary scientists showed that the tidal forces were much too weak to account for the drifting of continents and thus his theory was abandoned. Wegener did not live long enough to witness the revival and acceptance of many aspects of his theory, as he perished in 1930 at age 50 during a fourth expedition to Greenland while on a rescue mission. Indeed, it was not until the late 1950s and early 1960s that scientists solved the puzzle of the moving continents and Wegener's theory was revived.

In this chapter we demonstrate that the overall physical character of the edge of continents for several thousands of kilometers of coast, such as the west coasts of North and South America, is a function of plate tectonic processes. This includes whether a coast is seismically and volcanically active, whether it is bordered by mountain ranges, a broad coastal plain, or something in between, and whether it is fronted by a narrow or wide continental shelf. Plate tectonics are also shown to have an important influence on the supply of sediment to a coastline and the extent of depositional landforms such as deltas, barrier chains, marshes, and tidal flats.

2.2 The Earth's interior

In order to grasp why continents have "drifted" to new positions, why ocean basins have opened and closed, how mountain systems and ocean trenches have formed, as well as the dimensions of these systems, it is necessary to understand the composition, layering, and processes that occur within the Earth's interior. In its early beginning, the Earth was essentially a homogeneous mass consisting of an aggregate of material that was captured through gravitational collapse and meteoric bombardment. The heat generated by these processes, together with the decay of radioactive elements, produced at least a partial melting of the Earth's interior. This melting allowed the heavier elements, especially the metals of nickel and iron, to migrate toward the Earth's center, while at the same time the lighter rocky constituents rose toward the surface. These processes, which have decreased considerably as the Earth has cooled, ultimately led to a layered Earth containing a core, mantle, and crust. The layers differ from one another both chemically and physically (Fig. 2.5). Most of our knowledge of the Earth's interior comes from the study of seismic waves that pass through the Earth and are modified by its different layers.

The center of the Earth is divided into two zones: a solid inner core (radius 1250 km) with a density almost six times that of the crust and with a temperature comparable to the surface of the sun; and a viscous molten outer core (width 2220 km) some four times as dense as the crust. The mantle is 2880 km thick and contains over 80% of the Earth's volume and over 60% of its mass. It is composed of iron and magnesium-rich silicate (silica–oxygen structure) minerals, which we have observed at locations where material from the mantle has been intruded into the overlying crust and subsequently exposed at the Earth's surface. The chemical composition is consistent throughout the mantle, but due to the increasing temperature and pressure the physical properties of the rocks change with increasing depth.

Compared to the inner layers, the outer skin of the Earth is cool, very thin, much less dense, and rigid. The crust is rich in the lighter elements that rose to the surface during the Earth's melting and differentiation stage, forming minerals such as quartz and feldspar. There are two types of crust: the relatively old (up to 4.0 billion years) and thick granitic crust of the continents (20–40 km thick) and the thin and geologically young (less than 200 million years) basaltic crust beneath the ocean basins (5–10 km thick). Continental crust can be up to 70 km thick at major mountain systems. The less dense and greater thickness of the continental crust (2.7 g cm^{-3}) as compared to the thinner more dense (2.9 g cm^{-3}) ocean crust has important implications for how these two crusts have formed and the stability of the crusts.

The divisions described above are based mostly on the different chemical character of the layers. However, in terms of plate tectonic processes, especially in understanding the movement of plates, the crust and mantle are reconfigured into three layers on the basis of physical changes in the nature of the rocks:

1 Lithosphere. This is the outer shell of the Earth and is composed of oceanic and continental crust and the underlying cooler, uppermost portion of the mantle (Fig. 2.5). The lithosphere is approximately 100 km thick and behaves as a solid, rigid slab. In comparison to the diameter of the Earth, the lithosphere is very thin and would represent the shell of an egg. The outer shell is broken up into eight major plates and numerous smaller plates. The Pacific Plate is the largest plate, encompassing a large portion of the Pacific Ocean, whereas the Juan de Fuca Plate off the coasts of Washington and Oregon is one of the smallest. The tectonic or lithospheric plates, as they are called, are dynamic and are continuously moving, although very slowly, with an average rate varying from a few to several centimeters per year. Whereas once this movement was calculated through indirect means, such as age determinations of oceanic crust, rates can now be measured directly from satellites orbiting the Earth.

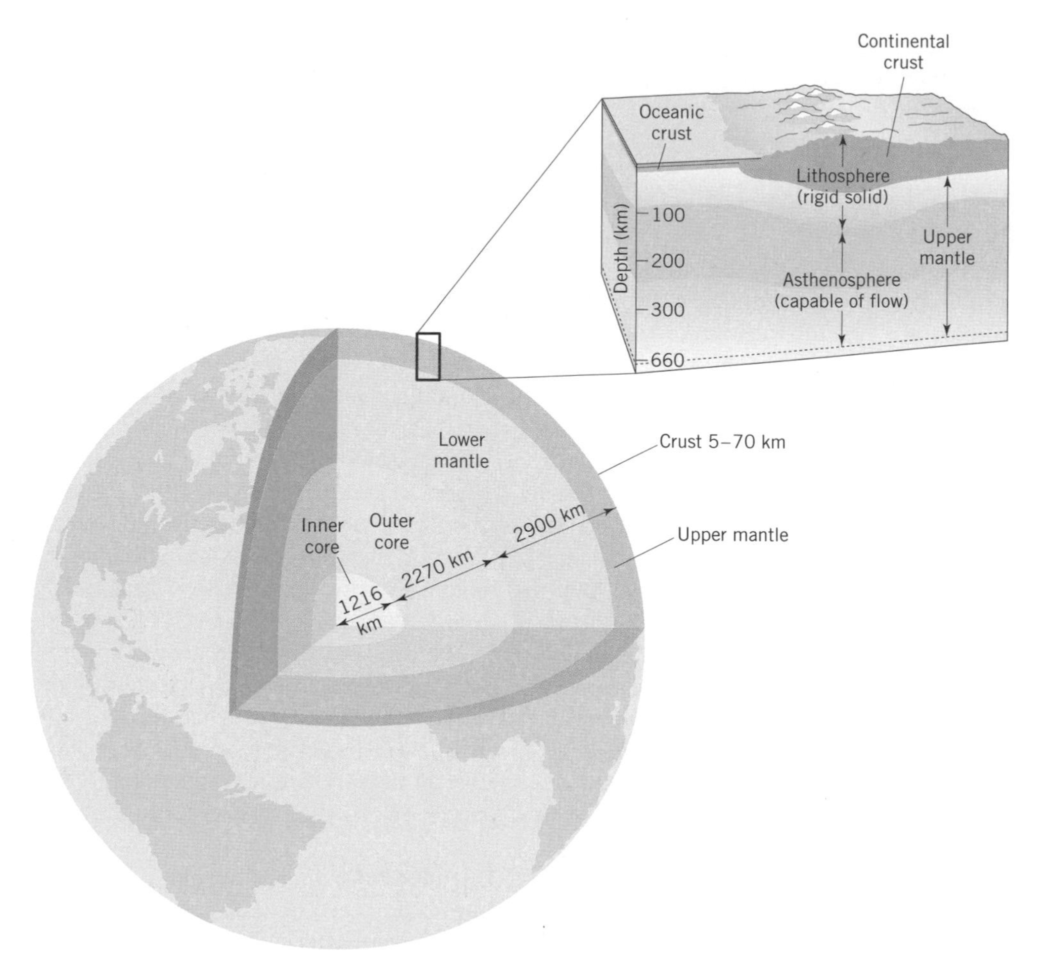

Fig. 2.5 The layered Earth reflects gross chemical and physical changes of the rocks with depth. These changes (which are really gradual changes) are responsible for the major boundaries in the Earth, including the inner and outer core, lower and upper mantle, and thin crust.

2 Asthenosphere. The lithospheric plates float on top of a semi-plastic region of the mantle called the asthenosphere, which extends to a depth of about 350 km. In this part of the mantle the high temperature and pressure causes the rock to partially melt, resulting in about 1–2% liquid. The partially melted rock allows the asthenosphere to be deformed plastically when stress is applied. Geologists have compared this plasticity to cold taffy, hot tar, and red hot steel. One way of illustrating this concept to yourself is by putting a chunk-sized piece of ice between your back teeth and applying slow constant pressure. You will see that the ice will deform without breaking in a manner similar to how the asthenosphere behaves when stressed. In terms of plate tectonics, the semi-plastic nature of this layer allows the lithospheric plates to move.

3 Mesosphere. Below the asthenosphere is the mesosphere, which extends to the mantle–core boundary. Although this layer has higher temperatures than the asthenosphere, the greater pressure produces a rock with a different, more compact mineralogy. This portion of the mantle is mechanically strong.

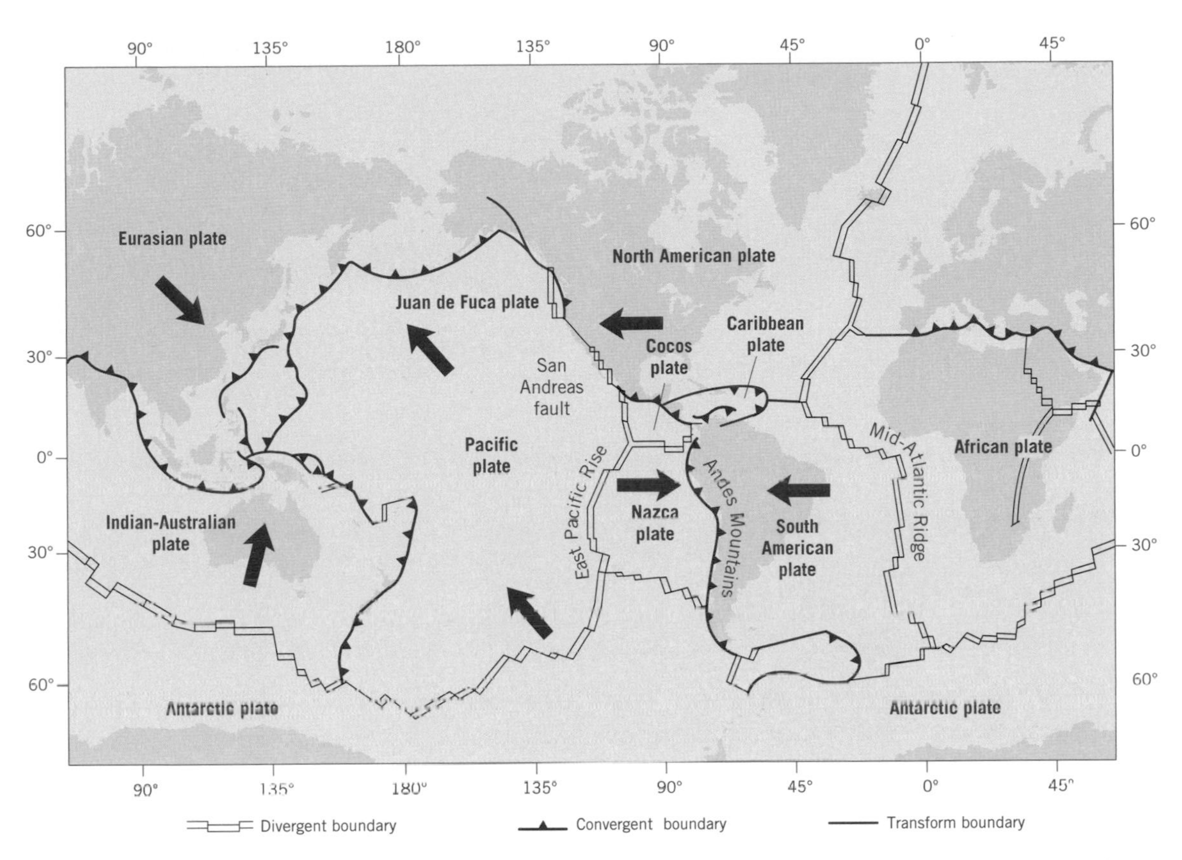

Fig. 2.6 The outermost rigid portion of the Earth, termed the lithosphere, is broken into eight major and several smaller lithospheric plates. Plates are separated from adjacent plates by divergent, convergent, and transform boundaries. Arrows indicate directions of plate movement.

2.3 Plate boundaries

In a general sense, the edges of the eight major lithospheric plates as well as those of the smaller ones are defined by a number of tectonic processes and geological features (Fig. 2.6). Most of the significant earthquakes and volcanic eruptions coincide with plate boundaries, as do the major mountain systems, ocean trenches, and mid-ocean ridges. As we discuss in more detail later, the movement of plates is a consequence of the heat transferal from interior of the Earth toward the surface. Because the plates are rigid slabs, when they move over the surface of the Earth they interact with other plates. This produces three types of plate boundaries (Fig. 2.7): plates that move apart from one another (divergent boundary), plates that move toward one another (convergent boundary), and plates that slide past one another (transform boundary). It should be noted that a single lithospheric plate can contain many different plate boundaries. For example, the Arabian Plate is separating from the African Plate along the Red Sea (divergent boundary), which is causing a convergent zone with the Eurasian Plate at the Zagros Mountains (along northern Iraq and Iran). Finally, the northwestern edge of the Arabian Plate is slipping past the African Plate (transform boundary) along the Dead Sea Fault, which coincides with the Gulf of Aqaba, the Dead Sea, the Sea of Galilee, and the border between Israel and Jordan.

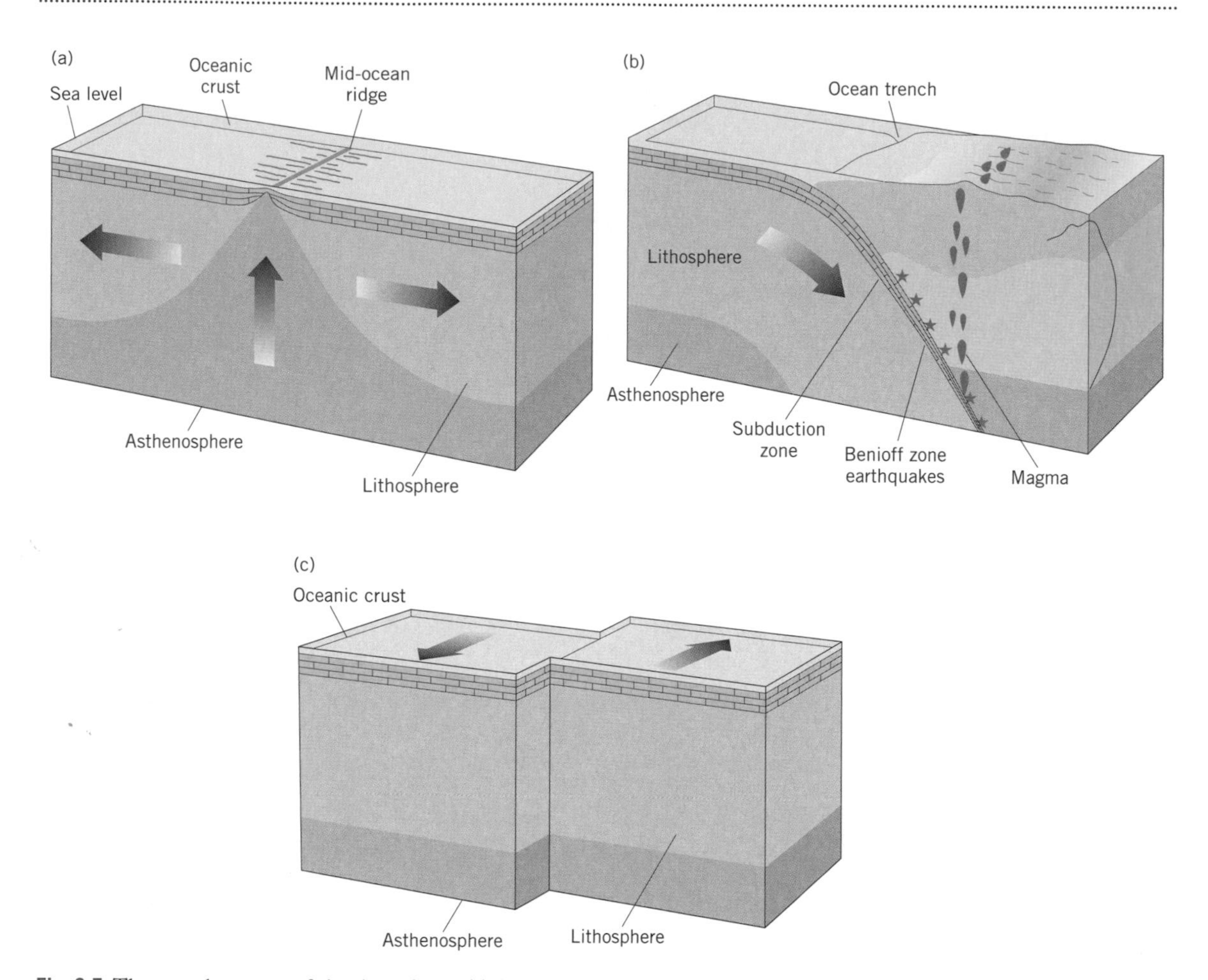

Fig. 2.7 There are three types of plate boundaries: (a) divergent, (b) convergent, and (c) transform boundaries.

2.3.1 Divergent boundaries

This type of boundary, also called a spreading zone and a rift zone, occurs in the middle of ocean basins or in the middle of continents (Fig. 2.8). Today most divergent boundaries are found in ocean basins; however, 200 million years ago extensive rift zones on land produced the breakup of Pangaea. In ocean basins, as the two lithospheric plates move apart, the asthenosphere wells up between the diverging plates, forming new oceanic crust. The decrease in pressure of the upwelling mantle produces partial melting of the asthenosphere and the formation of molten rock called magma. Some of the magma ascends to the sea floor, producing submarine volcanoes. The combination of volcanism, intrusion of magma in the overlying ocean crust, and the doming effect of the upwelling mantle creates a submarine ridge that extends along the length of the divergent boundary. The ridge, which rises approximately 2 kilometers (6600 feet) above the sea floor, rivals some mountain systems on land in size and stature. It is the single most continuous feature on the Earth's surface. The Mid-Atlantic Ridge, which marks the location where North America and South America separated from Europe and Africa, is only part of an extensive mid-ocean ridge system that winds it way through the world's oceans for some 65,000 kilometers (41,000 miles).

The exact boundary between the two plates is defined by a central rift valley. A view of this type of valley can be seen in the northern Atlantic, where

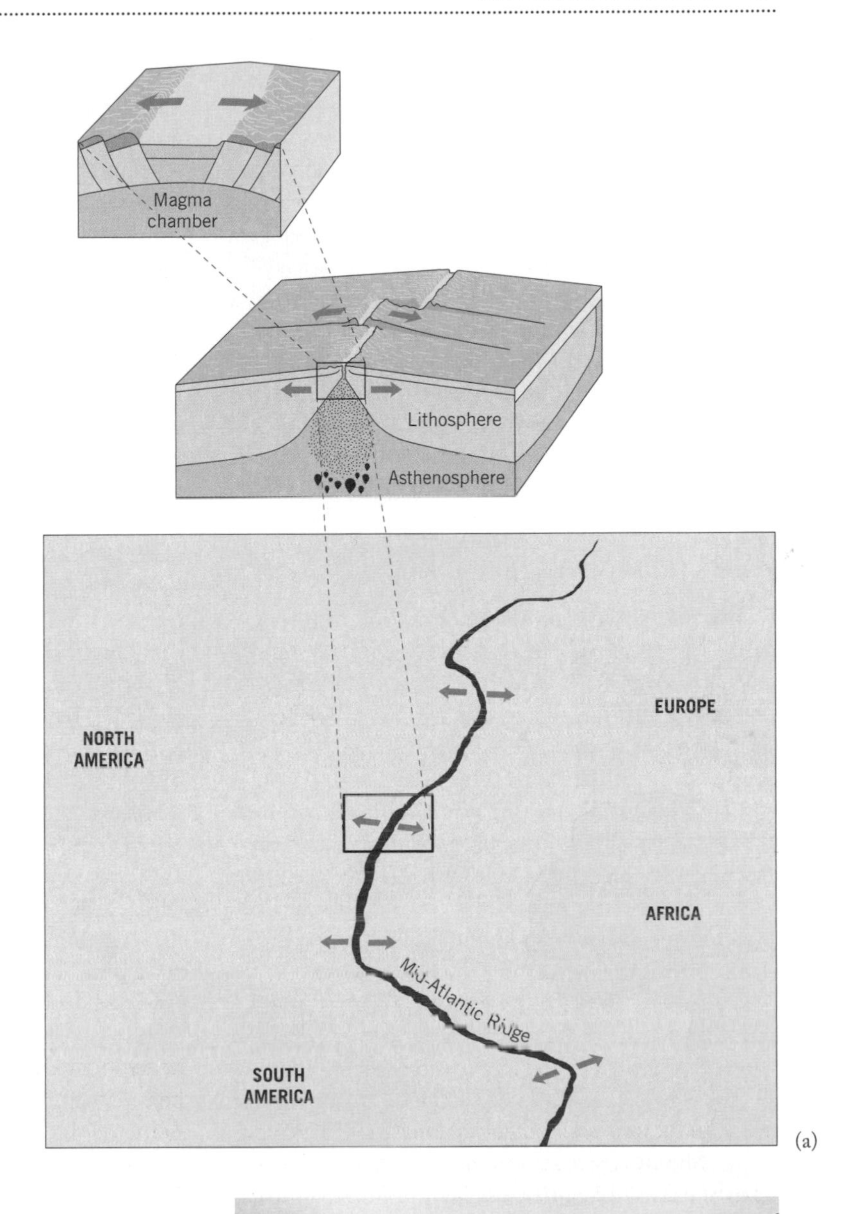

Fig. 2.8 The divergent boundary of the Mid-Atlantic Ridge system is (a) conceptualized in cross section and (b) shown as it appears on land where the divergent boundary moves onshore and bisects Iceland. (Photograph by Albert Hine, University of South Florida.)

the rift valley of the Mid-Atlantic Ridge is exposed in western Iceland. The valley is steep-sided and there is active volcanism. The volcanic rock comprising the large island of Iceland increases in age away from the rift valley. Likewise, the age of the oceanic crust on either side of the Mid-Atlantic Ridge gets older toward the bordering continents. This leads to greater sediment accumulation away from the ridge, resulting in a general smoothing of the once irregular, young seafloor and the formation of the relatively flat abyssal plains. Thus, divergent boundaries are like two conveyers moving newly formed oceanic crust away from a central ridge, where it cools, sinks, and becomes covered with sediment.

2.3.2 Convergent boundaries

Lithospheric plates moving toward one another are composed of the upper rigid portion of the mantle and topped by either the dense basaltic oceanic crust or the lighter granitic continental crust. This leads to three types of plate convergences (Fig. 2.9), each dominated by different tectonic processes and resulting landforms: (i) oceanic plate colliding with an oceanic plate; (ii) oceanic plate colliding with a continental plate; and (iii) continental plate colliding with a continental plate. The contact between the plates is not always head on but rather the two plates commonly meet in an oblique convergence.

Ocean–ocean plate convergences occur throughout the margins of the northern and western Pacific Ocean, where the Pacific oceanic plate, moving northwestward, collides with the oceanic crust of the Eurasian and North American plates. This type of plate boundary is also found along the Caribbean Islands and in the southern Atlantic (South Sandwich Islands). When two oceanic lithospheric plates converge, the older and hence cooler and denser plate descends beneath the younger and more buoyant ocean plate. Thus, the Pacific Plate is sliding under the Eurasian and North American lithospheric plates and is descending into the semi-plastic asthenosphere. This process whereby plates that are created at mid-ocean ridges descend into the mantle and are consumed at convergent zones is called subduction (Fig. 2.10). The depth to which the plate descends into the mantle and the geometry of the downgoing slab are known from the numerous earthquakes that are produced during the subduction process.

One of the major features associated with subduction zones is deep ocean trenches, which are caused by the flexure of the downgoing plate. Trenches are the deepest regions in the oceans, being 8–11 km in depth or 3–5 km deeper than the surrounding ocean floor. They can be thousands of kilometers long. The deepest region on Earth is found in the Mariana Trench (11 km) in the western Pacific, which is over 2 km deeper than Mount Everest (8.8 km) is high. Ocean trenches are relatively steep on the descending plate side, whereas the overriding plate margin has a shallow slope. As the oceanic plate is subducted, some of the sediment that has accumulated on the ocean floor is scraped off and plastered against the adjacent plate margin. This produces an accretionary sedimentary prism that manifests itself as a low submarine ridge along the length of the ocean trench.

Paralleling the overriding plate margin is a chain of volcanic islands referred to as an island arc. Examples include the Philippine Islands in the western Pacific and the Aleutian Islands, which extend 2500 km westward from the Alaskan Peninsula (Fig. 2.11). As the subducted plate descends into the mantle, water is released from the downgoing slab. This water lowers the melting point of the already hot surrounding rock, facilitating the generation of magma at a depth of about 120 km in the overlying asthenosphere. Being less dense than the mantle rock, the magma rises toward the surface. Some of the magma intrudes and solidifies within the overlying ocean crust. A small portion of the magma reaches the surface and erupts on the ocean floor. As this process proceeds, the volcanic pile coupled with the thickening ocean crust produces the island arc. It should be noted that the volcanic islands are the surface expression of this arc and comprise only a small portion of the system. As the island arc increases in elevation, more and more sediment is shed from the volcanic arc and some of the sediment is transported to the ocean trench, where it is metamorphosed and deformed by the compressive forces of the converging plates. The

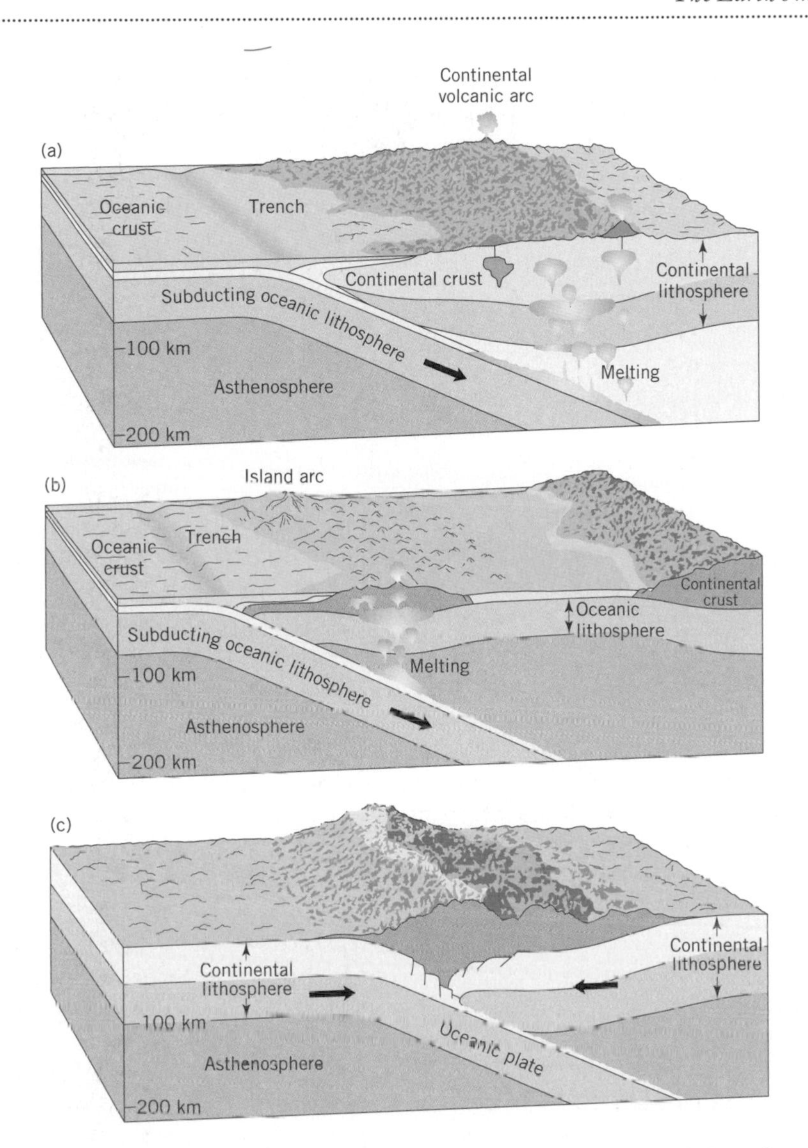

Fig. 2.9 Convergent zones involve three types of lithospheric plate collisions: (a) oceanic–continental plate collision (Nazca and South American plates); (b) oceanic–oceanic plate collision (Pacific and Philippine plates); (c) continental–continental collision (Arabian and Eurasian plates).

mature island arcs, such as the Japanese arc, consists of a complex mix of volcanic rocks, deformed sedimentary and metamorphic rocks, and intruded igneous rocks. Because of the varying angle of the descending plate, the formation of the island arc along the adjacent plate margin occurs 50–200 km from the ocean trench.

Ocean–continent plate convergences occur where a lithospheric plate containing relatively thin and dense oceanic crust is subducted beneath a lighter, thicker continental crustal plate. Similar to the ocean–ocean plate convergences, the flexure of the downgoing plate causes a deep ocean trench offshore of the continent. Likewise, dewatering of the subducting slab produces partial melting of the asthenosphere. Although most of the rising magmatic

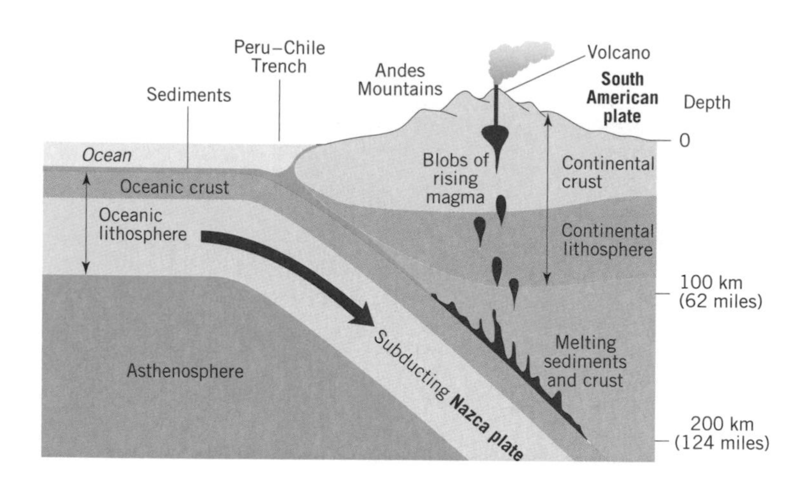

Fig. 2.10 Subduction is the process whereby an oceanic lithospheric plate descends into the mantle at a zone of plate convergence. Earthquake activity, volcanism, mountain building, and oceanic trench formation characterize subduction zones.

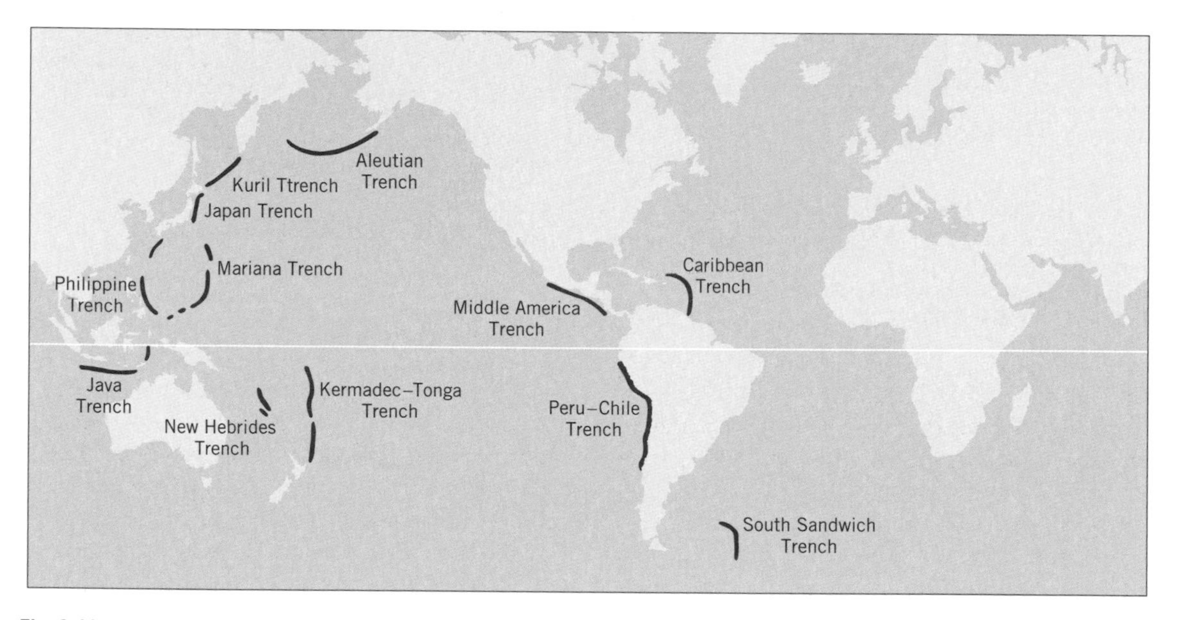

Fig. 2.11 Distribution of oceanic trench systems. Trenches are sites of the deepest depths in the oceans.

plume solidifies within the overlying crust, some of the magma reaches the surface, causing explosive volcanic activity. The combined processes of magma intrusions and volcanic eruptions produce thick continental crust and high mountain systems.

One example of this type of convergent boundary occurs where the Nazca Plate, moving eastward, is being subducted beneath the South American Plate moving westward. The convergent boundary is marked by the 5900 km long Peru–Chile Trench. The thickening of the South American Plate margin is evidenced by the immense Andes Mountains, which reach over 6 km in elevation and are the site of frequent volcanic and earthquake activity. Another ocean–continent convergent boundary is found where the Juan de Fuca and Gorda plates are descending beneath North America along northern California, Oregon, and Washington. The devastating eruption of Mount Saint Helens in Washington state on May 18, 1980, in which a cubic kilometer of rock

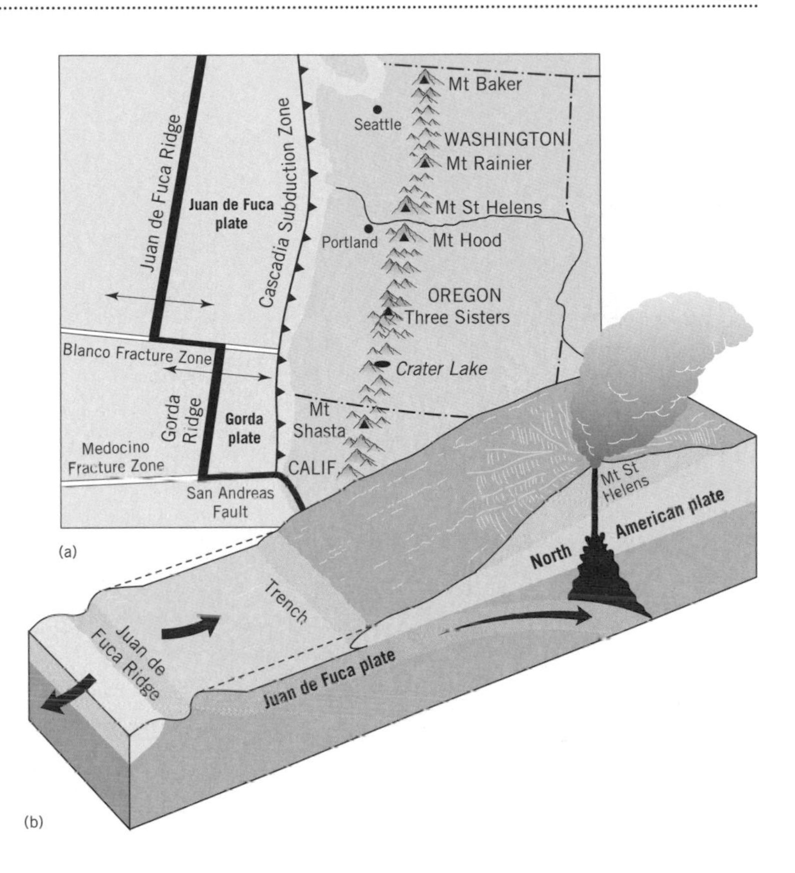

Fig. 2.12 Subduction of the Juan de Fuca and Gorda plates beneath the North American Plate has produced the Cascade Range and is responsible for the 1980 eruption of Mount St Helens.

was ejected, lowering the mountain by 410 m (14% of its height), is a product of this subduction process (Fig. 2.12). During the initial eruption 59 people were killed.

In some ocean–continent plate convergences, such as the former western margin of North America 180 to 80 million years ago, the long-term subduction of oceanic crust led to the "docking" of numerous microcontinents. These terrains, such as island arc systems, rode the ocean plate eastward toward North America but were not subducted into the mantle because they were too light. Instead, they are added to the continental margin, widening the continent up to 1000 km.

Continent–continent plate convergences are responsible for the closure of ocean basins and the formation of majestic mountain systems. A geologically young and spectacular example of this type of convergence occurred as a result of the breakup of Pangaea when India rifted away from Antarctica and rode a northward-moving plate toward Eurasia. The leading edge of this plate consisted of oceanic crust. Several thousand kilometers of oceanic lithosphere were subducted beneath Eurasia before the collision with the subcontinent of India, approximately 45 million years ago. During the period of oceanic–continental convergence, the margin of Eurasia was greatly thickened by magmatic intrusion and volcanic accumulation. As the subcontinent of India approached and the ocean basin closed, the great pile of sediment that had been shed by rivers draining Eurasia and deposited into the adjacent sea was bulldozed, along with some of the oceanic crust, onto the continental margin (Fig. 2.13). The formation of this accretionary sedimentary prism in combination with the aforementioned crustal thickening produced the highest mountains on Earth, the Himalayas and the Tibetan Plateau. Although we know that the Indian–Australian Plate continues to collide with Eurasia, geologists are uncertain

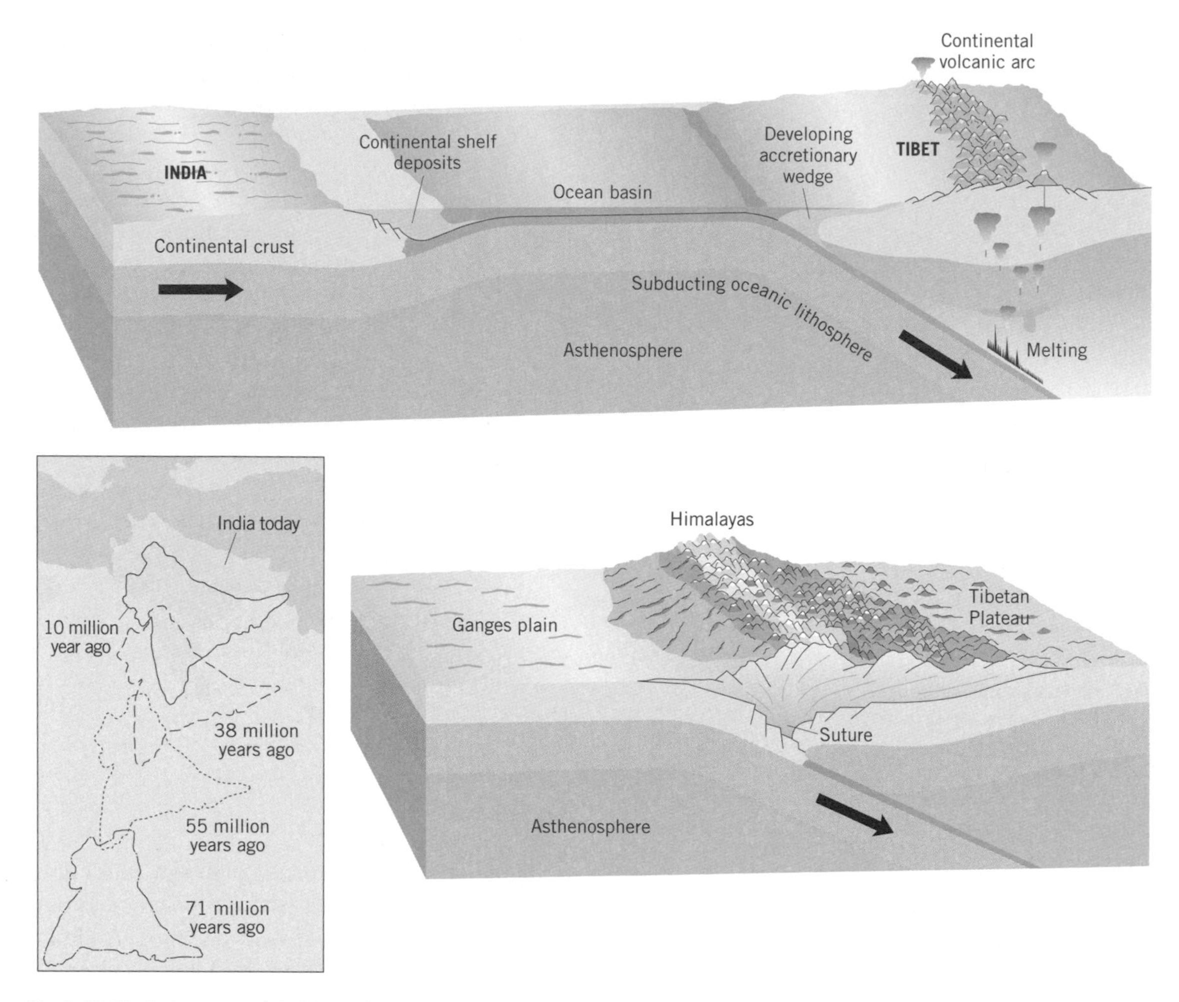

Fig. 2.13 Evolutionary model of the Indian–Eurasian continental collision. (a) The actual collision of the two landmasses, which began approximately 45 million years ago, was preceded by the subduction of extensive oceanic lithosphere beneath Eurasia. (b) During this period of oceanic and continental convergence, the margin of Eurasia was greatly thickened through magmatic intrusions, thrust sheets, and an accreting volcanic arc. (c) Ultimately, the collision with India further uplifted the margin of Eurasia, producing the majestic Himalayas and Tibetan Plateau.

what is happening to the downgoing portion of the plate. Continental crust is too buoyant to be subducted and, therefore, most geologists believe that the oceanic part of the Indian–Australian Plate has broken from the rest of the plate and continues to descend and be consumed into the mantle.

In addition to the formation of the Himalayas, other continental–continental convergences have occurred in the geological past, including the collision between North America and northern Africa (during the formation of Pangaea), which was responsible for the development of the Appalachian Mountains. The suture of Europe to Asia formed the north–south trending Ural Mountains.

2.3.3 Transform boundaries

Transform boundaries occur where crustal plates shear past one another. Unlike at the other plate boundaries, lithosphere is neither created nor

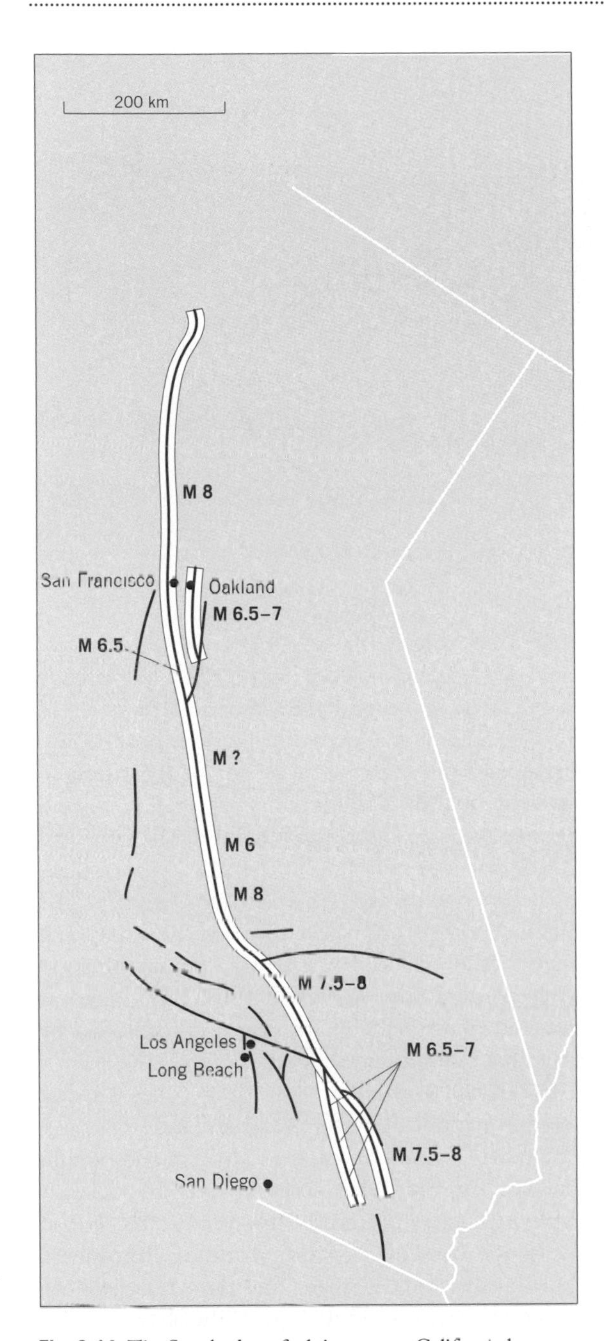

Fig. 2.14 The San Andreas fault in western California has been the site of some of the largest recorded earthquakes in the northern hemisphere. Land on the western side of the transform is part of the Pacific Plate, moving in a northwesterly direction, whereas the land on the eastern side of the fault is attached to the North American Plate.

destroyed at transforms. The most common site of transforms is along mid-ocean ridges. The spreading that occurs at ridges is produced by upwelling magma from many different magma chambers. These magma sources are not aligned and may not even be connected. Consequently, the ocean ridges and spreading that occur at these sites are not aligned. A close inspection of mid-ocean ridges, such as the Mid-Atlantic Ridge, shows that while the ridge is continuous, it is offset along its axis. On either side of the offset the newly formed ocean crust moves in opposite directions, but only between the two ridges. Beyond the ridge axis the crust spreads in the same direction. These shear zones are called transforms because the motion along the boundary can terminate or change abruptly. They are sites of shallow earthquake activity.

Transforms lie not only between spreading ocean ridges but also between other types of plate boundaries. One of the most famous is the San Andreas fault, which is a transform boundary between the Pacific and North American plates (Fig. 2.14). This transform connects the spreading zone of the Juan de Fuca ridge and a spreading zone centered in the Gulf of California. It is almost 1300 km (780 miles) long and encompasses a large segment of western California. Because the boundary is located on a continent, the sideways grinding movement between the two plates involves a much greater thickness of crust than oceanic transforms. Hence, the energy released when the plates abruptly slide past one another can be large and catastrophic, such as in the great San Francisco earthquake of 1906. The San Andreas fault has been active for approximately the past 30 million years and lateral movement along the fault has amounted to more than 550 km (340 miles).

2.3.4 Plate movement

While satellite technology can now accurately measure the exact rate at which North America and northern Africa are moving apart from one another and how fast the Hawaiian Islands are approaching the Aleutian Trench, where they will be consumed, the mechanism that drives the

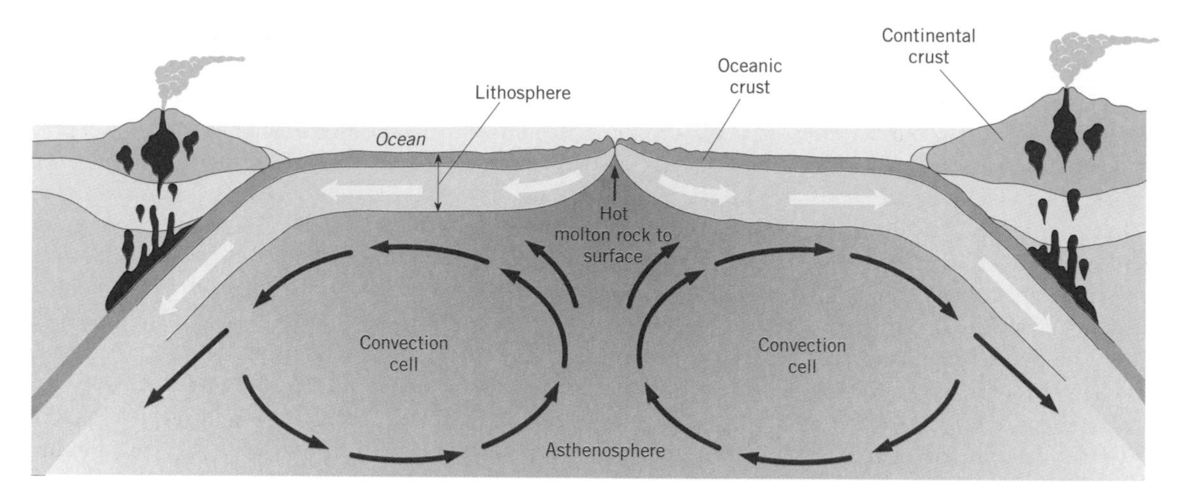

Fig. 2.15 Simplified view of the convection process that drives plate motion and is involved with the ascension of magma at mid-ocean ridges and the descent of oceanic lithosphere at subduction zones.

lithospheric plates, and the continents and island chains that they carry, is much less well known.

Most scientists believe that plate movement is due to some type of convection process produced by the unequal distribution of heat within the Earth. Although considerable heat has been given off since the Earth's original melting phase, additional heat is still being conveyed to the surface. This heat is derived from the molten core and from heat that is produced in the mantle and crust through radioactive decay of unstable atoms as they change to more stable atomic configurations. The expulsion of this heat likely involves convection cells, whereby hot material rises toward the surface and cool material descends (Fig. 2.15).

Convection cells operate in a manner similar to the way in which a beaker full of liquid reacts to being heated at the bottom along a single side. As the liquid heats, it expands and rises upward. Upon reaching the surface, the warm liquid moves out laterally toward the opposite side of the beaker and cools. At the same time, the void left by the upwelling heated liquid is replaced by cool liquid moving along the bottom. The circulation cell is completed as the moderately cool liquid at the surface descends along the other side of the beaker.

There is still considerable debate concerning the dimensions of convection cells within the Earth and the processes that produce them. Proposed models range from convection cells that are relatively shallow and circulate material within the upper mantle to cells that encompass the whole mantle down to the core boundary. One synthesis of various theories envisions large masses of very hot rock forming at the core–mantle boundary and ascending toward the lithosphere. Upon reaching the relatively cool lithosphere the mantle plumes, as they are called, are believed to mushroom and spread laterally. The doming process and shear force of the expanding plume ultimately lead to a rifting apart of the overlying plate. Once spreading (divergent zone) is initiated, movement of the plates is facilitated by a number of forces, including:

1 Gravitational sliding. Spreading zones are hot elevated regions due to upwelling mantle material. The newly formed lithosphere at these sites is thin and hot. As the plates spreads from the ridge, it thickens, cools, and sinks as the doming effect is lost. Thus, the base of the newly forming lithosphere slopes away from the ridge. This slope may produce a condition whereby movement of the plate is caused by gravitational sliding.

2 Slab pull. Lithospheric plates float on top of the asthenosphere because they are less dense. As stated above, when a plate spreads away from the mid-ocean ridge where it was formed, it gradually cools and thickens. At some distance, the contraction of the lithosphere due to cooling produces a slab more dense than the underlying asthenosphere. At this point, the plate descends into the mantle and a subduction zone is formed. It is thought that the subducting plate may pull the rest of the plate with it. These descending slabs have been traced by earthquake activity as deep as 700 km into the mantle.

In trying to assess the importance and viability of the convection processes described above, geologists have made a number of important observations that help to constrain the various models. The North American Plate, which is spreading westward from the Mid-Atlantic ridge, contains no subduction zone, indicating that the slab pull is not a required force in the movement of all plates. Similarly, when ocean basins are first being formed, such as in the Red Sea, there are no topographically high mid-ocean ridge systems to produce gravity sliding. Earthquake studies have also shown that the magma chambers responsible for the formation of new lithosphere at mid-ocean ridges appear to be no deeper than 350 km. Therefore, while deep mantle plumes may produce the rifting apart of continents, they may not sustain the spreading process along the entire length of the plate. Submersible investigation of the axial valleys of the mid-ocean ridges reveals evidence of giant cracks and fissures, suggesting that these sites are not a product of magma pushing the plates to the side; instead, the upwelling magma is a passive response to the plates moving apart by some other force. Thus, we are left with a working hypothesis that plates are formed, move laterally, and are subducted into the mantle as a result of some type of convection cell encompassing the lithosphere and mantle, but the details of the process are still being discovered!

2.4 Continental margins

Continental margins are the edges of continents and the container sides of the deep ocean basins. Geologically, they represent a transition zone where thick granitic continental crust changes to thin basaltic oceanic crust. The margin includes the physiographic regions known as the continental shelf, continental slope, and continental rise (Fig. 2.16).

Continental shelves are the submerged, shallow extensions of continents stretching from the shoreline seaward to a break in slope of the seafloor. Beyond the shelf break, water depths increase precipitously. The average shelf break occurs at 130 m but this depth ranges widely. Continental shelves are underlain by granitic crust and covered with a wedge of sediment that has been delivered primarily

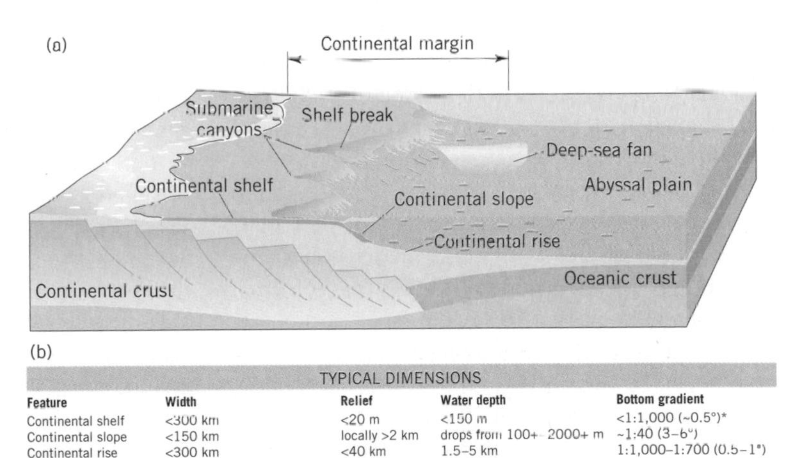

(b)

TYPICAL DIMENSIONS				
Feature	**Width**	**Relief**	**Water depth**	**Bottom gradient**
Continental shelf	<300 km	<20 m	<150 m	<1:1,000 (~0.5°)*
Continental slope	<150 km	locally >2 km	drops from 100+–2000+ m	~1:40 (3–6°)
Continental rise	<300 km	<40 km	1.5–5 km	1:1,000–1:700 (0.5–1°)
Submarine canyon	1–15 km	20–2,000 m	20–2,000 m	<1:40 (3–6°)
Deep-sea trench	30–100 km	>2 km	5,000–12,000 m	–
Abyssal hills	100–100,000 m (100 km)	1–1,000 m	variable	–
Seamounts	2–100 km	>1,000 m	variable	–
Abyssal plains	1–1,000 km	0	>3 km	1:1,000–1:10,000 (>0.5°)
Midocean ridge flank	500–1,500 km	<1 km	>3 km	–
Midocean ridge crest	500–1,000 km	<2 km	2–4 km	–

* A bottom gradient of 1:1,000 means that the slope rises 1 m vertically across a horizontal distance of 1,000 m.
Source: Adapted from B.C. Heezan and L. Wilson, Submarine geomorphology in *Encyclopedia of Geomorphology*, R.W. Fairbridge, ed. (New York: Reinhold, 1968); and C.D. Ollier, *Tectonics and Landforms* (New York: Longman, 1981).

Fig. 2.16 (a) Physiographic provinces of continental margins including the continental shelf, continental slope, and continental rise. (b) Depths and dimensions of continental margins and other ocean regions.

by rivers to the shore, where it has been redistributed by tides, waves, and shelf currents. In glaciated regions much of the sediment may have been derived through glacial deposition (Georges Bank east of Cape Cod). In equatorial areas the sediment cover may consist of calcium carbonate (which makes up sea shells) that is biogentically derived (i.e. from the exoskeletons of various organisms including coral) or precipitated directly from seawater (particles called oolites, a calcium carbonate concretion).

Although all shelves are relatively flat, their gradients and widths vary considerably. The widest shelves occur in the region surrounding the Arctic Ocean and in a band extending from northern Australia northward toward Southeast Asia. Here the shelves may be more than 1000 km wide. At other sites, such as the eastern margin of the Pacific Ocean, continental shelves are comparatively narrow. The average shelf is 75 km wide and has a slope of 0° 07′. This is equivalent to the slope of a football field (100 yards long) in which one goal line is 6 inches (15 cm) higher than the other goal line; such slopes appear flat to the human eye. However, the topographic expression of continental shelves is not always flat. There can be tens of meters of relief in the form of valleys, hills, sand ridges, and other features that can be attributed to the erosional and depositional processes associated with glaciers, rivers, tidal currents, and storms.

Continental slopes mark the edge of the continental shelves. Here the sea-floor gradient steepens dramatically as the continental crust thins and is replaced by oceanic crust. Continental slopes descend to depths ranging from 1500 to 4000 m but may extend much deeper at ocean trenches. They are commonly only 20 km in width. Slopes are composed of sediment that forms the outer portion of the tilted sedimentary layers comprising the continental shelf. In regions where oceanic lithospheric plates are being subducted offshore of a continent (e.g. the west coast of South America) or an island arc (Aleutian Islands off Alaska), the slope may also consist of oceanic sediment that is scraped off the downgoing slab. The gradient of continental slopes varies greatly from 1 to 25°. The average gradient is slightly steeper along the margin of the Pacific Ocean (5°) than in the Atlantic or Indian Oceans (3°) due to the number and extent of deep ocean trenches that are associated with the Pacific margin's subduction zones.

Continental rises are formed from the transport of sediment down continental slopes and its accumulation in an apron-like fashion at the base of the slope. The thickness of these deposits can be more than several kilometers. The slope of the rise decreases toward the flat abyssal plains and is typically less than 1°. Their widths vary greatly but average a few hundred kilometers. Continental rises are not found where the continental borderlands coincide with subduction zones. In these regions, continental rises are replaced by deep sea trenches (e.g. Mariana Trench, 11 km deep).

Submarine canyons are the conduits through which sediment is delivered to continental rises (Fig. 2.17). As the name suggests, they are V-shaped, usually steep-walled valleys, and they resemble river-cut canyons on land in both size and relief. Most commonly, they are incised into the edge of continental shelves and extend down the continental slope to the rise. They occur ubiquitously throughout the world's oceans. Their origin has been the subject of controversy since they were discovered and studied by the late Francis P. Shepard (1898–1985) of Scripps Institute of Oceanography in California. Because of their likeness to river canyons and the fact that some submarine canyons extend across portions of the continental shelf toward major rivers, they were once believed to have formed by river erosion when sea level was much lower than it is today and the shelf edge was exposed. However, some canyons begin at water depths much lower than sea level is ever thought to have dropped. Additionally, the fact that canyons extend across continental slopes requires some mechanism of submarine erosion. It is important to note that the relationship between rivers and submarine canyons cannot be totally discounted. For example, Hudson Canyon off the eastern seaboard of the United States can be traced to the mouth of the Hudson River.

It is now generally accepted that submarine canyons are formed due to erosion by turbidity currents.

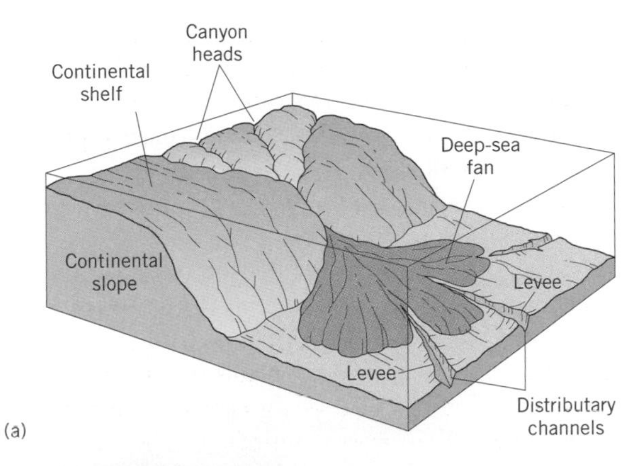

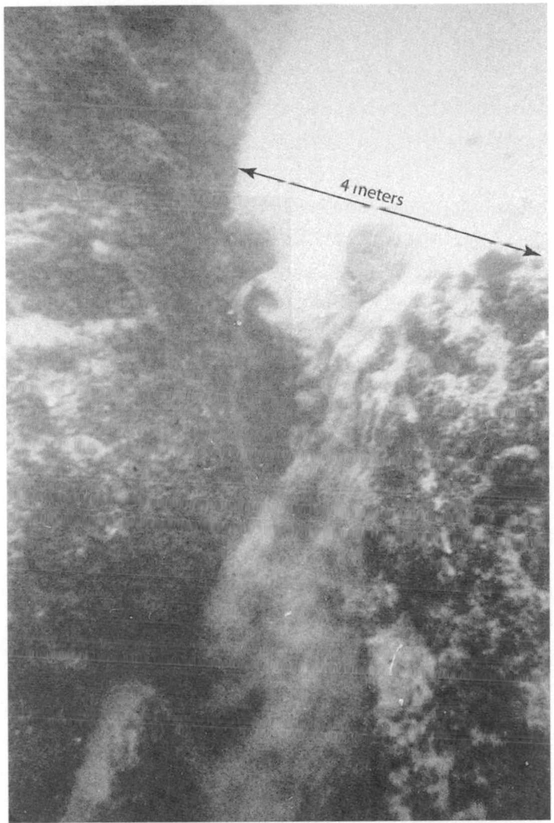

Fig. 2.17 (a) Conceptualized drawing of a submarine canyon. Submarine canyons occur along the shelf–slope break and are found ubiquitously on continental margins throughout the world. (b) Photograph of sand cascading down the head of a submarine canyon. (From Robert Dill.)

A turbidity current is a sediment-laden torrent of water moving downslope under the influence of gravity. Turbidity currents are initiated along the shelf edge by sediment being put into suspension by a turbulent event, such as an earthquake or underwater landslide. As the suspended sediment moves downslope, the sedimentary particles entrain water, forming the turbidity current. In a manner identical to rivers, the moving sedimentary particles, including sand and silt, act as abrading agents and erode the valley, producing submarine canyons. At the base of the continental slope the sediment carried by turbidity currents through canyons is deposited, creating a submarine fan. It is the coalescence of these fans that forms the continental rise. The largest submarine fans in the world coincide with some of the greatest sediment discharge rivers of the world. The Bengal Fan (2500 km long) and Indus Fan (1600 km long) are direct products of the Ganges Brahmaputra and Indus rivers, respectively, draining huge quantities of sediment from the Himalayan Mountains and the gradual transport of this sediment into deep water by turbidity currents. The Amazon Cone is another very large submarine fan forming from a high sediment discharge river.

2.4.1 Tectonic evolution of continental margins

Although the morphological character of margins ranges widely throughout the world, much of the variability can be traced to different stages of two evolutionary tectonic styles: the Atlantic and Pacific models (Fig. 2.18).

Atlantic margin

The Atlantic margin, which is also referred to as a passive margin, occurs where the edge of a continent coincides with the middle of a lithospheric plate, and hence there is no tectonic plate interaction. Because of this location, there is little seismic activity and no volcanism. As the name implies, the eastern margin of North and South America and the western margin of Africa and Europe are examples of this margin type. The margin surrounding India and most of Australia fits into this category too.

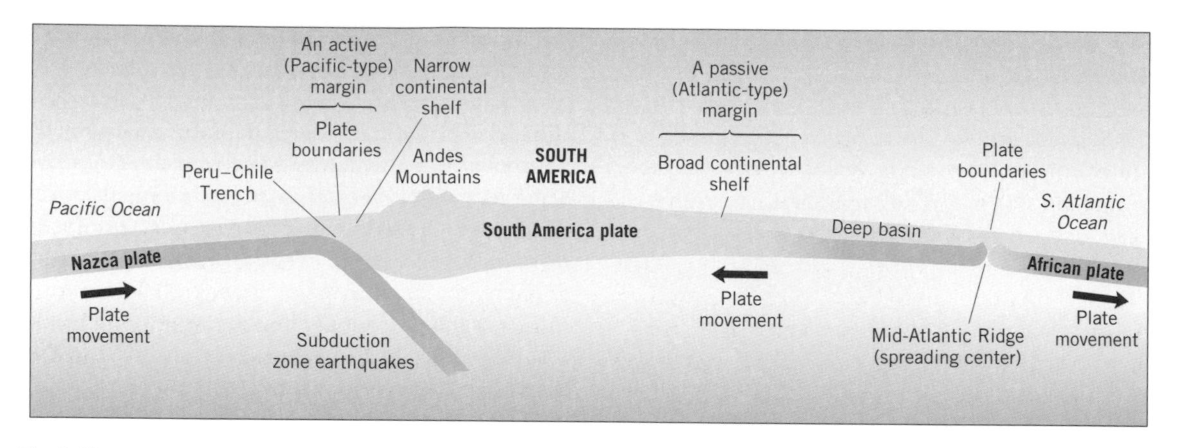

Fig. 2.18 Cross-sectional view of Atlantic (passive) and Pacific (active) continental margins using South America as a model.

During its early history, the evolution of Atlantic-type margins involves the rifting apart of large landmasses, such as the breakup of Pangaea. In the initial phase the continental crust is bulged upward by a rising mantle plume, perhaps ascending from the core–mantle boundary, as discussed earlier in this chapter, or from elsewhere within the mantle. This stage may be represented by Yellowstone National Park in northwestern Wyoming and the Rio Grande region of Colorado and New Mexico, where high heat flow values have been measured and uplift and crustal thinning are occurring. As the plume mushrooms and spreads laterally, the overlying crust is stretched and thinned until it ruptures, producing down-dropped continental crustal blocks that form a series of elongated basins. The central basin of this system is called the rift valley, which marks the separating landmasses. In the early stages, basaltic lava derived from the mantle plume may flow onto the valley floor. As the central valley widens and its elevation lowers, the ocean invades forming a linear sea. The Red Sea and Gulf of California are examples of this stage of Atlantic margin development. Continued spreading of the landmasses through the creation of oceanic crust leads to the formation of a new deep-ocean basin and mid-ocean ridge system.

As the ocean basin evolves, so too do the continental margins. The linear basins that were formed by the down-dropped continental blocks during the initial rifting phase become depocenters for sediment delivered chiefly by river systems. As the underlying granitic crust subsides due to a deflation of the mantle plume, the rift valley is ultimately transformed into a juvenile ocean. The land-based, river-lain deposits are now covered by shallow marine sediments. The incipient shelf continues to develop and the sedimentary wedge slowly thickens as sediment is shed from the adjacent continent and dispersed by waves, tidal currents, and storm processes. During this phase the slowly subsiding shelf provides accommodation space for continued sedimentation, while maintaining a shallow marine platform. Subsidence of the shelf results from cooling and contraction of the mantle plume as well as from the weight of the accumulating sedimentary deposits. In warm water environments, coral reef formation along the edge of the shelf aids in trapping sediment by providing a barricade to the deep sea. The end product of these sedimentation processes is the formation of a broad, shallow, flat continental shelf.

Along the eastern margin of the United States, oceanographic studies have revealed that some of the sedimentary deposits are more than 10 km thick and their average thickness is 4–5 km. The outermost portion of the shelf from Florida to New England contains the framework of a buried coral reef that is several kilometers in height. It appears that this extensive coral reef system lived in relatively shallow water (10–30 m) and was able to grow vertically as the shelf platform slowly subsided. Sometime, approximately 100 million years ago, the coral reefs died and were subsequently buried by shallow marine sands, silts, and clays. Not all Atlantic-type margins are identical to that of the east coast of the United States. The widths and gradients vary greatly, as do the sediment thicknesses and the topography of the shelf. For example, along the northwestern margin of Africa the continental shelf is very narrow (30 km wide) and only about one-sixth that of the eastern United States. The variability of Atlantic margins is related to the original processes of rifting, the extent of crustal subsidence, the supply of sediment to the shelf, climatic factors (e.g. controlling reef development), the strength of oceanic currents, and other factors.

Pacific margin
The Pacific margin type occurs at the edge of lithospheric plates and thus it is also called an active margin. These margins are confined primarily to the rim of the Pacific Ocean, where oceanic plates are being subducted beneath continental plates. Because they coincide with subduction zones, they are tectonically active and are characterized by earthquake activity and onshore volcanism. Pacific margins have narrower continental shelves and steeper continental slopes than Atlantic margins. Continental slopes of Pacific margins descend deep into adjacent oceanic trenches and thus continental rises are usually absent. An exception to this trend occurs along much of the west coast of North America, south of Alaska, where there is no subduction zone today but where one existed 25 million years ago. Here, sediment transported across the narrow shelf drains through submarine canyons and is building large sediment fans on the floor of the deep ocean.

Differences in the dimensions and morphology of Pacific versus Atlantic margins are the result of contrasting tectonic histories. Whereas Atlantic margins develop wide continental shelves due to rifting, the slow subsidence of broken up continental blocks, and the accumulation of great thicknesses of sedimentary deposits (several kilometers), Pacific margins are narrow and are a product of the compressive forces of an oceanic plate being subducted beneath a continent. As the oceanic plate flexes downward, forming the seaward margin of the ocean trench, some of the deep sea sediment that overlies the basaltic ocean crust is scraped off and plastered onto the adjacent continental margin. The forces involved in this process chemically alter the sedimentary layers and physically disrupt the sedimentary prism through folding and faulting. The shelf region receives some sediment from material that is eroded from adjacent continental hinterlands and delivered to the coast via river systems. However, the sedimentary wedge comprising Pacific shelf margins is relatively thin when compared to most Atlantic margins.

2.5 Tectonic coastline classification

In the early 1970s, plate tectonic theory, which had served to enlighten and transform many subdisciplines of geology, was applied to the field of coastal geology. Two scientists from Scripps Institute of Oceanography in California, Douglas Inman and Carl Nordstrom, produced their now classic work "On the tectonic and morphologic classification of coasts." This scheme provides a first-order characterization of the morphology and tectonic processes of 1000 km long stretches of continental margin, including a description not only of the coastline but also of the continental shelf and the uplands bordering the coast. The classification is based primarily on the tectonic setting of the coast (Pacific- versus Atlantic-type margins). Secondary factors dictating coastline subclasses include: (i) tectonic setting of the opposite side of the continent; (ii) geological age of the coast; and (iii) exposure of the coast to open ocean conditions.

It is important to note that this classification is meant as a first-order characterization of a coastline along the length of a continent. There will be many exceptions to the general trends presented here due to secondary factors such as the presence of a major river, the effects of glaciation, or climatic influences, which may have widespread effects too.

The Inman and Nordstrom classification consists of the following (Fig. 2.19):

I Collision coasts
 A Continental collision coasts
 B Island arcs collision coasts

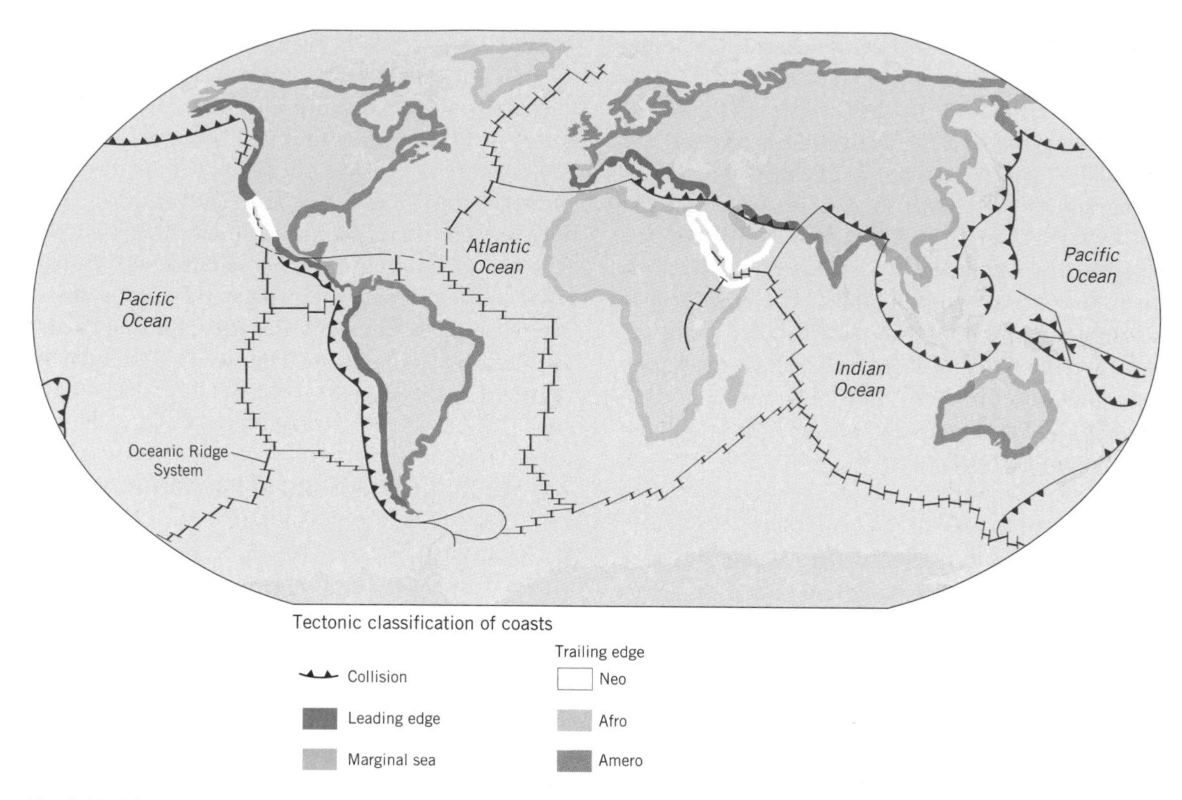

Fig. 2.19 Global tectonic coastal classification. (After D. L. Inman & C. E. Nordstrom, 1971, On the tectonic and morphologic classification of coasts. *Journal of Geology*, **79**, 1–21.)

II Trailing edge coasts
- **A** Neo-trailing edge coasts
- **B** Afro-trailing edge coasts
- **C** Amero-trailing edge coasts

III Marginal sea coasts

2.5.1 Collision coasts

Continental collision coasts

Continental collision coasts (IA) coincide with convergent, Pacific-type margins where oceanic lithospheric plates are being subducted into the mantle beneath the continent. Characteristically, these coasts have narrow continental shelves that terminate next to deep ocean trenches. The coastlines are backed by high mountain systems such as the Andes, which contain many peaks over 6000 m (Fig. 2.20). Mountain building is a product of crustal thickening and volcanism caused by upwelling magma emanating from the subducted slab. In

Fig. 2.20 Mountainous coast along a tectonically active continental margin (Alaska).

some locations, including the Coast Range along northern California, the mountains adjacent to the coast are much lower in elevation because they consist of ocean sediments and pieces of ocean crust that were scraped off the subducted plate and thrust onto the continental margin. Located much further inland are the considerably higher Sierra Nevadas (>6000 m), which are similar in origin to the Andes but now consist primarily of granite. Because the subduction process along most of California ceased tens of million of years ago, most of the Sierra's volcanic pile has been removed through erosion, exposing the underlying granite plutons. These rocks are evidence of the magma that solidified within the crust during the subduction process.

Continental shelf widths along collision coasts are a function of how steeply the ocean plate descends into the mantle. Along Chile and Peru there is almost no continental shelf because the Nazca Plate dips very steeply beneath South America. In contrast, the continental shelf along the coasts of Oregon and Washington is comparatively wide because the Juan de Fuca Plate descends at a shallow angle beneath North America and material has scraped off the Juan de Fuca Plate, forming a narrow shelf.

Collision coasts tend to be rocky due to the scarcity of sediment and, where sediment does exist, gravel-sized material is often a major component. This is particularly true at high-latitude coasts due to the effects of glaciation and frost weathering. The lack of major rivers leads to localized sediment sources that produce isolated accumulation forms such as small barrier spits and pocket beaches. Extensive barrier development is normally absent. Exceptions to this trend occur in regions where glacial meltwater streams transport large quantities of sand and gravel to the coast or near the mouths of the few moderate to large rivers that exist along collision coasts. The south-central coast of Alaska and the mouth of the Columbia River, respectively, are examples of these conditions.

Alaska Collision coasts are not only majestic and ruggedly beautiful, they also experience some of the Earth's most dramatic processes. The southern coast of Alaska exhibits many of the features and processes that typify collision coasts. The Pacific Plate is being subducted beneath North America along the south-central margin of Alaska continuing along the Aleutian Islands. This tectonic setting produces volcanism, frequent earthquakes, and large crustal displacements. The coastal mountains of the region are young and their uplift has been rapid. Coastal Alaska is the site of some of the world's largest earthquakes and volcanic eruptions. On March 27, 1964 at the head of Prince William Sound (130 km west of Anchorage), the largest earthquake ever recorded in the northern hemisphere, registering 8.6 on the Richter scale,[1] shook much of the central Alaskan coast along an 800 km-long tract (Fig. 2.21). The effect on Anchorage and many other coastal communities was one of complete devastation, including the loss of 131 lives. The primary shock lasted from three to four minutes and produced large-scale slumps, landslides, and avalanches. The earthquake also created tsunamis (seismic sea waves) that completely wiped out native communities in Prince William Sound and large sections of several seaports, including Valdez and Seward. These giant waves are not only triggered by movements of the ocean floor, but in the case of some Alaskan fjords they also may be caused by earthquake-induced gigantic rock falls that crash into the heads of deep, elongated water bodies sending walls of water toward the mouth of the fjords (see Box 2.1). The great 1964 Alaskan earthquake produced permanent changes along much of the central coast due to uplift and subsidence (down-faulting). In the Prince William Sound region, the Island of Montague rose as much as 9 m, and there are many other areas along the coast in which results of the uplift can be seen today in the form of raised beaches and intertidal platforms. Contrastingly, 200 km southwest of Anchorage, the earthquake caused a general subsidence of 2–3 m along the Kenai Peninsula, resulting in drowned forests.

In addition to being earthquake prone, coastal Alaska experiences continual volcanism, particularly

1 Scientists have re-evaluated the 1964 Alaskan earthquake and determined that it was magnitude 9.2, second only to the 9.5 Chilean earthquake of 1960. The new calculations are used for very large earthquakes whose signal saturates the recording instruments. The moment magnitude, as it is called, is based on the amount of fault slippage and the surface area of the plates involved.

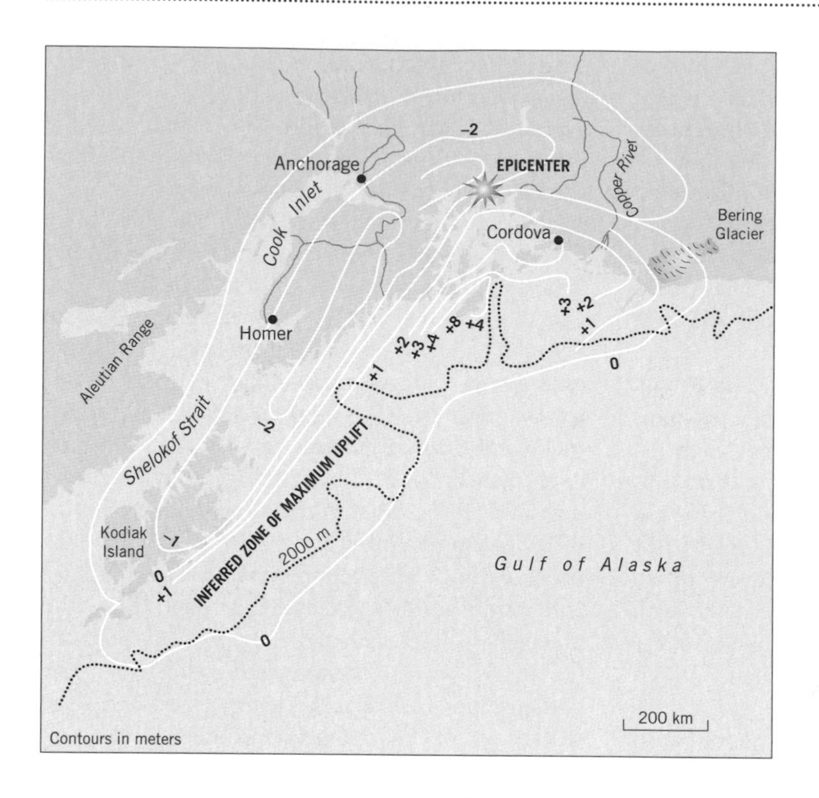

Fig. 2.21 Map of displacement contours illustrating the regional effects of the 1964 Great Alaskan Earthquake. Note that portions of the Alaskan coast east of Anchorage experienced uplift, whereas areas southwest of Kodiak Island underwent subsidence. (From G. Plafker, R. Kachadoorian, E. B. Eckel & L. R. Mayo (1969) Effects of the earthquake of March 27, 1964, on various communities. USGS Professional Paper 542-G.)

along the Alaska Peninsula, where at least 17 volcanoes have erupted during the past 10,000 years. Some of these volcanoes have erupted with a tremendous ferocity. In 1912, Mount Katmai erupted, ejecting 15 km^3 of volcanic debris. Much of the ash was blown southwestward, blanketing portions of Kodiak Island, 150 km away, with up to half a meter of ash. This enormous eruption is the third largest during recorded history, with only Tambora and Krakatoa in Indonesia being larger. Truly, Alaska is a very active collision coast.

Island arc collision coasts

Island arc collision coasts (IB) are similar to continental collision coasts and are active tectonically, with characteristic volcanism and earthquakes. They differ from continental collision coasts because the convergence is between two oceanic plates. These coasts are backed by low to moderately high mountains and are fronted by narrow continental shelves that are bordered by deep ocean trenches. Most island arc collision coasts are located in the northern and northwestern Pacific and include the ocean coasts of the Aleutian Islands and the Philippines. Japan is another collision island arc coast, and has recently experienced major earthquakes (Kobe, Japan) and volcanic eruptions.

2.5.2 Trailing edge coasts

Neo-trailing edge coasts

Neo-trailing edge coasts (IIC) are geologically young coasts (<30 million years old) that have formed as a consequence of continental rifting (Fig. 2.22). Examples include the coasts surrounding the Red Sea, Gulf of Aden, and Gulf of California. These **linear seas** are a stage in the evolution of new ocean basins, which is summarized below:

Stage 1a Mantle plume causes doming and stretching of overlying continental crust.

Stage 1b Alternatively, plate motion causes passive rifting and ensuing mantle upwelling.

Stage 2 Thinning crust ruptures and down-faulted blocks create a topographically low rift valley.

Box 2.1 Lituya Bay, Alaska: site of the largest waves ever seen

Lituya Bay is located on the rugged southeast coast of Alaska along the southwest side of the Fairweather Mountain Range (Fig. B2.1). Frenchman Jean François La Perouse first explored the region in 1788. He noted that the bay provided good harborage, although navigating the strong currents at its mouth was hazardous. He was unaware, however, of the apocalyptic waves that frequent this site. The bay extends 13 km inland from the coast, where it is bordered by mountains and glacier-filled valleys. In his study of the region, Don J. Miller of the US Geological Survey concluded that the bay was once occupied by the confluence of at least two major glaciers, which had deepened the valley and deposited a moraine at its mouth. Slight reworking of the terminal moraine created a spit-like landmass, La Chaussee Spit, which narrows the bay's entrance, forming a tidal inlet about 300 m across. Two inlets, Gilbert Inlet and Crillon Inlet, open onto the head of Lituya Bay. Gilbert and Crillon inlets are currently enlarging as Lituya Glacier retreats to the northwest and North Crillon Glacier recedes to the southeast, respectively (Fig. B2.2).

The mountainous landscape of southeast Alaska is the result of the collision between the Pacific and North American plates. The plate boundary in the Lituya Bay region is a transform fault that has caused considerable vertical motion as well as horizontal slip. It is known as the Fairweather fault and it coincides with the overall trend of Lituya and Crillon glaciers and the inlets that form the upper T-shaped portion of the bay. On the evening of July 9, 1958 at 10:16 p.m. a section of the fault ruptured approximately 23 km southeast of upper Lituya Bay, producing an 8.3 magnitude earthquake. Measurements taken along the fault revealed vertical ground displacements of 1.05 m and horizontally slip of 6.5 m. The earthquake was felt throughout southeastern Alaska, and as far south as Seattle, Washington, and as far east as Whitehorse, Yukon Territory, Canada, covering an area of over 1,260,000 km^2.

On the evening of July 9, 1958, three fishing boats were anchored at the entrance to Lituya Bay. The boatsmen were enjoying the late evening sunlight that accompanies Alaskan summers. Only four people lived to tell the story of a giant wave that filled the bay. At the onset of the earthquake the ground around Lituya Bay shook violently and continued doing so for about a minute. Eyewitnesses from the fishing vessels indicate that about 1–2.5 minutes after the earthquake began there was a tremendous explosion in the upper bay. The noise they heard came from the mountain slope at the head of Lituya Bay giving way, sending 30 million m^2 of rock into the northeast side of Gilbert Inlet. Failure of the mountainside was caused by ground tremors, which weakened the high-angle bedrock slope that had been oversteepened by glacial erosion. Due to the steepness of the mountain slope (75–80°) the slab of rock fell almost directly into the inlet from an average height of about 610 m (2000 feet). The

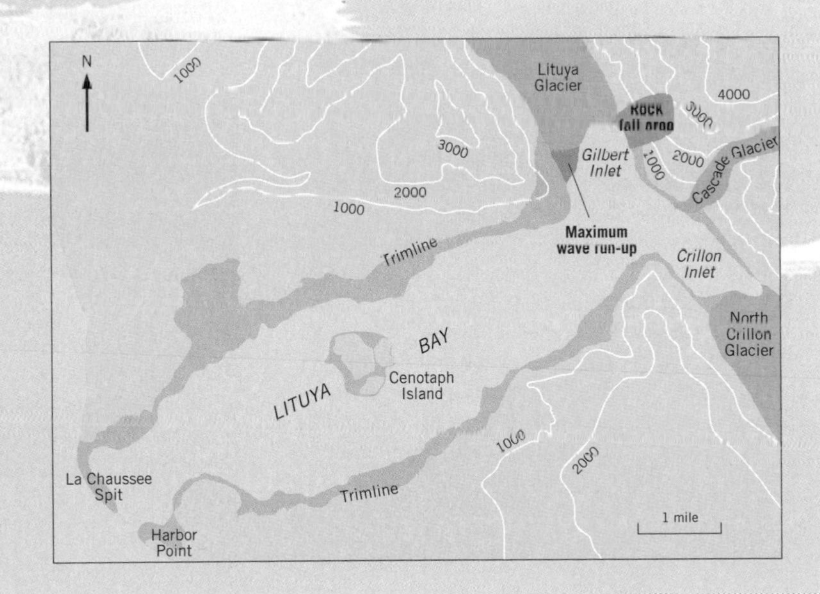

Fig. B2.1 Location map of Lituya Bay.

Box 2.1 *(cont'd)*

Fig. B2.2 Aerial photograph of Lituya Bay. A terminal moraine known as La Chaussee Spit narrows the entrance to the bay. The island of Cenotaph is situated in the middle of the bay. (Photograph by D. J. Miller.)

huge mass of rock plunged through the water column, cratering the sediment bottom of the bay. The displaced water produced a massive wave that propagated across Gilbert Inlet, traveling up the opposite cliffside to a maximum height of 530 m (1740 feet). Here the force of the wave stripped away the vegetation and the underlying soil down to the bedrock surface. Where trees were rooted deeply in bedrock, the trunks, some of which were more than a meter in diameter, were sheared off as if they were small twigs. Scientists have theorized that a wave of this magnitude would be similar to that generated by the impact of a small asteroid.

After carving away much of the southeast mountainside of Gilbert Inlet the giant wave swept down Lituya Bay with a speed of between 156 and 209 km h^{-1} (97–130 m.p.h.), advancing toward the fishing boats anchored at the mouth. As the wave approached the first fishing boat, Mr Ulrich and his young son described the onslaught of a black wall of water that crested over at 30 m (100 feet). The giant wave snapped the boat's anchor chain and propelled the craft and occupants toward the bay's south shore, where they were saved by the wave's backwash, which carried the boat and passengers back to the middle of the bay. There they encountered several smaller but still terrifying waves, some reaching over 6 m (20 feet) in height. Miraculously, the Ulrichs sailed out of Lituya Bay next day under the boat's own power. A second boat, with Mr and Mrs Swanson aboard and anchored closer to the northern bay shoreline, was not so lucky. The Swansons' fishing trawler was picked up by the giant wave and carried out of the bay like a swimmer on a surfboard. By the Swansons' account, as the trawler surfed over La Chaussee Spit the tops of the trees were 25 m (82 feet) below them. Once deposited in the ocean their boat immediately began to sink. Fortunately, they were able to climb aboard their dinghy and were picked up by a passing fishing boat a few hours later. The third boat was engulfed by the monstrous wave and neither crew nor boat was ever found.

The immense wave of July 9, 1958 was not the first extraordinarily large wave to sweep through Lituya Bay. On October 27, 1936 four people witnessed three waves traveling down the bay ranging in height from 15 to 30 m (50 to 100 feet). These waves cut a neat trimline along the tree-covered slopes surrounding the bay. Other trimlines at even higher elevations have been reported by Don J. Miller of the US Geological Survey, and these indicate that giant waves repeatedly crashed along the bay shoreline. Trimlines have been found at elevations of 60–150 m (200–490 feet) (Fig. B2.3). It is interesting to note that the 1936 waves were not related to an earthquake and no bedrock scar corresponding to a rockfall was ever found. Scientists have suggested that other mechanisms, such as the drainage of a subglacial lake or the frontal collapse of a glacier, may also trigger giant wave formation. Lituya Bay is a beautiful anchorage site. However, boaters should be aware of the roar in the upper bay that portends the coming of the great wave!

Fig. B2.3 Aerial view of trimline cut by the seismic wave. The northern end of Cenotaph Island is in the foreground. (Photograph by D. J. Miller.)

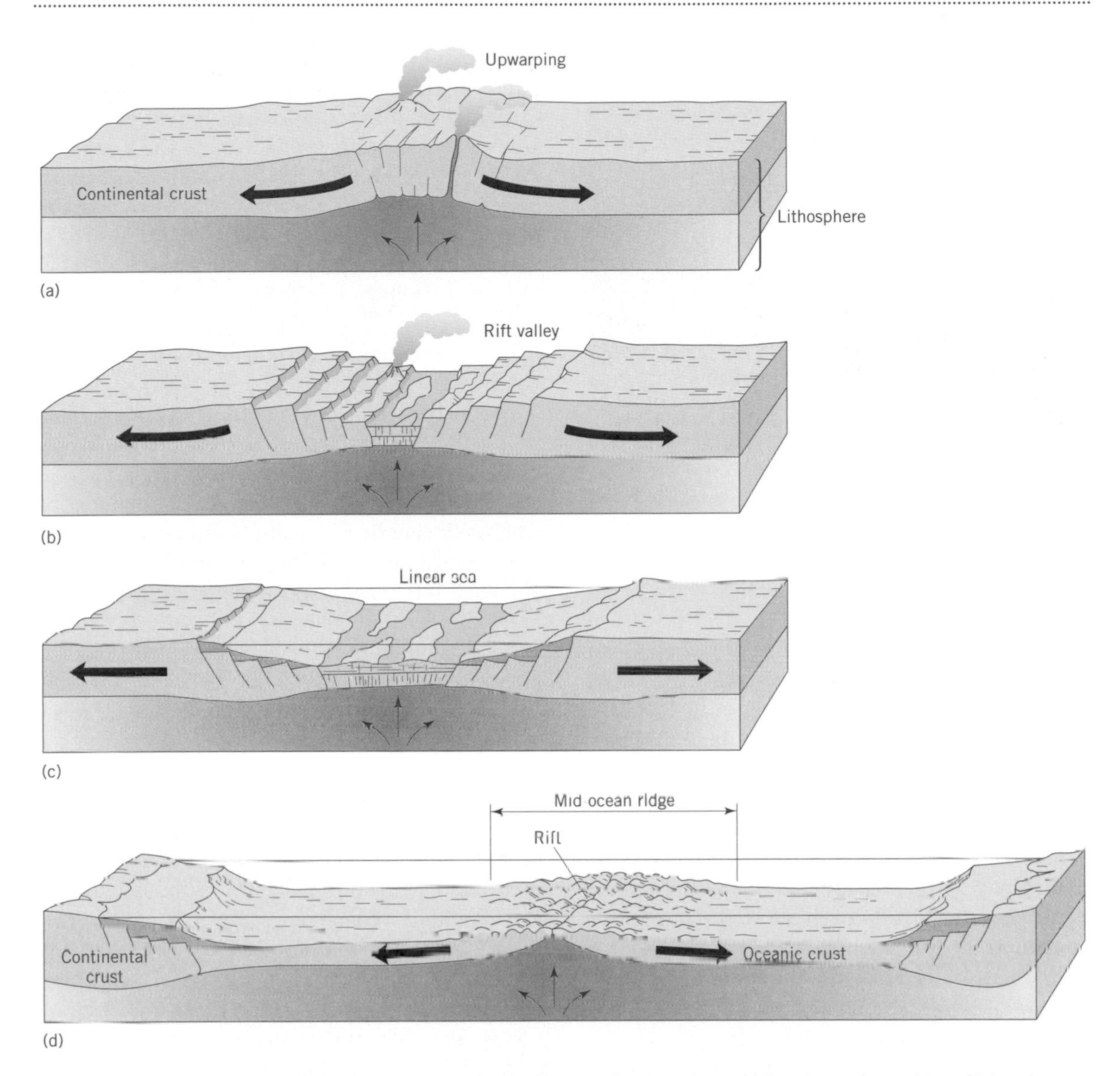

Fig. 2.22 Four-stage model of rifting of a continent and establishment of an ocean basin. (a) Doming and stretching of lithosphere produced by upwelling magma. (b) Spreading causes rupture of lithosphere and creation of a rift valley due to block faulting. (c) Continued spreading, subsidence, and ocean crust formation causes flooding of the rift valley and the creation of a linear sea. (d) Long-term spreading creates an ocean basin.

Stage 3 Subsidence and widening of rift valley causes invasion of ocean and formation of a linear sea.

Stage 4 Long-term sea-floor spreading leads to ocean basin formation.

Neo-trailing edge coasts are rugged and coastal borderlands tend to have narrow to non-existent coastal plains with adjacent mountains. Sediment tends to be scarce along these coasts and there is little barrier development (Fig. 2.23). Where sand is locally abundant, thin mainland beaches or pocket beaches are present. A major exception to this trend occurs in the Gulf of California along the mainland coast of Mexico, where there are extensive coastal lowlands and several barrier island chains.

Fig. 2.23 Neo-trailing edge coasts tend to be rocky and mountainous, with few beaches and barriers (Baja California).

Neo-trailing edge coasts experience frequent low-magnitude earthquakes due to crustal adjustment associated with past rifting. Volcanism is absent.

Sinai Peninsula The Gulf of Suez is a relatively narrow body of water that connects the northwestern end of the Red Sea to the Suez Canal and the Mediterranean Sea. The Sinai Peninsula, which forms its northeast shoreline, is typical of neo-trailing edge coasts. The coast is bordered by a narrow hilly region that gives way to high mountains. Little sediment reaches the coast and thus depositional landforms are mostly absent. The arid climate of the region and the lack of rivers contribute to this condition. In the Ras Mohammed area, at the southern tip of the Sinai, the barren landscape along the coast belies the abundant sea life immediately offshore. Not more than a 100 m from the shoreline, a robust coral reef community provides some of the best scuba diving in the world.

Afro-trailing edge coasts
Afro-trailing edge coasts (IIB) differ from Amero-trailing edge coasts (see below) in that the opposite side of the continent is not a collision zone. Instead, the opposite continental margin coincides with the middle of a plate and thus it is also a trailing edge coast. Afro-trailing edge types are found along the east and west coasts of Africa, the southwestern half of the Australian coast, and the coast of Greenland. The lack of a collision zone along the east or west coast of Africa means that there is no large-scale organization of river drainage within the continent. Consequently, sediment delivery to the margin of Africa as well as other the Afro-trailing edges has only been local and generally does not compare in magnitude to quantity that has been transported to Amero-trailing edges. This condition is also a function of differences in climate. Similar to Amero-trailing edges, Afro-trailing edges are tectonically inactive, with minor earthquake activity and no volcanism.

Due to the sedimentation history of Afro-trailing edges, they exhibit a great deal of morphological variability. For example, in some regions, such as most of the northeast coast of Africa, the continental shelf is quite narrow (<25 km wide), whereas along much of its southwest coast, shelf widths approach 100 km. Coastal borderlands along Africa range from coastal plains and hilly settings to cliffs and low mountains. Likewise, there are estuarine, barrier, and deltaic coasts where sediment is abundant and other long stretches of coast that are rocky and barren of sediment.

South Africa The southern portion of South Africa is a good example of the variability of Afro-trailing edge coastlines. The port of Cape Town is a mixture of lowland areas surrounded by several flat-topped low mountains composed of layered sedimentary rocks. The coastline south of Cape Town extending to the Cape of Good Hope is rugged, with high cliffs and low mountains. The Indian and Atlantic Oceans meet at this site and their unlimited fetches combine to produce huge 6 m and higher waves that crash upon the rocks and send salt spray 40 m high to the top of the cliffs (Fig. 2.24). East of this region is False Bay, where wave abrasion of sandstones has led to extensive sand accumulation. At the western end of the inner bay the beach is 300 m wide and fronted by a very wide surf zone. The abundance of sand at this locality is most clearly demonstrated by the extensive dune system backing the beach, where individual dunes reach 10 m in height and the dune field extends several kilometers inland. Contrastingly, a few tens of kilometers from

Fig. 2.24 Coast of South Africa where large waves carve away at sandstone cliffs. In this region beaches only occur in embayments, where sediment can accumulate under high wave energy conditions.

this site the coast consists of steep rocky cliffs devoid of any sediment.

Amero-trailing edge coasts
Amero-trailing edge coasts (IIC) occur along passive, Atlantic-type margins in which the opposite side of the continent is a collision coast. The mountain chains associated with collision coasts organize the drainage of the continent such that the major rivers flow from the mountains and away from the collision coast, across the continent, and discharge their loads along the passive margins. Amero-trailing edge coasts include the eastern margins of North and South America, the Atlantic coast of Europe, and the coast of India. These are geologically old, and long-term sedimentation at these sites has led to the development of wide flat margins, which encompass the low profile coastal plain and seaward continental shelf. The division between the coastal plain and shelf is simply a function of sea level. This boundary has changed dramatically during the past two million years as sea level has fluctuated in response to the growth and decay of the continental ice sheets. When the ice sheets advance and sea level falls, the coastal plains expands and, likewise, when the ice sheets shrink and sea level rises, the continental shelf widens landward.

Amero-trailing edge coasts tend to be tectonically inactive with relatively few major earthquakes; the ones that do occur are of low magnitude (<4 on the Richter Scale). The Charleston, South Carolina, earthquake of 1886, magnitude 7.3, is a major exception to this trend. These coasts also lack volcanic activity.

East coast of the United States The primary characteristics of Amero-trailing edge coasts include their depositional landforms, such as barrier island chains, broad sediment-filled lagoons, marsh systems, tidal flats, and river deltas. For example, almost the entire east coast of the United States, south of glaciated New England, is fronted by barriers chains that are interrupted by only a few major re-entrants such as Delaware and Chesapeake Bay (Fig. 2.25) and several mainland beaches, including Myrtle Beach, South Carolina, and Rehoboth Beach, Delaware. Some of these barrier systems, such as the Outer Banks of North Carolina, are separated from the mainland by wide shallow bays (Albemarle Sound and Pamlico Sound), whereas other barrier chains, including those along much of South Carolina and Georgia, are backed by an extensive region of marshes and tidal creeks. Although there are no active river deltas along the east coast today, many existed in the past, including the Santee River delta in South Carolina. This delta is no longer building and is now eroding because much of its river discharge, and hence its sediment source, was diverted into Charleston Harbor in the late nineteenth century.

The east and west coasts of the United States illustrate well the sharp contrasts in morphology that different tectonic settings produce (Table 2.1).

2.5.3 Marginal sea coasts

Along much of the western and northern Pacific Ocean, a series of island arcs, including the Aleutians, Kuril Islands, Japan, and Philippines, separates the

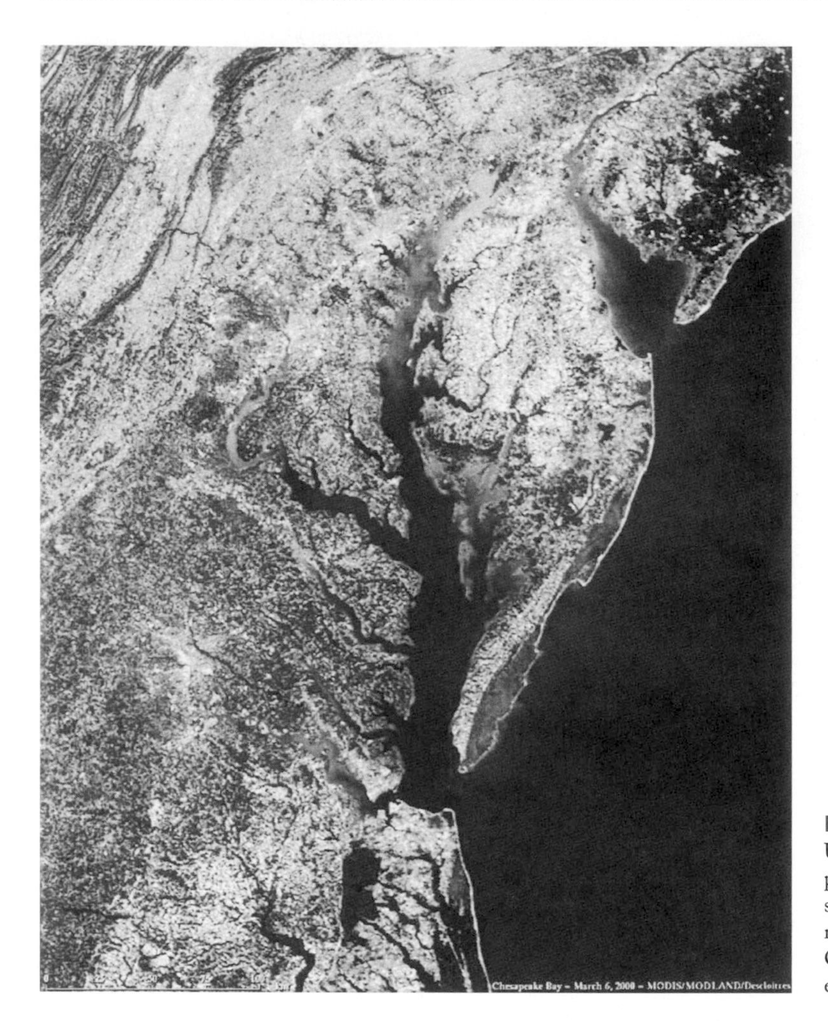

Fig. 2.25 Much of the east coast of the United States is characterized by a coastal plain setting fronted by lagoons, marsh systems, and barrier chains. The drowned river valleys of Delaware (top) and Chesapeake Bay (bottom) are major exceptions to this trend.

edge of continents from the open ocean. These protected shorelines of Alaska and Asia are defined as marginal sea coasts (Fig. 2.26). Characteristically, they are fronted by shallow water bodies, including the Bering Sea, Sea of Okhotsk, Sea of Japan, East China Sea, and South China Sea. In the Atlantic Ocean, the Gulf coast and east coast of Central America are also marginal sea coasts and are protected by the Caribbean island arc as well as by the landmasses of Florida and Cuba.

Marginal sea coasts exhibit considerable variability depending largely upon the geological history of that portion of the continent, but overall they tend to be more similar to Amero-trailing edges than any other class of coast. For example, many of the world's largest sediment discharge rivers occur along these coasts, including the Mississippi and Magdalena, rivers which empty into the Gulf of Mexico and the Caribbean Sea, and the Huang Ho, Yangtze, and Mekong rivers, which discharge along the Asian continent. Interestingly, despite this apparently abundant supply of sediment, long barrier chains are sparse along the marginal sea coast of Asia. This condition may be related to the lack of coastal plain development in the region, which does provide the continuous platform upon which barrier island chains can develop. Contrastingly, the Gulf coast is backed by an extensive coastal plain and barriers occur ubiquitously in this region. Along the marginal sea coasts of Asia, much of the sediment

Table 2.1 Comparison of the east and west coasts of the United States.

Feature or process	East coast	West coast
1 Tectonic class	Amero-trailing edge coast	Collision coast
2 Earthquakes	Small magnitude earthquakes	Site of large earthquakes
3 Volcanism	None	1980 eruption of Mount St Helens. Other mountains in Cascade Range also active
4 Shelf width	Wide and flat	Narrow and steep
5 Coastal borderland	Coastal plain	Mountains, some very high
6 Sediment supply	Long-term erosion of Appalachian Mountains provided sediment to the coast through numerous moderately sized rivers	Coming mostly from short steep gradient rivers
7 Coastal morphology	Depositional coasts with extensive barrier island development. Numerous bays, lagoons, marshes, tidal flats, tidal deltas, and estuaries	Rocky coast with mostly pocket beaches and small spits
8 Wave energy	Low (0.7–0.8 m)	Moderate (1.5–1.7 m)
9 Tidal range	Micro- to macro-tidal (1–6 m)	Micro- to meso-tidal (1–3 m)
10 Description of car trip along the seaboard	South of New York exceedingly boring due to flatness of coastal plain. Only excitement is "south of border" signs	Magnificent coastal vistas and view of mountain ranges

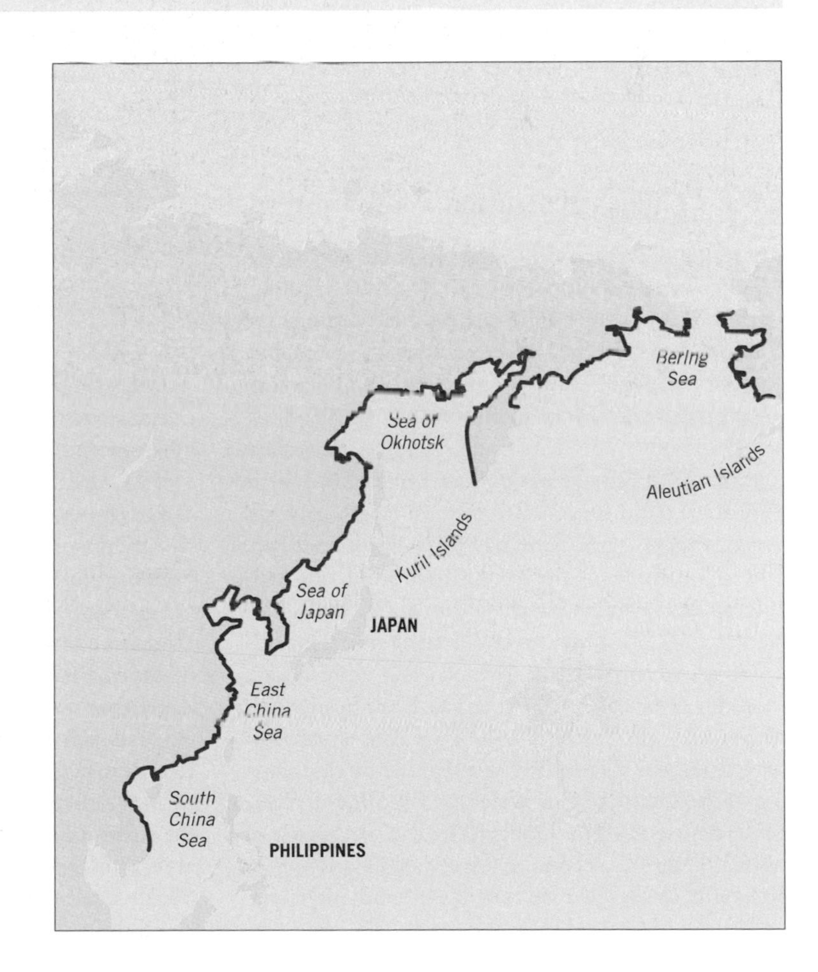

Fig. 2.26 Most of the marginal sea coasts of the world are found along the borders of the northern and western Pacific Ocean, where the Pacific Plate is being subducted beneath the North American and Eurasian plates.

may go into filling valleys rather than building barriers.

Generally, marginal sea coasts experience low to moderate wave conditions due to the semi-protected nature afforded by the shallow seas and offshore island arcs. However, these coasts can suffer damaging short-term high wave energy conditions during the passage of major storms, such as hurricanes.

2.6 Tectonic effects on coastal sediment supply

The tectonic history of a continent strongly influences the distribution and abundance of sediment along coastlines, which in turn control the size and extent of depositional features, such as barriers, deltas, marshes, and tidal flats. These influences can be on a scale of continent-wide drainage and the widths of continental shelves or they can be more regional such as dictating the position of a river mouth.

2.6.1 Continental drainage

Drainage of a continent is determined by the distribution of major mountain ranges, which is a product of the continent's tectonic history. For example, the collision between the eastward moving Nazca Plate and the westward moving South American Plate produced the Andes Mountains, which span the entire length of South America's Pacific margin. The eastern margin of South America has experienced little mountain building activity since the opening of the south Atlantic, approximately 130 million years ago. Thus, the Andes is the major drainage divide separating rivers that flow east from those that drain west (Fig. 2.27). The eastward flowing rivers, including the Amazon, Parana, Orinoco, and others, drain more than 90% of South America. Their headwaters begin along the eastern flank of the Andes and flow long distances across the continent before discharging large quantities of water and sediment along the Atlantic margin. Conversely, the west coast of South America receives relatively small volumes of sediment delivered by short, steep gradient rivers.

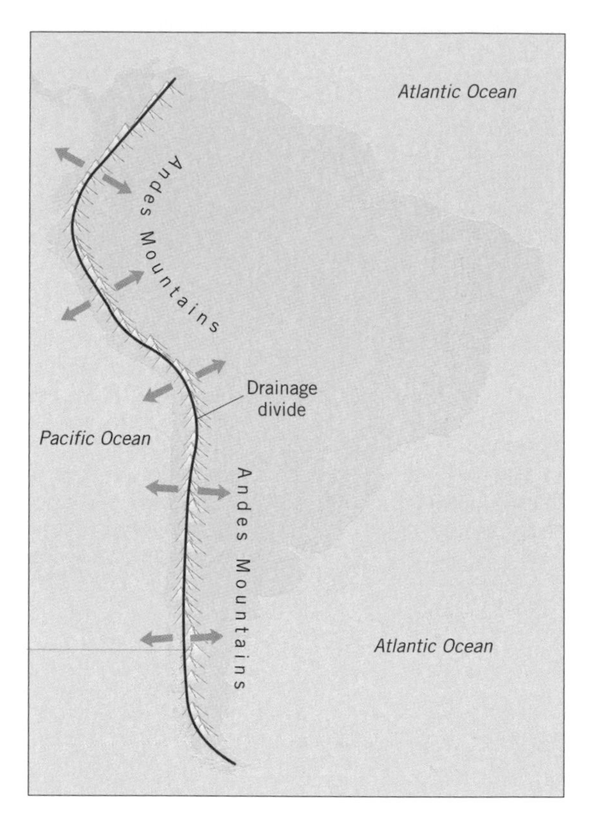

Fig. 2.27 Physiographic map of South America illustrating that the Andes Mountains produce a drainage divide along the entire length of the continent.

The diminutive size of the rivers reflects their small drainage basins, resulting from the proximity of the adjacent Andes drainage divide.

The African continent illustrates a very different drainage condition. Because most of Africa is in the middle of a plate, there is little active mountain building and most of the existing mountain systems are geologically very old. Unlike in South America, there is no large-scale organization of river drainage in Africa. Instead, Africa has moderate to small-sized rivers that discharge along its Mediterranean coast (e.g. Nile River), Atlantic coast (e.g. Niger and Congo rivers), and Indian coast (e.g. Zambezi River). Due to the relatively small volume of sediment reaching the African coast there are fewer depositional features along this coast as compared to the east coast of South America. Certainly the effects of climate and

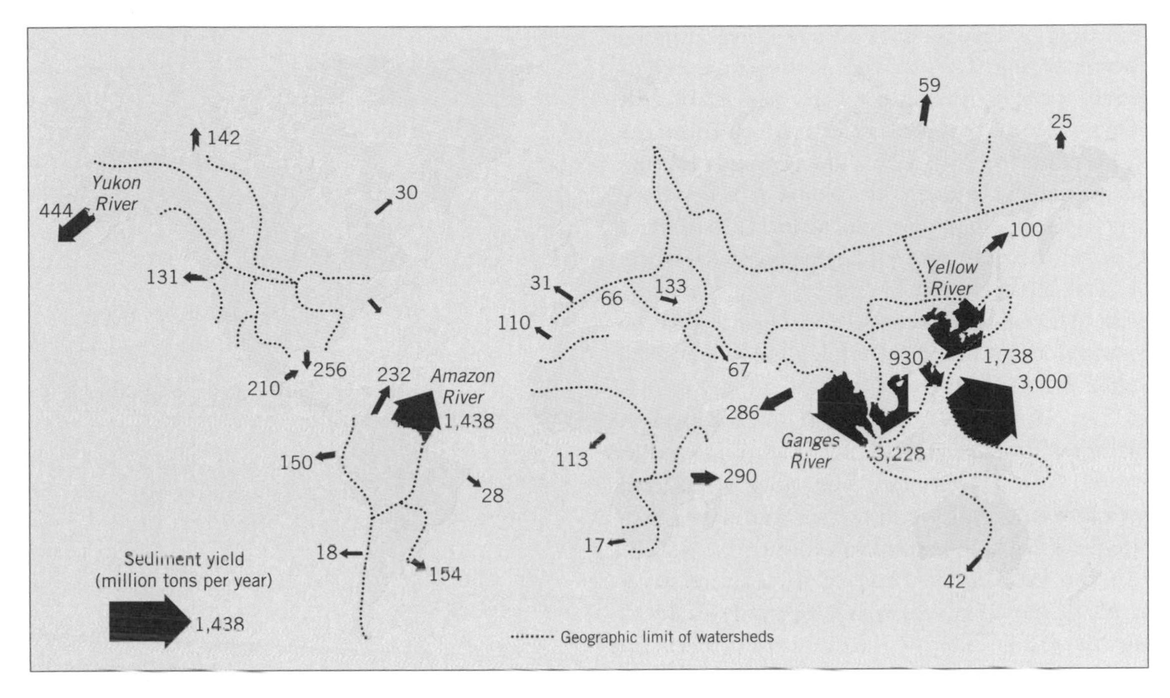

Fig. 2.28 Pattern of sediment discharge to the world's ocean based on suspended sediment loads. Note that most of the suspended sediment discharged to the ocean by rivers occurs along Amero-trailing edge and marginal sea coasts. (From Milliman & Meade 1983.)

the overall lesser amounts of precipitation falling on Africa than South America contribute to this condition.

River drainage in the United States is strongly affected by the tectonic history of North America. Similar to South America, the western margin of the United States was a collision zone until 30 million years ago, resulting in the formation of the Rocky Mountains, Sierra Nevadas, and Basin and Range. It is the Rocky Mountains that form the western drainage divide of the Mississippi River, whose drainage basin comprises two-thirds of the United States. The eastern divide is defined by the Appalachian Mountains, which were formed when the landmasses forming Pangaea were being assembled.

As demonstrated for three different continents, the delivery of sediment to the coast is governed by the size of the rivers discharging along the coast. This, in turn, is a function of climatic factors and drainage basin size, which is largely controlled by the tectonic history of the continent (Fig. 2.28). One other factor that influences the supply of sediment to a coast is the geology of the drainage basin. For example, the collision between the Indian and Eurasian plates created the Himalayan Mountains and the Tibetan Plateau, some of the highest landscapes on the surface of the Earth. The elevation of the Himalayas, coupled with ongoing glaciation of this region, produces huge quantities of sediment. The rivers draining the Himalayas, including the Indus, Ganges, and Brahmaputra rivers, rank as some of the largest sediment discharge rivers in the world.

2.6.2 Location of rivers

In addition to the quantity of sediment delivered to the coast, tectonic processes dictate the location of many large river systems. As described previously, the breakup of a continent involves several plumes of magma rising from the mantle that dome, stretch, and eventually rift apart the overlying lithosphere. If we consider a single mantle plume, the doming and rifting process produce a three-arm tear in the lithosphere (Fig. 2.29). As the mantle plume spreads laterally beneath the lithosphere, two of the rift valleys widen and deepen, while the third arm of the rift becomes inactive. The third arm fails because the pressure of the mantle plume is relieved by the continued spreading of the other two arms of the rift. An example of this process is seen where the Arabian Peninsula has rifted away from northeast Africa. The Indian Ocean has invaded the two active arms of the rift, forming the Red Sea and Gulf of Aden. The failed third arm of this system is the East African Rift, which is a low-lying valley, portions of which are below sea level, that extends into the African continent. The valley is bordered by escarpments and is still tectonically active, with infrequent earthquakes and volcanism.

In the breakup of Pangaea there were many mantle plumes involved, each having its own three-arm rift system. The two active arms of each rift widened, elongated, and connected to one another, eventually producing the separation of entire continent-sized landmasses, such as North America separating from Africa. In many cases the failed third arms of these systems coincide with sites where the internal drainage of a continent empties into the sea. It is believed that because failed third arms are topographic lows that extend into continents, either major rivers evolve at these locations or existing rivers migrate to these sites. The Niger River valley in western Africa and the Amazon River valley in eastern South America are examples of these processes. The Mississippi River valley may also be the location of a geologically very old rift valley.

Another more unusual form of tectonic control of river drainage and sediment supply to a coast occurs in the Indian Ocean. The triangular shape of the Indian subcontinent dictated that, when it collided with Eurasia 50 million years ago, not only was a sizable mountain system produced with a large potential sediment supply, but the shortest route for drainage to the sea was on either side of India. Thus, the high sediment discharge rivers of the Indus draining into the Arabian Sea and the Ganges–Brahmaputra emptying into the Bay of Bengal are predictable.

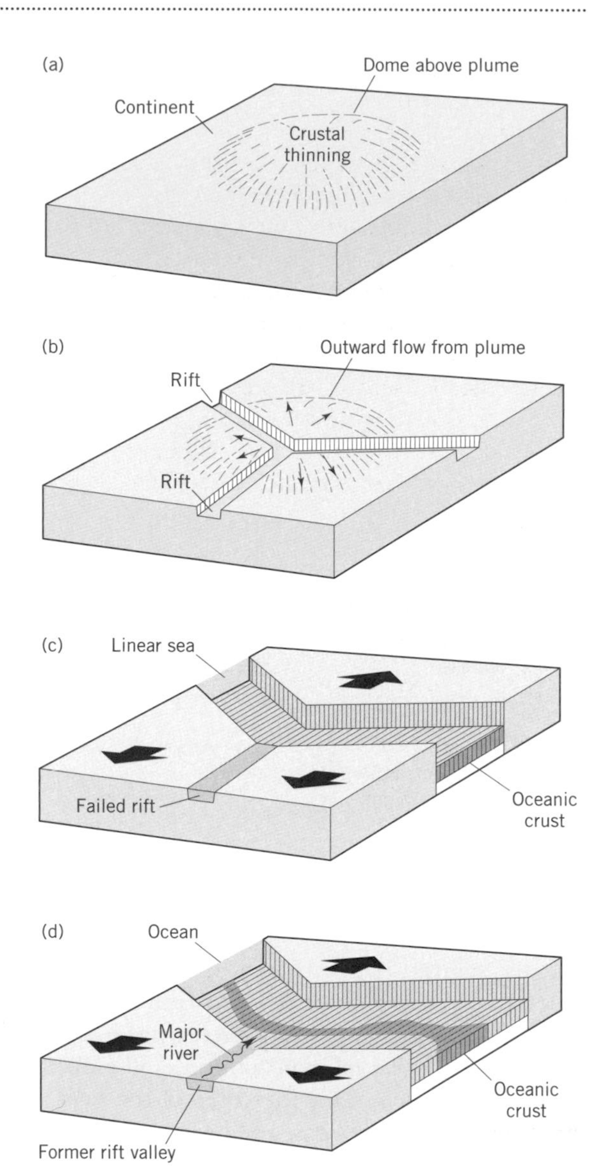

Fig. 2.29 Four-stage model of the development of internal drainage of a continent. (a) Time 1, doming and thinning of lithosphere due to upwelling magma. (b) Time 2, rupture of the lithosphere and development of a three-arm rift system. (c) Time 3, linear sea development while the third arm of the rift system becomes inactive. (d) Time 4, evolution of ocean basin, with interior of continent draining through old rift valley.

2.6.3 Continental shelf width

To a large extent, plate tectonic setting dictates the width and slope of the continental shelf, with Pacific (active) margins tending to have narrow, steep shelves and Atlantic (passive) margins having relatively wide, shallow gradient shelves (Fig. 2.30). Shelf width and slope, in turn, influence the effects of storms, wave energy, and sediment dispersal, as well as cliff abrasion rates along rocky coasts. Along collision coasts, such as the west coasts of North and South America, the narrow shelf widths result in large wave energies and large potential sediment transport rates. Remember too that these margins have localized sediment supplies derived by relatively small river systems. Thus, sediment accumulation forms along these types of settings are dependent on nearby sediment input and embayments where sediment can collect.

Fig. 2.30 Physiographic map of southern South America illustrating the difference in shelf width between the active western margin and the passive eastern margin.

A very different pattern exists along Amero-trailing edges and marginal sea coasts that are fronted by wide continental shelves. In these regions, much of the deep water wave energy is attenuated by friction imparted by the shallow gradient continental shelf. Longshore and offshore sediment transport along these coasts occurs during the passage of major storms. Accretionary landforms are common along these coasts and are coincident with abundant supplies of sediment.

Continental shelf width also influences sediment supplies along continental margins by dictating the proximity of submarine canyons to the coast. For example, along the coast of California the narrow shelf has led to canyons being located only several hundred meters seaward of the beach. During storms and other high wave energy events sand is eroded from beaches at these sites and transported offshore. Some of this sand makes its way into the heads of submarine canyons, such as La Jolla and Scripps canyons, where eventually it will transported to deep sedimentary basins. In contrast, this process does not occur along the eastern margin of North America because the submarine canyons of this region are too far from shore to capture inner shelf sands.

2.7 Summary

Plate tectonics provides a means for explaining the overall features and major processes on the surface of the Earth, including the character of the world's continental margins and its coastal settings. We are able to understand why coastal regions on either side of a continent can be so markedly different and why the west coast of the United States is bordered by mountains and experiences volcanic eruptions and frequent earthquakes, whereas the tectonically inactive east coast abuts a flat coastal plain and is fronted by an almost uninterrupted chain of barriers. It has also been shown that the tectonic history combined with climatic factors determines the drainage of a continent and how much sediment is delivered to a particular coast. Even the location of individual majors rivers has been shown to be a function of plate tectonic processes. In subsequent chapters it is shown that coastlines are molded by numerous physical, chemical, and biological processes, but all these operate on the backbone of the plate tectonic setting of the region.

Suggested reading

Bird, E. C. F. (1993) *Submerging Coasts*. Chichester: Wiley.

Bird, J. M. (ed.) (1980) *Plate Tectonics*. Washington, DC: American Geophysical Union.

Cox, A. & Hart, R. B. (1986) *Plate Tectonics: How It Works*. Palo Alto, CA: Blackwell Science.

Davies, J. L. (1980) *Geographical Variation in Coastal Development*. New York: Longman.

Glaeser, J. D. (1978) Global distribution of barrier islands in terms of tectonic setting. *Journal of Geology*, **86**, 283–97.

Hayes, M. O. & Kana, T. (1978) *Terrigenous Clastic Depositional Environments*. Columbia: Department of Geology, University of South Carolina.

Inman, D. L. & Nordstrom, C. E. (1971) On the tectonic and morphologic classification of coasts. *Journal of Geology*, **79**, 1–21.

Moores, E. M. (ed.) (1990) *Shaping the Earth: Tectonics of Continents and Oceans*. New York: McGraw-Hill.

Shepard, F. P. (1977) *Geological Oceanography*. New York: Crane, Russak.

3 Sediments and rocks: materials of coastal environments

3.1 Introduction

We learned in Chapter 2 that the Earth's crust has a wide range in composition; that is, it is comprised of many different rock types. Igneous rocks comprise most of the volume of the crust but metamorphic rocks and sedimentary rocks are also important. All of these rock types and their contained assemblages of minerals may occur in the coastal zone. Rocky coasts are widespread along the shorelines of the world and are generally characterized by erosion (see Chapter 20). In this book, we are more interested in the depositional coastal environments where various types of sediment accumulate. The sediments that characterize these depositional environments, such as beaches, dunes, estuaries, and deltas, are products of the destruction of bedrock through various physical and chemical processes we call weathering.

These weathering products will be transported on or near the earth's surface, eventually coming to rest as an accumulation of particles that either settle due to a change in surface conditions or are precipitated from solution (physicochemical) or by an organism (biochemical). Even in the coastal zone, some of these particles become buried, then cemented into sedimentary rocks, but most remain as unconsolidated particles of various sizes.

In this chapter both the textural and compositional aspects of sediments are discussed, along with a general treatment of the rocks from which they came.

3.2 Rock types

The igneous rocks that comprise most of the Earth's crust, and that form from magma that originates in the mantle, are composed primarily of what are called silicate minerals. These minerals are produced by a variety of cations of various elements combined with silicon and oxygen. Common elements involved in these silicates are iron, magnesium, potassium, calcium, sodium, and aluminum, though there are others. Depending upon where these minerals form within the Earth's crust they may result in granite, diorite, gabbro, basalt, or other igneous rock types.

(a)

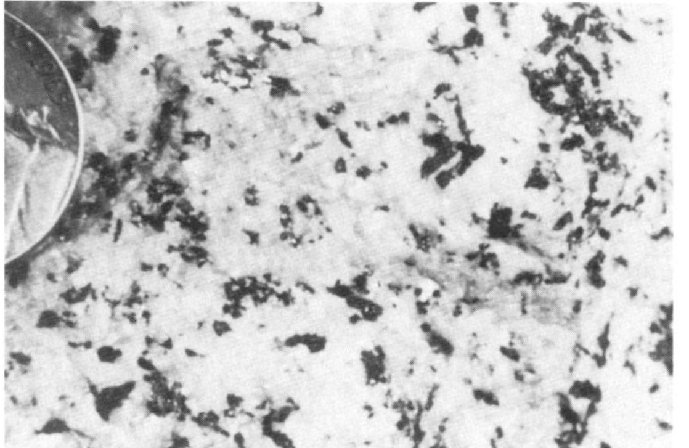
(b)

Fig. 3.1 (a) Outcrop and (b) close-up of granite, an example of igneous rocks that form from the crystalization of molten magmas at very high temperatures beneath the Earth's surface.

Weathering of igneous rocks (Fig. 3.1) can produce sediment particles of varying sizes that eventually become buried and cemented, forming sedimentary rocks (Fig. 3.2) such as arkose, quartz sandstone, or mudstone. Heat and pressure may be applied to these rock types to produce metamorphic rocks (Fig. 3.3), or "changed rocks." These include gneiss, schist, slate, and others.

In some cases sedimentary rocks can form from shells (Fig. 3.4) and other skeletal material, typically composed of calcium carbonate, the mineral calcite. This produces limestone, and when metamorphosed that becomes marble.

All of these rocks, as well as other less common

(a)

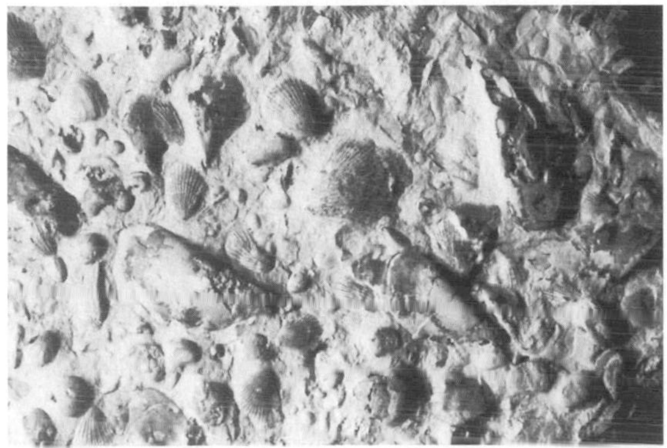

(b)

Fig. 3.2 (a) Outcrop and (b) close-up of limestone, an example of sedimentary rocks that form at the Earth's surface. Limestone has a composition of calcium carbonate, which is commonly shell material.

(a)

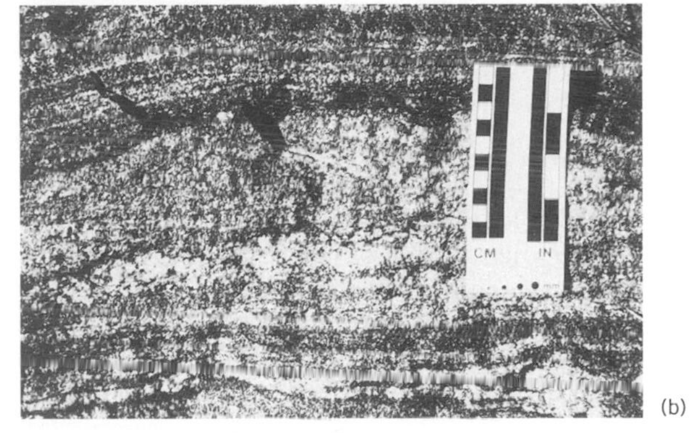

(b)

Fig. 3.3 (a) Outcrop and (b) close-up of gneiss, an example of metamorphic rocks that form from the application of heat and pressure on granite. Metamorphic rocks are commonly associated with mountain ranges.

types, may occur in various combinations to produce rocky coasts. More importantly in the development of depositional coastal environments, they produce sediment as the result of physical and chemical weathering and abrasion. These sediment products, regardless of where they form, may eventually find their way to the coast, where they accumulate in various coastal environments. The bulk of our discussion in this chapter focuses on these sediments, and their size, composition, and arrangement within various coastal settings.

3.3 Sediment texture

The production of sedimentary particles during erosion can be related to many phenomena, including climate, organisms, gravity processes such as rocks falls, stream transport, and others. In this discussion no attention is paid to the composition of the sediment particles. Instead we focus on the size, shape, and distribution of these particles as they accumulate in various coastal environments.

Fig. 3.4 Typical shells that accumulate in coastal environments. Some beaches are almost exclusively composed of sediment that is shell. Notice that in this example there are few different shells and most are the same size.

3.3.1 Grain size

When we determine the grain size of sediment particles, it is common to treat a sample of the sediment that includes many individual grains. This causes potential problems if we think about how we are going to try to measure these many grains. There are multiple ways whereby one might assign a size to a sediment particle. Three come to mind: the diameter, the volume, and the mass of the particle. The mass is rather easy to deal with because shape is not a factor and it can easily be related to the density of the material. Because of density differences mass is not practical as a grain size parameter. Measurement of the volume of large numbers of grains is both difficult and time consuming because we would have to treat each one individually. The diameter of a particle is far easier to measure, although there is an infinite range of particle shapes that may be exhibited, ranging from a sphere to a flat disk shape. Even though relatively few perfect spheres are produced as erosion products, many sediment particles exhibit a rough approximation to a sphere.

The range in grain size of sedimentary particles is almost infinite, from less than 1 μm to several meters in diameter. Obviously, several techniques must be utilized for measuring such a wide spectrum of grain sizes, especially because grain size analysis involves extremely large numbers of grains, which places further constraints on size analysis techniques. Commonly, grains larger than a couple of centimeters or so in diameter are measured individually with calipers or by a similar method. Grains ranging from a centimeter down to about 50 μm in diameter are measured either by passing them through several sieves that separate the grains into rather narrow size classes or by allowing the grains to settle through a column of a fluid, usually water. The rate at which particles settle is proportional to their size, shape, and density, although certain assumptions are made. Settling tubes in which this type of analysis is conducted have become the standard in recent years and may be quite sophisticated, with direct connection to computers for rapid data processing. Particles smaller than 50 μm may also be measured by settling rates, although such methods are time consuming and not very accurate. Recently, instruments utilizing photoelectric sensors, lasers, or X-ray beams have been developed for measuring these small grains.

Because sedimentary particles display such a wide range of sizes, it is practical to utilize a geometric or logarithmic scale for size classification. About 100 years ago a scale was devised based on a factor of 2 such that as one moves from unity there is a multiplier or divisor of 2: one category is either twice or half the diameter of the next one. Later modifications resulted in the grain size scale as it is used today (Table 3.1). Each size class possesses a name as well as a size range.

3.3.2 Statistical analysis of grain size data

After sediment has been subjected to one of the standard methods for measuring grain size, these measurements are then used to calculate various statistical parameters that describe the nature of the population of particles in the sediment sample. Data obtained from the size analysis are usually in the form of weight percentages for each size class. These data may be plotted as a histogram or as a frequency curve (Fig. 3.5).

The distribution of grain sizes in a sediment population typically follows or approaches a log-normal distribution resulting in a normal or bell-shaped

Table 3.1 The grain size scale.

Particle diameter (mm)	Size class	
	Boulders	Gravel
256		
	Cobbles	
64		
	Pebbles	
4		
	Granules	
2		
	Very coarse	Sand
1		
	Coarse	
1/2		
	Medium	
1/4		
	Fine	
1/8		
	Very fine	
1/16		
	Silt	Mud
1/512		
	Clay	

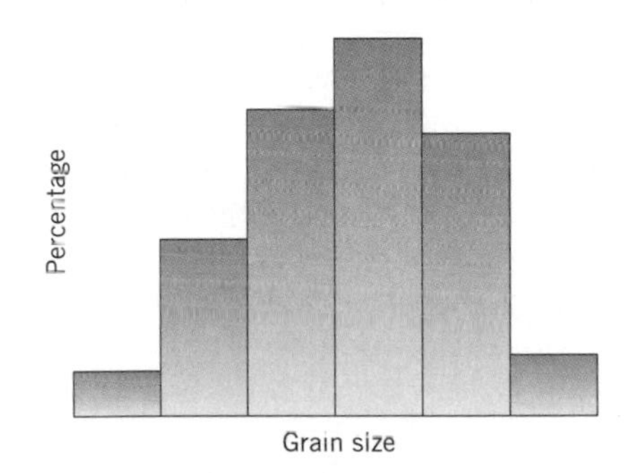

Fig. 3.5 Distribution plot of the grain size of a typical beach sediment. This is a histogram that shows the amount of sediment in each grain size class. It can be smoothed to form a distribution curve.

curve. This plot provides data from which statistical values are calculated. Two statistical parameters are commonly used to describe grain size distribution: the mean and the standard deviation. Such parameters are easily determined by various spreadsheet software. The **mean** is the statistical average, determined in just the same way as a teacher determines the average score on an examination. All values are

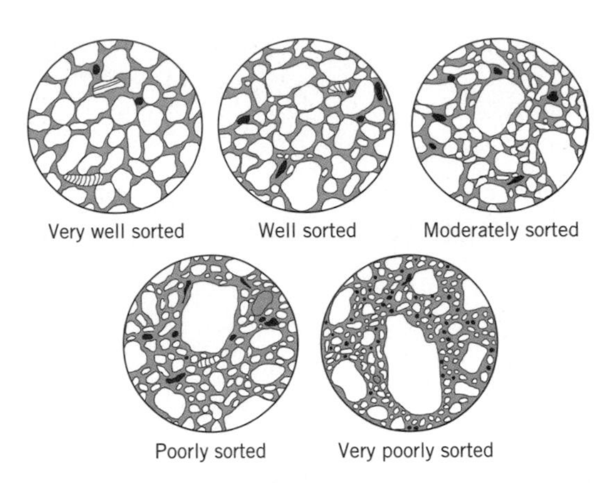

Fig. 3.6 Diagrams showing the various levels of sorting of sediment particles. Although the categories here are distinguished by qualitative terminology, they can also be categorized numerically.

summed and the total is divided by the number of values in the population. The **standard deviation** is the statistical determination of the deviation from the mean. In other words, were all of the test scores bunched together (a small standard deviation) or were they spread out along a wide range (a large standard deviation)? When dealing with sediments we generally refer to this measurement as the sorting. A well sorted sediment has most of the grains about the same size, whereas a poorly sorted sediment has a wide range in particle sizes that comprise the population or sample (Fig. 3.6).

Another statistical condition that commonly occurs in natural accumulations of sediment is the mixing of populations. Examples include the combination of quartz sand with shells or pebbles (Fig. 3.7), which is common on many beaches. Each is derived from a different source and each has a very different grain size; the quartz is sand and the shells are gravel. Such a sediment is said to be bimodal. Each population represents a mode; that is, a large number of particles of about the same size (Fig. 3.8).

3.3.3 Grain shape

Sediment particles display a great variety of geometries. This variation is due to a combination of the internal structure of the minerals that comprise the

Fig. 3.7 Photograph of sediment that is a combination of quartz sand and shell gravel. Such a sediment would be near the poorly sorted end of the spectrum shown on Fig. 3.6.

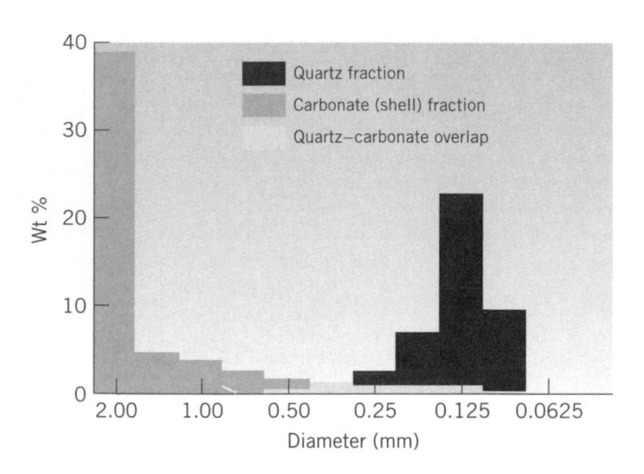

Fig. 3.8 Distribution plot of a bimodal sediment such as shown in Fig. 3.7. This histogram shows a large amount of coarse sediment on the left and a large amount of fines on the right, with little sediment of medium-size grains.

particles with the origin and history of the particle. Some have simple and symmetrical shapes, whereas others are extremely complex.

The roundness of a particle refers to the sharpness or smoothness of its edges and corners. Both physical abrasion and chemical reactions contribute to this characteristic, although abrasion is generally the most important of the two. Descriptive names for this characteristic range from "very angular" to "well rounded" (Fig. 3.9).

The term sphericity refers to the degree to which a particle approaches a sphere. Although many ways of determining sphericity are available, it is most common to compare the lengths of three mutually perpendicular axes. As this ratio approaches unity the particle is becoming more spherical (Fig. 3.10). Sphericity is more strongly influenced by the origin of the particle than is roundness. Some grains are inherently elongate because of their crystallographic or biogenic makeup. Examples are such minerals as mica, which tends to be flat or disk-shaped, and many shell types, such as bivalves or branching corals.

3.4 Mineralogy

Although the crust of the Earth is composed of a wide variety of minerals, relatively few are present in great abundance. Those that are include the feldspars, iron–magnesium silicates such as pyroxenes and hornblendes, and quartz, with feldspars being most abundant. Sedimentary rocks are present as only a relatively thin skin near the surface of this crust. Although there is a wide variety of minerals present in these rocks, they contain primarily four mineral types – feldspars, quartz, clay minerals, and carbonates – with feldspars being the least abundant of the four.

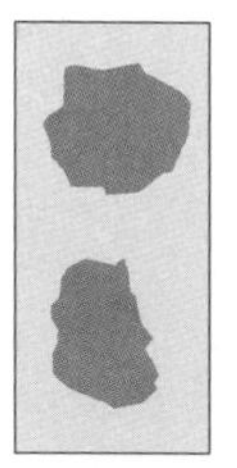

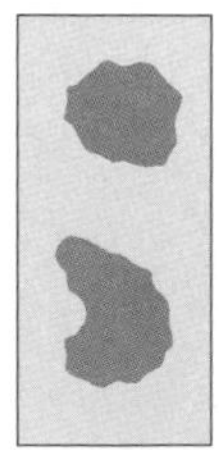

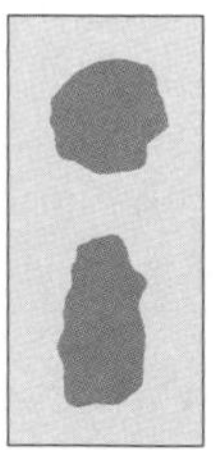

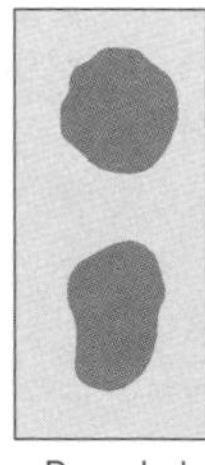

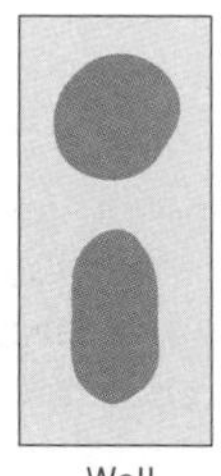

Fig. 3.9 Examples of rounded particles and angular particles. Although some have smooth edges, others have rough edges, the angular ones.

(a)

(b)

Fig. 3.10 Examples of (a) spherical particles and (b) nonspherical particles. Regardless of the roundness or angularity, the three-dimensional shape ranges from elongate particles to nearly equidimensional ones, the spherical particles.

This difference in abundance between the crust as a whole (mostly igneous) and sedimentary rocks is largely a reflection of the relative chemical stability of the minerals present. The various iron–magnesium silicate minerals weather readily and are not common in sedimentary rocks. Most feldspars are relatively unstable, and they weather to produce clay minerals. Quartz is quite stable chemically and is physically very durable. Thus, during the cycle of weathering and transport, quartz tends to persist because it is hard and chemically inert at surface conditions. Carbonate minerals are dominantly the result of physicochemical or biochemical precipitation from solution at or near the surface of the earth. Most of what we see in coastal sediments is in the form of skeletal material, typically shells. The result is that sediments and sedimentary rocks are composed largely of a few minerals that are the relatively stable products of erosion.

3.4.1 General origin and distribution of sediments

Having discussed the broad aspects of the composition of the Earth's crust and some aspects of sediment textures, we can now turn our attention to where sediments on the coast come from and what controls them in their specific environment of deposition. There are several factors that determine how sediments are distributed throughout the coastal zone. Some of the important ones are:

1 Composition of coastal sediments is a direct reflection of the composition of the source materials that produced these sediments.

2 The dispersal of sediments is related to the processes that transport them from their source and throughout the coastal zone.

3 The grain size of the sediments is a reflection of the rigor of the processes acting in the environment in which they accumulate.

If we were to ask someone what kind of sediment is present on the coast, the answer would probably be light-colored sand. This simply reflects the experience of most people, whose coastal visits are typically to the beach, and most beaches are composed of light-colored sand. This sand is mostly quartz but could have some other minerals or rock fragments included as well. There are, however, lots of beaches that don't look like this and there are also many other coastal environments that might have other types of sediment.

3.4.2 Composition

Minerals and rocks that comprise the sediments of the coastal zone reflect the composition of the rocks and sediments that are in the source area from which these sediments come. In many situations, these sediments are produced by some type of weathering and erosion, then transportation by a river to the coastal zone, where they may end up in a river delta, be carried to the open coast and then distributed by currents and waves, or be dumped

Box 3.1 Beach nourishment borrow areas

The construction of a beach nourishment project is a very complicated undertaking. It is much more than putting some sand on the beach to make it wider. One of the most important aspects of such a project is the location of appropriate borrow material: the sediment that will be used to build the beach. The basic objective is to find material that is similar to that on the natural beach at a price that falls within the amount determined by budget analysis. It is really a long and detailed undertaking to locate such material, and to determine its suitability.

Obviously we are looking for sand-sized material as a first requirement. We also want something that is durable, with a grain size that matches or is a little coarser than the natural beach at the location in question. We do not want too much shell debris because that is not comfortable for sunbathers, and tourism is a big factor in beach nourishment. Color is another important consideration: some projects have caused residents and tourists alike to complain because of a mismatch in color of the sand.

Appropriate material can be located underwater in the nearby ocean or in upland areas where old sand dunes have been preserved. For a variety of economic and logistical reasons, the shallow marine environment tends to be the typical location of most nourishment material. The shoals at the mouths of tidal inlets, large sand bars, and old beaches from lower sea levels tend to be the most commonly targeted sites.

The exploration first requires detailed bathymetric maps to show where these mounds of sand are located. Then various surveys are conducted throughout the prospective area. The bottom community must be investigated to determine if there are endangered species, organisms that would be destroyed by the dredging, and other factors. Detailed high-resolution seismic surveys provide knowledge of the thickness of the sand body and the depth to solid rock. Because of concerns for antiquities, magnetometer surveys are conducted to determine if there are shipwrecks buried in the sand. Generally if these assessments prove to be positive toward using the sand in question, then cores are taken through the potential borrow area to determine the grain size, sorting, and composition of the sediment. All of these efforts take months to a year or more and are very expensive.

After all of this has taken place, and assuming that positive results have been obtained, then applications are completed and filed with the appropriate state and federal agencies in order to obtain a permit for dredging and construction.

into an embayment along the coast. Fine sediments are commonly dominated by clay minerals, sand is generally mostly quartz but might also have feldspar, rock fragments, and other minerals, and gravel could be similar to sand in composition, although rock fragments tend to be the most common.

A good example of how the local source materials dictate the composition of coastal sediments is found in many of the tropical islands of the Pacific Ocean. Here the islands are typically volcanic in origin, with a composition of basalt, and most of these islands are surrounded by coral reefs. The sediments of the beaches and surf zone on these islands are typically black basaltic sand and gravel, with scattered fragments of white coral reef debris (Fig. 3.11).

Coasts in glaciated areas show a similar match to source materials. The glacial sediments of these areas contain a wide range of grain sizes and sediment particle compositions. The coastal sediments derived from these materials also have this spectrum of textures and compositions. The New England coast is a good example of this situation (see Chapter 18).

Regardless of how they originate or are transported to the coast, all of the coastal environments of the Bahamas are composed of a single mineral: calcite (calcium carbonate). That is because there is nothing but calcium carbonate in the entire Bahamas Platform; it is isolated from a landmass that could contribute other mineral and rock compositions.

Fig. 3.11 A beach on Tahiti that is dominated by black basaltic gravel and has scattered pieces of white coral. The dominant black gravel particles are eroded from volcanoes, which form the core of the island. They are smoothed by wave action. The white particles are eroded from the nearby reef and transported to the beach.

3.4.3 Texture

The processes that dominate a particular environment in the coastal zone are responsible for the grain size of the sediments that accumulate there. Mud carried by rivers ends up in the river delta or in bays where the rivers discharge their load, or it may be carried offshore onto the continental shelf and beyond. In general, the mud and other fine sediments tend to accumulate in places where waves and currents are absent or weak. Sand tends to accumulate where wind, waves, and currents are relatively strong. This includes beaches, dunes, tidal inlets, and other open coastal environments. In some places this sand is brought to the coast by rivers, or it may erode from rocks along the coast or be carried to the beach from offshore by waves.

In general, we can say that the grain size of the sediment along the coast is directly proportional to the rigor of the processes operating in the environment of sediment deposition.

3.5 Summary

The materials that make up the coast are fundamental in determining the nature of the coast. Obviously, one of the most important factors is whether the coast is dominated by bedrock or by sediment. Because sediments are transported by a wide variety of processes, with there being a wide range in the intensity of these processes, they are important factors in determining the morphology and scale of the various coastal environments. Low-energy processes tend to accumulate sediments that are fine-grained and/or those that are not well sorted. More energetic conditions result in sand and gravel that tends to be well sorted. Regardless of the nature or intensity of the processes operating along a particular coastal location, the sediment is the product of the source materials in that vicinity.

Suggested reading

Boggs, S. N. (1998) *Principles of Sedimentation and Stratigraphy*, 2nd edn. Englewood Cliffs, NJ: Prentice Hall.

Davis, R. A. (1992) *Depositional Systems: An Introduction to Sedimentology and Stratigraphy*, 2nd edn. Englewood Cliffs, NJ: Prentice Hall.

Greensmith, J. T. (1978) *Petrology of the Sedimentary Rocks*, 6th edn. London: George Allen & Unwin.

Selley, R. C. (1976) *An Introduction to Sedimentology*. London: Academic Press.

4 Sea-level change and coastal environments

4.1 Introduction

When sea level changes, for whatever reason, the coast responds with a change of its own. Some changes in sea level are very abrupt and can be recognized easily, others are so slow and of such a small magnitude that we cannot tell that they occur without sophisticated instruments. There are sea-level changes that take place globally but there are also those that are local or regional in extent. In this chapter we look at the causes of sea-level change, the extent over which they occur, and the rates at which they take place.

Sea level is changing throughout the world but it is doing so very slowly as a consequence primarily of climatic change and plate tectonics. It is also changing locally by a wide variety of means. In just a few moments, an earthquake can lower sea level by lifting the earth at the shore, or it can raise sea level by causing the surface to sink. All of these phenomena cause the position of sea level to change in comparison to the adjacent land. Sea level regularly rises and falls over short time periods as wind patterns shift over the seasons and as a result of the shifting positions of ocean currents.

Slow and long-term changes in sea level on scales of centuries to millennia take place as glaciers enlarge and subsequently melt in response to global changes in climate. The removal of the great mass of these glaciers permits the crust to rise, thus producing a lowering of sea level, resulting in old shorelines being many meters above existing sea level.

These kinds of sea-level change are all relative changes; that is, the position of sea level has changed relative to the surface of the crust. They also occur locally or regionally as opposed to globally. These local relative changes may be limited in extent to hundreds of meters, but they could be regional, extending over a thousand kilometers or more. A eustatic sea-level change takes place throughout the world. In order for such a change to occur there must be a change in the volume of the world ocean or there has to be a change in the overall size or shape of the ocean container, i.e. the crustal basins in which the oceans occur. For example, the melting of glaciers due to global warming will add water to the ocean system and thereby cause sea level to rise globally. Similarly, sea-floor spreading will change the shape of the ocean basins and thus cause sea level to rise everywhere.

Details of these and many other causes of sea-level change are considered in the following sections. The regular and predictable change in sea level caused by astronomical conditions that we call tides is not treated here because tides do not result in any net sea-level change. They are discussed in a separate chapter.

4.2 Changing the size and shape of the container: tectonic causes

Movement of the Earth's crust may take place over a wide range of rates and scales: from meters per hour to millimeters per year and from local events such as volcanic eruptions to plate movements that extend through entire ocean basins. Some of these changes take place at or near the coast and cause the crust to shift, and thereby cause sea level to change. Others are so large that they cause the entire ocean basin or even multiple ocean basins to change in size.

4.2.1 Local and regional changes

Many times in the past a seismic event resulting from crustal movement will take place at the coast, especially if the coast is at or near the leading edge of a plate boundary.

On March 27, 1964, a severe earthquake took place on the coast of the northeastern Pacific Ocean in Alaska. Here the Pacific and North American crustal plates are colliding and the movement along this crustal boundary produced some spectacular changes in local, relative sea-level positions (Fig. 4.1). The earthquake had its epicenter along the north shore of Prince William Sound between Anchorage and Valdez, the area more recently made famous by the Exxon Valdez oil spill disaster. The event registered about 8.3 on the Richter scale, making it one of the highest ever recorded in North America. Numerous locations along the shoreline

Fig. 4.1 Photograph of coast in the Anchorage, Alaska area showing uplift and property damage associated with the earthquake of March 27, 1964. Here the displacement is more than 2 m. (Official US Navy photograph.)

displayed shifts of meters upward and downward relative to sea level. Extreme examples of elevation change showed movement of 6 m above sea level, which caused harbors and docks to rise well above sea level and their docks to become useless. More extreme examples took place offshore, where islands were uplifted over 10 m and a portion of the sea floor on the adjacent continental shelf was lowered 15 m. It is important to remember that these sea-level changes took place in only minutes, but they were quite local in their extent.

Other local sea-level changes can be associated with volcanic eruptions. These phenomena are common along plate boundaries and will result in significant uplift of coastal areas affected by volcanic eruptions that push up the shoreline on many islands. Some subsidence could also take place during this type of phenomenon. Any of these conditions will cause local sea level to change, in some cases by several meters.

These extreme, event-related sea-level changes are not rare along collision coasts at plate boundaries. Such situations are present along the west coasts of both North and South America, on many of the volcanic island arcs of the Pacific Ocean, and in parts of the eastern Mediterranean, such as Italy, Turkey, and Cyprus.

4.2.2 Eustatic tectonic changes

Sea-floor spreading and the evolution of crustal material at spreading centers on the ocean floor can also have an influence on sea level. As new crust moves up from the asthenosphere the older crust above moves away from the oceanic ridge, causing the plates to diverge. Subduction takes place at the collision zones, causing the descent of one plate under the other. This combination of slow uplift and slow downwarping on the plate margins, along with movement of land masses on the conveyor belt called the crust, produces changes in the basins that change their volume. This set of conditions causes sea level to change on a global scale. The amount of change produced by these circumstances is quite small, on the order of millimeters per decade or more.

We therefore have both local and eustatic sea-level changes that can be produced by tectonic activity. Some of these are very rapid, such as those resulting from volcanic activity or earthquakes, and others are very slow, caused by sea-floor spreading. Also, we can generalize to say that local changes in sea level are generally more rapid and greater in magnitude than eustatic changes.

4.3 Climate and sea-level change

Sea level is affected by changes in climate. These changes may be due to seasonal or other short-term fluctuations in climate or they may be due to long-term changes, some of which are cyclic and others are not.

4.3.1 Seasonal changes

On a global scale, mean sea level typically shows a seasonal difference ranging from 10 to 30 cm depending upon location (Fig. 4.2). Such sea-level differences are due to the changes in wind patterns

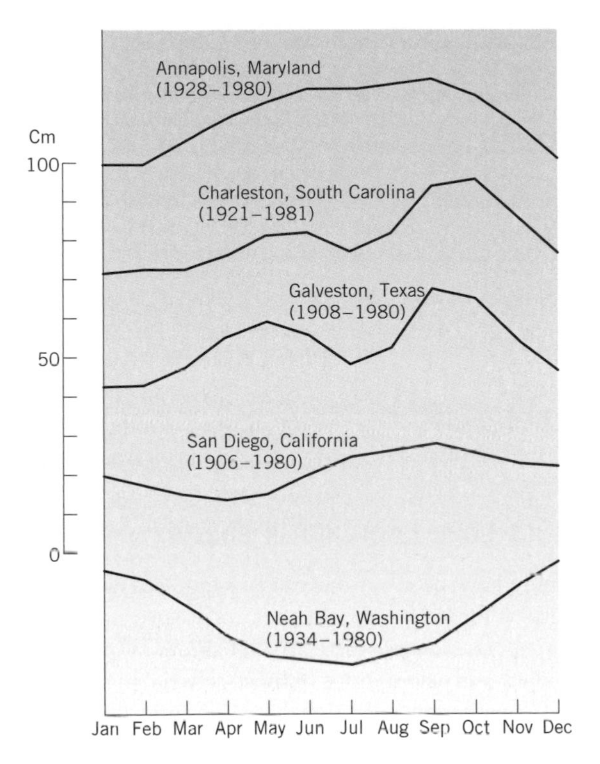

Fig. 4.2 Examples of annual cycles of sea-level changes as obtained from smoothing of tide gauge records. These data are taken from southern California. (From P. D. Komar & D. B. Enfield, 1987, Short-term sea-level changes and coastal erosion. In D. Nummedal, O. H. Pilkey & J. D. Howard (eds), *Sea-level Fluctuations and Coastal Evolution*. Tulsa, OK: SEPM Special Publication 11, p. 18.)

and velocities, and persistant, large-scale high or low pressure systems. For example, the Bermuda High, a large high pressure system in the central Atlantic Ocean, depresses sea level due to the high atmospheric pressure. Sea level in the Atlantic is typically lowest in the spring and highest in the fall, alternating back and forth between the northern and southern hemispheres as the seasonal wind patterns change. In the low latitudes the seasons are more subtle and, as a consequence, mean sea level also shows less seasonal change than in the mid-latitudes.

The most basic and most easily predictable seasonal sea-level fluctuation is caused by changes in weather patterns as the Sun and the Earth shift in their positions relative to each other (Fig. 4.3). As the sun moves through the latitudes with the seasons there are changes in water temperatures, which produce changes in wind patterns and velocities. A sea level change of 10–30 cm is produced as a response to these seasonal wind differences. This change is caused by water being pushed in different directions as the result of seasonal differences in atmospheric pressure and wind direction.

Seasonal temperature changes can change the water volume and thereby produce changes in sea level because water volume increases as it warms. The amount of change resulting from this phenomenon is so small that it is difficult to measure. In some situations, however, this type of seasonal change is combined with seasonal change produced by winds to show typical annual patterns of sea level change. Sea level is highest in the summertime on the east coast of the United States because the warm, expansive Gulf Stream flows very near the coast at that time of year. On the Pacific side of the continent, sea level is at its lowest on the Washington coast during the summertime because of the cold coastal currents that flow from the north at that time of year. The highest mean sea level on this coast is in winter, when Arctic winds generate coastal storms that blow onshore and pile up water.

4.3.2 Nonseasonal cyclic changes

Probably the most famous of the weather-related causes of sea level change is the El Niño phenomenon. This climatic change influences most of the world. It is caused by the warm current that is produced by changes in wind conditions off the west coast of Peru in South America and it has occurred every four to seven years over most of the south Pacific Ocean. Fishermen have named the current "the child," referring to the birth of Christ, because the El Niño phenomenon tends to begin in late December. Normally during this middle part of the southern hemisphere summer, the Peru Current, which is cold, flows to the north along the west coast of South America. As the current flows, the Coriolis effect produced by the Earth's rotation (explained in Chapter 5) causes a deflection of the current to the left or west, resulting in upwelling of

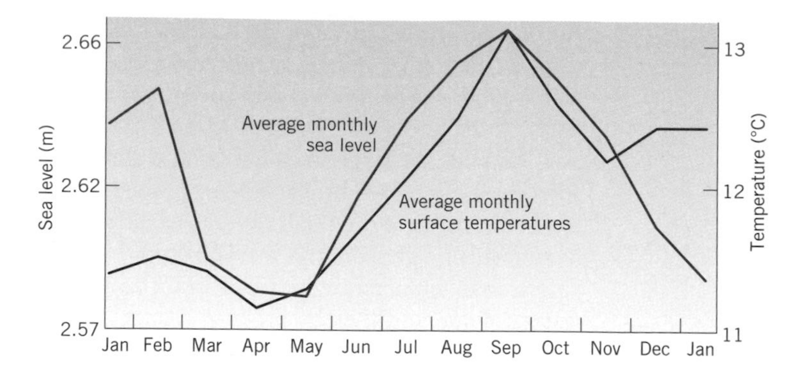

Fig. 4.3 Annual cycle of both sea level and water temperature showing their coincidence and their relationship to the dominant oceanic conditions of the time. (From P. D. Komar & D. B. Enfield, 1987, Short-term sea-level changes and coastal erosion. In D. Nummedal, O. H. Pilkey & J. D. Howard (eds), *Sea-level Fluctuations and Coastal Evolution*. Tulsa, OK: SEPM Special Publication 41, p. 18.)

even colder, nutrient-rich deep water. This upwelling provides the nutrients that support a huge population of plankton, which provides the food for one of the world's most important fishing grounds.

At four- to seven-year intervals, the west-to-east blowing trade winds diminish in their intensity, allowing a warm current to move south and force the Peru Current into the Pacific at a higher than normal latitude. This relatively warm current is the result of the El Niño conditions. When the current enters South American coastal waters, its thick, surface layer of warm water inhibits the upwelling of the cold nutrient-laden waters below. This situation keeps the plankton population from reaching its typical high density and, therefore, the fish populations are smaller than normal. The current may stay to the south for a year or more, which causes fish populations to either migrate to other feeding grounds or die, thereby disrupting the economics of the adjacent coastal areas. The warmer water of the sea warms the air above it, so that the rate of evaporation increases. This moist air mass moves landward, bringing torrential storms that produce flooding and coastal erosion in some areas and a large influx of sediments in others. Changes in other wind and current patterns are also associated with these air masses.

There is a rise in sea level associated with El Niño along the entire western South American coast and extending up to California (Fig. 4.4), due to both the wind effects and the fact that the warm water takes up more space than the colder Peru Current. The El Niño condition also causes anomalous weather conditions, including storms on the South American side of the basin and some drought conditions in places like Indonesia on the western side.

4.3.3 Long-term climatic effects

There have been various causes suggested as explanations for long-term changes in climate, including sun spots, changes in relative positions of celestial bodies, and orientations of those celestial bodies.

Climate-induced global or eustatic sea-level change can occur in two ways: by increasing or reducing the amount of water in the entire world ocean, or, more subtly, by changing the temperature of the world ocean, which causes its volume to increase or decrease as water expands or contracts. In either case, when the volume of water changes it causes an absolute, or eustatic, change in sea level. A worldwide and long-term change in climate can bring about both of these conditions simultaneously. A cold climate results in glaciers, which incorporate huge volumes of water while at the same time lowering the temperature of the ocean. Both of these situations cause the volume of the world ocean to be reduced. A warmer climate melts the glaciers and frees the water to return to the ocean; simultaneously it raises the ocean's temperature, causing an increase in ocean volume and, therefore, a rise in sea level.

Drastic temperature changes are not required to produce major changes in sea level. A rise or fall in the mean annual global temperature of only 2–3°C has a profound effect on both the volume of ice retained on the surface of the earth and the volume of the water in the world ocean. During the last ice

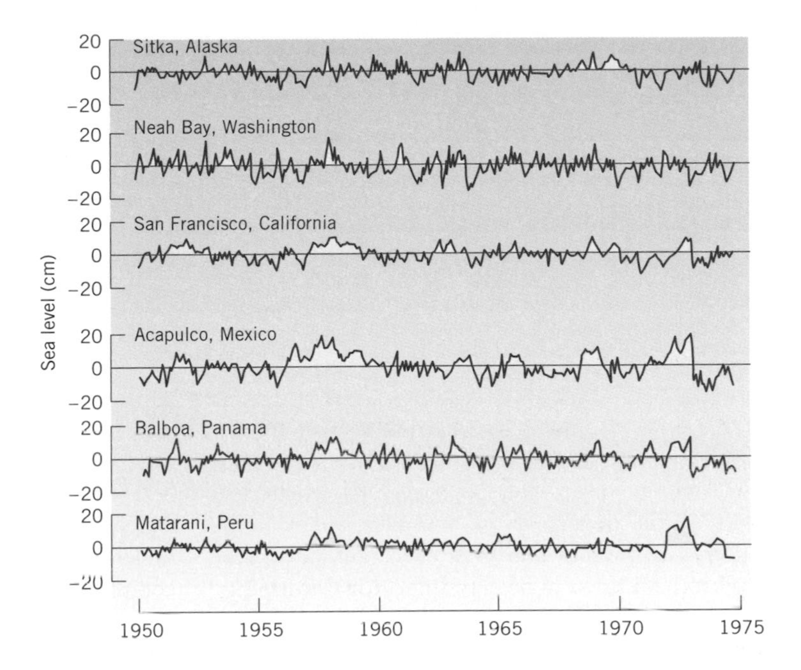

Fig. 4.4 Sea-level variations from the east coast of the Pacific Ocean over a 25-year period showing anomalies that have been attributed to an El Niño effect. (Modified from P. D. Komar & D. B. Enfield, 1987, Short-term sea-level changes and coastal erosion. In D. Nummedal, O. H. Pilkey & J. D. Howard (eds), *Sea-level Fluctuations and Coastal Evolution*. Tulsa, OK: SEPM Special Publication 41, p. 24.)

age, thick ice sheets covered much of the land mass of the northern hemisphere. The volume of water in the world ocean was greatly reduced; water was incorporated in ice and that remaining in the ocean became contracted with the cold. This combination resulted in the exposure of nearly all of the continental shelf. These conditions were produced with a reduction in global mean annual temperature of only about 2–3°C as compared to the present. If the present apparent trend toward a warmer global climate continues and the mean temperature increases by only a few degrees, the entire process will reverse – the ice sheets will melt and the ocean will encroach upon the continents until many of the port cities are at least partly under water.

Insufficient data have made it difficult to assess the pace and direction of global climate and sea-level changes over the past few decades. Accurate records of sea level have been kept for little more than a century and weather records in most parts of the world do not extend back any further. The great sea-level changes of the past, as recorded in the layers of sediment accumulation, occurred in cycles of tens of thousands of years or more. Our hundred-year-old records, therefore, cannot be superimposed on past changes in order to make any sort of valid prediction. Nevertheless, the recent rise in global temperature has forced us to take note of a possible human-generated cause for the worldwide increase in the rate of sea-level rise taking place today. The still-accelerating release into the atmosphere of carbon dioxide and other greenhouse gases has some climatologists projecting a global warming of 3°C by the year 2030 (Fig. 4.5). This increase in mean annual temperature could melt a large portion of the ice cover in Greenland and Antarctica; enough to raise global sea level by as much as 5 m in only a few centuries. This is too long for a person to worry about but still a short time in terms of human occupation of the coast as we know it. Even here, however, we may be trying to superimpose a short-term data base on very long periods of cyclicity. The current increase in carbon dioxide might only be part of a longer cycle that predates civilization and will decline by itself as the cycle proceeds. It is hard, however, to discount the obvious contribution to global warming being made by our current high levels of combustion and our destruction of photosynthesizing plants that take up huge amounts of carbon dioxide and release large quantities of oxygen. It is important for us to

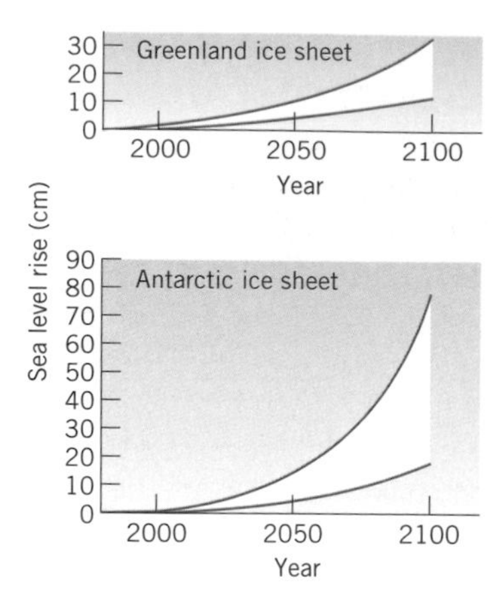

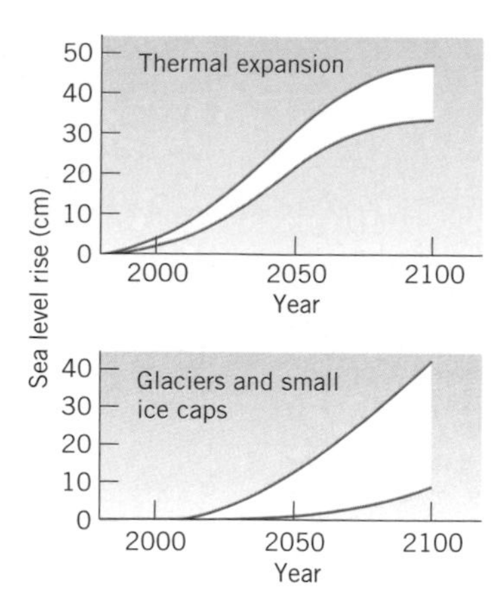

Fig. 4.5 Contributions to sea-level rise during the next century. Notice that nearly all of the rise in sea level is predicted to come from the melting of glaciers as the result of global warming. (From M. Ince, ed., 1990, *The Rising Seas*. Proceedings of the Cities on Water Conference, Venice, Italy.)

examine the details of the warming and cooling patterns of the past and their effects on global sea level to help us to understand the global warming pattern and sea-level changes that we are presently experiencing.

4.4 Sea-level rise due to sediment compaction and fluid withdrawal

In some of the coastal areas associated with river deltas sea level is rising locally because the land itself is sinking, or subsiding. This is particularly apparent along east Texas and within the Mississippi delta in Louisiana. Here local sea-level rise is much greater than the global rate. The highest rates of local sea-level change in the United States are in the Mississippi River delta area, where it is rising by 9–10 mm per year, about four times the global average. As much as 6–7 mm of that rise is due to land subsidence caused by a combination of isostatic loading of the delta sediments, fluid withdrawal primarily by the petroleum industry and compaction of water-laden deltaic sediments.

Huge quantities of fine sediment are transported by the river and deposited at its mouth, an average of up to 1.6 tonnes per day at the active lobe. This sediment is deposited so rapidly that it traps much water as it settles. The resulting mud, at places up to 90% water by volume, accumulates as thick sequences through the active delta region. As the weight of the new mud layers compresses the underlying ones, the water is squeezed out, thereby compacting the sediment to a lesser volume, which causes the land surface to subside, resulting in a relative rise in sea level.

While compaction of sediments has occurred for several thousand years, subsidence along a significant part of the northern coast of the Gulf of Mexico caused by the withdrawal of large volumes of fluid is a recent phenomenon that is the result of recent human activities. Nearly 100,000 wells have extracted large quantities of oil and natural gas from the Mississippi delta and also from the nearby coast and shelf of the Gulf of Mexico. Large volumes of ground water have also been taken for domestic and industrial use. This results in a dewatering effect similar to that of squeezing of the water from the delta muds, resulting in land subsiding and sea level rising. An example of this phenomenon is near Galveston, Texas, where land has sunk nearly 2 m in the past century (Fig. 4.6). To reduce, or possibly eliminate, further subsidence, the major domestic and industrial uses are being shifted to surface water, and water is now being pumped back into the ground to replace what had previously been withdrawn.

Another human impact on the coastal zone is not as easily remedied. The building of large cities on thick accumulations of unstable sediments has been an environmental disaster. The water of the nearby Gulf has responded by flooding parts of the inland

Fig. 4.6 Photograph of flooding in southeast Texas near Galveston Bay, which is, in part, due to a rise in sea level resulting from withdrawal of oil and gas in this area. (Courtesy of Orrin H. Pilkey.)

port cities of Houston and New Orleans. In fact, the city of New Orleans is a few meters below sea level and even more than that below the level of the Mississippi River.

If sea level in the lower Mississippi delta continues to rise at a rate of 9–10 mm yr^{-1}, and if this rate is sustained for 50 years, it could amount to 0.5 m of sea-level rise, enough to submerge large areas of coastal wetlands, which would destroy the ecosystem that is so important to the delta area.

Compaction is not only an important contributor to land subsidence in delta regions throughout the world, it also affects peat bogs, marshes, and other organic-rich sediment accumulations that hold large volumes of water. The influence of compaction on these environmentss is quite small, however, because their overall thickness is less than a few meters in most places.

4.5 Isostasy

The Mississippi River delta is one example of regional subsidence, but broad regions of the continental plates can also sink under a heavy load. Over a period of thousands of years, the mass of a huge ice sheet can cause the continental lithosphere to become depressed by about 100 m, with the amount dependent on the thickness and density of both the ice mass and the underlying lithosphere. The vertical movements of the lithosphere are accommodated by the semi-plastic portion of the upper asthenosphere. As the glacier melts, removing the overlying pressure of the ice from the land mass, the lithosphere rises, or rebounds. Such vertical adjustments of the lithospheric crust are called isostatic adjustments; isostasy is the condition of equilibrium of the Earth's crust that takes place as the forces that tend to elevate the lithosphere are balanced by those that tend to depress it.

The depression and subsequent rebounding of the lithosphere as mass is added and removed is only one kind of isostasy. A change in density of the lithosphere will cause a similar isostatic crustal response and produce changes in sea level as a consequence. When the lithosphere is young and still hot as it is produced at a rift zone of the oceanic ridge system, it is relatively low in density. As this lithosphere cools over several millions of years, it is reduced in volume, becoming more dense. This causes subsidence over the asthenosphere. Because most of this activity takes place in the ocean basin, the resulting rise in sea level over the subsiding sea floor is so small as to be imperceptible. There are a few places on the globe where coastlines are close to sites of plate divergence. At these places the local rise of the young lithosphere can produce an increase in sea level and slow inundation of land. The coasts of both the Red Sea and the narrow Gulf of California are on such diverging oceanic plates and are places where this type of local sea-level rise can be expected.

Another reason for isostatic adjustment of the Earth's crust is related to the thick accumulation of sediments and volcanic material in the lithosphere. This currently takes place along thick prograding coastal plains such as the Gulf and Atlantic coasts of the United States. In both cases, thousands of meters of sediment accumulate over tens of millions of years, causing crustal subsidence, which produces a relative rise in sea level. The same phenomenon, but in a more rapid scenario, is associated with thick accumulations of volcanic crust, such as the islands in the Pacific Ocean. One of the best examples of this phenomenon is the Hawaiian Islands, where about 5000 m of volcanic material is piled up on the thin oceanic crust. This enormous mass causes subsidence and a relative rise in sea level.

4.6 Changes in the volume of the world ocean

4.6.1 Advance and retreat of ice sheets

During the Quaternary Period, the most recent period in the geologic time scale, which is from about two million years ago to the present, eustatic sea level has changed very rapidly as compared to most times in the Earth's history. It is also likely that we have this opinion about rates of sea-level change simply because we do not have the detailed information about sea-level change throughout geologic time that we do for the past couple of million years. The cause of these geologically recent and rapid sea-level changes has been the advance and retreat of continental ice sheets and the polar ice caps that formed the extensive glaciers of the northern hemisphere and the Antarctic. The Pleistocene Epoch of the Quaternary Period, also called the "Ice Age," came to a close 10,000 years ago. At this time the last of the major ice sheets had melted, except in Greenland and Antarctica. For the previous million or two years, there were thick ice sheets in the northern hemisphere that repeatedly covered and withdrew from most of Europe, northern Asia, and North America down to the Missouri and Ohio River valleys (Fig. 4.7). The development of similar ice sheets in the southern hemisphere was limited due to the general absence of land in the high latitudes, except in Antarctica and the southern tip of South America.

Fig. 4.7 Map of the northern hemisphere showing the distribution of glacial ice sheets during the peak condition of the Pleistocene as compared with the present. (From J. T. Andrews, 1975, *Glacial Systems*. North Scituate, MA: Duxbury Publishing Co.)

The record of the Pleistocene Epoch that is preserved on the continents is generally considered to include four cycles, in each of which there was glaciation alternating with interglacial melting. Each of these four cycles has a name for both the glacial advance and the following melting portion of the cycle. For example, the last portion of the cycle characterized by glacial advance is called the Wisconsinan in North America. It was preceded by the Sangamonian interglacial period, when sea level was high and glaciers were probably smaller and covered less area than they do at the present time. In Europe and other areas where similar glacial cycles took place there is a different terminology for each of the cycles.

For the first half of the twentieth century this interpretation of four glacial cycles was accepted and taught throughout not only North America but also the rest of the world. As we developed more capability for studying the ocean basins through sediment cores, the climatic history depicted showed that there were many more Pleistocene glacial cycles than the four shown in the land-based stratigraphic records. The evidence for these numerous glacial cycles has been masked on land by deposits of the four larger and longer-lasting glacial cycles.

The discovery from the oceanic record of these numerous glacial cycles is primarily due to techniques for investigating ancient climatic conditions on the Earth that did not become available until the late 1940s, just after the Second World War. Probably the most important of these is the use of oxygen isotopes to help to interpret past climatic conditions. Isotopes are varieties of a given element that have different atomic masses due to variation in the number of neutrons. Oxygen, for example, typically has an atomic number of 16, but there is also a heavier isotope, oxygen-18. Both isotopes behave the same chemically and both are incorporated in the skeletons of organisms like corals and mollusks as part of the compound calcium carbonate ($CaCO_3$). The discovery was made at the University of Chicago in 1947 that the relative concentration of the two oxygen isotopes taken into the skeletons of organisms was a function of the temperature of the ocean water in which the animals were living. This led researchers to believe that the temperature of the water at the time when the organism was living could be determined by analyzing the $^{16}O/^{18}O$ content of the calcium carbonate in the skeleton. This relationship was not applied to global climate interpretations in any significant fashion until 1955, when the skeletons of floating single-celled animals (foraminifers) from several deep sea cores were analyzed using these techniques. These floating, single-celled animals are very common in the upper few hundred meters of the water column throughout the ocean. Their calcium carbonate skeletons settle to the bottom when the individuals expire. The sediments in cores taken from the ocean floor have many of these skeletons. Oxygen isotope ratios determined for samples taken from these cores provided data that demonstrated that there were numerous periods of significant temperature fluctuation in ocean waters in only 300,000 years. These changes in ocean temperature were interpreted to be the response to climatic changes associated with glacial activity. These data demonstrated that the record preserved in the sediments of the ocean floor was much more complete and complex than that on the continents.

Perhaps more importantly, the temperature cycles shown by the oxygen isotopic data in these sediment cores are in agreement with the cycles that had been predicted many years earlier by Milutin Milankovitch, a Serbian astronomer. He developed a theory of climatic changes that was based on cycles of radiation received by the Earth as it tilts relative to the Sun. Three different astronomical conditions produced cycles (Milankovitch cycles) associated with variation in tilting and its effect on climate:

1 Variation in eccentricity of the Earth's orbit around the Sun, with a periodocity of 90,000–100,000 years.

2 Changes in the obliquity of the Earth's plane of orbit and the angle it makes with the plane of the ecliptic, with a period of 41,000 years.

3 The precession or wobbling of the Earth's axis, with a period of 21,000 years.

The coast as we see it now is the product of the most recent of these cycles. The last of the great advances of glaciers, the Wisconsinan Ice Age, began about 120,000 years ago and lasted for more than 100,000 years. The formation of these glaciers took most of this time; only about 20,000 years were involved in the melting of them to their present size. Since the end of that period we have been in what is called an interglacial period, characterized by global warming, glacial melting, and rapid sea-level rise. The Antarctic and Greenland ice sheets of today are remnants of the last Wisconsinan advance. The period of overall melting and related warming has been interrupted by a few "little ice ages," which are extended periods of abnormally cold weather, some of which have occurred during recorded history. The most prominent was chronicled in Europe between about 1450 and 1850. Sea level actually dropped during these short cold periods that occurred during the overall warming trend.

By calculating the difference in volume between the Wisconsinan ice sheets and the ones that remain today, it is possible to determine the volume of ice that has melted and how this volume has changed global sea level. The next step is to extrapolate what would happen to sea level if the present rate of global warming continued until the remaining ice sheets melted. In order to make these determinations, it is necessary to estimate the area and thickness of the previously existing ice sheets. The surface area covered by the Antarctic ice sheet during the Wisconsinan advance was almost 14 million km^2 and today it is about 12.5 million, which is not a lot of difference. By way of contrast, the North American ice sheet once extended over more than 13 million km^2 of land and now covers only 147,000 km^2, a loss of about 99%. There were also ice sheets covering Greenland, much of Europe, and the northern part of Asia. The combination of these and other smaller ice sheets covered more than 44 million km^2 about 12,000–15,000 years ago, of which just under 15 million now remain, one-third of the former extent.

Whereas we have geologic and geographic information that enables us to determine the areal extent of these enormous ice sheets, it is much more difficult to determine the volume of the Wisconsinan ice sheets. A reasonable approach is to use the Greenland ice sheet as a model. This ice sheet is well known because of extensive petroleum exploration surveys, as well as scientific and military studies. Glaciologists have estimated that the Greenland ice sheet holds about 2.5 million km^3 of ice. By extrapolation of surface areas and by estimating average thicknesses, we interpret that 75 million km^3 of ice were contained in the vast glaciers of the Wisconsinan Ice Age. Given the ratios mentioned above, then about 50 million km^3 of ice have melted since the Wisconsinan Ice Age – roughly equivalent to about 20 times the volume of the present Greenland ice cap. Considering that there is a 10% decrease in volume when ice turns to water, then 45 million km^3 of water were returned to the ocean. This obviously caused a major change in volume and therefore a dramatic increase in sea level. The mass of this volume of water was enormous and caused isostatic adjustment of the crust, both on the continents where mass was removed and on the oceanic areas where mass was added. This tremendous shift of mass from the continent to the ocean had, and is still having, a pronounced influence on sea level.

As the ice sheets were removed from the continents, the continental lithosphere began to rebound in an isostatic adjustment. If we assume that some areas were covered with 3000 m of ice at peak glacial conditions, the isostatic rebound after melting would have reached about 1000 m. This is determined by assuming that the 3000 m of ice had a density of 0.9 $g\ cm^{-3}$, which is close to one-third of the density of the continental lithosphere. This would be equivalent to about 1000 m of lithospheric crust. The rate of the rebound has been slow, and in some places, such as Norway and Sweden, it is still ongoing (Fig. 4.8).

While this was taking place on the continents, the ocean was experiencing the opposite condition. The added mass coming from the continents in the form of glacial meltwater caused a sinking of the ocean floor, which has a much younger and thinner crust than the continents. This combination of continental rebound and oceanic crustal depression with added volume of the ocean still produced an increase in sea level in most places despite the sinking floor of the ocean. There were, however, regional decreases in sea level due to rapid rates of

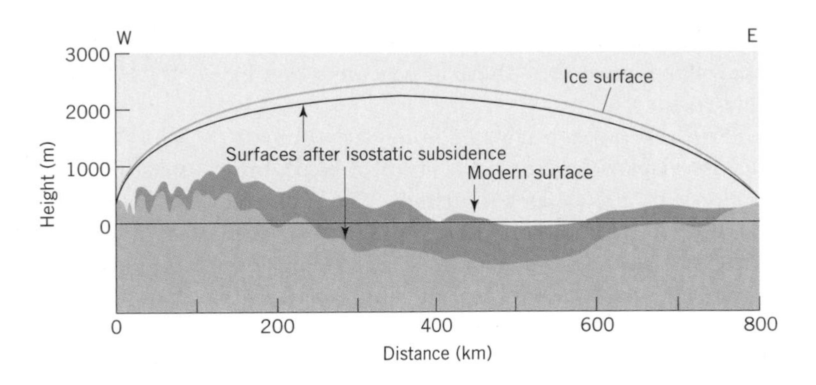

Fig. 4.8 Map and cross section showing the subsidence and rebound associated with glacial development and melting across the northern part of Norway and Finland, where the greatest isostatic adjustments have taken place and are still under way. (From A. L. Bloom, 1978, *Geomorphology*. Englewood Cliffs, NJ: Prentice Hall, p. 405.)

continental rebound. Considering the combination of all of these phenomena, the overall eustatic increase in sea level over the past 18,000 years has been about 100 m, with local variations due to differences in continental rebound.

4.6.2 Continental rebound

Because the Wisconsinan glacial ice melted slowly, the isostatic adjustment in the form of continental rebound was also slow. Most of the crustal rebound took place before the melting of the ice sheets reached near their present size, because the remaining ice was no longer sufficient to cause significant subsidence. In some places the rebound is still ongoing, such as in Scandanavia, in the Hudson Bay area of Canada, and in Argentina and lower Chile.

Along the coast of Sweden and Norway, the mean annual sea level is still dropping a few millimeters a year as a result of the rebound of continental crust. If the annual global sea-level rise of 2–3 mm is added, the rate of uplift is up to 9 mm a year. If we assume that this has been going on for about 5000 years, the rebound has been a net uplift of about 30–35 m. It is even more in the Hudson Bay area of north-central Canada. Aerial photographs of these areas show parallel ridges marking the old shorelines and abandoned beaches as the sea level dropped, causing the ocean to recede from the land (Fig. 4.9).

The amount of isostatic adjustment diminishes toward the low latitudes because the ice formed later, it was thinner, and it melted more rapidly toward the warmer climate. This can be shown along the

Fig. 4.9 Photograph of beach ridges around Hudson Bay, Canada, providing evidence for isostatic rebound after glaciation. As relative sea level fell, the waves formed these numerous beach ridges. (Photograph courtesy of A. Hequette.)

New England coast, where old shorelines are somewhat higher than those of the same age in New York.

The Great Lakes originated from glacial activity and also display clear signs of a north-to-south tilt, as shown by the ancient shorelines. These shorelines are several meters higher on the northern part of the lakes than on the south sides. Lake Michigan is gradually getting shallower at the north end and deeper in the south. If the present rate of rebound continues, after about 3000 years the lake will drain into the Mississippi River system rather than through the other lakes into the Saint Lawrence Seaway as it does now.

Wherever the high-latitude land mass is still responding to the melting of an ice sheet, coastal uplift leaves the shorelines well above present sea level (Fig. 4.9), abandoned by the receding water level. Eventually the rebound will slow and the water level will rise again relative to the land mass.

4.7 Holocene rise in sea level

The Holocene Epoch is arbitrarily defined as beginning 10,000 years ago. It is characterized by the rapid and large rise in global sea level as a response to the melting of the Wisconsinan glaciers. Geologists, oceanographers, and climatologists who study sea-level fluctuations cannot agree among themselves on the details of the rise in sea level over this period, especially during the past few thousand years.

It is generally agreed that sea level was at its lowest position about 18,000 years ago, when the ice sheets of the Wisconsinan had reached their maximum extent. The lowest position of sea level, called the lowstand, is deduced by uncovering evidence of the oldest drowned shoreline now located beneath the waters of the continental shelf. This evidence might take the form of beach sand, marsh deposits, drowned wave-cut platforms, drowned river deltas, or almost any sort of indication of an old shoreline. Other factors, such as tectonic uplift and subsidence, must also be taken into account

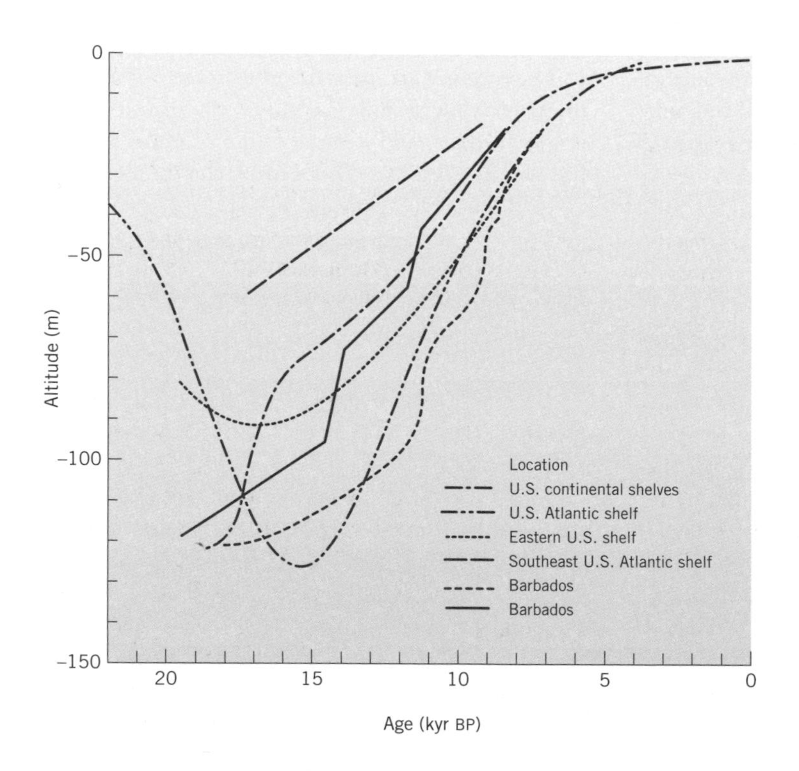

Fig. 4.10 Plot of sea-level curves for the late Quaternary for numerous locations in the Atlantic Ocean and Caribbean Sea. The differences are in part due to tectonic factors but all show a significant slowing in the rate of rise about 6000–7000 years ago. (From Fairbanks, 1987.)

before deciding the vertical position of the low-stand. Scientists know that sea level was more than 100 m below its present position; perhaps as much as 130 m. There is general agreement that sea level rose very rapidly for the first several thousand years and that it slowed about 6000–7000 years before present (BP) (Fig. 4.10). The annual rise during this period was near 10 mm yr^{-1}, a rate fast enough for the sea to cover parts of many of the present coastal cities in a century.

During this period of rapid rise in sea level the shoreline moved so quickly that the sand bars and barrier islands that protect so many of our coasts today had no time to build vertically. In order for these barriers to develop, it is necessary for the shoreline to be either stable or very slowly moving, to give waves and currents enough time to construct these long sand bodies. The lack of a stable shoreline for any length of time coupled with moderate to high tidal ranges along the irregular coasts produced tide-dominated coasts with widespread estuaries and tidal flats. Some areas, such as Australia and New Zealand, saw relative sea level reach its present position about 6500–7000 years ago, at the end of this period of rapid rise because of land subsidence in these areas of the globe. Most land masses of the world experienced continued but much slower sea-level rise due to the decrease in the rate of glacial melting. This slower rate of sea-level rise permitted shorelines to become more stable in their position and resulted in waves becoming the dominant coastal process. These conditions led to the formation of beaches and barrier islands.

There is no general agreement about the position of sea level along the coast of North America during the past 3000 years or so. There are three scenarios that have been proposed during this period: (i) it has been stable at the present position; (ii) it has changed about a meter or so above and below its present position; and (iii) it has been gradually rising during the period, over only about 3 m (Fig. 4.11). It is possible that each of these situations has prevailed at various places. Because of the resolution necessary to determine which of these three situations might have existed at any given coastal location, we are still working on this question.

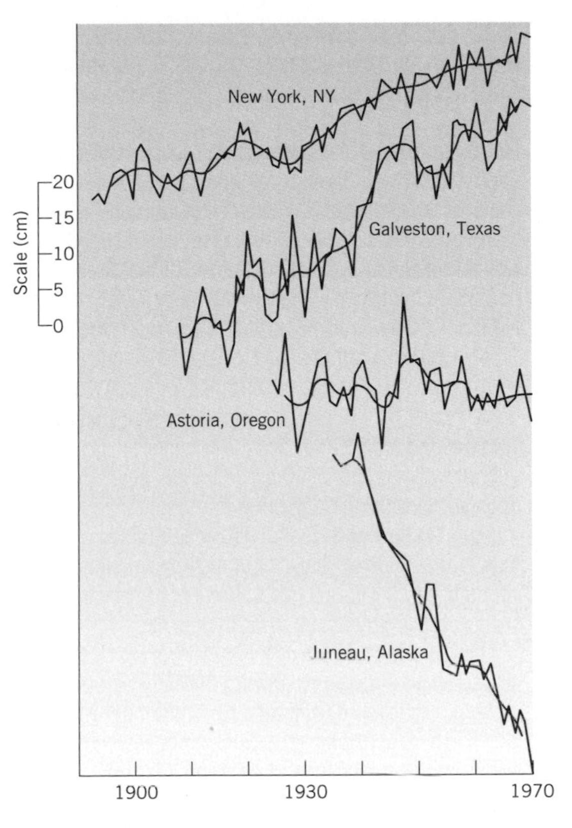

Fig. 4.11 Tide gauge records for the past several decades as compiled by Hicks and colleagues. These show well the differences around the USA based largely on tectonic conditions; Alaskan sea level is subsiding locally but it is stable in other areas, as it is in Oregon. The east coast is showing a rise of about 2 mm yr^{-1} (Adapted from S. D. Hicks, J. A. Debaugh Jr & L. E. Hickman Jr, 1983, *Sea Level Variations for the United States, 1855–1980*. Rockville, MD: US Department of Commerce, NOAA, National Ocean Service.)

4.8 Current and future sea-level changes

The impact of the rapid increase in the human population of the Earth has produced elevated levels of carbon dioxide, which, according to many authorities, has led to global warming (see Box 4.1). The warming of the atmosphere over a long period of time causes an increase in the rate of melting of the ice sheets and also causes the ocean to warm up. Both of these phenomena result in an increase in

Box 4.1 Global warming and the future of sea-level rise

During the past decade or so, there has been much speculation about the future trends in sea-level rise. Conservative estimates give a rise of about 50 cm over the twenty-first century, whereas the most liberal predictions are in excess of a meter for that time period. Let us take a close look at the situation; then make your own decision about "global warming."

It is well known that sea level rose tremendously at various rates during the past 18,000 years as extensive glaciers melted thoughout many parts of the earth. This melting took place because of an increase in global temperatures of about 2–3°C per year. That does not seem like much but it is enough to cause melting of snow and ice that had accumulated over many thousands of years. Obviously this melting added huge amounts to the volume of the oceans and sea level rose greatly: nearly 130 m in all.

The important aspect of this sea-level rise is the rate at which it occurred. During the early part of the melting, sea level rose about a centimeter per year, which is a catastrophic rate. Such a rate lasted for several thousand years and accounted for the vast majority of the rise. Although a centimeter a year (an inch each 2.5 years) seems like a pretty slow change, it causes great erosion of the coast and drowning of wetlands (see Box 4.2). The addition of sediment to compensate for these processes is generally insufficient to prevent the sea from encroaching on the land. In a lifetime of this rate of rise sea level will rise more than 60 cm (2 feet), enough to cause problems for many low-lying coastal communities. Recall that this rate of change occurred for thousands of years.

During most of the time in which significant numbers of people have inhabited the coast, sea level has been rising very slowly: only a millimeter or so per year. This goes unnoticed by almost everyone except those studying the position of sea level. Why did sea level rise rapidly for an extended period and then rise only slowly for another extended period? The simple answer is that melting of the ice caps throughout the world slowed as a response to global temperature: it stopped warming up so fast.

Why this happened provides the more complicated and complete answer to the question. Part of it can be found in the various perturbations of the Earth–Sun system as recognized by Milankovich in the early twentieth century. He showed that there are various cycles to the way the Earth moves in its orbit, the shape of the orbit, and the wobble and tilt of the earth on its axis. These cycles range from about 20,000 years to nearly 100,000 years. As a consequence of these fairly long cycles and their shorter-term variations, the amount of insolation that is received on the Earth as a whole, and on various parts of the Earth, changes with time, resulting in fluctuations in the mean annual temperature.

These are the well documented natural changes in the Earth's climate. What about the possible influence of humans on the global climate? Since the industrial revolution in the nineteenth century, we have been emitting large amounts of carbon dioxide into the atmosphere. This carbon dioxide is concentrated in the upper atmosphere, where it acts as an insulator, preventing heat loss from the Earth into the stratosphere. It acts like a greenhouse: thus the term "greenhouse effect." Since the advent of the automobile this effect has been exacerbated. Most of the problem originates in North America and Europe – the so-called developed countries.

During the past 100 years, about the time for which we have good records, it is evident that the rate of sea-level rise has increased. It is now about 1.5 mm per year. Is it coincidence that this increase in sea level parallels the great increase in CO_2 production, or is CO_2 the cause of increased global temperatures and therefore the rise in sea level? Actually there is still some disagreement among scientists as to the answer. There are more favoring anthropogenic factors as very important but many believe that the changes in sea level are a result of natural changes in the Earth's climate. The case is not yet closed on global warming.

the volume of water in the world ocean, and thus an increase in sea level. Recent data from tide gauges around the world show that the rate of rise in eustatic sea level is increasing. Because tide gauges are available from all over the world this is one of the most prevalent types of sea-level data and must be studied very closely. The records from many of these stations must be discarded because of instability

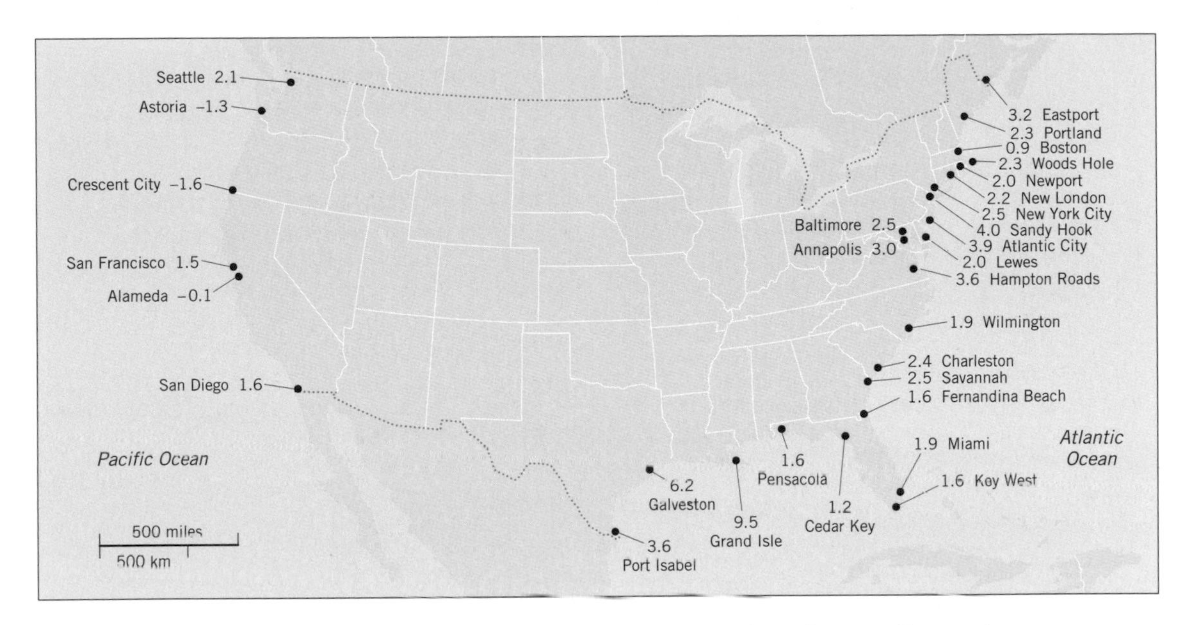

Fig. 4.12 Map showing rates of sea-level rise around the continental USA as shown from tide gauges. Compare the east coast, Mississippi delta area, and west coast. (From National Research Council, 1987, *Responding to Changes in Sea Level.* Washington, DC: National Academy of Sciences, 1987, p. 10, fig. 1.1.)

of the location due to subsidence or compaction, or because of tectonic activity. For most locations, we have an average of about one hundred years of reliable data for the position of sea level.

Most of the data for the United States are collected through the efforts of the National Ocean Survey (NOS), which is part of the National Oceanographic and Atmospheric Administration (NOAA). Their personnel have studied and analyzed data from hundreds of tide stations. Similar efforts are being extended throughout many other countries to provide a decent global coverage of these data. Examples from studies of various parts of North America serve to illustrate the nature of these data and the trends that are present (Fig. 4.11). The east coast of the United States shows a general increase in sea level, with considerable short-term variation. In New England the same general pattern is present but with a slightly lower general rate of rise. The west coast of the United States shows more variation from one location to another but there is little overall increase in sea level for the period of record. This is due to the tectonic activity on this leading edge coast. The more stable east coast shows a rise that ranges from relatively slow in the north, where isostatic rebound is still going on, to more rapid rates of rise to the south, where rebound is absent and some compaction is taking place. The west coast, a crustal plate collision area, experiences great variety in sea-level conditions. In Alaska many locations show a decrease in sea level due to tectonic conditions of uplift: up to 14 mm yr^{-1} (Fig 4.11).

The Gulf coast of the USA shows a great range in sea-level rise due to differences in the geologic setting. The Florida peninsula is a carbonate platform, one of the most stable and least compacting geologic provinces. Its rate of rise is only about 1.5 mm yr^{-1}, which is probably a good reflection of the actual sea-level rise due to the increase in the volume of the ocean. By contrast, the Mississippi delta region is experiencing an annual sea-level rise of 10 mm (Fig. 4.12). This is due mostly to isostasy with contributions by thick deltaic mud and the large amount of fluid removal by the petroleum industry.

On a global scale, the tide gauge records show a distinct but variable increase in sea level everywhere except for the Pacific Ocean (Fig. 4.13). The coasts

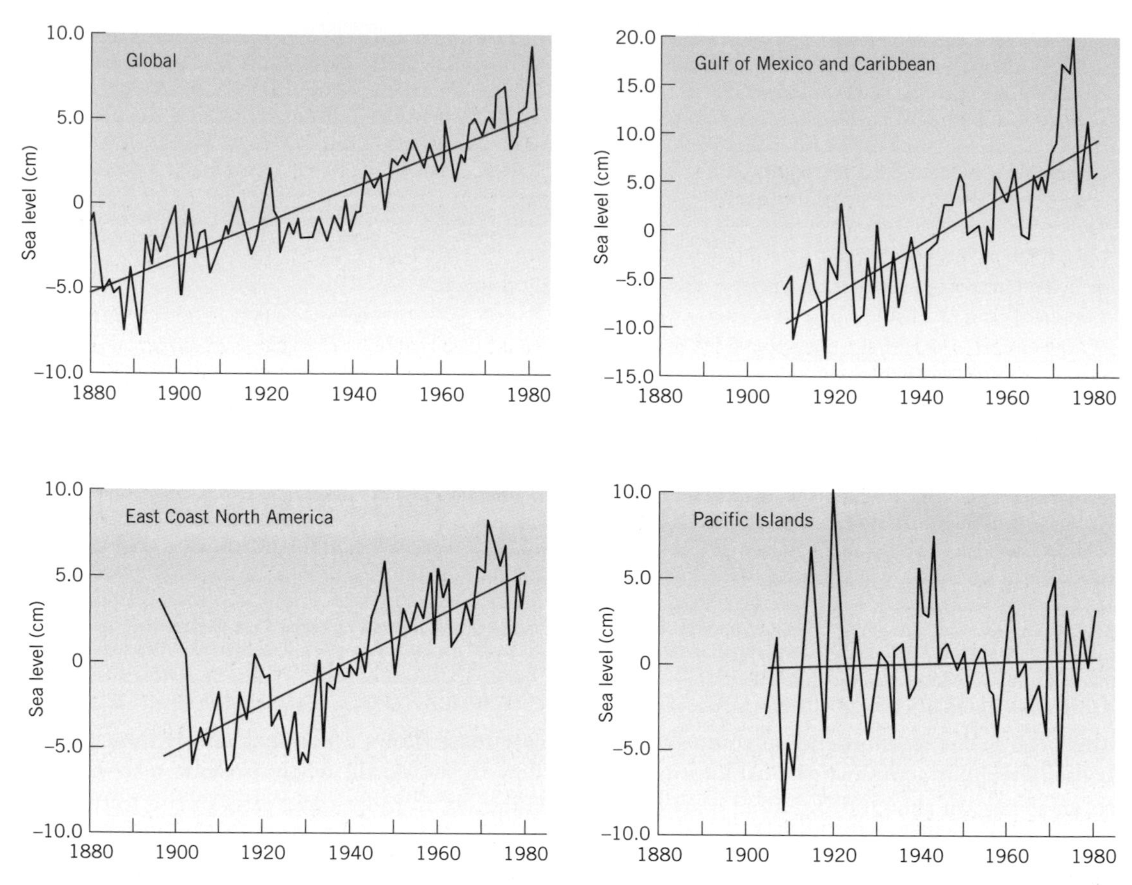

Fig. 4.13 Summary of sea-level curves for the past century from around the world. Note that the only place where a noticeable increase has not occurred is the Pacific Ocean, the area where tectonic activity is most active around the edge of the Pacific Plate, which closely corresponds to the continental coasts. (From V. Gornitz & S. Lebedeff, 1987, Golbal sea-level changes during the past century. In D. Nummedal, O. H. Pilkey & J. D. Howard (eds), *Sea-level Fluctuation and Coastal Evolution*. Tulsa, OK: SEPM Special Publication 41, p. 10.)

of the Pacific Ocean are mostly collision coasts where tectonic conditions override eustatic changes in sea level.

We must remember that 100 years of data for sea level changes represents an insignificant period of time in the cyclic systems of the Earth. It is inadequate to accurately predict long-term sea-level trends. During the past several thousand years, large changes in sea level have occurred as climates displayed smaller cycles within the long-term warming and melting trends of several millennia. These short reversals on the long-term trend were sometimes a century or more long. One of the best examples of this took place a few hundred years ago in the aforementioned "little ice age." During that time the existing ice sheets increased in size, and there was a slight reversal of the rise in sea level. This reversal of a long-term trend was best documented in Europe and China in various historical records. We expect that similar short-term reversals are typical of most long-term trends in climate. Factors that can contribute to these climatic changes include variations in sun spots, the shifting of oceanic currents and slight changes in Earth–Sun positions.

Box 4.2 Sea-level rise in the Mississippi delta area of Louisiana

The combination of decreased discharge of sediment from the Mississippi River, isostatic loading, compaction of sediment and fluid withdrawal by the petroleum industry has caused sea level to rise at catastophic rates – about four times the global average. This is resulting in the loss of tens of square kilometers of land each year; Louisiana is literally drowning in the Gulf of Mexico.

Most river deltas are the sites of very rapid rates of sediment accumulation. This is typically fine sediment, much of it clay minerals. These clay minerals have shapes like small sheets of paper or small sticks, about 2–4 μm in maximum dimension. When these particles come to rest they trap a tremendous amount of water in the sediment: it is very soft mud! As hundreds, or in some cases thousands, of meters of this sediment pile up along the continental margin there is significant compaction because of the mass of the sediment involved. This compaction causes water to be driven from the sediment and the particles to be better organized or layered. The result is that there is a significant decrease in the volume of the sediment, which causes the surface of the delta to sink. When the surface sinks that means that the water is getting deeper: sea level is rising.

In the early days of human habitation of the Mississippi delta area, such as at New Orleans and other nearby communities, this sinking did not occur because the river was continually providing new sediment to fill the areas where compaction was taking place. This became even more true during the mid-nineteenth century when agriculture first became extensive througout the Midwest and Great Plains, the primary drainage area of the Mississippi. Removal of grasses and trees for agriculture caused increased rates of erosion and sent increased volumes of sediment down the river.

As commerce developed in this area, boat traffic on the river increased greatly. Flood control and navigation gave rise to numerous dams along the entire Mississippi system, especially during the first half of the twentieth century. These dams are quite helpful for shipping but they stop sediment from moving down the river. The result was that the amount of sediment being delivered to the mouth of the Mississippi has been about cut in half over the past century. The result has been the "starvation" of the delta.

In the middle of the twentieth century the petroleum industry discovered the great oil and gas resources of the Mississippi delta and exploration took off. Now there are thousands of wells on and adjacent to the delta, from coastal lands out to many hundreds of meters of water. Billions of barrels of oil and trillions of cubic feet of gas have been removed from under the delta surface. The removal of such volumes caused tremendous compaction of the delta sediments. This has added to the rate of sea-level rise.

When combined, the compaction of sediments (about 4 mm yr^{-1}), the withdrawal of oil and gas (about 4 mm yr^{-1}), and the global sea-level rise (about 2 mm yr^{-1}) causes a centimeter of sea-level rise each year. This occurs on a coast that is dominated by wetlands that are intertidal to less than a meter above mean sea level.

Small and low barrier islands along this coast have been destroyed or greatly reduced during historical times. Predictions are that some will be completely gone by the end of the next century. Marshes, which are among the most productive environments, are being drowned because of a lack of sediment to nourish them. In order to sustain the marsh environment, it is necessary for the rate of sediment being delivered to the marsh to be the same as the rate of sea-level rise. That is not even close to happening on the Mississippi delta at the present time.

One positive note that give some hope is that the Corps of Engineers is planning to return significant portions of the river to its natural state, especially in the delta area. This will permit floods to take place and thereby provide the wetlands with the sediment needed to maintain them during this time of rapid sea-level rise.

4.9 The impact of increasing sea-level rise on modern coastal environments

The present coastal morphology is very young, even in the context of the Holocene sea-level rise. Virtually all of the coastal features we see have developed in less than 6500–7000 years. The interaction of coastal processes with sediments and rock causes rapid changes in erosion and deposition. When these conditions, including sea level, remain constant for a period of time, an equilibrium situation develops. Changes in sea level can upset this equilibrium and affect virtually all of the environments in the coastal zone. A rising sea level causes the shoreline to move, which brings about changes in the coast.

Since the time of widespread human occupation on Earth, eustatic sea level has risen only a few meters. Nevertheless, there have been important changes to various coastal environments over this period. Examples include the development of the present lobe of the Mississippi delta, the formation of most of the barrier islands on the Florida Gulf coast, and considerable erosion of bluffs on the west coast of the United States. The possibility of nearly doubling the eustatic rate of sea-level rise from about 1.5 to 3 mm yr^{-1} means that there will be about 30 cm of increase in a century. Such a rate over only a thousand years would produce an increase in sea level of about 3 m, enough to cause major changes in the nature of the shoreline and flood portions of many of the world's formost cities, like New York, London, Amsterdam, and Los Angeles. This is a reasonable forecast.

4.10 Summary

Perhaps more than any other coastal process, sea-level change influences the entire world. Global and slow changes are always taking place, and regional or local changes may be rapid and catastrophic. Sea-level change is, in reality, everywhere! The current concern about global warming and its influence on sea-level change is a front page story in many parts of the world. Rates of rise have increased over the past century but the long-term future is still a matter of speculation.

Ranging from changes on the order of meters in less than a day associated with earthquakes to only a millimeter or so in a year, sea-level changes can impact coastal management in many ways. The location of development, types of construction, density of occupation, and other factors must all be factored in when considering the local sea-level situation. We are still wrestling with the causes for these increases in global sea-level rise: are they natural or anthropogenic?

Suggested reading

Eisma, D. (ed.) (1995) *Climate Change: Impact on Coastal Habitation*. Boca Raton, FL: CRC Press.

National Research Council (1990) *Sea-level Change*. Washington, DC: National Academy Press.

Titus, J. G. & Narayanan, V. K. (1995) *The Probability of Sea-level Rise*. Washington, DC: US Environmental Protection Agency.

Williams, M., Dunkerley, D., DeDeckker, P., Kershaw, P. & Chappell, J. (1998) *Quaternary Environments*, 2nd edn. London: Arnold Publishers (Chapters 5 and 6).

Wind, H. G. (ed.) (1987) *Impact of Sea-level Rise on Society*. Rotterdam: A. A. Balkema.

5 Weather systems, extratropical storms, and hurricanes

5.1 Introduction

Storms are a frightening but fascinating manifestation of the Earth's surface energy (Fig. 5.1). They have a dramatic influence on coastlines, in terms of both sediment movement along the coast and damage to homes and other structures. Storm waves and storm-induced currents are the primary agents responsible for removing sand from beaches and for causing long-term shoreline recession. In a matter of hours storms can produce dramatic changes in coastal morphology, as evidenced by the breaching of barriers, relocation of tidal channels, formation of dune scarps, and deposition of extensive aprons of sand along the landward side of barriers.

(a)

(b)

Fig. 5.1 (a) Photograph of Hurricane Elena taken in 1985 from Space Shuttle Discovery (from http://www.wvwi.net/hurricane.htm). (b) In late August 1995 five major storms were present simultaneously in the Atlantic Ocean, while another storm, tropical storm Gil, existed in the Pacific Ocean (from http://obs-us.com/people/karen/storms/stormstuff.htm). Most Atlantic storms are born off the coast of northwest Africa and are steered westward by the easterlies or trade winds.

Property damage resulting from major hurricanes can be enormous. In 1988, Hurricane Gilbert cut a wide swath through the Caribbean Islands, resulting in US$6.5 billion of damage in Jamaica alone. In the wake of Hurricane Hugo, which hit Charleston, South Carolina, in 1989, $3 billion were spent in reparations. Three years later Hurricane Iniki (1992) swept through the Hawaiian Islands, causing a billion dollars in damage. During the same year Hurricane Andrew produced a staggering $30 billion in property damage as it crossed over southern Florida into the Gulf of Mexico. Eventually it made a second landfall in western Louisiana, where it wreaked devastation totaling another $1.5 billion.

As recorded in history and well chronicled during the past decade, the passage of major hurricanes does not only result in substantial property loss; these storms have also proved to be "killers" to the unprepared coastal population. The greatest loss of life in the United States due to natural causes is attributed to the 1900 hurricane that killed at least 6000 people on Galveston Island along the Texas coast. Most of these deaths were due to drowning resulting from a 6 m high storm surge that covered the island, most of which was less than 2.5 m in height. The storm surge is the super-elevation of the water surface produced by the hurricane's strong onshore winds and low pressure (Fig. 5.2). In response to this Galveston disaster a substantial seawall was constructed along the ocean shoreline and at the same time much of the island was raised some 1.5 m using sediment dredged from a nearby site.

The storm-related deaths that occurred in Galveston pale in comparison to the loss of life that has taken place in Bangladesh, located in the apex of the Bay of Bengal. Similar to the state of Louisiana, the southern part this country has been built through deltaic sedimentation. Much of the lower portion of the Ganges–Brahmaputra delta

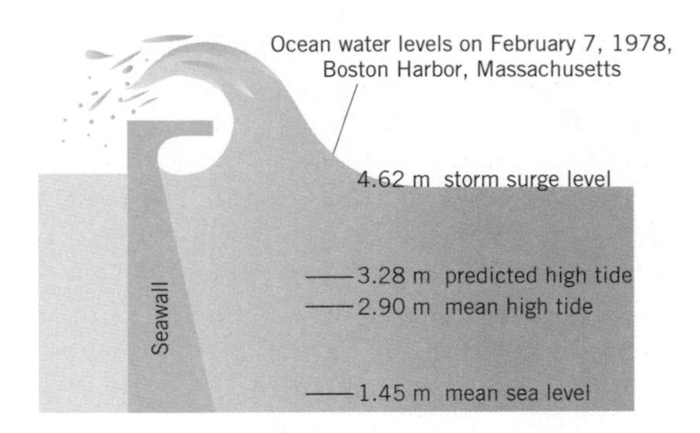

Fig. 5.2 During the Blizzard of 1978, the storm of record for much of northern New England, large astronomic tidal conditions combined with record storm surges caused significant lowland flooding and extensive coastal damage. The high tide levels allowed large storm waves to break over seawalls, against fore dune ridges, and across barrier systems.

in Bangladesh is highly populated. Most of these people live on land that is less than 4 m above sea level. An intense cyclone in 1970 moved through the northern Bay of Bengal and pushed a dome of water onshore that reached 10 m above normal water levels. An estimated half million people lost their lives as a result of this storm. A view of the area after floodwaters subsided revealed a land almost completely stripped of human occupation. A second tropical cyclone in 1991 (Cyclone Gorky) took another 100,000 lives. What makes barriers and other depositional coasts so vulnerable to intense storms is the combination of storm-induced elevated water levels and the inherent low elevation of these coastal regions.

This chapter describes fundamental atmospheric circulation because global wind patterns largely control the pathways of storms and major weather fronts. A basic discussion of air flow associated with high and low pressure systems provides a foundation for understanding the two primary types of coastal storms: those that form above the tropics, called extratropical cyclones, and those that are generated in the tropics, named tropical cyclones. Typically, these storms have different strengths and modes of formation. Similar factors, such as storm path, speed, and magnitude, as well as the configuration of the coast, govern the extent and type of damage resulting from these storms. Case histories of various storms are useful in illustrating these points. Classifications of the hurricanes and extratropical storms are presented so that the effects of various magnitude storms can be compared. The discussion emphasizes hurricanes that impact the Gulf of Mexico and eastern Atlantic Ocean, although similar storms elsewhere in the world are also included. Likewise, primary attention is given to the northeast storms that affect the northeast coast of the United States.

5.2 Basic atmospheric circulation and weather patterns

5.2.1 Wind

Wind is defined as the horizontal movement of air. It is caused by differences in atmospheric pressure. In turn, pressure gradients are produced by differential heating or cooling of air masses. On weather maps the distribution of atmospheric pressure is presented by isobars (Fig. 5.3). Isobars are lines connecting points of equal pressure. As seen on weather maps, wind flows at a slight angle to the isobars toward the central low pressure. Wind velocity increases as isobars become more closely spaced. The pressure gradient can be thought of as the slope of a hill. A ball rolling down the hill will attain a greater velocity the steeper the slope. Likewise, the greater the contrast in pressure (pressure gradient) the stronger the wind velocity.

5.2.2 Atmospheric circulation

The movement of air masses over the surface of the Earth represents one of the processes whereby temperature differences between the equator and poles are balanced. The Sun's preferential heating of the equatorial region compared to polar areas causes the strong temperature gradient that exists between the low and high latitudes. A simple heat budget of the Earth reveals that between approximately 35° N and 40° S of the equator, more energy is absorbed from incoming solar radiation than is

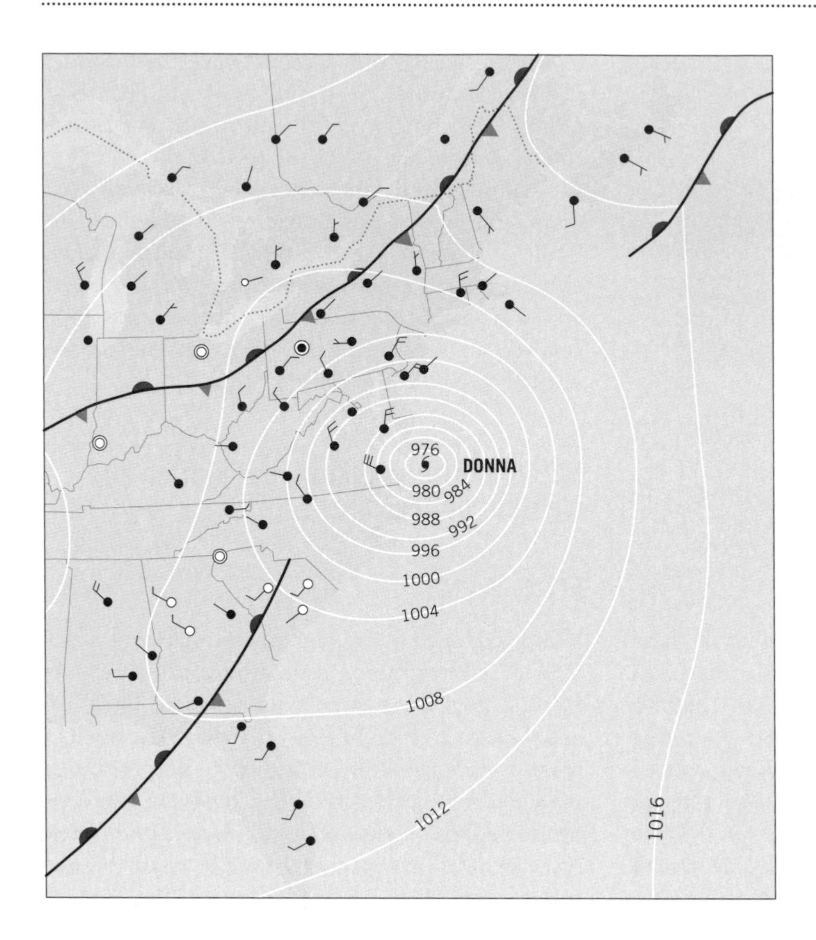

Fig. 5.3 Atmospheric pressure is measured in millibars (equal to 100 newtons per square meter). Standard sea-level pressure is 1013.25 millibars. Individual dots indicate the amount of total sky cover (black dot means totally cloudy). The arrow shaft shows the direction from which the wind is blowing. A full barb on the arrow equals 10 knots. Commonly, weather maps show concentric isobars around low and high-pressure systems.

radiated back into space (Fig. 5.4). Above these latitudes the curvature of the Earth produces a deficit of heat caused by a combination of less incoming radiation per unit area and greater reflectance of solar radiation by the ice-covered poles. While the uneven heating of the Earth has always existed, the equator does not grow warmer with time, nor do the poles grow colder. Offsetting the heat imbalance are mechanisms of heat transfer involving both the atmosphere and oceans that move heat poleward and cold toward the equator. The California Current, which transports cold water southward along the Pacific coast, and the Gulf Stream, which moves warm water northward along the Atlantic coast, are examples of this heat transfer. The north–south movement of air masses accomplishes the same task of equilibrating energy over the Earth's surface as seen when polar air invades southern Canada and the continental United States during winter or when hurricanes move into the north Atlantic during late summer and early fall.

The global wind patterns are also a product of differential heating. The Sun's concentrated incoming radiation in the equatorial latitudes heats the surrounding air and evaporates water from the oceans. As this warm moist air rises, it expands and cools, resulting in water condensation and rainfall in equatorial regions. In a highly simplistic case involving a non-rotating Earth one would expect that the dry air would flow outward from the equator toward the poles, resulting in further cooling of the air. At the poles the now dense cold dry air would sink and then flow southward along the Earth's surface toward the equator. In this ideal model two convection cells would operate, one each in the northern and southern hemispheres

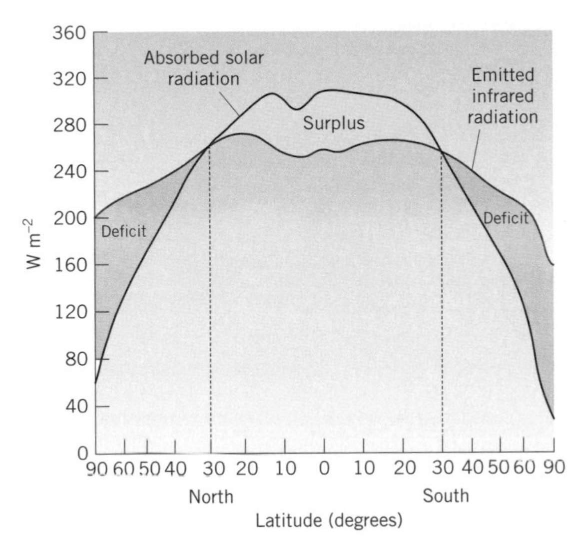

Fig. 5.4 Over the Earth's surface there is an uneven distribution of incoming solar radiation and outgoing infrared radiation. Ocean currents and weather systems are responsible for equalizing the geographic heat imbalance.

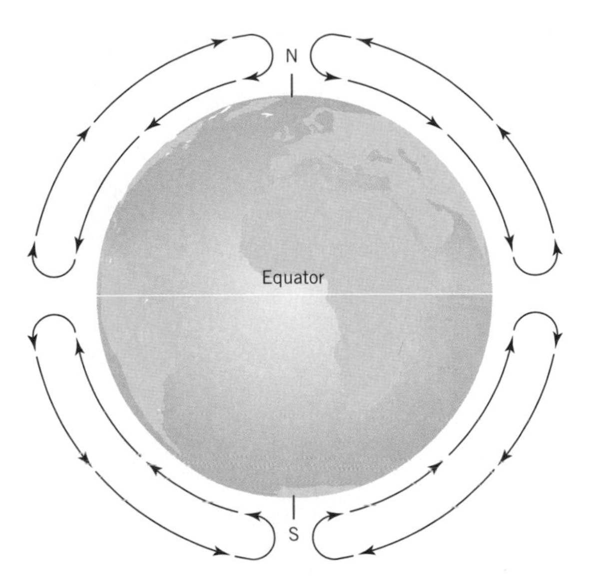

Fig. 5.5 Hypothetical model of air convection due to uneven solar heating on a non-rotating Earth.

(Fig. 5.5). However, due to the Coriolis effect, the actual global air circulation is much more complex. The Coriolis effect is produced by the Earth's rotation and causes all moving objects to be deflected to the right in the northern hemisphere and to left in the southern hemisphere. The Coriolis effect applies to ocean currents, moving air masses, and even jet planes flying across the globe.

Differential heating of the atmosphere coupled with the Coriolis effect produces a global air circulation model with six convection cells: three each in the northern and southern hemisphere (Fig. 5.6). As described in the simplistic model, preferential heating causes warm moist air to rise in the equatorial region. The rising air is replaced by surface air flowing toward the equator from the northern and southern latitudes. This rising air expands and cools, producing precipitation. Thus, the equatorial region is noted for hot temperatures, low pressure, water-laden clouds, and rain. High above the equator the air mass flows northward and southward where it continues to cool and lose moisture. At about 30° latitude the air mass descends, thus completing an atmospheric circulation cell known as the Hadley cell. As the air sinks at 30°, it compresses, producing high pressure, dry air, and variable winds. These are known as the horse latitudes. Folklore tells us they are so named because ships sailing into these latitudes became becalmed. As the ship expended its animal feed and began running out of water, the crew was forced to throw horses and other farm stock overboard. A second convection cell exists between 30 and 60° latitude, which is called the Ferrel cell. This cell is formed because some of the air that sinks at 30° latitude flows northward upon reaching Earth's surface. At the same time a secondary low pressure system stationed at about 60° latitude coincides with a rising air mass and precipitation. Aloft, this air mass cools and moves southward. The polar cell is the third circulation cell and occurs between 60 and 90°. It results from upper air masses moving northward and descending at the poles, while at the same time surface air flows south.

5.2.3 Prevailing winds

Now that the atmospheric circulation has been established, we can use air flow along the Earth's surface to understand the global prevailing wind

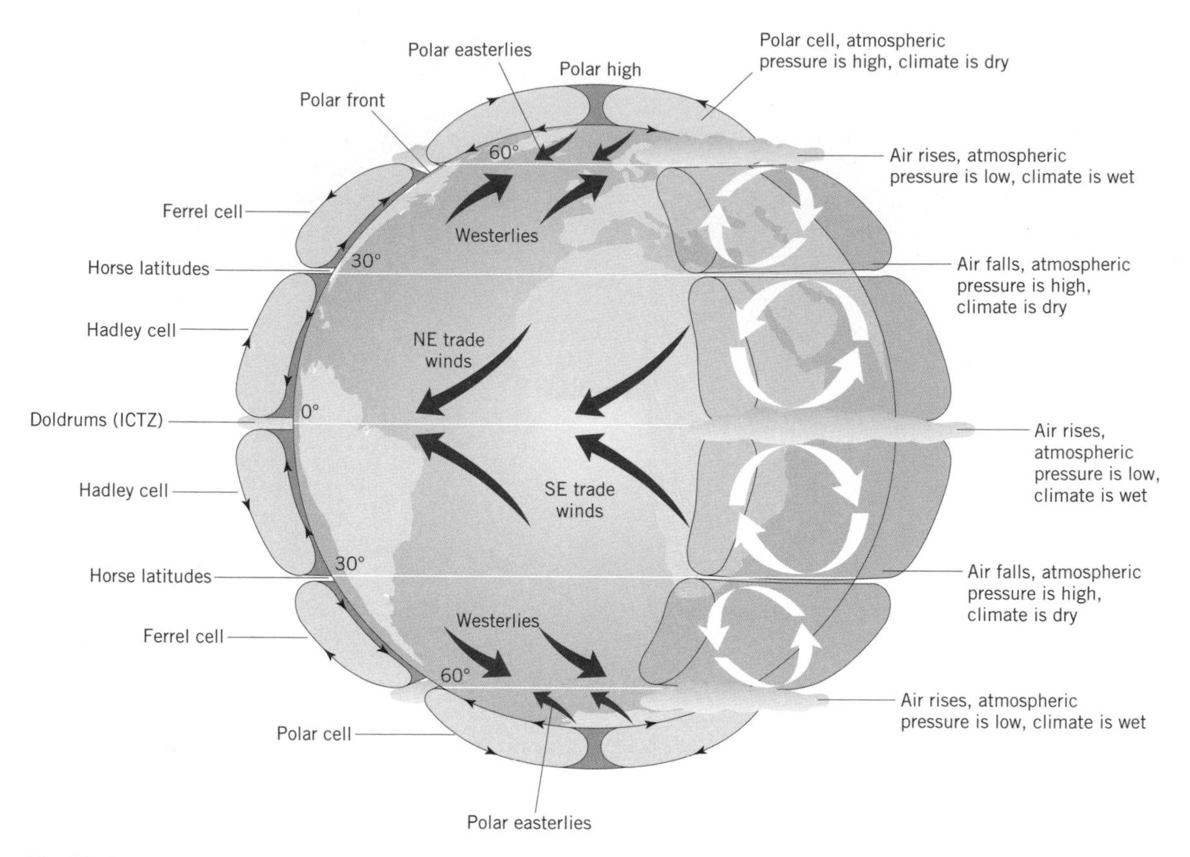

Fig. 5.6 Simplified air circulation model for a rotating Earth containing the Hadley, Ferrel, and polar convection cells north and south of the equator.

patterns (Fig. 5.6). In the northern hemisphere circulation in the Hadley convection cell would seem to indicate that air should flow from north to south between 30 and 0° latitude. Remembering, however, that the Coriolis effect causes all moving masses to be deflected to the right (in the northern hemisphere), the flow of air is actually from the easterly quadrant toward the west. These winds are called the easterlies because the winds blow from the east. They are also referred to as the "trade winds." The early merchants who sailed from Europe bound for the New World gave them this name. The word "trade" was used by the English of that day to mean constant and steady, and this is how they described the winds that helped them sail across the Atlantic. In the southern hemisphere the pattern is repeated, resulting in the convergence of the northeast and southeast trade winds at the equator. This region is called the intertropical convergence zone (ITCZ) and scientists have shown that it profoundly affects ocean currents and weather patterns in the equatorial area.

In the mid-latitudes, surface winds associated with the Ferrel circulation cell are deflected to the east by Coriolis. These winds are known as the westerlies and affect most of the continental United States. The west coast is known as a windward coast because the westerlies blow onshore, augmenting the wave energy in this region. Conversely, the east coast is a leeward coast because the westerlies blow offshore. The prevailing winds diminish wave energy, partially explaining why average shallow water wave heights are more than twice as large on the west coast as they are along the east coast. The westerlies are also responsible for steering the weather systems, including hurricanes.

In the northern hemisphere, air flowing southward from the pole forms the polar easterlies. At about 50–60° latitude the polar easterlies meet the westerlies, establishing the polar front. This convergent zone produces a near permanent boundary separating the polar cold dense air from the warm tropical air mass. The variable weather that characterizes much of the United States reflects a latitudinal wandering of the polar front.

5.2.4 Cyclonic and anticyclonic systems

Although exceptions do exist, low atmospheric pressure is most often associated with rainy or stormy weather, whereas high pressure is a sign of fair weather. A well developed low pressure system is characterized by a gyre of air that rotates in a counterclockwise direction (in the northern hemisphere; clockwise in the southern hemisphere) around a central low pressure cell (Fig. 5.7). These are called **cyclonic systems** and may be hundreds of kilometers is diameter. The counterclockwise movement of air is produced by the convergence of surface currents. Low pressure systems contain ascending air masses, which many people have observed in film footage of tornadoes. As air streams in to replace the rising air, the currents are deflected to the right by the Coriolis effect, producing the counterclockwise circulation. Hurricanes, extratropical storms, and tornadoes are all types of cyclonic systems, although tornadoes are considerably smaller systems.

High pressure systems occur where air masses are sinking. As the air mass descends toward the ground, it flows outward from a central high pressure cell. In the northern hemisphere the outward flowing currents are deflected to the right due to Coriolis, resulting in a rotating air mass that circulates in a clockwise direction. These are called anticyclones. The Bermuda High is an example of an anticyclonic system. This high pressure system stabilizes over Bermuda during mid-summer and is responsible for transporting the uncomfortable hot, hazy, humid air from the Gulf of Mexico to the northeastern Atlantic seaboard.

5.2.5 Land breezes and sea breezes

Anyone who has spent the summer along the seashore or has sailed along the coast is familiar with the systematic breezes that characterize the coastal zone. The onshore and offshore winds are a result of differential heating of the land surface versus that of the ocean. Under fair weather conditions when the sun has just risen, the air is usually calm because the air over the ocean and land has similar temperatures. However, as the Sun ascends in the sky, the land surface preferentially warms compared to that of the ocean. In turn, the land surface warms the overlying air, causing it to rise. Over the ocean the air remains cool and dense. Thus, a pressure gradient develops between the relatively low pressure over

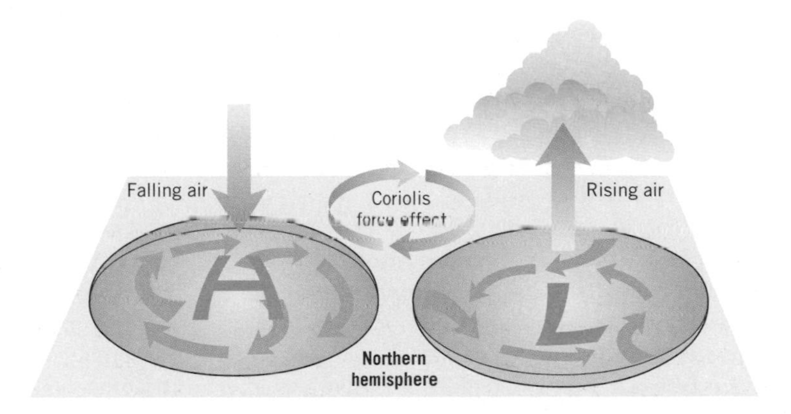

Fig. 5.7 Gyres of moving air are created around low- and high-pressure systems due to the Coriolis effect. Air masses in the northern hemisphere flow in a counterclockwise direction in low-pressure systems (cyclonic circulation) and a clockwise direction in high-pressure systems (anticyclonic circulation).

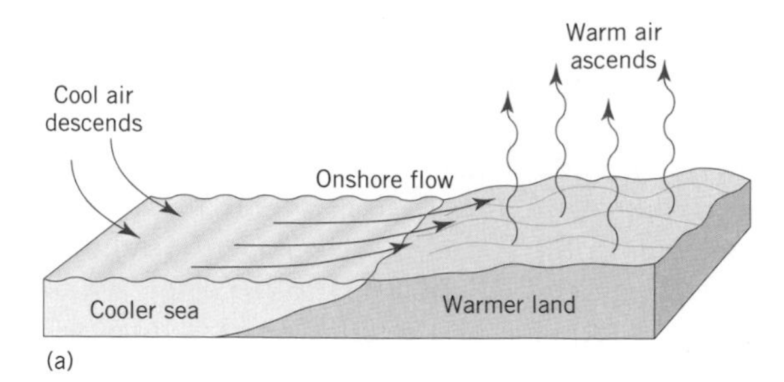

(a)

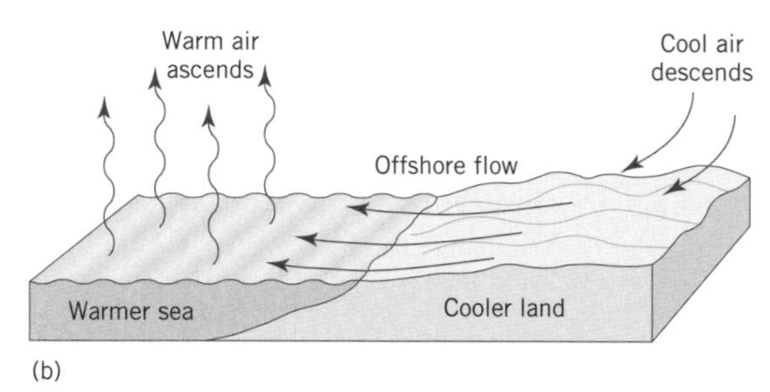

(b)

Fig. 5.8 (a) Sea and (b) land breezes are a common diurnal pattern of winds along many coasts. They are formed due to temperature differences between the land and adjacent seawater.

the land and the higher pressure over the ocean. This results in an onshore flow of air that is referred to as a sea breeze (Fig. 5.8). The sea breeze strengthens during mid-day and reaches a maximum during mid to late afternoon, before diminishing by late evening. During the night, the land surface radiates heat back to the atmosphere at a much higher rate than the ocean, which reverses the temperature and pressure gradients along the shore. This results in a flow of air from the land toward the ocean, which is called a land breeze. The extent and magnitude of the land and sea breezes are a function of ocean and daytime temperatures, coastal morphology, vegetation, and other factors.

5.3 Mid-latitude storms

Whereas tropical storms and hurricanes dominate coasts of low latitudes, extratropical storms and weather fronts are the major weather systems impacting mid-latitudes coasts. A transition zone exists between these two regions where both weather systems are common. As the name implies, extratropical cyclones form north of the Tropic of Cancer or, in the case of the southern hemisphere, south of the Tropic of Capricorn. These storms are associated with low pressure systems and affect the Pacific, Gulf, and Atlantic coasts of the United States and other mid-latitude coastlines of the world. Like all cyclonic weather systems, extratropical storms are air masses that rotate around a central low pressure with a counterclockwise circulation (Fig. 5.7). Another important type of mid-latitude weather system is the **front**. The passage of frontal systems along the Gulf coast strongly influences the coastal processes of this region. Frontal systems are discussed first, as they are the precursors to extratropical cyclone development.

5.3.1 Frontal weather

A front is defined as a narrow transition zone (25–250 km wide) between two air masses with different densities (Fig. 5.9). Fronts may extend for more than 1000 km. One air mass is usually warmer and more humid than the adjacent air mass. The boundary between the two is usually inclined, with the warmer lighter air rising over the colder, denser air. A cold front is one in which the cold air mass advances, thereby displacing the position of the warm air. The opposite occurs during the passage of a warm front. Generally, the two air masses travel with nearly the same velocity and in about the same direction. Cold fronts usually advance at a slightly more rapid rate (35 km h^{-1}) than warm fronts (25 km h^{-1}). In addition, cold fronts tend to be accompanied by more energetic weather and often contain concentrated precipitation and severe wind.

In North America, cold fronts commonly are initiated by cold air (polar air mass) sweeping down from Canada meeting warm air from the south (Fig. 5.10). Fronts travel west to east across the country and may extend all the way to the Gulf of Mexico, where they can produce strong winds and surf. The northern Gulf region experiences more than twenty cold fronts each year, lasting from 12 to 24 hours depending on the speed of the storm and whether it becomes stalled or not. Due to the overall low tidal

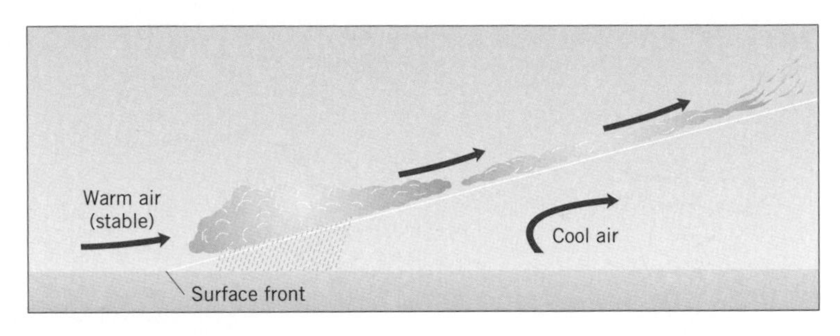

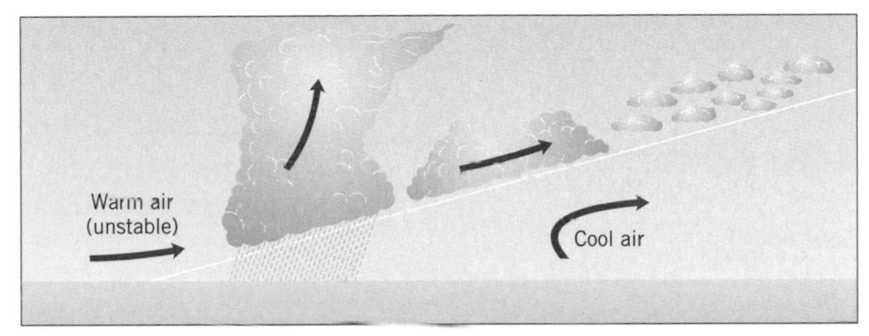

Fig. 5.9 Weather fronts are formed at the boundary between two air masses that have contrasting densities, usually due to differences in temperature. Fronts mark a change in the weather and are generally associated with moderate to intense precipitation.

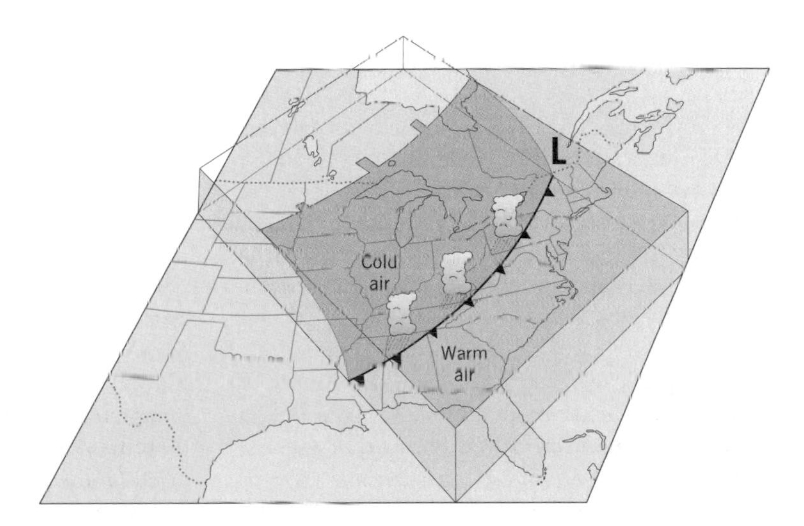

Fig. 5.10 Cold fronts in the central United States are caused by polar air moving south from Canada, displacing a warm air mass. The front moves eastward across the country and may stretch from the Gulf of Mexico to the Great Lakes.

range of most of the Gulf coast (<1 m), cold fronts can be effective agents in substantially augmenting or diminishing tidal elevations. Along the Louisiana coast cold fronts create westerly to southwesterly winds, which cause higher tide levels than would be expected due to astronomic forcing alone. This condition allows storm waves to break higher along the beach, leading to the overwashing of low barriers. At the same time, elevated tides increase the flow of ocean water into the bays and marsh systems that back the barrier islands. This generates strong landward currents through the tidal inlets during the passage of the front and strong seaward currents when the floodwaters exit the inlets. As the front passes, there is a dramatic shift from southerly to northerly winds that occurs within a few hours or less. Strong northerly winds surpress the waves in the nearshore of the Gulf-facing beaches, while at the same time generating substantial waves in the larger bays behind the barriers. In fact, scientists

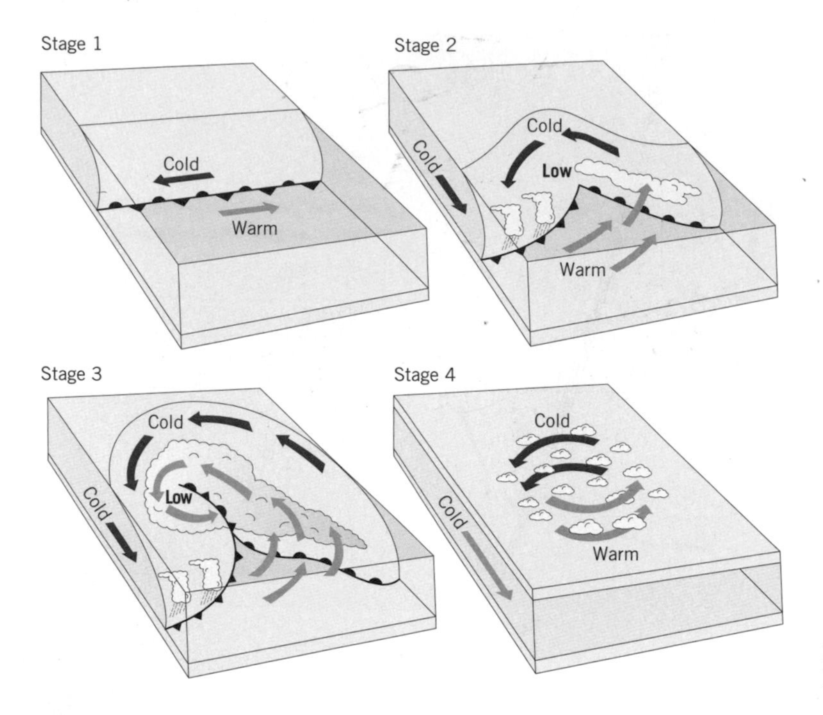

Fig. 5.11 Model of cyclogenesis as proposed by Norwegian scientists and published by J. Bjerknes.

have shown that these bay waves, which can approach 3 m in height, are responsible for chronic shoreline erosion on the backside of barriers along long stretches of the northern Gulf of Mexico. Thus, cold fronts are important agents in modifying beaches and tidal inlets in the Gulf coast region.

5.3.2 Cyclogenesis

The process of extratropical cyclone formation, cyclogenesis, was first described by Norwegian scientists during the First World War and eventually the theory was published by J. Bjerknes in 1918. Even though it was devised using limited ground observations, the basic tenets of the model are still deemed acceptable today despite major advances in the collection and analysis of weather data. Cyclones develop along advancing frontal systems in which the two air masses have a slight component of differential movement. At the surface of the front this condition is manifested by the two air masses moving in opposite directions to one another (Stage 1, Fig. 5.11). The next stage of cyclogenesis coincides with a disturbance along the front that is produced by topographic irregularities, such as mountain ranges, temperature differences, or other factors. The end result of the disturbance is a wave-like form, similar to a breaking ocean wave, in which the warm low-density air penetrates into the cold air mass (Stage 2, Fig. 5.11). Extratropical cyclone formation also appears to be strongly linked with upper air circulation where currents flow west to east in long meanders. This pattern serves to initiate or reinforce the counterclockwise rotation of air around a central low pressure (Stage 3, Fig. 5.11). The demise of a cyclone occurs when the cold air supplants the rising warm air, the sloping gradient between the two air masses ceases, and the storm runs out of energy (Stage 4, Fig. 5.11).

5.3.3 Extratropical storms

Occurrence and storm tracts

Although extratropical cyclones can occur in the mid-latitudes at any time of the year, they happen most frequently in the hemisphere that is going through winter, because this is when the temperature contrasts between the polar air masses of high

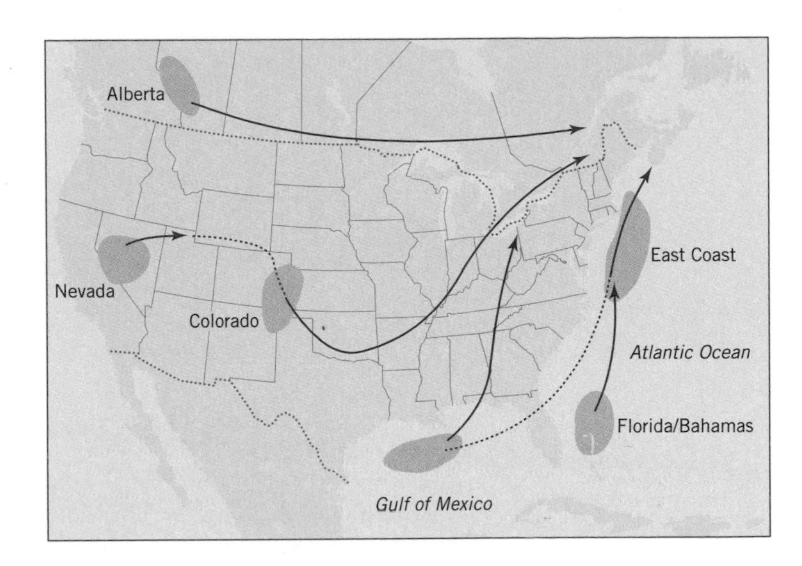

Fig. 5.12 Extratropical cyclones develop in many regions across the United States. Regardless of their origin, their pathway of these storms is towards northeast Canada and the northern Atlantic Ocean.

latitudes and the warmer air of the lower latitudes reach their maximum. In the United States extratropical storms are most common from late fall through early spring. The weather conditions that lead to extratropical cyclone formation often originate in the Pacific Ocean. Full development of the storm is associated with several locations throughout North America due to the complexity of the factors governing the weather there. These sites include the continental Southwest, Midwest, and Southeast United States, southern Alberta, Canada, and in the Gulf of Mexico and the Florida–Bahamas regions (Fig. 5.12). The continental storms travel in an eastward path across the country, eventually moving northeastward into the northern Atlantic, where they dissipate. Those that pass through the Gulf of Mexico and in the Florida–Bahamas region move northward along the eastern seaboard. Other storms track through the Great Lakes and into the Gulf of Saint Lawrence. Because most extratropical storms track along the east coast, they are major storm producers of this region, particularly north of North Carolina. Hurricanes have a greater frequency and influence along the coastal states south of Virginia.

Northeasters

Northeaster is the name given to the extratropical cyclones that pound the northeast coast of the United States and Canada with driving rain, strong winds, elevated tide levels, and storm waves. During the winter, northeasters can produce blizzard conditions, blanketing the northeast with more than two feet (60 cm) of snow. They are called northeasters because the winds associated with these storms come from the northeast (Fig. 5.13). Remember that winds are named for the direction from which they blow. In fact, it is the path of the extratropical cyclone that determines the wind direction as well as the type and severity of the storm. Northeast storms occur when the eye of the storm tracks in a northeasterly path offshore of the coast, eventually moving east of Cape Cod and Nova Scotia. Under these conditions the counterclockwise air circulation associated with the cyclone generates onshore winds that blow out of the easterly quadrant. The eastern seaboard of the United States is particularly susceptible to these storms because most of this coast faces eastward and is directly exposed to northeast winds and waves.

The Ash Wednesday Storm of 1962 was one of the most powerful and damaging northeasters ever to strike the east coast of the United States during historic times. The storm lasted for more than five tidal cycles and impacted 1000 km of mid-Atlantic coastline stretching from North Carolina to Long Island, New York. Wind gusts exceeded hurricane

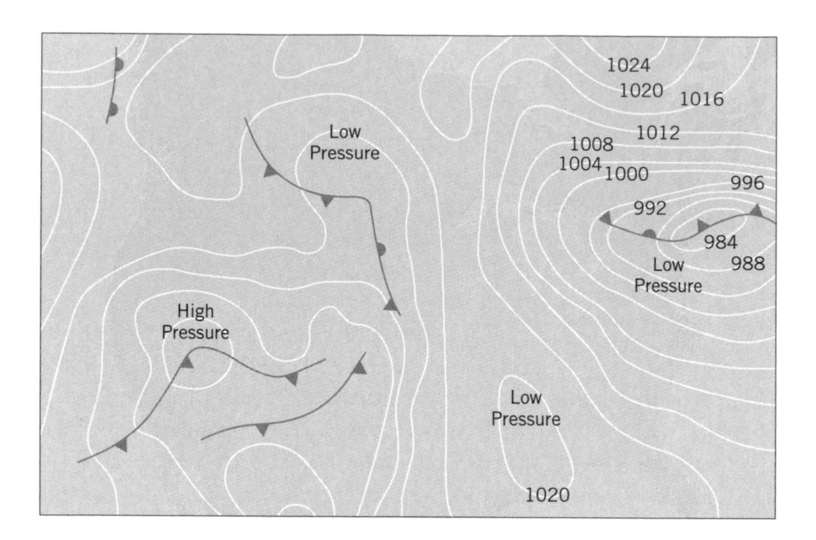

Fig. 5.13 Along the eastern seaboard of the United States, particularly in New England, extratropical cyclones travel offshore of the coast and generate strong winds and storm waves from the northeast. The weather map illustrates the conditions that existed during a northeast storm on January 25, 1979. In the Boston, Massachusetts, area winds reached 60 km h^{-1} (40 m.p.h.) and the region was blanketed with over two feet (60 cm) of snow. Pressure measured in Millibars. Contour interval 4 millibars.

force, producing deepwater wave heights greater than 10 m. The storm hit during perigean spring tides, which contributed to the elevated water levels. The large storm surge and erosive waves removed beaches and carved deeply into adjacent dunes. Along many shorelines storm waves carried beach sands across barriers, forming extensive overwash deposits. The storm breached numerous barriers, creating more than a dozen ephemeral tidal inlets. Along populated sections of shoreline, rows of houses and other buildings were destroyed. Ultimately the "storm of the century" caused more than $300 million in damage and accounted for numerous deaths.

A classification of northeasters has been devised by Robert Dolan and Robert Davis, who have studied 1347 storms affecting the North Carolina coast over a 42-year period. Their scheme is based on the power of the northeaster, which is calculated by multiplying the storm's duration by the square of the maximum significant wave height (Table 5.1). They have divided storms into five classes, with the weakest storms, Class 1 and 2, comprising about 75% of all northeasters impacting the Outer Banks of North Carolina. As seen in Table 5.1, Class 4 northeasters have a frequency of 2.4%, with a significant deepwater wave height of 5 m and an average duration of 63 hours. The extreme northeaster, Class 5, has a recurrence interval of 67 years. These storms have a significant wave height of 7 m, an average duration of 96 hours, and are agents

Table 5.1 The Dolan–Davis classification of northeast storms.

Storm class	Frequency of storms: Number	Frequency of storms: Percentage	Significant wave height (m)	Duration (h)	Power (m^2 h): Mean	Power (m^2 h): Range
1 Weak	670	49.7	2.0	8	32	≤71.63
2 Moderate	340	25.2	2.5	18	107	>71.63 to 163.51
3 Significant	298	22.1	3.3	34	353	>163.51 to 929.03
4 Severe	32	2.4	5.0	63	1455	>929.03 to 2322.58
5 Extreme	7	0.1	7.0	96	4548	>2322.58
	1347					

of permanent change along the coast, such as the Ash Wednesday Storm of 1962.

Shoreline vulnerability

Many of the factors that control the severity of hurricanes also apply to the impact of northeasters. The storm's size and intensity are of paramount importance, as these elements control wave height, magnitude of the storm surge, and to some extent the storm's duration. Astronomic tidal conditions can significantly augment or reduce the effects of the storm surge, thereby affecting the elevation where waves break along a shoreline. For example, the Blizzard of 1978, which is the storm of record along much of the New England coast, occurred during extraordinarily high tidal elevations in which the predicted mean high tide was 60 cm higher than mean high water levels. The high astronomic tides coinciding with the blizzard were a major factor contributing to coastal flooding, beach erosion, overwashing of barriers, inlet formation, and wave-induced damage (Fig. 5.14).

Although the majority of extratropical cyclones travel in a northeasterly path offshore of the coast, producing northeasterly winds and waves, some low-pressure systems take an inland track, such as through the Hudson or Connecticut valleys. In some instances, storms move through the Gulf of Saint Lawrence. In these circumstances the cyclonic wind regime produces wind from the southerly quadrant. In New England the shorelines have a variety of orientations and exposures to wave energy. In this region the storm track is particularly important in determining the impact of storm processes. The passage of the common northeaster primarily affects the eastward facing shorelines of outer Cape Cod and the north and south shores of Massachusetts Bay as well as the New Hampshire and southern Maine coasts. Storms that travel west of New England generate southerly wind and waves that impact the southward shores of Rhode Island, Cape Cod, and the central peninsular coast of Maine.

Another factor that influences the vulnerability of beaches to storm erosion and damage to adjacent structures is the interval between storms. The width and elevation of the berm strongly affects erosion of the abutting dune during storms. Beaches with wide accretionary berms can withstand the onslaught of an intense northeaster because the berm constitutes a large quantity of sacrificial sand that can be eroded before the dune or adjacent dwelling is destroyed. If a previous storm has removed the sand buffer, then the impact of the next storm will be much greater than that of the first. One of the reasons why the Halloween Eve Storm of 1991 caused significant damage to New England was that Hurricane Bob, which passed through the region a month and half earlier, had left many of the shorelines with little sand on the beaches.

Fig. 5.14 View of houses destroyed in Scituate, Massachusetts, during the Blizzard of 1978. Storm waves and the onshore movement of gravel toppled foundations, leading to the collapse of numerous houses along this section of coast.

5.4 Hurricanes and tropical storms

5.4.1 Low latitude storms

Hurricanes and their forerunners, tropical storms, are the major storms of the tropics. Similarly to other weather systems, they commonly move beyond their typical latitudinal range given the right set of circumstances. Most of us are familiar with tropical storms and hurricanes due to the wide media coverage they receive, particularly when a major coastal region in the United States is impacted.

Tropical storms and hurricanes are known by different names in other parts of the world. In the western Pacific they are called typhoons, in the Indian Ocean they are known as cyclones, and in Australia they are given the name "willi-willis." Tropical storms are low-latitude cyclonic systems that may intensify to a hurricane if wind velocities surpass 119 km h^{-1}. During an average year, approximately 20 tropical storms form in the equatorial Atlantic and of these eight to ten reach hurricane strength. Most hurricanes that make landfall in the United States do so in the Gulf of Mexico or along the Florida and North Carolina coasts. In a study of hurricane frequency for the southeast United States coast, Robert Muller and Gregory Stone of Louisiana State University showed that Morgan City, Louisiana, and the Florida Keys have the shortest recurrence of major hurricanes (wind velocity >179 km h^{-1}, 111 m.p.h.) (Table 5.2). The Florida Keys (including Key West and Key Largo) have survived over 32 hurricanes during the past century and seven of these were major hurricanes.

Table 5.2 Hundred-year summary of hurricane strikes and return periods.

	Hurricane strikes	Major hurricane strikes	Hurricane return period (years)	Major hurricane return period (years)
South Padre Island, TX	8	3	12	33
Port Aransas, TX	9	2	11	50
Port O'Connor, TX	13	2	8	50
Galveston, TX	13	3	8	33
Cameron, LA	5	2	20	50
Morgan City, LA	10	7	10	14
Boothville, LA	16	5	6	20
Gulfport, MS	10	2	10	50
Dauphin Island AL	15	4	14	25
Pensacola Beach, FL	14	2	7	50
Destin, FL	10	3	10	33
Panama City Beach, FL	10	2	10	50
Apalachicola, FL	11	0	9	100
Cedar Key, FL	3	1	33	100
St Petersburg, FL	5	1	20	100
Sanibel Island, FL	8	1	12	100
Marco Island, FL	12	3	8	33
Key West, FL	17	2	6	50
Key Largo, FL	15	5	7	20
Miami Beach, FL	15	2	7	50
Palm Beach, FL	13	4	8	25
Vero Beach, FL	10	1	10	100
Cocoa Beach, FL	4	0	25	100
Daytona Beach, FL	3	0	33	100
Jacksonville Beach, FL	2	0	50	100
St Simons Island, GA	1	0	100	100
Tybee Island, GA	5	0	20	100
Folly Beach, SC	3	1	33	100
Myrtle Beach, SC	7	2	14	50
Wrightsville Beach, NC	11	2	9	50
Atlantic Beach, NC	10	3	10	33
Cape Hatteras, NC	15	3	7	33

TX, Texas; LA, Louisiana; MS, Mississippi; AL, Alabama; FL, Florida; GA, Georgia; SC, South Carolina; NC, North Carolina.
Source: Study of hurricane frequency of the southeastern USA by Robert Muller and Gregory Stone of Louisiana State University.

5.4.2 Origin and movement of hurricanes

Formation

Tropical weather is generally considered to be that occurring between the Tropic of Cancer and the Tropic of Capricorn (23.5° N to 23.5° S). Here there is little change in daylength, seasons are subtle to nonexistent, year-round temperature is warm to hot, and major changes in weather patterns are linked to the dry and wet seasons. In these latitudes the winds typically blow from the southeast, east, or northeast depending on the latitude. Weather systems of the tropics, such as hurricanes, are steered by the trade winds, in contrast to those of the mid-latitudes that move west to east due to the westerlies.

The first indication of the potential development of a storm in low latitudes is the presence of a tropical wave. This feature is identified on weather charts as a bending of the streamlines, which show pathways of airflow within the wind system. Tropical waves form over western Africa and move westward into the Atlantic Ocean, where they gain strength over the warm water of the tropical latitudes (Fig. 5.15). The warm water reduces pressure along the wave, transforming it into a trough. In a typical year, about 60 of these develop during the hurricane season or about one every three to four days. They have a wavelength of about 2500 km and travel westward with speeds of 10–40 km h^{-1}. Some of these troughs intensify and develop into tropical disturbances, which are the infancy stage of a tropical cyclone. Tropical disturbances are characterized by a line of thunderstorms that maintain their identity for a day or so. These weather systems have a rotary circulation, which is counterclockwise in the northern hemisphere and clockwise in the southern hemisphere. Further strengthening of these storms produces a tropical depression, which is a weather system having maximum cyclonic wind velocities up to 61 km h^{-1} (38 m.p.h.). Storms with wind velocities greater than 61 km h^{-1} but less than 119 km h^{-1} are classified as tropical storms.

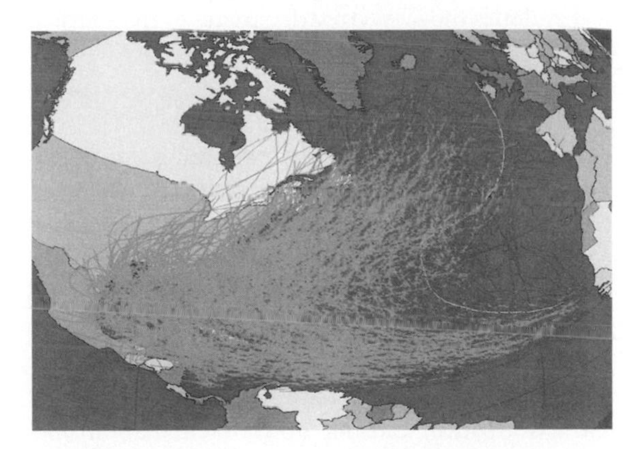

Fig. 5.15 Pathway of hurricanes in the central Atlantic during approximately the past 100 years. (Courtesy of Gregory Stone, Louisiana State University.)

In about 10% of the cases, the developing tropical storm receives sufficient energy from the warm ocean waters to reach hurricane strength. Hurricanes are one of Earth's largest weather systems, with wind velocities of at least 119 km h^{-1} (74 m.p.h.) but some exceeding 250 km h^{-1} (155 m.p.h.). The conditions necessary to produce a hurricane include:

1 Warm ocean temperatures (>26°C) occur from the beginning of June until the end of November in the northern hemisphere and during the opposite time of year in the southern hemisphere. The conditions extend late in the season because ocean waters cool slowly in the fall. Most hurricanes form in August and September when ocean waters are at their warmest.

2 Vertical movement of warm moist air rising within the storm from the ocean surface upward to a height of 10–20 km. As the air rises it cools. Eventually the water vapor contained in the air condenses, releasing huge quantities of energy in the form of heat.

3 The Coriolis effect produces the spinning of the hurricane. Air that flows toward the center of the low-pressure system to replace the air that is rising in the hurricane is deflected to the right in the northern hemisphere. This causes the air mass to rotate. Hurricanes generally do not form within 5° of the equator because the Coriolis effect is very weak in this region.

Hurricane pathways

Once formed, hurricanes and developing tropical storms move in a variety of pathways, all having a

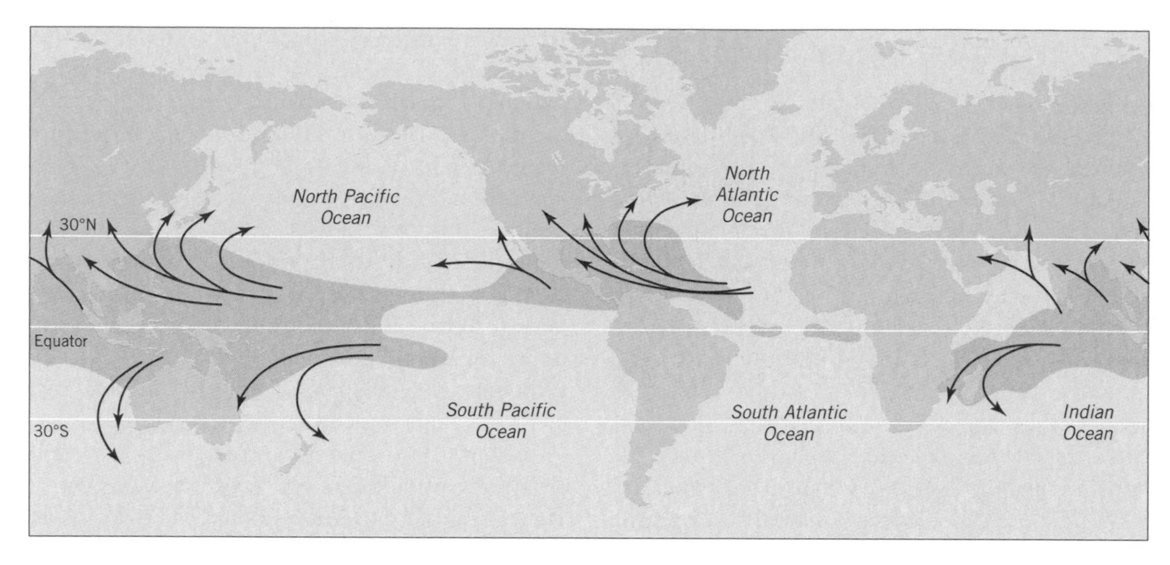

Fig. 5.16 Hurricanes and tropical storms form in the tropics above 5° north and south of the equator. Trade winds steer these storms westward where they degrade over continental areas, and move east and poleward, where they die in the colder waters of the mid-latitudes.

general westerly direction (Fig. 5.16). South Atlantic hurricanes usually travel across the Caribbean Sea and then either enter the Gulf of Mexico or move northward up the western margin of the Atlantic. Storms that move into the Gulf of Mexico continue west or northwest, making a landfall encompassing the shoreline between the Yucatan peninsula and the western Florida panhandle. It is rare for these storms to swing northeast and cross the western Gulf coast of Florida. The last hurricane to do so was Donna in 1960.

Atlantic tropical storms that do not enter the Gulf generally move northward with a slight westerly component. Although they do not ordinarily cross the Florida peninsula, they may travel close enough to impact coastal communities with high surf, strong winds, and possible lowland flooding. These storms commonly make a landfall in the Carolinas or, in rare instances, as far north as New England. Recent examples are Hurricane Hugo, which hit South Carolina in 1989, and Hurricane Fran, which came across the Outer Banks of North Carolina in 1996.

Hurricanes and tropical storms in other ocean basins of the northern hemisphere have pathways that are similar to the Atlantic systems (Fig. 5.16). In the western Pacific storms travel westward toward the Philippines and Southeast Asia, as well as swinging northward, where they impact the coasts of China, Japan, and Korea. In the northern Indian Ocean cyclones move northwestward, making landfalls along Bangladesh, India, Pakistan, and along the Arabian Gulf. The mirror image of this pattern takes place in the southern hemisphere, where tropical cyclones move westward, curving to the south.

5.4.3 Anatomy of a hurricane

Conditions that lead to the formation of a full-blown hurricane start with what is called a seedling. In addition to warm water, there must be weather conditions that enhance the vorticity of the storm, which is the upward spiraling of winds. High humidity, lack of vertical wind shear, and wind surge are also conducive to hurricane formation. Wind surge adds bursts of high velocity flow to the center of the disturbance, causing upward circulation and intensification of the storm.

Although hurricanes vary greatly in size, intensity, speed, and path, they have many common character-

Fig. 5.17 Satellite view of Hurricane Andrew on August 23, 24, and 25, 1992 as it moved westward across Florida toward Louisiana. (From http://rsd.gsfc.gov/rsd/images/andrewSequence_md.jpg)

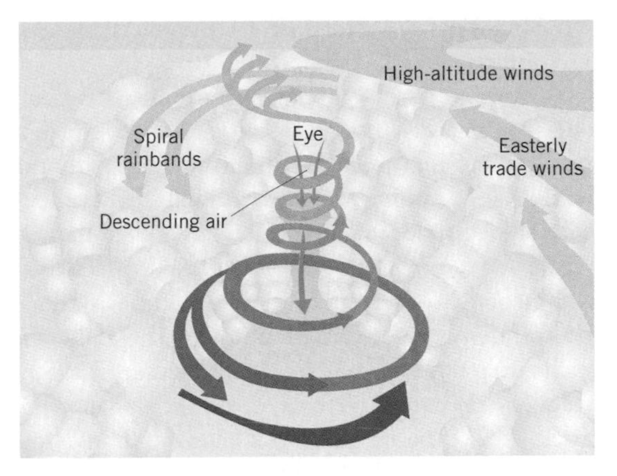

Fig. 5.18 Internal structure of a hurricane illustrating the overall counterclockwise flow of air at the base of the storm.

istics. A satellite view of a hurricane reveals spiraling bands of thunderstorm-like clouds (Fig. 5.17). Some of these cloud systems contain abundant moisture and some do not, which explains why the intensity of rainfall is so variable during the passage of a hurricane. Hurricanes may have a single cloud band or more than seven, each extending from the center of the storm outward to a distance of about 80 km. The storm itself can range in diameter from about 125 km to more than 800 km; the average is generally 150–200 km.

Most people know that the center of the hurricane is called the **eye** and that within this zone winds are weak to perfectly calm. The eye is 5–60 km wide, averaging about 20 km in diameter. This part of the hurricane is cloudless and many people comment about seeing the Sun or stars as the eye passes over them. It is also common for birds to be trapped within the eye; the strong winds beyond the storm's center make escape impossible (Fig. 5.18).

The fact that the strongest wind occurs near the center of the storm is due to conservation of angular momentum. The **law of conservation of momentum** states that the product of an object's velocity around its center and its distance from the center squared is constant. The law can be stated mathematically as:

$$MVD^2 = \text{constant}$$

where M is the mass of the object, V is its velocity around the center, and D is its distance from the center. This principle can be illustrated by thinking of a figure skater who is doing a spin on the ice. As skaters pull their arms in toward their body, they decrease the diameter of the rotation. To maintain the same angular momentum the skater's rotational velocity increases drastically. By doing the opposite and extending her arms straight outward, the skater stops. Likewise, the winds in a hurricane gradually diminish toward the periphery of the storm.

Adjacent to the eye of the hurricane is a wall of clouds, which may reach nearly 20 km in height. The eye wall, as it is called, contains abundant water vapor. This vapor moves upward and eventually condenses, releasing vast amounts of energy in the form of heat, which strengthens the storm. During a single day, a hurricane releases energy equivalent to that needed to supply electrical power to the United States for an entire year.

Because of the numerous conditions affecting the development of hurricanes, their ultimate size and intensity vary considerably from one storm to another. The primary factors determining the intensity of a given storm are wind velocity and barometric pressures. Storm surge, which is largely responsible for determining the amount of damage resulting from a hurricane, is difficult to predict because of the variations in the speed of the storm and diversity in the bathymetry of the inner continental shelf and the configuration of coast.

Most people are aware that hurricanes are named. This practice was initiated because of the confusion surrounding situations of multiple hurricanes at one time. Prior to naming, storms were identified by their location: latitude and longitude. The first names were given during the Second World War and were in alphabetical order, such as Able, Baker, Charlie – the commonly used designations for the alphabet by the military. This practice continued until 1953, when female names were used, also in alphabetical order. This style of designating storms lasted for 25 years, and then in 1978 the policy was changed to include both male and female names. A set of names is chosen years in advance, with the first one alternating between male and female and continuing through the alphabet. Names of hurricanes that have severely impacted the United States are permanently retired from the list. Hurricanes that have achieved this status include Camille (1969), Hugo (1989), and Andrew (1992). Different names are used for the north Atlantic and eastern Pacific storms. The name is applied from the time it achieves tropical storm level until it has completely dissipated.

Hurricanes lose their power when they move over cool ocean waters or onto land. Generally by 40° latitude the waters are too cold to supply the large amounts of moisture needed to fuel the storm. Likewise, when a hurricane moves over land, sources of water vapor are greatly reduced. The ability of the storm to take up moisture is further lessened by the cooling effect of the land. Finally, the friction imparted by the land surface rapidly diminishes the low-level storm winds. These factors contribute to an increase in barometric pressure and a spreading out of the storm, leading to its general unraveling and loss of identity. Hurricanes typically last about a week to ten days but some have been known to last as long as a month.

5.4.4 Hurricanes at the coast

Many coastal regions around the world, including numerous sites along the Gulf coast and eastern seaboard of the United States, are low-lying and moderately to densely populated, with numerous dwellings, buildings, and other infrastructure. This combination presents a highly vulnerable situation to a major storm. Devastation to natural environments, destruction of property, and injury or even death to people are all typical hurricane impacts to the coastal zone. This section considers what happens when a hurricane approaches and passes over a coastal area.

Factors affecting their severity

The strength, speed, and size of the storm are major elements in determining how the hurricane will affect the coast. In addition, the gradient and width of the inner continental shelf are important considerations. In some regions, such as New York Bight, the configuration of the coast is also a critical factor. All of these variables contribute to the size of the waves, magnitude of the storm surge, and overall impact of the hurricane. Remember that storm surge, or storm tide as meteorologists frequently call it, is the super-elevation of the ocean water surface above the predicted tide level. It is the storm surge that allows the high-energy storm waves to break high against a dune ridge, across a barrier, or over a seawall. We can better assess the impact and behavior of different hurricanes, and compare those of increasingly higher category, if first we understand the factors that influence them:

1 Magnitude. Encompasses both hurricane size and intensity. Hurricane size governs the length of coast that is affected by the storm as well as the duration of high velocity winds and high-energy waves. The greater the intensity of the hurricane, the stronger the wind velocities and the larger the wave heights and storm surge.

2 Speed of storm. Determines the amount of time over which storm winds can transfer energy to the water surface waves and pile water onshore. Generally, slower moving storms produce higher waves and larger storm surges than faster moving storms of equal magnitude.

3 Path of storm. Determines the landfall of a hurricane and areas along the coast of greatest storm impact. In the northern hemisphere, when a hurricane moves onshore, areas to the right of its landfall

Box 5.1 The fury of Hurricane Camille

Hurricane Camille slammed into the Mississippi coast late on August 17, 1969. Despite hurricane warnings and evacuations, 143 people lost their lives along the Gulf coast. Another 113 people drowned in Virginia floods caused by intense rainfall spawned by Camille. The birth of this destructive storm began in early August when a tropical wave traveled westward off the coast of Africa. By August 14 it reached tropical storm status and later that day intensified to a hurricane 100 km southeast of Cuba. When it passed over the western tip of Cuba on August 15, atmospheric pressure had dropped to 964 millibars and wind velocities reached 185 km h^{-1} (115 m.p.h.). Once in the Gulf of Mexico the hurricane traveled northwestward at 23 km h^{-1} and intensified dramatically (Fig. B5.1). By the afternoon of August 16 an Air Force plane measured a low pressure of 905 millibars and wind velocities of 260 km h^{-1} (160 m.p.h.). The last flight into the storm was made on the afternoon of August 17. By that time minimum pressure had dropped to 901 millibars (26.61 inches of mercury; 30 inches is normal) and surface winds had increased to more than 322 km h^{-1} (200 m.p.h.). This was the second lowest pressure ever recorded in the United States and was only surpassed by the great Labor Day Hurricane of 1935 (892 millibars; 26.35 inches), which swept through the Florida Keys, killing 408 people.

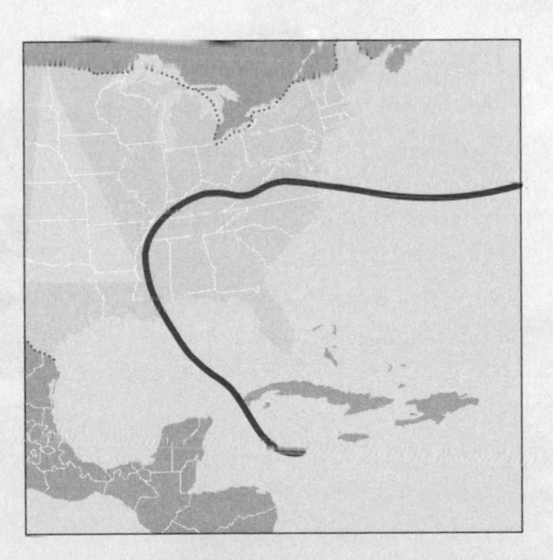

Fig. B5.1 Storm track of Hurricane Camille, which slammed into the Mississippi coast late on August 17, 1969, killing 143 in the Gulf coast region.

Early on August 17 police and civil defense officials used television and radio messages to call for an immediate evacuation of coastal regions along the Mississippi shore, knowing that a category 5 hurricane was located just 400 km south of Mobile Bay and that its landfall was imminent. That night at 10:30 p.m. one of the strongest storms ever witnessed by mankind came onshore. Maximum wind velocities are unknown because all instruments stopped working before the storm reached its greatest intensity. However, estimates of wind velocity based on surface pressures and previous flights into the hurricane were calculated at 324 km h^{-1} (202 m.p.h.). While wind of that magnitude can rip a structure apart, it is the accompanying wall of water, known as the "storm surge," that accounts for most of the destruction and loss of human life during a storm. The extreme low pressure and high velocity winds of Camille produced a wall of water of that measured 24.6 feet (7.5 m) above mean sea level at Pass Christian, located 20 km east of the landfall. Not only does a storm surge of this magnitude flood areas far inland from the coast, but it also allows monstrous waves to break against and over anything along the coast, including dune ridges, seawalls, buildings, and other structures. On the offshore barriers of Ship and Cat Island, debris marks indicated that hurricane waves broke across the tops of trees covering the island.

Hurricane Camille ranks as one of the deadliest and costliest storms in United States history. Certainly the amount of damage caused by the storm is attributed to its category 5 status, but the loss of life was also due to the fact that some of the coastal residents failed to respect the danger of an intense hurricane and did not respond to repeated warnings. Stories chronicled by the National Hurricane Center help to demonstrate the magnitude and destructive force of Hurricane Camille. Perhaps the fate of the Richelieu Apartments and its occupants best illustrates the storm's immense power. The Richelieu was a three-story, brick-front building located about 100 m from the ocean. Standing between the building and the beach was a 2.4 m high seawall and a four-lane highway. Civil defense personnel pleaded with the occupants of the Richelieu to evacuate, but 25 people planned a "hurricane party" instead. The entire structure was destroyed by the storm and only

Box 5.1 *(cont'd)*

Fig. B5.2 Before and after photographs of Richelieu Motel, which was dismantled during Hurricane Camille. (Photographs by Chauncey Hinman.)

two people survived (Fig. B5.2). One of the survivors was Mary Ann Gerlach, who remained in the Richelieu with her husband because they had lived through previous hurricanes in Florida and thought that this one would be no more challenging. When the storm surge and waves began dismantling the Richelieu, Mary Ann managed to jump out a second story window. Miraculously, she washed ashore 7.2 km (4.5 miles) from the apartment after being in the water for almost 12 hours. A small boy from the Richelieu also survived the hurricane. He was saved by a father of the family who lived next to the apartment building. After the man's house was broken apart by gigantic waves he swam onshore and found temporary safety by clinging to the top of a tree. He succeeded in saving his son and was also able to grab the ten-year-old boy as he floated from the Richelieu. His wife and two daughters perished. Another group of people went to the Trinity Episcopal Church seeking shelter; of the 21 at the church only 13 survived the storm (Fig. B5.3). These stories and experiences with other major storms have contributed to a growing public awareness that intense hurricanes can strike with tremendous force, causing death and injury to those who don't heed their fury.

Fig. B5.3 Before and after photographs of the Trinity Church, which was destroyed by Hurricane Camille. (Photographs by Chauncey Hinman.)

will experience the strongest winds and greatest storm damage.

4 Coastline configuration. This is an important factor in large deeply embayed coastlines. In this setting certain hurricane tracks can significantly amplify storm surge levels as water is forced into the funnel-shaped embayment.

The circulation of wind within a hurricane and its speed and pathway toward land dictate the relative intensity of the storm and the amount of damage the hurricane will inflict along a given stretch of coast. In the northern hemisphere the wind orbiting the central low pressure of the hurricane travels in a counterclockwise direction. At the same time that this wind is blowing in a circular fashion, the hurricane is also moving over the ocean surface with a velocity that ranges from 5 to 40 km h^{-1}. This forward movement of the hurricane has an additive effect on wind blowing in the same direction as the path of the storm and a subtractive effect where the storm wind is blowing in the direction opposite to the storm's forward motion.

To illustrate this point, consider a moderate-sized hurricane in the Gulf of Mexico with an average wind velocity of 180 km h^{-1} (Fig. 5.19). The storm is moving northward with a speed of 30 km h^{-1}. East of the storm's eye the hurricane wind blows in a northerly direction with a velocity of 210 km h^{-1} (180 + 30 km h^{-1}). At the same time, wind west of the eye blows southward with an effective velocity of only 150 km h^{-1} (210 – 30 km h^{-1}). Not only is the effective wind velocity different on either side of the storm, but also as the storm passes the coast the shoreline east and west of the eye experiences very different wind and wave conditions. East of the eye there is a wind "set-up" where wind continues to blow onshore building the storm surge. Conversely, west of the eye there is a "set-down" producing lower water level and waves due to the wind blowing offshore. Damage from a hurricane is always greater to the right of the landfall than to the left of the landfall.

Hurricane Frederick was a moderate-sized storm that struck the Mississippi coast in 1979. Meteorologic and oceanographic data collected during the storm illustrate the relationships among the storm's

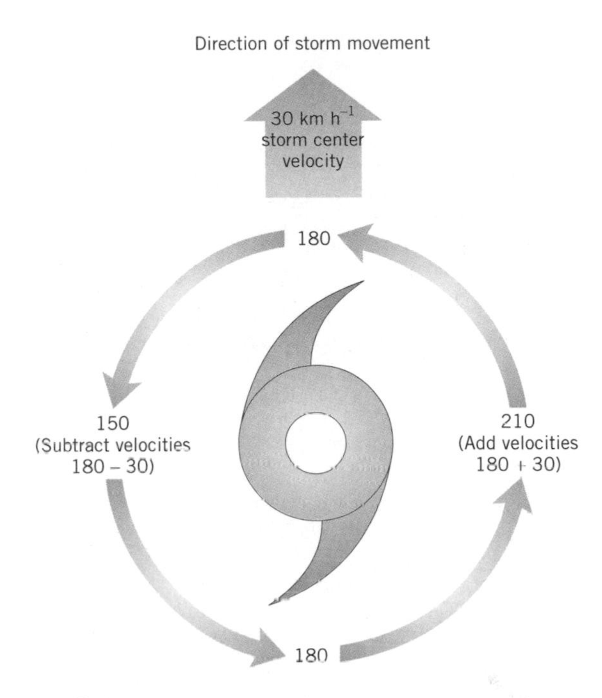

Fig. 5.19 The forward speed of a hurricane increases wind velocities on the advancing right side of the storm while diminishing wind velocities on the advancing left side of the storm.

low pressure, its wind velocity, and the storm surge pattern (Fig. 5.20). The hurricane moved onshore from the Gulf of Mexico at an average forward speed of 15 km h^{-1}. The eye of the storm made landfall at Dauphin Island, which is located along the western flank of Mobile Bay. As seen in Fig. 5.21, the highest wind velocity that was recorded at the coast, 126 km h^{-1}, coincided with the period of lowest pressure. Generally the lowest pressure is found near the eye of the hurricane and it gradually increases toward the perimeter of the storm. The strongest wind corresponds with the steepest pressure gradient, which occurs just beyond the eye of the storm. During the passage of Hurricane Frederick maximum tidal elevations recorded along the coast ranged from 1.0 to 3.8 m above mean sea level. As illustrated in Fig. 5.22, water elevations were much greater east of Dauphin Island than in the coastal regions to the west. Note that the greatest storm surge took place approximately 30 km east of the eye. This pattern of higher water level, flooding,

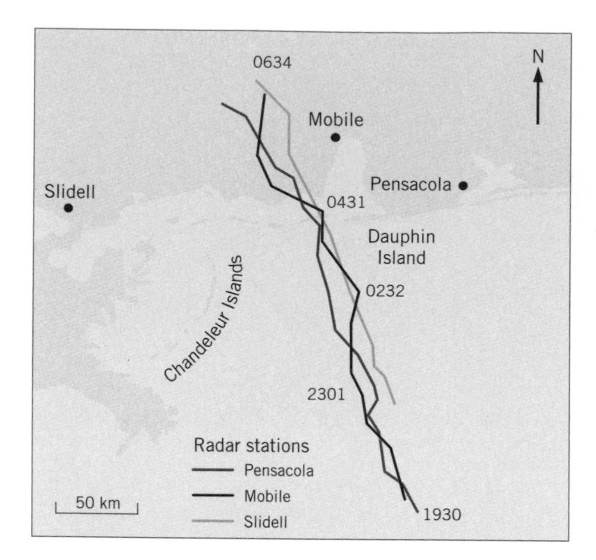

Fig. 5.20 Storm track of Hurricane Frederick, which struck Dauphin Island along the Mississippi coast on September 13, 1979. (Courtesy of Shea Penland, University of New Orleans.)

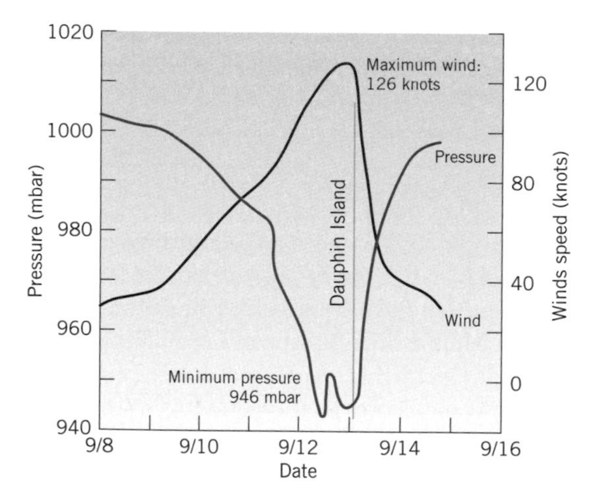

Fig. 5.21 Graph of atmospheric conditions at Dauphin Island, Mississippi, during the passage of Hurricane Frederick. Maximum wind velocities coincided with lowest atmospheric pressures. (Courtesy of Shea Penland, University of New Orleans.)

and damage that occurs east of the storm center is due to the hurricane's counterclockwise wind circulation. When the hurricane was centered over Dauphin Island wind was still blowing onshore east of the island. At the same time, wind was

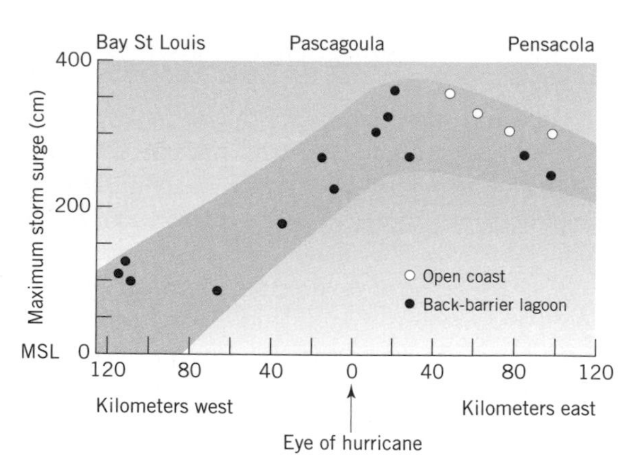

Fig. 5.22 Storm surge values along the Gulf coast from Mississippi to the panhandle of Florida resulting from Hurricane Frederick. (Courtesy of Shea Penland, University of New Orleans.)

already blowing offshore along the coasts of eastern Louisiana and western Mississippi. The longer the time during which hurricane winds blow onshore, the greater the storm surge and resulting damage.

The importance of coastline configuration has been clearly demonstrated in a study of New York Bight by Nicolas Coch of Queens College (New York). Eastern Long Island and northern New Jersey meet at near right angles to one another, forming a funnel-shaped embayment. Coastlines with this type of configuration are susceptible to amplification of the storm surge as waters are constricted by the adjoining landmasses. Dr Coch has shown that if a moderate-sized hurricane tracked in a northwesterly direction across northern New Jersey, the storm surge would be heightened from 1 m at the inner shelf to more than 6 m at the mouth of the Hudson River (Fig. 5.23). It is obvious that if (when) this hurricane scenario occurs, New York City would be devastated! The large storm surges associated with extensive loss of life in Bangladesh have been caused in part by the same funneling effect that is produced in the upper Bay of Bengal.

Hurricane categories

The strength of a hurricane is formally classified using the Saffir–Simpson Scale and is based on the maximum wind velocity, barometric pressure, storm

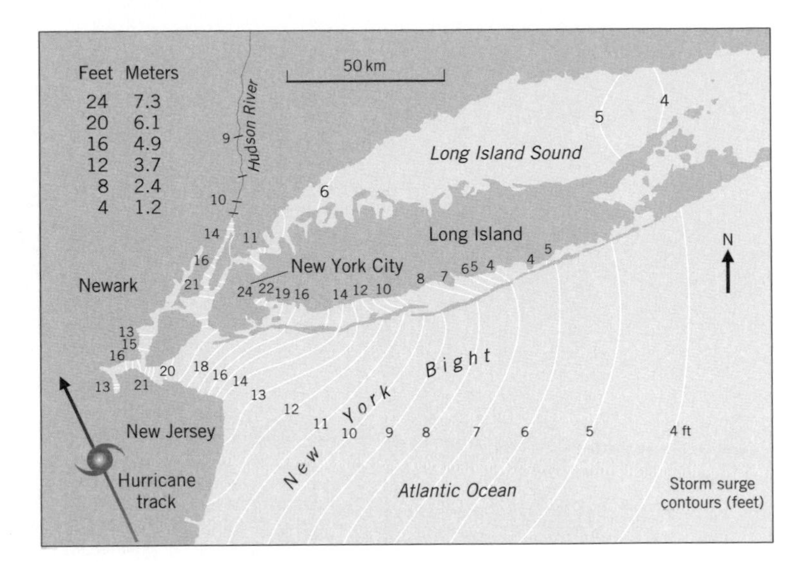

Fig. 5.23 Predicted storm surge values for the New York Bight. The contours are based on the SLOSH hurricane computer model. (From N. K. Coch, 1995, *Geohazards Natural and Human*. Englewood Cliffs, NJ: Prentice Hall, p. 444, Fig. 16.19.)

Table 5.3 Storm genesis.

Tropical disturbance	Prevalence of thunderstorms, infancy stage of tropical cyclone
Tropical depression	Cyclonic wind pattern, wind velocities up to 61 km h^{-1} (38 m.p.h.)
Tropical storm	Highly organized ocean storm visible in satellite imagery, velocities from 61 to 119 km h^{-1} (38 to 73 m.p.h.)
Hurricane	Spiraling cloud pattern around a central eye, wind velocities exceed 119 km h^{-1} (73 m.p.h.)

surge level, and expected damage (Table 5.4). The scale has five categories and each higher level represents a substantial increase in hurricane intensity (Table 5.5). A list of the most intense hurricanes in the United States in terms of loss of life and property damage is provided in Table 5.6. It is evident from this table that death tolls have decreased in recent times as meteorological forecasts and evacuation procedures have improved.

Most hurricanes fall within categories 1 to 3. One of the difficulties of accurately determining the maximum wind in severe storms such as a category 4 or 5 hurricane is the inability of instruments to withstand the storm. As an example, there has been considerable disagreement about the maximum wind velocity of Hurricane Andrew as it struck south of Miami in August 1992. Instrumentation at the National Hurricane Center in Coral Gables measured sustained wind of 210 km h^{-1} (125 m.p.h.), with gusts of about 275 km h^{-1} (165 m.p.h.) at the time that it became disabled. Calculations indicate that peak sustained wind velocities were probably greater than 235 km h^{-1} (140 m.p.h.). Inland from the shoreline high-velocity wind exerts the greatest influence on vegetation and manmade structures. The path of a hurricane is often evidenced by broken and uprooted trees, destroyed homes, and public infrastructure in complete disarray.

Damage to structures due to hurricanes is associated with lowland flooding, wave attack, and strong wind. Because of the current stringent rules and regulations regarding construction in the coastal zone, there are usually few problems with wind damage. In hurricane-prone areas, most damage is now confined to roofs, where shingles are commonly torn from roofs. Under extreme conditions such as during a category 4 or 5 hurricane, damage can be much greater and the building itself can be

Table 5.4 The Saffir–Simpson hurricane intensity scale.

	Central pressure		Wind speed		Storm surge		
Scale number (category)	**Millibars**	**Inches**	**Miles per hour**	**Kilometers per hour**	**Feet**	**Meters**	**Damage**
1	≥980	≥28.94	74–95	119–154	4–5	1–2	No real damage to building structures. Damage primarily to unanchored mobile homes, shrubbery, and trees. Also some coastal road flooding and minor pier damage.
2	965–979	28.50–28.91	96–110	155–178	6–8	2–3	Some damage to roofing material, and door and window damage to buildings. Considerable damage to vegetation, mobile homes, and piers. Coastal and low-lying escape routes flood 2–4 hours before arrival of hurricane eye. Small craft in unprotected anchorages break moorings.
3	945–964	27.91–28.47	111–130	179–210	9–12	3–4	Some structural damage to small residences and utility buildings with a minor amount of curtainwall failures. Mobile homes are destroyed. Flooding near the coast destroys smaller structures, with larger structures damaged by floating debris. Terrain continuously lower than 1.5 m (5 feet) above sea level may be flooded inland as far as 9.6 km (6 miles).
4	920–944	27.17–27.88	131–155	211–250	13–18	4–6	More extensive curtainwall failures with erosion of beach areas. Terrain continuously below 3 m (10 feet) above sea level may be flooded, requiring massive evacuation of residential areas inland as far as 9.6 km (6 miles).
5	<920	<27.17	>155	>250	>18	>6	Complete roof failure on many residences and industrial buildings. Some complete building failures, with small utility buildings blown over or away. Major damage to lower floors of all structures located less than 4.5 m (15 feet) above sea level and within 457 m (500 yards) of the shoreline. Massive evacuation of low areas on low ground within 8–16 km (5–10 miles) of the shoreline may be required.

Table 5.5 Comparative destructive level of hurricanes, by Saffir–Simpson category.

Category	Relative hurricane destruction potential
1	1 (reference level)
2	4 times the damage of a category 1 hurricane
3	40 times the damage
4	120 times the damage
5	240 times the damage

Source: Based on empirical analysis over the past 42 years by Dr William M. Gray, Colorado State University meteorologist.

destroyed. A good example of how recent construction codes have protected dwellings during the passage of moderate hurricanes is illustrated by the impact of Hurricane Opal in 1995. This severe storm made landfall in the panhandle of Florida just east of Pensacola. Of the nearly 1500 homes subjected to the storm's fury, approximately half were constructed prior to the new regulations and the other half were built in compliance with new building codes. There were hundreds of homes severely damaged or destroyed during the storm, but all of those built under present guidelines received only minor damage.

Storm surges and the accompanying large waves can, however, cause major damage, including the destruction of structures. For this reason, current zoning in Florida requires that the first level of occupancy must be above the 100-year storm surge level. This level is based on existing data and predicted frequency of storm surge. For example, the Gulf coast of central and southern Florida is predicted to experience a 4 m storm surge on a 100-year return interval. This is based in part on historical data of previous storms and in part on computer modeling of storm conditions using variables such as shelf gradient, wind velocity, and wave size, among others. Qualification for federally supported insurance in the coastal zone requires compliance with these regulations.

Table 5.6 The deadliest and costliest hurricanes in US history.

Hurricane	Year	Category	Pressure (millibars)	Damage (US$ billion)*	Deaths
Camille (Mississippi, Louisiana)	1969	5	909	6.10	256
Florida (Keys)	1935	5	892		408
Andrew (SE Florida, SE Louisiana)	1992	4	922	30.48	15
Audrey (SW Louisiana, N Texas)	1957	4	945	0.80	390
Florida (Keys), S Texas	1919	4	927		600†
Florida (Lake Okeechobee)	1928	4	929		1836
Florida (Miami, Pensacola), Mississippi, Alabama	1926	4	935	4.83	243
Hugo (South Carolina)	1989	4	918	7.00	57
Louisiana (Grand Isle)	1909	4	931		350
Louisiana (New Orleans)	1915	4	931		275
Texas (Galveston)	1900	4	931	0.81	6000†
Texas (Galveston)	1915	4	945	1.35	243
Mississippi, Alabama, Pensacola	1906	3			134
NE USA	1944	3	947	1.06	390
New England	1938	3	946	4.14	600
SE Florida	1906	2			164
Diane (NE USA)	1955	1	969		184

*Adjusted to 1996 dollars on the basis of US Department of Commerce implicit price deflator for construction.
†Minimum estimates.
Source: National Hurricane Center (http://www.nhc.noaa.gov/pastall.html).

5.5 Summary

Hurricanes and extratropical cyclones are a dramatic expression of the Earth's weather system. Through recorded history severe storms have accounted for permanent changes to the coastal landscape, billions of dollars of damage, and the unfortunate loss of many lives. Hurricanes are tropical storms with wind velocities exceeding 119 km h^{-1}. The largest hurricanes have storm surges greater than 7 m and winds attaining 220 km h^{-1}. They form in the tropics and are steered by the prevailing global wind patterns. Hurricanes affecting the United States are initiated off the west coast of Africa. They intensify over the warm Atlantic Ocean waters and travel in an easterly direction until they make landfall along the Gulf of Mexico and along the east coast, most commonly from Florida north to the Outer Banks of North Carolina. Rarely do hurricanes strike the west coast of the United States. They degenerate after moving over land or cold water.

Extratropical cyclones form above the tropics and are commonly associated with cold fronts. These storms are usually weaker than hurricanes, but occur more frequently. They develop over the continental United States, in southwestern Canada, in the Gulf of Mexico, and off the east coast of Florida. These storms generally move eastward and northward, eventually traveling offshore of New England and passing east of Nova Scotia. They generate northeasterly winds and waves and therefore are called "northeasters." Northeasters last for one to two days and are accompanied by wind velocities of 40–65 km h^{-1} and storm surges ranging from 0.2 to 1.2 m.

Both hurricanes and extratropical storms are low-pressure systems with counterclockwise wind patterns. The strength, speed, and size of these cyclonic systems control the severity of the storm, including the amount of erosion and structural damage. Storm track, gradient of the shelf, configuration of the shoreline, and astronomic tidal conditions are other important factors governing storm processes.

Suggested reading

Abbott, P. L. (1996) *Natural Disasters*. Dubuque, IA: W. C. Brown Publishers (Chapters 9 and 10).

Coch, N. L. (1995) *Geohazards: Natural and Human*. Englewood Cliffs, NJ: Prentice Hall.

Dunn, G. E. & Miller, B. I. (1960) *Atlantic Hurricanes*. Baton Rouge: Louisiana State University Press.

Elsner, J. & Kara, A. B. (1999) *Hurricanes of the North Atlantic: Climate and Society*. Oxford: Oxford University Press.

Henry, J. A., Portier, K. M. & Coyne, J. (1994) *The Climate and Weather of Florida*. Sarasota, FL: Pineapple Press.

Pielke, R. A. Jr & Pielke, J. A. Sr (1997) *Hurricanes: Their Nature and Impacts on Society*. New York: Wiley.

Simpson, R. H. & Riehl, H. (1981) *The Hurricane and Its Impact*. Baton Rouge: Louisiana State University Press.

Williams, J. M. & Duedall, I. W. (1997) *Florida Hurricanes and Tropical Storms*, rev. edn. Gainesville: University Press of Florida.

6 Waves and the coast

6.1 Introduction

Waves are a surface disturbance of a fluid (gas or liquid) in which energy is transferred from one place to another. In the case of the coast, we are concerned with the interface between the ocean and the atmosphere, but waves also occur between different liquid masses and between different gaseous masses. For example, waves occur between water masses of different densities caused by temperature and/or salinity contrasts, and these are termed internal waves. Waves also occur within the atmosphere, such as between a warm, light air mass and one that is cold and heavy.

When the surface of the fluid is disturbed, the perturbation is transferred from one location to another but the medium itself, water in the case of the coast, does not move with the propagating disturbance. Waves occur in a wide range of sizes (Fig. 6.1) and may be caused by various phenomena. The primary type of wave that influences the coast is what is called a progressive, surface wave that is produced by wind. In this type of wave, energy travels across and through the water in the direction of the propagation of the wave form. The movement of these waves is due to restoring forces that cause an oscillatory or circular motion that is basically sinusoidal in its form (Fig. 6.2). That is, it is shaped like a sine curve: perfectly symmetrical and uniform. Both gravity and surface tension are important restoring forces that maintain waves as they propagate. Surface tension is important in very small waves called capillary waves. These waves are less than 1.7 cm long. Larger waves are called gravity waves because the primary restoring force is gravity. It is these gravity waves that are the typical waves we see at the coast or on any water surface (Fig. 6.1).

Each individual wave has several components that are important for describing the wave and its motion (Fig. 6.2). The wavelength (L or λ) is the horizontal distance between two like locations on

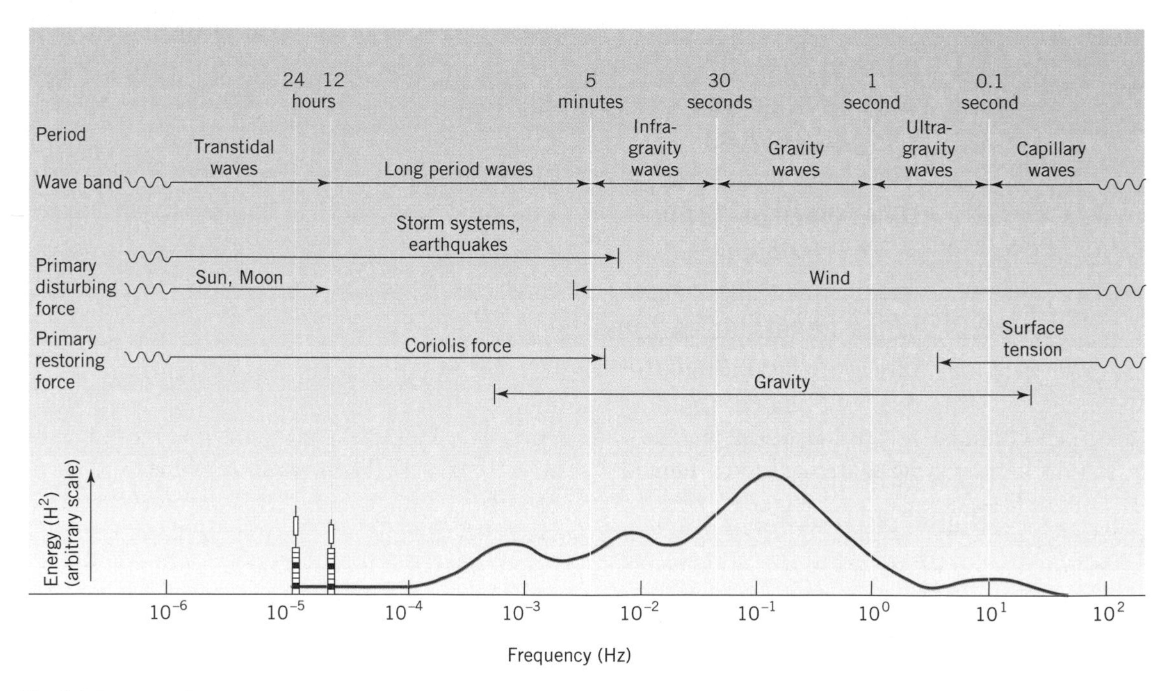

Fig. 6.1 Diagram showing various wave types and their respective frequencies. The large and broad peak is the typical surface waves of the open ocean or the coast and the peaks near 24 and 12 hours represent the tidal wave. (From B. Kinsman 1965, *Wind Waves*. Englewood Cliffs, NJ: Prentice Hall, p. 23, Fig. 1.2–1.)

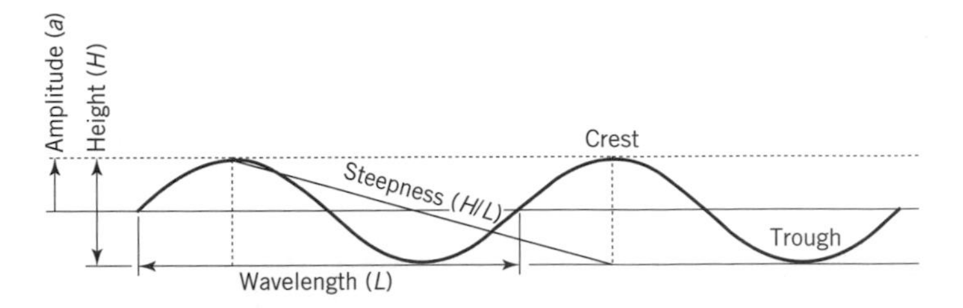

Fig. 6.2 Typical wave form showing the various elements of a gravity wave. This is a conceptual model. The natural waves tend to be peaked instead of sinusoidal.

the wave form: crest to crest, trough to trough, etc. The wave height (H) is the vertical distance between the base of the trough and the crest. The steepness of the wave is the ratio of the height to the length (H/L). Another important characteristic of gravity waves is the **period** (T), the time in seconds that it takes for a complete wavelength to pass a reference point. The velocity that a wave propagates across the water surface is called the celerity (C) and is expressed as

$$C = L/T \tag{6.1}$$

Another way of measuring wave propagation is by the frequency (f); that is, the number of wavelengths passing per second. Therefore, a 10 s wave would have a frequency of 10^{-1}.

The actual transfer of energy from the wind to the water surface is complicated and not completely understood. It is well known, however, that the size of waves depends upon three primary factors: (i) how fast the wind is blowing (velocity); (ii) the length of time the wind blows (duration); and (iii) the distance over the water that the wind blows (fetch). Any one or a combination of these factors can be limiting in wave size. In many situations the theoretical limitation of wave size is caused by the fetch; these are typically termed fetch-limited basins. Any location can have any wind velocity and the wind can blow for any length of time. What cannot change, however, is the size of the water body that is being subjected to these winds, and therefore the fetch. Good examples are Lake Michigan and Gulf of Mexico, both large bodies of water to be sure, but also ones that are not nearly as big as an ocean. Huge waves cannot form in these water bodies even during hurricanes or other severe storms.

Fig. 6.3 Photograph of complicated sea conditions during a storm on the North Sea.

Although the wave form and its components are discussed in relatively simplistic terms, the actual wave conditions in nature are extremely complicated. Typically there are several families of waves of different sizes (Fig. 6.3) moving in different directions all superimposed at a given location in the sea (Fig. 6.4). These waves combine to produce a wave field that can be recorded and analyzed. The data are in the form of a wave spectrum, which can be separated into its component wave forms, each with its own period or frequency and height (Fig. 6.5). One of the most important aspects of the spectral analysis of a wave field is the determination of the significant wave height, the most commonly used wave-measuring category. This value (H_s) is the average height of the highest one-third of the waves

Fig. 6.4 Photograph of complicated sea conditions showing distinct but different wave sets. The different sizes of waves are the result of different weather in proximal areas and the movement of waves from one place to another.

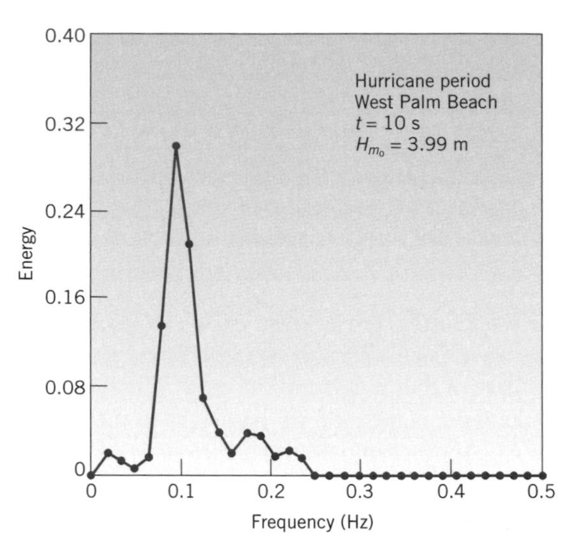

Fig. 6.5 Plot of a wave spectrum showing the range of frequencies recorded at this particular location in a brief time series. The peak represents the dominant frequency for the wave spectrum, in this case 0.1 Hz or periods of 10 seconds.

occurring during the time period being analyzed. The significant wave height is commonly used as an index of the wave energy.

6.2 Water motion and wave propagation

Recall from the previous section that only the wave form is propagated in gravity waves, not the water itself. That being the case, we need to consider how the water actually moves within gravity waves. In this wave type, the water moves in an orbital path, with the circulation in each orbit being in the direction of wave propagation (Fig. 6.6). As the wave moves toward the coast, the surface water is moving landward on the crest of the wave and seaward in the wave trough. The wave form moves toward the shore but the water itself moves only in circles. This orbital motion in the wave extends well below the water surface. The diameter of the orbital motion at the water surface is equal to the wave height, and this diameter decreases with depth. At a depth of approximately one-half the wavelength of the surface wave, the orbital motion is very slow and the orbits are very small, and can no longer move sediment on the bottom (Fig. 6.6). Anyone who has gone scuba diving knows that even if it is quite rough on the surface, there is a depth below which

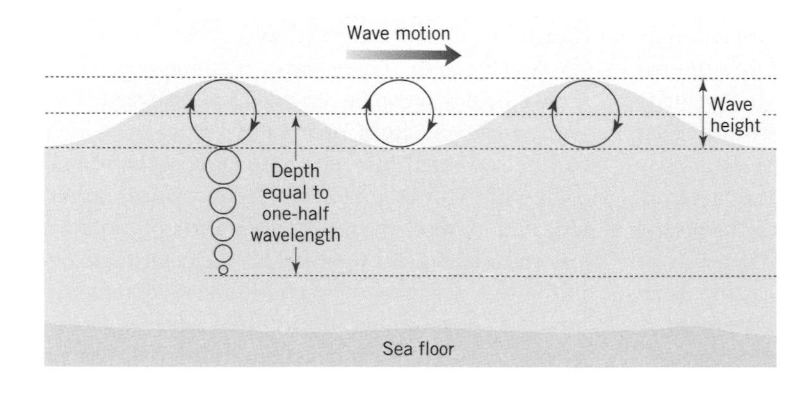

Fig. 6.6 Diagram of water motion in waves showing the decrease in the diameter of the orbital paths with depth. You must remember that this is a very simplified diagram. There is an infinite number of particles moving in this same pattern.

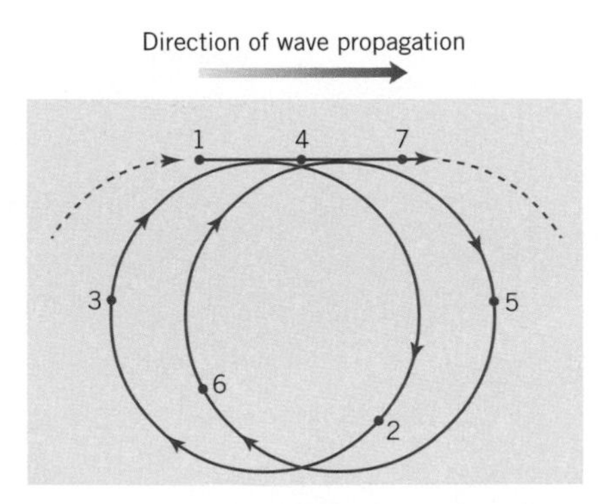

Fig. 6.7 Slight landward movement of water as wind friction causes orbital paths to move toward the shore. This motion is superimposed on the motion shown in Fig. 6.6.

you do not feel any wave motion. Because this orbital motion is forward on the crest and backward on the trough, with vertical motion halfway between, a fishing float or a ball appears to actually move up and down as the waves pass, but without progressing significantly itself. Where wind is present, close inspection of the actual path of water particles shows that there is slight net movement of water in the direction of propagation due to friction between the wind and the water surface (Fig. 6.7). At the coast, this wind may push the water landward and produce what is called setup, a temporary elevation of water level. Setup is a phenomenon that is the major factor in undertow, a process that is discussed later in this chapter.

When the depth of water is more than half of the wavelength of the surface wave, the orbital motion of water within the wave is not influenced by the ocean floor as the wave propagates. As the wave moves into increasingly shallow water during its approach to the coast, these orbital motions begin to interact with, or "feel," the bottom at a depth about equal to half the wavelength. This is called wave base and this interference causes the orbits of water particles to become deformed and to slow down. At first the circular motion becomes squashed into an oval shape and eventually becomes simply back and forth motion. This condition progresses as the wave moves into increasingly shallow water. The slowing of the wave causes it to steepen because it is being compressed somewhat like an accordian. At the same time that this is happening, bottom friction causes the wave to slow down at the bottom, but at the surface the wave is traveling faster in the absence of this friction. These conditions cause the wave to eventually become so steep that it is no longer stable and its shape collapses. This is the breaking of waves that we see in the **surf** along the coast. This instability due to excessive steepness takes place when the inclusive angle of the wave form seeks to be less than 120° or the steepness exceeds 1:7. After breaking, restoring forces cause the wave to reform and continue its progression, perhaps to break again before reaching the shoreline.

Multiple breaking of waves in shallow water at the coast is produced by a bar and trough topography of the seabed (Fig. 6.8). As the wave moves into shallow water and begins to feel bottom, slow its forward speed, and steepen, it will eventually break. This first break is commonly over the crest of a sand bar, which may rise a meter or so above the gently sloping nearshore gradient. Landward of this bar is a trough of deeper water and it is over this trough that the wave reforms after initial breaking. The water shallows again and breaking takes place again,

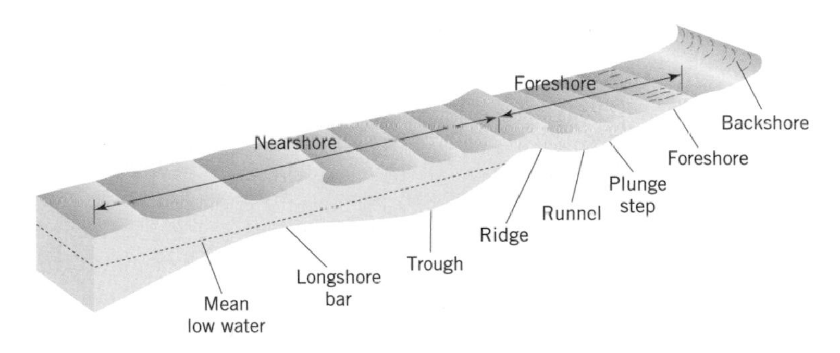

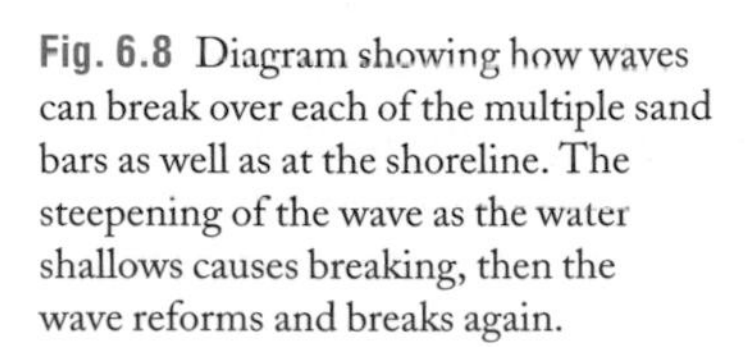
Fig. 6.8 Diagram showing how waves can break over each of the multiple sand bars as well as at the shoreline. The steepening of the wave as the water shallows causes breaking, then the wave reforms and breaks again.

Fig. 6.9 Aerial photograph showing waves breaking over multiple bars in the nearshore area along Mustang Island on the Texas coast.

Fig. 6.10 Waves that have moved away from the wind or where the wind has stopped are called swell waves. They are most common along the Pacific coast, as shown here.

over a second sand bar. It will reform again and finally break at the shore. This situation with two sand bars and waves breaking over them (Fig. 6.9) is typical of many of the gently sloping nearshore areas of the world. It is also possible for a coast to have just one bar or three bars. Under conditions of small waves, breaking will take place only over the shallow bar and not the deeper one. The fairly narrow coastal area where wave breaking occurs is called the surf zone.

6.3 Wind wave types

Although the theoretical and simplistic wave form is a sinusoidal curve, that form is not common in nature. Wave shape depends on the conditions of the wind, water depth, and the progression of the wave itself. Although wind is responsible for the production of most gravity waves, it is common for the waves to travel well beyond the area where the wind blows or for waves to continue long after the wind stops blowing. Waves that are directly under the influence of wind are called sea waves (Figs 6.3 and 6.4). They typically have relatively peaked crests and broad troughs. In nature, sea waves tend to be complicated by multiple superimposed sets of various sized waves. Whitecaps occur when the wind blows off the tops of sea waves.

A common wave type is the swell wave, which actually has a wave form that approaches a sinusoidal shape (Fig. 6.10). Swell develops after the wind stops or when the wave travels beyond the area where the wind is blowing. Swell waves commonly have a long wavelength and small wave height, thus having a very low steepness value, much lower than sea waves.

The breaking of waves as they enter shallow water takes on different characteristics depending upon the type of wave that is present in deep water, and the conditions of breaking as the waves move into the surf zone. For example, swell waves are long and low, and therefore they begin to feel the bottom in relatively deep water. They gradually slow and steepen until they break as plunging breakers (Fig. 6.11). This type of breaking wave is typified by a large curling motion, with an instantaneous crashing of the wave characterized by a sudden loss of energy. It is typical of gently sloping nearshore zones.

The other common type of breaking wave is the spilling breaker (Fig. 6.12). This type is most common as sea waves with their shorter wavelength and higher steepness enter shallow water. As these waves break they do so slowly over several seconds and some distance. The wave looks like water spilling out of a container. Surging breakers and collapsing breakers (Fig. 6.13) are other types of breaking waves; they look about the same. They have an appearance that is similar to spilling breakers. Surging breakers typically develop on or near the beach as the wave runs up

Box 6.1 The Duck Pier

In 1977, the US Army, Corps of Engineers established a Field Research Facility (FRF) at Duck, North Carolina, near the northern end of the Outer Banks. The central element of this facility is a 180 m (590 feet) long research pier that extends to a depth of 7 m. This pier was constructed to permit monitoring of a variety of coastal processes across the inner shoreface and surf zone. Other elements of the facility include research laboratories, sophisticated computer capabilities, and a conference room. An observation tower rises 13 m above the adjacent dunes and various specialized vehicles for taking measurements in the rigorous conditions of storms are available. The staff of 12 includes coastal scientists and engineers, computer specialists, and technicians.

The specialized vehicles and other equipment allow the FRF to make observations and take measurements that are typically not possible. A specially constructed motorized vehicle called a CRAB (coastal research amphibious buggy) can move across the surf zone to survey with centimeter accuracy to a depth of 9 m. It can also deploy instruments and provide a stable platform for other activities, such as vibracoring, side-scan sonar surveying, and sediment sampling. The sensor insertion system (SIS) is a large crane that can extend out to 24 m from the pier and can carry wave gauges, current meters, suspended sediment sampling devices, and other instruments. The SIS can be moved on a track along the entire length of the pier.

One of the main tasks of the FRF is to continuously monitor coastal process and change in order to provide a large database. Wave height, period, and direction are monitored by pressure transducer arrays beyond the pier, and wave height and period are measured at three locations along the pier. The water current profile is measured, as are various meteorological parameters. Various water parameters, such as temperature, salinity, and light penetration, are also included in this database. The changes in bathymetry and shoreline position are measured and related to the processes. All of these data are compiled into monthly and annual reports and are available to the public.

In addition to the regular collection of data by the FRF staff, the facility also hosts visiting researchers from other Corps of Engineers locations and from universities. There have been huge experiments at the FRF that have involved more than 100 researchers each. Such events as "Super Duck" and "Sandy Duck" were typically held in the fall of the year and lasted at least two weeks. Investigators brought their own instruments and personnel to interact with the total group. In order to participate in these large and complex experiments, a principal investigator had to submit a proposal for research and relate it to the overall objectives. These events have provided a tremendous amount of comprehensive data for a selected time period.

One criticism of the FRF is that there has been too much time, effort, and money expended at this single location, instead of having multiple locations where a wealth of data are collected. Funding is always a limiting factor, so such shortcomings are realistic given the budget of the supporting agencies.

Fig. 6.11 Waves that steepen in the nearshore and break almost instantaneously are called plunging waves.

Fig. 6.12 Waves that steepen in the nearshore and break over some time and distance are called spilling breakers.

Fig. 6.13 In very shallow water waves break as surging breakers.

to the shoreline. It steepens, and just as the wave begins to break, it surges up the beach.

6.4 Distribution and transfer of wave energy

As waves move into shallow water and are influenced by the bottom and by various natural features or structures made by humans, they may experience changes in their energy distribution and/or direction. This occurs in three primary ways: (i) refraction; (ii) diffraction; and (iii) reflection. Therefore, waves act much like light. Waves typically approach the shoreline at some angle; the crest may have an orientation ranging from nearly perpendicular to parallel to the shoreline. As the wave enters shallow water and begins to be slowed by interference with the bottom, this slowing of the rate of movement of the wave takes place at different times and places along the crest of the wave. The result is a bending or **refraction** of the wave as it passes through shallow water on its way to the shoreline (Fig. 6.14). As the waves bend or refract, they cause a vector of energy to move along the shoreline in the form of a longshore current (discussed later in this chapter), and they cause energy to be distributed according to the relationship between the bending wave and the shoreline. The distribution of wave energy can be represented by orthogonals, lines constructed perpendicular to the wave crests. This type of construction enables us to show where wave energy is concentrated and where it is dispersed.

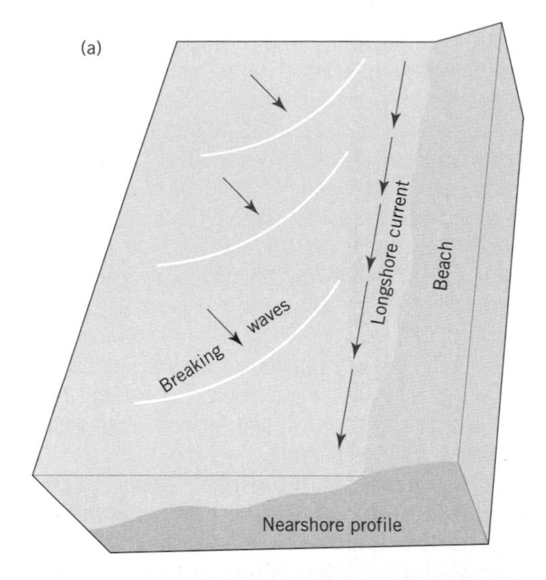

Fig. 6.14 (a) Schematic diagram of waves refracting across the surf zone and (b) photograph of the same phenomenon. This happens when offshore waves approach the shoreline at an angle. The shorter the wavelength the more the angle at which the waves approach the shore. Long-period waves tend to refract completely and approach the shore essentially parallel.

If we have a uniformly sloping nearshore with wave approaching a straight shoreline at an acute angle, we would expect the wave energy to be uniformly distributed along the shoreline (Fig. 6.14). If the shoreline and/or the nearshore topography is irregular, then the refraction of the waves will reflect these irregularities and the distribution of wave

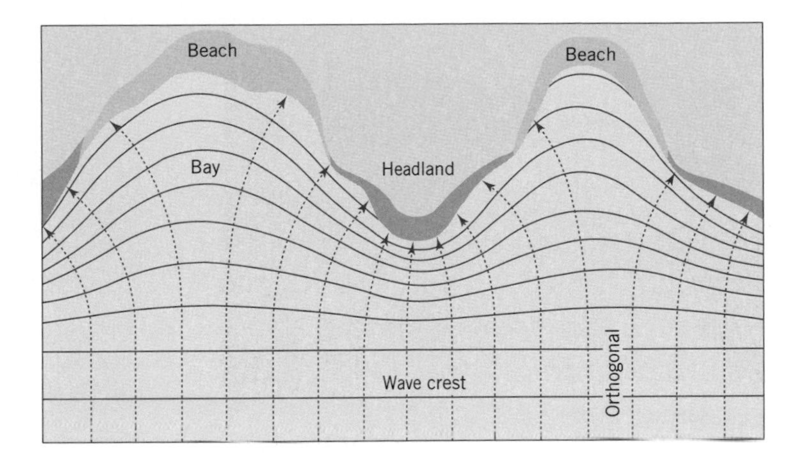

Fig. 6.15 Diagram showing wave refraction with erosion at the headlands and deposition in the embayments. The lines perpendicular to the wave crests are called orthogonals. They converge to concentrate energy at headlands and they diverge to dissipate energy at embayments.

energy will be complicated. As the result of refraction, wave energy is concentrated at headland areas along the coast and is dissipated in embayments (Fig. 6.15). This combination of conditions causes headlands to erode and embayments to accumulate sediment.

As waves pass an impermeable obstacle such as a jetty, breakwater, or other type of structure, the wave energy is spread along the crest behind the obstacle – it is **diffracted**. Part of the wave crest is stopped by the structure and the rest of it passes by (Fig. 6.16). As the wave passes the structure, energy is transferred laterally along the wave crest to the sheltered areas behind the obstacle. In this way waves gradually progress behind the obstruction, although they will have smaller heights than the wave that is unaffected by the structure. This is the reason why boats behind breakwaters or other shelters still feel waves. It should be noted that the obstacle also influences wave energy beyond the structure itself. The same phenomenon occurs when waves approach and pass an island or prominent headland (Fig. 6.17).

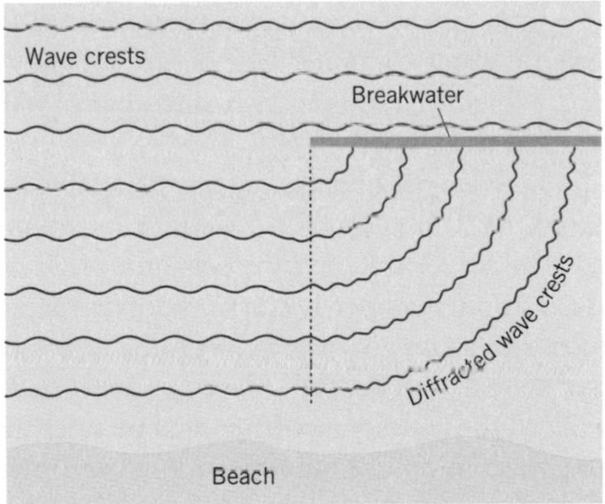

Wave crests
Breakwater
Diffracted wave crests
Beach

Fig. 6.16 Diagram showing how wave energy is distributed as the result of diffraction around a structure. The energy essentially spreads along the crests to extend behind barriers.

Fig. 6.17 Aerial photograph showing wave refraction and diffraction around a headland. Such a pattern results in a very complicated distribution of wave energy.

Fig. 6.18 Wave reflecting from a steep beach along the California coast. The amount of energy reflected depends on both the size of the wave and the steepness of the beach.

Waves can also have some or all of their energy reflected as the wave meets the beach (Fig. 6.18), a structure, or rock exposures. The amount of reflected energy and the direction in which the reflected energy is directed are dependent upon the amount of energy absorbed by the coast or structure, and the angle of approach. If a wave approaches and hits a vertical and impermeable structure without any interference by the bottom, 100% of its energy will be reflected. This would be the situation if a seawall or breakwater presented a solid vertical surface such as one of poured concrete. A vertical rock outcrop would present the same situation. A structure that is sloping and/or has an irregular surface will absorb some of the wave energy and will reflect only part of it. The same is true of a steep beach, where some of the wave energy is reflected and some dissipated or absorbed. Typically, only a small percentage of incident wave energy will be reflected at a beach.

The direction along which this reflected energy proceeds is in accordance with the law of **reflection**: the angle of reflection is equal to the angle of incidence. In other words, waves approaching a vertical sea wall act just like light rays on a mirror. A wave that approaches a sea wall at an angle of 45° will be reflected at the same angle (Fig. 6.19). Incident waves on a rocky coast may experience just this sort of condition.

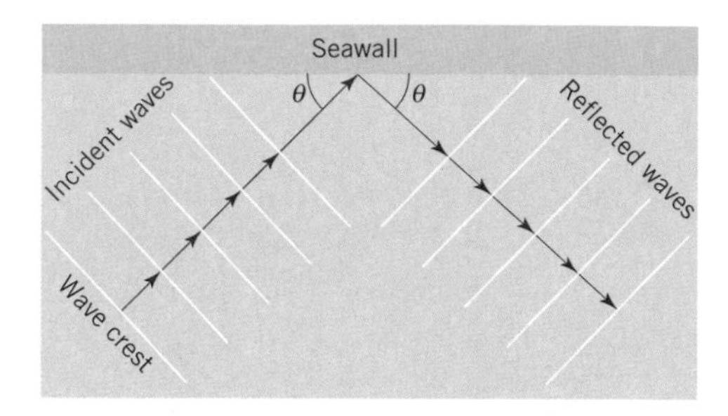

Fig. 6.19 Diagram showing how incident waves reflect from a seawall according to Snell's Law; that is, the angle of reflection is equal to the angle of incidence.

6.5 Other types of waves

6.5.1 Tsunamis

A tsunami is a large, rapidly moving wave that can be very destructive when it reaches the coast. The word tsunami is Japanese for very long harbor wave. This name has been applied to these waves because it is in harbors that much of the damage occurs. The name is typically applied to waves produced by seismic or other events that cause major movements on, or of, the ocean floor. Such events cause a disturbance of the sea surface, a tsunami, commonly referred to as a seismic sea wave. Earthquakes, volcanic eruptions, landslides, and possibly even strong, deepwater currents may trigger tsunamis. A good way to envision the generation of a tsunami is that the water container (the ocean floor) is disturbed, which in turn causes a disturbance of the water surface.

The wavelength of a tsunami is generally up to hundreds of kilometers, wave height may be about a meter in deep water, and the wave travels at hundreds of kilometers per hour. Because of the great wavelength, a tsunami will begin to be affected by the sea floor at great depth. This causes a very marked slowing of the velocity, accompanied by steepening of the wave, with the result being a wave height of several meters at the coast. Many catastrophic tsunamis have influenced various coastal areas of the world; for example, in Japan and many of the small islands of the Pacific, especially Hawaii. The

Fig. 6.20 Destruction of property along the coast of Central America caused by a tsunami. Current abilities to predict tsunamis enable people to avoid the huge waves, but property damage can be extensive. (Courtesy O. H. Pilkey.)

west coast of North America has also experienced large tsunami waves. Not only do these huge waves destroy property (Fig. 6.20) and erode the beaches, but several have caused a high loss of life. Among the recorded disasters of this type were the tsunamis of 1692 in Jamaica, 1755 in Portugal, 1896 in Japan, and 1946 in Hawaii. The eruption of Krakatoa in the southwest Pacific in 1883 produced a tsunami that carried a large ship 3 km inland to an elevation of 9 m above sea level. The 1964 earthquake in Alaska produced a tsunami that caused severe damage to Crescent City, California.

Much of the loss of life has been greatly reduced since the establishment of a network of seismic monitoring stations that covers the entire Pacific Ocean. This monitoring system was constructed in the 1950s and 1960s, and now makes it possible to predict the development, movement, landfall, and size of tsunamis. This does not make them less dangerous, nor does it do anything to prevent damage or loss of property. It only provides warnings of minutes or a few hours to permit rapid evacuation of coastal areas.

6.5.2 Standing waves

All of the previously discussed wave types are progressive gravity waves. The wave form moves forward and gravity is the restoring force. There are special conditions under which waves do not propagate but can influence the coast. Such waves are called standing waves. These waves are trapped in a container or a restricted body of water such as an embayment, a harbor, or a lake. The wavelength of a standing wave is equal to the diameter or length of the water body in which the wave develops. There is a node in the middle about which there is no motion, and an alternating up and down motion at each end of the water body (Fig. 6.21). The best example of a small-scale standing wave is the water motion that occurs when you walk with a cup of coffee or a pan of water.

Most common and most noticeable among the standing waves in the natural environment is the seiche. Seiches are most commonly caused by weather conditions that create rapid changes in barometric pressure and/or wind conditions. Water is piled up on one side of the basin and then is released as the wind stops quickly or when the barometric pressure rises quickly. As a consequence of these phenomena, there is an alternating extremely high and extremely low water level at the shore as the standing wave sloshes back and forth across the water body. The Great Lakes tend to experience the most dramatic seiches, with changes in lake level of over a meter with the passage of a strong frontal system. Seiche waves have swept people from the beach and drowned them in Lake Michigan. Lake Erie is the most susceptible due to its shallow depth and its orientation along the dominant wind direction as storms pass

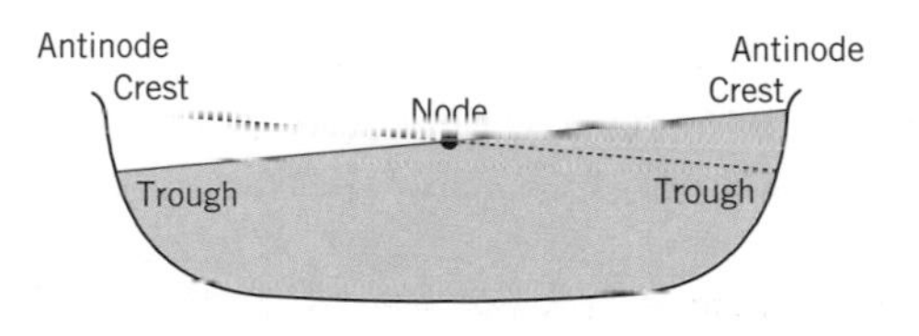

Fig. 6.21 Diagram of a standing wave showing a node at the midpoint of the vessel. This is what happens when you walk with a cup of coffee.

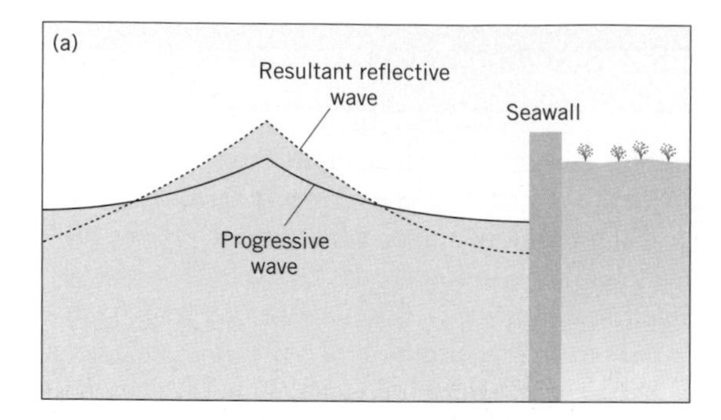

Fig. 6.22 (a) Diagram showing how waves reflect from a vertical obstruction and (b) photograph of the same phenomenon along the Georgia coast. A vertical and impermeable structure will reflect essentially all of the incident wave energy.

through this area. It has experienced seiches of over 2 m.

Wave reflection can also create a type of standing wave that is instantaneous in nature. As waves approach a sea wall, steep beach, or other obstacle with the wave crests parallel to the obstruction, the crest is reflected back and causes an instantaneous increase in the height of the next incoming wave (Fig. 6.22).

6.6 Wave-generated currents

There are three types of coastal currents that are produced by waves: (i) longshore currents; (ii) rip currents; and (iii) undertow. All three types of currents develop as the result of shoreward progression

Fig. 6.23 Photograph showing how waves approach the shoreline at an angle and produce a longshore current. As the wave refracts a current is produced in the direction of the open angle.

of waves, and all three can play a role in transporting sediment within the surf zone.

6.6.1 Longshore currents

There is a slow landward transport of water as gravity waves move landward and break in the surf zone. The rate at which this occurs is related to the refraction pattern of the waves as they move through the nearshore and surf zone. This refraction produces a current that flows essentially parallel to the shoreline: the longshore current, sometimes called the littoral current (Fig. 6.23).

As the wave refracts during steepening and breaking, there are both shoreward and shore-parallel vectors to its direction. This condition produces the longshore current with a velocity that is related to the size of the breaking wave and to its angle of approach to the shoreline. The longshore current is essentially confined to the surf zone and has an effective seaward boundary at the outermost breaker line and a landward boundary at the shoreline (Fig. 6.23). It acts much like a river channel, with the greatest velocity near the middle. Under some conditions, wind blowing along-shore will enhance the speed of the longshore current. These currents typically move at a few tens of centimeters per second, but during even modest storm conditions they commonly travel more than a meter per second.

Longshore currents work together with the waves to transport large volumes of sediment along the shoreline. As the waves interfere with the bottom when the waves steepen and break, large amounts of sediment are temporarily put into suspension. The sediment is most concentrated at the bottom of the water column but is present throughout. This suspended sediment and bed-load sediment is then transported along the shoreline by longshore currents.

Because waves may approach the coast in a wide range of directions depending upon the angle of wave approach, the longshore current may flow in either direction along the shoreline. As a result of this back and forth transport of sediment there might be a large amount of sediment flux over a designated period of time, but the net littoral transport will be in one direction and may be only a small portion of the total. We typically speak of **littoral transport** in terms of the annual amount in a particular length of coast. For example, it may be 100,000 $m^3 yr^{-1}$ from north to south. Remember, this is the net transport, and the total or gross amount may be many times that in each direction. Longshore currents and littoral sediment transport occur in any coastal environment where waves are refracted as they move into shallow water and approach the shoreline. Whereas large volumes of littoral drift take place on open ocean beaches, longshore currents and resulting sediment transport are also common in bays, estuaries, and lakes.

6.6.2 Rip currents

The previously mentioned landward transport of water as waves move to the shoreline produces an increase in water level at the shore called setup. This is largely the result of friction between the wind and the water surface as the progressive waves move through the surf zone. Setup is basically similar to storm surge but it is smaller in scale and is limited to a narrow zone along the shore. This elevation of the water is typically a few centimeters. The setup water is essentially piled up against the shoreline in an unstable condition because of the inclination of the water level. That situation cannot persist.

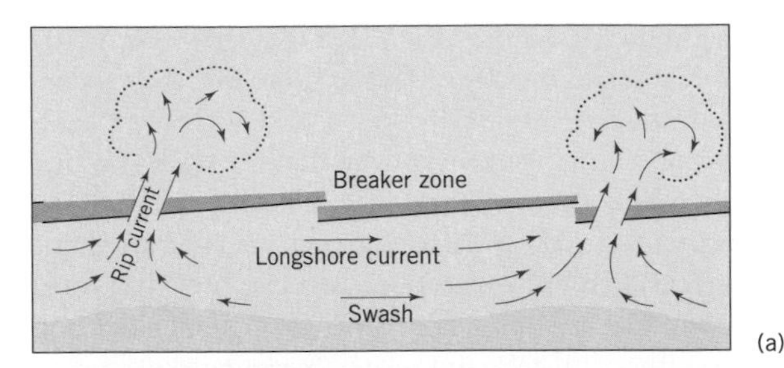

Fig. 6.24 (a) Diagram and (b) photograph showing the nature of circulation in a rip current system. These circulation systems are important in transporting sediment and may be dangerous to swimmers.

If this unstable condition exists along a barred coast or along some of the steeper coasts, the setup produces seaward flowing currents that are rather narrow and that create circulation cells (Fig. 6.24) within the surf zone. These narrow currents are called rip currents. They commonly flow at speeds of a few tens of centimeters per second and they can transport sediment. The rip currents are connected with feeder currents that are part of the longshore current system. The rip current is the phenomenon that is commonly and mistakenly called a "rip tide," even though it is not related to tides. Rip currents can be a danger to swimmers and cause several drownings each year at locations where they are common. In actuality they should not be a problem to swimmers because they are narrow and the swimmer simply needs to move a short distance along shore to escape their seaward path.

The size, orientation, and spacing of rip currents are related to wave conditions and nearshore bathymetry. In the absence of longshore sand bars and along a smooth, embayed shoreline, the rip current spacing is typically regular and fairly persistent. If incoming waves are essentially parallel to the shoreline they will produce a nearly symmetrical circulation cell with feeder currents that are symmetrical in each direction. Rip currents that form when waves approach at an angle generally develop asymmetrical cells and related feeder currents.

6.6.3 Undertow

Another nearshore circulation phenomenon that is caused by shoreward movement of water and setup is called undertow. Here the water that piles up at the shoreline is returned along the bottom in order to alleviate the unstable condition produced by the setup. This is a type of circulation that is commonly confused with rip currents and that may also cause problems for swimmers. Water that is transported toward the shoreline by forward motion of water as waves progress landward and from friction between wind and water is returned seaward in a strong and essentially continuous current (Fig. 6.25). Flow is commonly up to 50 cm s^{-1} during strong onshore waves, the only time when undertow is noticeable.

Undertow is not only limited to rather high-energy wave conditions in the surf zone but is also common in places where longshore bars are absent or far offshore and relatively deep. In some areas rip currents will persist during low- to moderate-energy wave conditions and then during high-energy wave conditions the rips will lose their definition and undertow will be the primary mode of seaward return of water from unstable conditions of setup.

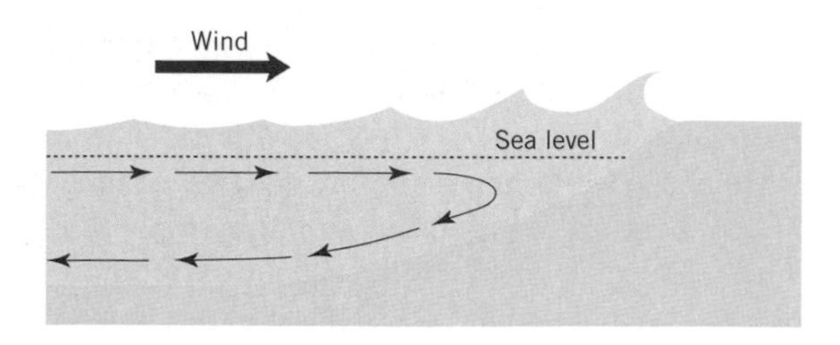

Fig. 6.25 Diagram of a section of the surf zone showing the circulation associated with undertow. This occurs when water is transported landward by wind and then returns seaward via a strong current along the bottom of the water column.

6.7 Summary

Waves are probably the most important factor in coastal development. They move sediment directly or through the generation of wave-driven currents. Wave activity can cause erosion and can also transport sediment to the coast. Although there are still many aspects of wave–sediment interaction that we do not understand, the basic characteristics of wave mechanics are well known. Part of the difficulty is that we are dealing with the relationships between fluid motion and a solid substrate. Another difficulty is that much of the interaction between waves and the coast takes place during severe conditions of storms. Making observations under such conditions is extremely difficult although great progress is being made.

Waves are primarily at work on the open coast and have only minor influence on more protected environments such as estuaries, tidal flats, and wetlands.

Suggested reading

Bascom, W. (1980) *Waves and Beaches*. Garden City, NY: Anchor Books.
CERC (1984) *Shore Protection Manual*, two volumes. Vicksburg, MS: Coastal Engineering Research Center, Waterway Experiment Station, Corps of Engineers.
Dean, R. G. & Dalrymple, R. A. (1998) *Water Wave Mechanics for Engineers and Scientists*, 2nd edn. Englewood Cliffs, NJ: Prentice Hall.
Weigel, R. L. (1964) *Oceanographical Engineering*. Englewood Cliffs, NJ: Prentice Hall.

7 Beach and nearshore environment

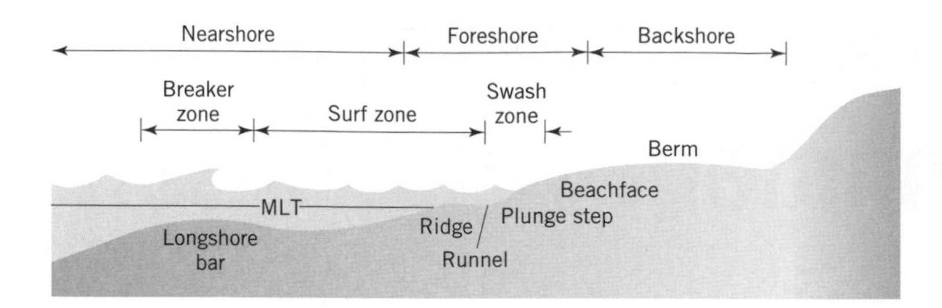

Fig. 7.1 Schematic diagram of a beach and nearshore profile showing specific sub-environments in profile view. There are variations on these themes depending on the nearshore slope, the tidal range, and the abundance of sediment.

7.1 Introduction

Most people visit or live along the coast to enjoy the beach. The constant change, interesting landscape, fascinating creatures, and beautiful scenery have captured the attention of people forever. The beach and related environments act as the seaward protection for the coast, whether it be a barrier island or the mainland. Coastal residents are familiar with many aspects of beach dynamics, especially the results of severe storms, and the seasonal or long-term changes. This chapter considers beaches in terms of the material that comprises them, their morphlogy, and their dynamics.

A proper understanding of beaches must include the adjacent shallow marine waters, generally called the **nearshore environment**. It includes the region that extends from the low tide line, which is the seaward extent of the beach, out across the surf zone, and includes the sand bars that are common along most coasts (Fig. 7.1). The width and depth to which the nearshore extends range widely depending upon its gradient, the wave climate, and the availability of sediment. In many places there are rather persistent sand bars that parallel the beach and over which waves break during storms (Fig. 7.2). Although some nearshore areas may be essentially smooth and gently sloping, most are comprised of a combination of these sand bars and intervening troughs.

Fig. 7.2 Waves breaking over a longshore bar in the nearshore environment caused by the shallowing of depth and the resulting steepening to the breaking point.

7.2 Beach and nearshore morphology

The beach itself extends from the low tide line landward across the generally unvegetated sediment to the next geomorphic feature in the landward direction, which may be a dune (Fig. 7.3), a cliff (Fig. 7.4), or nowadays a seawall (Fig. 7.5) or some other facet of human development. The beach is the most actively changing part of the coast; each wave causes sediment movement. The nearshore environment, where the beach ends, begins at the low tide line and extends to the outer limit of the bar and trough topography that typifies this innermost subtidal environment. This includes the surf zone under most conditions, although severe storms may develop breaking waves beyond the nearshore bars.

7.2.1 Nearshore

The overall profile of the beach and adjacent nearshore may tend to be steep or gently sloping. This depends upon a variety of factors, such as sediment

Fig. 7.3 Photograph showing where the backbeach merges with the dune environment, a rather modest change in the beach profile. This profile change varies with deposition and erosion. After a storm erodes the face of the dune, a near vertical slope will be present.

Fig. 7.4 Photograph showing where the backbeach merges with a steep bluff, an abrupt change in the beach profile. Such bluffs tend to be fairly permanent because of the cohesive sediment or rock that forms the bluff.

supply, wave climate, overall slope of the inner shelf, and perhaps tidal range. The steep beach and nearshore generally has poorly developed nearshore sand bars or none at all. Little energy is lost as the waves move across the nearshore and break for the first time at or near the shoreline. The general

Fig. 7.5 Photograph showing where the natural backbeach is interrupted by a seawall, an anthropogenic change in the beach profile. Commonly the beach will be removed in front of the seawall as erosion continues.

gradient and character of the profile across the nearshore and beach determines the way that waves behave as they move across this zone. These features also determine how the wave energy is dissipated across this zone and how much energy reaches the beach itself.

The most common morphology along the nearshore is the presence of one or two shore-parallel sandbars. The innermost bar is typically about 30–50 m from the shoreline and the outer one is at least twice that distance from the shoreline. These bars display a predictable pattern of sediment texture and sedimentary structures. The finest and best sorted sediment is on the seaward side of the bar crest where wave energy is highest, and the coarsest and worst sorted sediment is in the trough, which receives the lowest level of wave energy. Ripples form on the outer part of the bar and these give way to a plane bed surface on the upper slope and crest of the bar (Fig. 7.6).

The inner bar in this nearshore complex commonly includes rip channels where rip currents develop. The crest of longshore bars is rarely at exactly the same elevation. The low places, or saddles as they are commonly called, serve as a pathway of least resistance for water that is piled up between the bar and the shoreline. The rip currents (see Fig. 6.24) represent the primary circulation system in the nearshore zone.

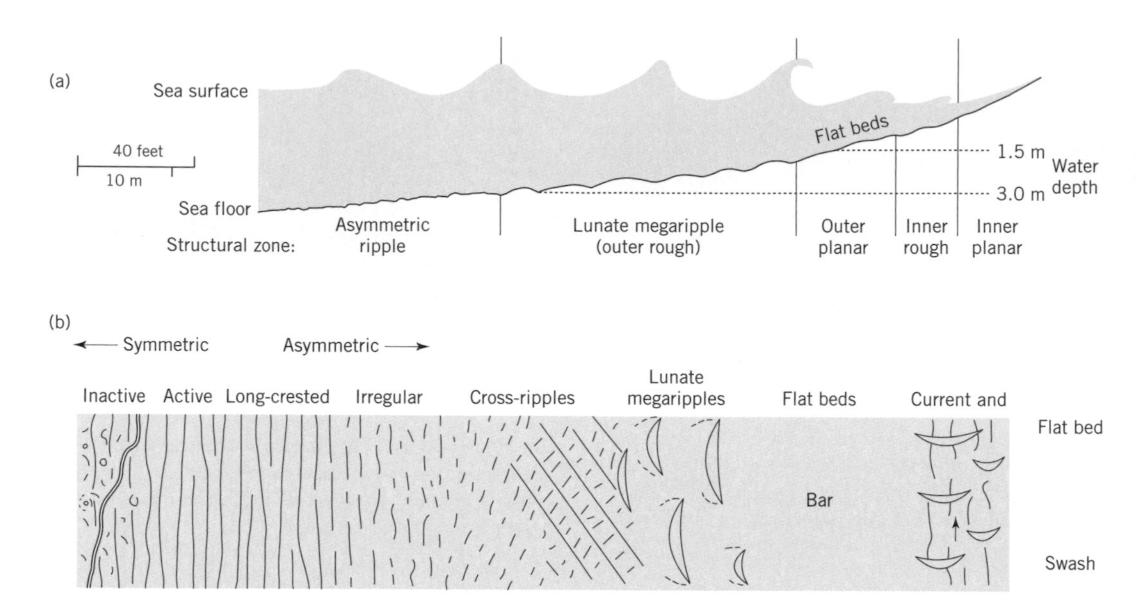

Fig. 7.6 (a) Profile and (b) plan view of the nearshore showing how surface features are distributed across the bar and trough topography. These features are the result of waves and wave-generated currents. (Modified from N. E. Clifton, R. E. Hunter & R. L. Philips, 1971, Depositional structures and processes in the non-barred high-energy nearshore. *J. Sed. Petrol.*, 41, 651–70.)

7.2.2 Beach

The typical sandy beach displays multiple profile configurations, of which two or three are common. These profile shapes are dependent upon the presence and configuration of the major elements that comprise the beach: the ridge and runnel if present, the foreshore, the backshore, and the storm ridge.

Ridge and runnel

The most seaward part of the beach is the ridge and runnel (Fig. 7.7), which is somewhat like a small-scale bar and trough. This lower intertidal part of the beach has a small bar called the ridge and a flat-bottomed trough called the runnel (Fig. 7.8). The ridge is initially symmetrical and then becomes asymmetrical as wave-generated currents move over its crest. As the tide rises, waves generate currents that cross the upper surface and transport sediment landward under high-energy conditions. This sediment reaches the landward edge of the ridge and the grains tumble down the steep landward side of it (Fig. 7.8). The seaward side is wide and gently sloping, and the height of the ridge is typically a few decimeters.

Fig. 7.7 Photograph of a large ridge and runnel topography in the seaward part of the intertidal beach along the Oregon coast. This is a coast with a large tidal range; hence the large and multiple ridges extending about 250 m from high tide. (Photograph by W. T. Fox.)

The runnel is broad and nearly flat. It is typically from 10 m or so to a few meters wide, and is a place where waves form small ripples (Fig. 7.9) at high tide when water covers this part of the lower beach. It is a fairly protected area where worms commonly burrow into the sediment, filter fine suspended particles from the water, and produce small fecal pellets

Fig. 7.8 Ridge and runnel where the ridge has migrated toward the beach. This process generally takes a few weeks but the rate of migration depends on the size of the ridge and the tidal range.

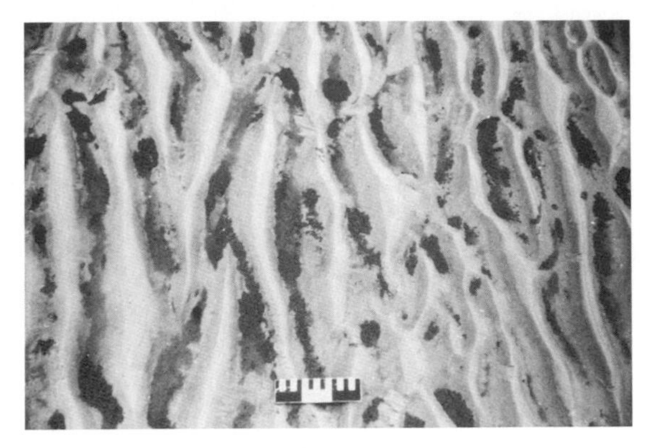

Fig. 7.10 Fecal pellets that have accumulated in the troughs of ripples in the runnel. These are commonly produced by worms and burrowing shrimp that live in the intertidal zone.

Fig. 7.9 Photograph of the leading edge of a ridge showing the steep, landward face. As sand is carried landward across the ridge surface it reaches the edge and rolls down the steep slope, causing the entire ridge to migrate landward.

that accumulate in the troughs of these ripples (Fig. 7.10). As the ridge migrates landward, it moves over the runnel. The rate of migration is commonly inversely related to the tidal range. In high tidal range situations the wave-generated currents move over the top of the ridge for only a few hours during each tidal cycle, whereas in areas of low tidal range it would be several hours. Eventually the ridge makes its way onto the beach and is reworked by wave swash into the beachface.

The ridge and runnel is not always present and, when it is, it only lasts for a period of weeks. These features are the result of abundant sediment in the shallow subtidal area adjacent to the beach. Formation of the ridge and runnel may result from low-energy waves building a shallow sand bar that then migrates shoreward by the same waves and their wave-generated currents. It also might result from waning storm conditions that erode the beach and make sediment available for the development of a shallow subtidal sand bar under low-energy wave conditions.

Foreshore

The foreshore part of the beach is essentially flat and slopes toward the sea (Fig. 7.11). In beach profiles that do not have a ridge and runnel, the foreshore includes the intertidal portion of the beach and extends to the landward change in slope. The foreshore, also called the forebeach or beach face, includes the swash zone, the part of the foreshore over which the waves uprush and backwash as each one meets the shore. The width and the slope of the swash zone change as the wave climate changes. Intense wave activity can also modify the foreshore dramatically, as is discussed in a later part of this chapter.

The base of the foreshore is typically marked by the plunge step, a small topographic break that is typically the site of the coarsest sediment on the foreshore. Most commonly this sediment is gravel, either shells or terrigenous lithic fragments. When

Fig. 7.11 Photograph showing the seaward sloping foreshore portion of a beach. It is on this part of the beach that the last breaking wave moves up and back, carrying sediment. (Photograph by L. Somers.)

Fig. 7.12 Opportunistic vegetation that rapidly colonizes the backbeach and helps to stabilize it. These plants require little moisture and few nutrients to grow and spread across the dry beach.

waves are large and the surge up the foreshore is great, this step tends to be smoothed out and the coarse particles spread over the lower part of the foreshore.

Another aspect of the foreshore texture is associated with the uprush and backwash of final breaking waves. On sandy beaches there is a strong uprush and most of the water returns with considerable energy. By contrast, on gravel beaches there is a strong uprush but the backwash is essentially absent because the water percolates into the coarse sediments.

Backshore

The backshore or backbeach extends from the berm crest, at the change in slope with the foreshore, landward across the remainder of the beach to the next physiographic feature. The backshore is dry and supratidal except during storm surge conditions when high water and large waves push water and sediment over the upper part of the beach. During these events, overwash of the beach and some wave activity may occur directly over the backshore. Most of the time this part of the beach is subjected to wind.

This portion of the beach is generally almost horizontal or slightly landward sloping. The backshore also includes the berm or, in some cases, multiple berms. These are the flat upper surfaces of the step-like features that may occupy the backshore. In some beaches the berm and the backbeach are the same; multiple berms are absent. The width of the backshore varies greatly in both space and time due to wave energy and sediment supply. The landward part of the backshore may support various types of opportunistic vegetation that help to stabilize the environment (Fig. 7.12). After major storms and the resulting erosion the backshore may be greatly reduced or even absent, especially along coasts where dunes or other factors prevent washover. In these circumstances, large storm waves scour the beach and carry the sediment seaward.

The surface of the unvegetated backbeach is typically covered with small ripples formed by wind and with a surface layer of shells. This shelly surface tends to become a stabilizing feature of the backbeach, in that the dry sand grains are blown away by the wind, generally to the dune areas, while the shells become more and more concentrated because they cannot be carried by the wind.

Storm ridge

The backbeach takes on a very different appearance if gravel is the dominant sediment present. Gravel beaches of shell and/or rock fragments commonly include a storm ridge that is just landward of the foreshore but that may, in some locations, also include an intertidal component. This feature can rise several meters above high tide and can totally replace the typical backbeach (Fig. 7.13). Its composition

Box 7.1 Daytona Beach, Florida: the fastest beach in the world

Nowadays Daytona Beach brings to mind thousands of college students on spring break or perhaps bike week, when thousands of motorcycles congregate for an annual rally. A hundred years ago, Ormond Beach, just north of Daytona, was a place where Stanley Steamers were run along the flat and hard-packed sand beach. It was called the speed capital of the world. The first race, in 1902, was between an Olds Pirate and a Winton Bullet No. 1, with the Bullet winning the mile race at a speed of 68.2 m.p.h. By 1919 the speed had increased to 149 m.p.h., achieved by a Mercedes. Indianapolis winners and other notables were racing on this coastal venue.

In 1928 a twin-engineer vehicle called a Slug claimed the first record at Daytona, with a speed of 203.8 m.p.h., which established this location as the new speed capital of the world. Bigger race cars with bigger engines kept coming to Daytona over the next several years to attempt to break the speed records. The record kept creeping up and up, until in March 1935 Bluebird V was clocked at 276.7 m.p.h. The main force behind the records and the development of these cars was Malcolm Campbell, an Englishman. His goal was to exceed the 300 m.p.h. barrier, but he never accomplished it at Daytona. In September 1935 he finally did it on the Bonneville salt flats of Utah. This accomplishment brought an end to the chase for speed records on the beach.

The next year Alexander E. Ulmann, a racing enthusiast, organized a race over a 3.2 mile oval that included a long straightway on the beach. The race was a financial bust and officials elected to halt beach racing. William France, a friend and colleague of Ulmann, organized a successful race at Daytona in 1938 and followed it up with a better one the next year. Then came the Second World War and racing stopped until 1946, when it resumed at Daytona. But now stock car racing had come about and was gaining popularity. France became involved in it and eventually helped to organize stock car racing. We can say that the beach at Daytona gave rise to what is rapidly becoming the biggest spectator sport in the USA – NASCAR.

Fig. 7.13 Beach profile showing high backbeach gravel ridge formed by storms. These may rise several meters above high tide. They form because the large storm waves carry large particles landward and then water percolates through the gravel, so there is no mechanism for the particles to return seaward.

Fig. 7.14 Storm ridge of mostly oyster shells that has formed in the backbeach along the western Louisiana coast. Storm ridges form from whatever coarse material is present.

depends upon the nature of the gravel material in the immediate area, and its size is typically directly related to the rigor of the storms that produce it. Shell ridges, commonly composed of oyster debris, occur along some of the Gulf of Mexico beaches, where they rise a meter or so above mean sea level (Fig. 7.14). By contrast, some of the terrigenous gravel beaches of the northeastern United States or adjacent Canada may rise several meters above sea level.

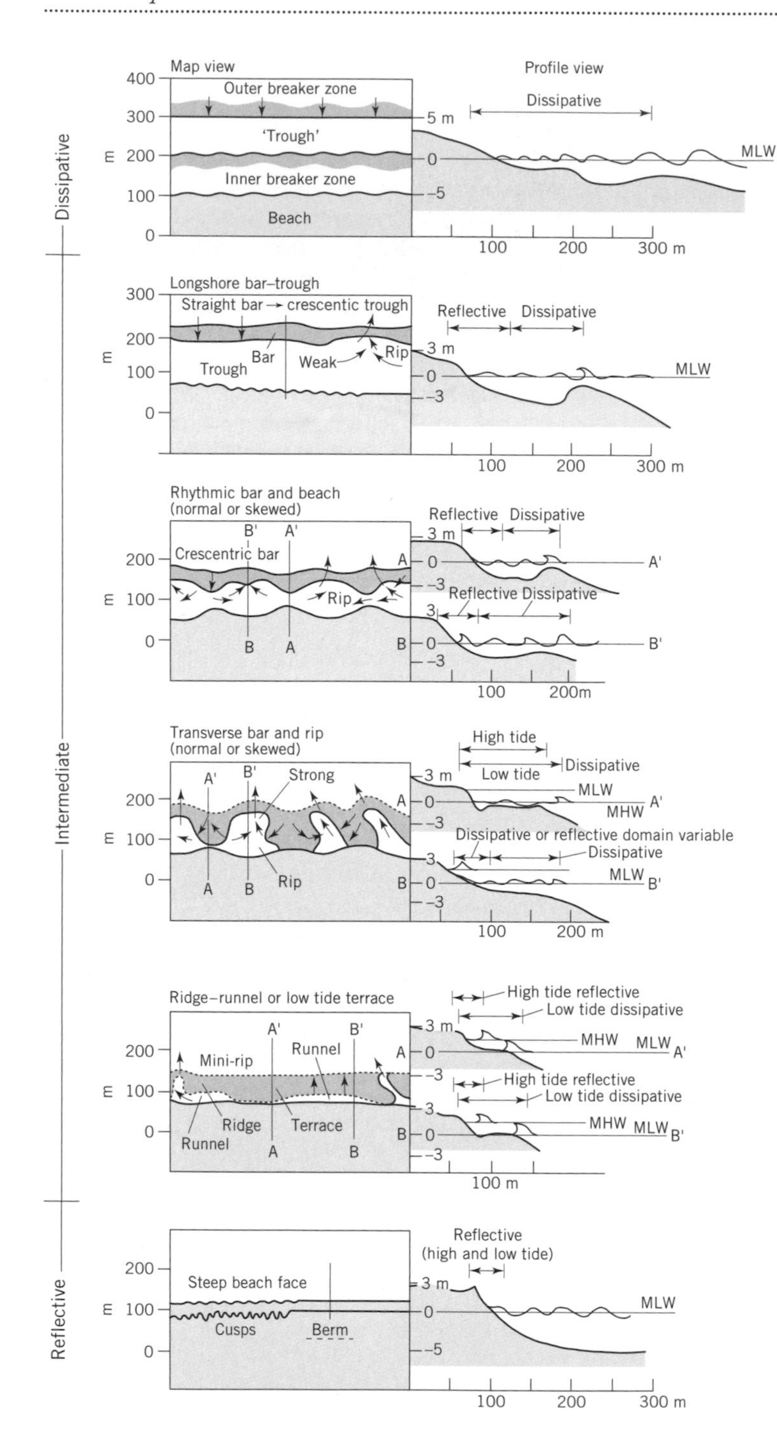

Fig. 7.15 Diagram showing various stages of beach and nearshore configurations, from highly reflective to quite dissipative profiles. This model is actually a continuum and is shown here in various stages. (From L. D. Wright, J. Chappell, B. G. Thom, M. P. Bradshaw & P. Cowell, 1979, Morphodynamics of reflective and dissipative beach and inshore systems; southeastern Australia. *Marine Geology*, **32**, 122.)

These ridges form as the result of great energy in the uprush of wave swash on the beach. This energetic uprush carries the gravel to the backbeach and piles it into a ridge. Because there is virtually no backwash due to the percolation of all of the water into the gravel, this ridge becomes large and stable in the back part of the beach.

7.3 Beach types

The overall gradient of the beach and nearshore is important in influencing both the amount of wave energy that reaches the beach and the configuration of the beach itself. Some are steep and others are not. An extensive analysis of these characteristics on Australian beaches has led to an important classification of beaches that is now applied worldwide. There are two end members to this classification: dissipative beaches and reflective beaches. A dissipative beach is one that has a gentle gradient in the shallow subtidal and intertidal areas where wave energy is gradually dissipated across this zone (Fig. 7.15). At the other end of the spectrum are reflective beaches that have relatively steep gradients, and where a significant amount of wave energy is actually reflected back toward open water (Fig. 7.15).

Dissipative beaches tend to have multiple sand bars that cause waves to break and wave energy to be dissipated as the waves progress to the shoreline. There are intermediate beaches that also have nearshore bars. These bars may have low areas that permit the development of rip channels and rip currents (Fig. 7.15). Reflective beaches generally do not have bars and have gradients that are so steep that little wave energy is lost as waves move across the nearshore zone to the beach. It is common for reflective beaches to display erosional conditions, whereas dissipative beaches are more commonly accreting sediment. The determining factor is the amount of wave energy that reaches the beach.

7.4 Beach materials

Beaches are composed of anything that can be transported by waves. Whatever is available will be incorporated as long as it can be carried to the beach and will be at least temporarily stable there. Mud is uncommon because the wave energy that characterizes beaches is typically too high to permit such fine sediment to accumulate and become stable there. It does accumulate in large quantities on some beaches; for example, along the coast of Surinam on the northern coast of South America (Fig. 7.16), where huge quantities of fine sediment are made available by the Amazon River. Wave energy is low and mud is the dominant sediment in the area; thus it accumulates along the shoreline.

Fig. 7.16 Photograph showing beaches comprised mostly of mud along the Surinam coast. Because beaches form from whatever sediment is available along a coast, there is a wide range in their texture. (Photograph courtesy of R. N. Ginsburg.)

Gravel particles of virtually all sizes and compositions are fairly common on beaches (Fig. 7.17) given the correct set of circumstances: a readily available source of gravel-sized particles, and generally a high wave climate or at least a location where severe storms occur. These beaches range from small pocket beaches between headlands on an otherwise rocky coast to beaches that extend for many kilometers. Those in the northern latitudes are typically composed of rock fragments from glacial deposits, stream accumulations, or directly eroded from bedrock. The gravel beaches that are associated with barrier islands are composed of shell gravel (Fig. 7.18), which is especially abundant in low latitudes where there is little other sediment available. Beaches on Pacific islands are commonly composed of a

Fig. 7.17 Photograph of gravel beach on the north coast of Alaska, where entire barrier islands are dominated by gravel. (Courtesy of E. Reimnitz.)

Fig. 7.18 Photograph of excavation in a beach dominated by shell debris.

Fig. 7.19 Photograph of a beach on Tahiti showing dark volcanic gravel.

combination of dark volcanic rock fragments (Fig. 7.19) and coral reef debris, the two primary types of material available.

When we think of beaches, we typically are thinking of sand beaches. This is the most common sediment size throughout the world, but the composition may extend over a wide range. The sand on beaches is commonly moved by the wind and wave and current action, but shells may represent a second, coarser population. The sand component is typically reworked from earlier accumulations within the coastal zone. In some areas, this deposition and reworking of older deposits may have taken place in many cycles over millions of years. As this reworking takes place there is both physical and chemical deterioration of the sediment particles. Eventually, it is common for the beach sediment to be dominated by quartz because it is very durable physically and is chemically inert at surface conditions.

The particle size of beach sediment ranges from mud to large boulders. The shapes vary depending upon their original composition and origin, and also upon the time and energy to which they have been subjected in the beach and nearshore environment. That is, some sediment particles like shells will always be flat or curved; they cannot become spherical like typical beach sand grains. Those particles that have been subjected to waves and currents for a long time tend to be smooth, whereas those particles "new" to the beach may be angular.

The interaction of waves and currents with beach sediments produces certain textural characteristics. It is possible to generalize about these characteristics, especially for sandy beaches. As a consequence of wave activity, a given section of beach displays a narrow range of particle size (Fig. 7.20). This is called good sorting. That is, the waves and their related processes arrange the sediment particles in such a fashion that, at any spot, one would expect to find sand grains that are nearly the same size. The same is true for gravel beaches: they are also well sorted even though the absolute particle size is much larger than on sand beaches.

Fig. 7.20 Beach materials of nearly all the same size displaying good sorting.

7.5 Beach dynamics

The changes on beaches may take place in only the time between individual waves striking the shore, or they may extend over decades, as shown by long-term trends in beach erosion or accretion. Likewise, the rate of change covers a very broad spectrum, up to the extreme erosional conditions caused by severe extratropical storms and hurricanes. Regardless of the rate or scope of the changes that take place along beaches, there is some general level of predictability that is associated with each level of change. There are various cycles that beaches experience, including those related to tides and to seasons of the year. The following discussion centers on the various levels of change, how they affect the beach, and what causes them. All these cycles tend to be present on beaches regardless of where they occur: mainland, barrier island, rocky coast, or wave-dominated delta.

7.5.1 Beach processes

Waves, tides, and the currents that they generate may influence the sediment and the morphology of beaches. Wind is also a factor, but it is limited primarily to the dry backbeach area. The breaking of waves produces obvious interaction with beach sediment: each crash of the wave places sediment in temporary suspension with the amount directly related to the size of the wave. This suspended sediment is then moved by currents, primarily wave-generated longshore currents (see Chapter 6).

The role of tides and tidal currents on beaches is subtle. There is no significant tidal current flowing in the nearshore zone. Indirectly, tides can influence wave activity as the tides produce a rise and fall in the water level. This causes the shoreline to move landward and seaward, which in turn affects how waves influence the underlying sediment. At a given location, there will be less wave energy transmitted to the sediment substrate during high tide than during low tide. The greater the tidal range, the wider the zone of change in incident wave energy.

Only near the mouth of tidal inlets are tidal currents important in moving sediment in the beach and nearshore area. The tidal currents are generally not strong enough to move sediment except with the assistance of waves. The waves suspend the sediment and the tidal currents carry it toward the inlet. These currents increase as one approaches the inlet mouth and their speed is directly proportional to the tidal prism, the amount of water that enters the inlet during the flood tidal cycle. These tidal currents adjacent to inlets are important only during the flood portion of the tidal cycle, when significant currents move laterally along the beach and nearshore toward the inlet mouth. During the ebb portion of the cycle the currents are very slow to nonexistent because ebb currents tend to exit the inlet channel in a jet-like configuration out into the open water.

Currents produced by waves are among the most important processes that generate change in the beach. They include: (i) combined flow currents; (ii) longshore currents; (iii) rip currents; and (iv) the onshore–offshore currents produced in the swash zone. All occur in the surf zone and adjacent beach, and all are responsible or partially responsible for transport of sediment.

Another important mechanism for sediment transport in this environment is in the swash zone, where the uprush and backwash of the final breaking wave carry sediment across the foreshore. Depending upon wave conditions, the slope of the foreshore, and sediment permeability, there may be a significant difference between the amount carried up the beach and that carried off the beach. Typically the grain size decreases up the foreshore as the result of waning energy in the uprush from the breaking wave.

7.5.2 Beach cycles

The general conditions of beach processes and their interaction with beach sediment can be considered in two distinctly different but common scenarios. The most prevalent condition is one where wave conditions are about equal to or less than the average energy conditions. At most beaches this includes a swell wave with a low wave height (generally <1 m) and a period of 8–12 s. Locally generated sea waves may be superimposed with a similar or smaller height and a period of 3–6 s. The sum of these conditions produces what is commonly called an accretionary beach (Fig. 7.21) because the dominant condition is one of deposition of sediment or stability; erosion is absent or quite limited. The profile assumed by an accretional beach is one where the backbeach is well developed and wide, with a relatively narrow foreshore. Nearshore sand bars display a relatively high amount of relief and are well formed. This is a common configuration on most low wave climate coasts, such as the Gulf of Mexico and the Atlantic coasts of the USA.

The other condition is one of an erosive or storm beach (Fig. 7.22). Storms, although short in duration,

Fig. 7.21 Beach showing accretion in the form of a wide backbeach and a wide foreshore.

Fig. 7.22 Storm beach showing a narrow and steeply sloping profile.

are the dominant process along many coasts. During such conditions there is typically an increase in wind wave size, resulting in their dominance. These relatively steep waves result in considerable entrainment of sediment through combined flow motion and longshore sediment transport. Large quantities of sediment are in suspension and the currents readily carry this sediment both offshore and along-shore. Additionally, the swash energy is high and uprush and backwash are more extensive due to the large waves.

The result is a removal of sediment from the beach, producing an erosional or storm profile. Such a profile is generally flat and featureless and has a narrow or nonexistent backbeach, with most or all of the beach being in the foreshore portion. The post-storm beach typically contains a ridge and runnel (Fig. 7.23) in the lower intertidal beach. A storm will cause the relief between the nearshore sand bars and related troughs to be reduced and the bars will move offshore up to a few tens of meters. The beach itself commonly is covered with a veneer of heavy minerals that accumulate as a lag deposit as the result of concentration during removal of the light fraction of the sand (Fig. 7.24). This is really a placer deposit in much the same fashion that gold is concentrated in streams.

The storm beach is a temporary condition and in the absence of successive storms the recovery period

Fig. 7.23 Oblique aerial photograph of ridge and runnel configuration and the adjacent beach after a storm, Hurricane Elena, in 1985 on the Gulf coast of Florida.

Fig. 7.24 Layer of heavy minerals concentrated on the storm beach surface after a storm.

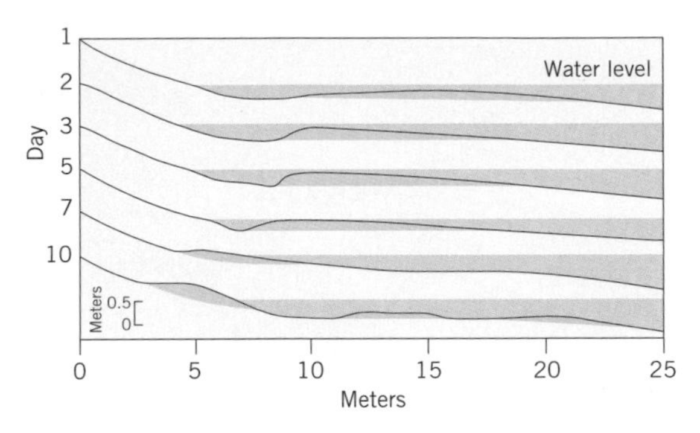

Fig. 7.25 Diagram showing the sequence of shoreward migration of a ridge as it makes its way to the beach.

begins immediately after the storm ends. This recovery process is initiated by the return of low-energy wave conditions and an absence of strong longshore currents. As swell and small wind waves persist there is significant landward transport of sediment. The nearshore sand bars return to their original position and to their original configuration. The ridge assumes a strongly asymmetrical profile (Fig. 7.25) and migrates landward as the result of washover during a flooding tide. Over a period that ranges from as little as a week to as much as three months, the ridge repairs the beach, with the result being an accretional profile that resembles the pre-storm situation. There may be some net loss or even some net gain depending upon how much sediment was lost or gained through longshore transport during the storm.

At some places where beaches are strongly influenced by waves that approach parallel to the coast, rip currents are strong during high-energy conditions and longshore transport is minimal. This condition is common along the Pacific coast of the United States and also occurs in northern New England.

The beach recovery process can be interrupted by storms that return the beach to the condition of an erosional beach. Because these storms are frequent during the winter season, the beach does not recover and assumes what is commonly called a winter beach profile, essentially the same as a storm profile. If storms are frequent and severe, such as is common along the west coast of the USA, the entire beach can be removed, leaving a bedrock bench, essentially devoid of sand (Fig. 7.26a). This condition may persist for a few months until the lower wave energy conditions of the spring and summer return sediment to the beach (Fig. 7.26b) and eventually build an accretionary profile. Good examples of this situation can be found along the central Oregon coast and also in southern California.

This sequence of beach conditions is distinctly seasonal and predictable. It produces the most obvious cyclic sequence in the dynamics of the beach environment. Other less pronounced cycles are associated with the diurnal tidal cycles and also with monthly lunar cycles. In each there is slight

(a)

(b)

Fig. 7.26 (a) Bare rock surface along the shoreline during the winter when storm waves have stripped all the sand away; (b) accretionary beach condition at the same location. The sand is returning and the beach is beginning to display its summer condition. (Photographs by W. T. Fox.)

change in the foreshore beach. As the tide rises and falls, the waves are permitted to impact on different parts of the foreshore. Likewise, during the lunar cycle, the neap tidal range produces a narrow range of wave impact on the swash zone as compared to the spring tidal range.

Sea level is another factor in beach conditions and is one of the primary coastal concerns over the next century. As discussed in Chapter 4, there has been a global increase in the rate of sea-level rise and there is much evidence to indicate that the rate will continue to increase. Global warming is the primary reason for this increase in sea level as a result of melting glaciers and expanding ocean volume due to rising temperatures.

This increasing sea level causes beaches to behave similarly to the way they do during storm conditions or during the winter season. Erosion is typical and there is little recovery or recovery is temporary. These conditions are produced by the combination of rising sea level with limited sediment supply. Statistics show that, depending on the source, from 60 to 80% of the sandy beaches in the United States are experiencing significant erosion. The addition of human interference in the form of various construction and development practices has added to the erosion problem.

7.6 Summary

The beach and adjacent nearshore environment display a wide range of morphologies, depending upon the overall geologic setting, wave conditions, and sediment available. Beaches are composed of all sizes, shapes, and compositions of sediment, not just the light-colored quartz sand that comes to mind when the beach is mentioned. The typical situation is that the beach reflects the type of sediment that is prominent in the area where the beach is located.

Steep beaches that lack well developed longshore bars are generally reflective and those that have gentle gradients with pronounced longshore bars are dissipative. These profiles reflect the interaction between wave energy and beach morphology.

Beaches also display cycles ranging from slight changes during diurnal tidal cycles and monthly lunar cycles to pronounced changes in storm/nonstorm cycles and seasonal cycles. The current extensive and chronic erosion of beaches has been attributable to global warming and sea-level rise. Whether or not this is part of a long-term cycle or whether it is part of a change in conditions is not known.

Suggested reading

Bird, E. C. F. (1996) *Coastal Management*. Chichester, John Wiley & Sons.

King, C. A. M. (1972) *Beaches and Coasts*, 2nd edn. London: St Martin's Press.

Komar, P. D. (1998) *Beach Processes and Sedimentation*, 2nd edn. Englewood Cliffs, NJ: Prentice Hall.

Short, A. D. (ed.) (1999) *Handbook of Beach and Nearshore Morphodynamics*. Sydney: John Wiley & Sons.

8 Barrier systems

8.1 Introduction

Barriers are the sites of some of the world's most beautiful beaches. They form due to the combined action of wind, waves, and longshore currents, whereby thin strips of land are built a few to several tens of meters above sea level (Fig. 8.1). They are called barriers because they protect the mainland coast from the forces of the sea, particularly during storms. They lessen the effects of storm waves, heightened tides, and salt spray. The bays, lagoons, marshes, and tidal creeks that form behind barriers provide harborages for pleasure boats and commercial craft, nursery grounds for fish and shellfish, and important sources of nutrients for coastal waters. Barriers occur throughout the world's coastlines but are most common along passive margins where sediment supplies are abundant and wave and tidal energy are conducive for onshore sand accumulation.

Barriers represent some of the most expensive real estate in many countries due to their development by the resort industry and because many people wish to live next to the sea. The lure of sandy beaches, salt air, water sports, and beautiful seascapes has always drawn vacationers and people wanting to erect summer cottages and permanent homes on barriers. Building lots on Figure Eight Island along the southern coast of North Carolina cost a million dollars and more. Some of the qualities that make barriers so attractive also make them vulnerable to storms, rising sea level, and erosion. Their sandy composition, general low elevation, and grass and shrubbery cover allow waves and storm surges to overrun barriers during the passage of intense hurricanes and major extratropical storms (northeasters). The greatest natural disaster in terms of lives lost in the United States occurred in 1900 when a hurricane swept over Galveston Island off the Texas coast near Houston, killing over 6000 people.

Fig. 8.1 Barriers may form in a variety of coastal settings if the supply of sediment is adequate and wave and tidal conditions are conducive to sand accumulation onshore. (a) Barrier system pinned to bedrock islands and promontories along the Rio de Janeiro coast in Brazil; (b) barrier chain along the New Jersey coast.

In this chapter we explore the different types of barriers, the factors controlling their worldwide distribution, and how they become modified by coastal processes. Depending upon their evolution, barriers may have a variety of architectures. We discuss the different components of barriers, what they look like in cross section, and the numerous theories that have been put forth to explain their formation. One of the major forces impacting barriers today, as well as in the past, is rising sea level. The response of barriers to this inundation and the dynamic processes that allow some of them to migrate onshore are also considered.

8.2 Physical description

Barriers are wave-built accumulations of sediment that accrete vertically due to wave action and wind processes. Most are linear features that tend to parallel the coast, generally occurring in groups or chains. Isolated barriers are common along glaciated coasts such as in northern New England and eastern Canada, and along high relief coasts such as those associated with collision coasts. Barriers are separated from the mainland by a region termed the backbarrier, consisting of tidal flats, shallow bays, lagoons, and/or marsh systems. An exception to this characterization occurs in instances where successive beach ridges have produced an outbuilding of the coast, such as along the flanks of a river delta.

Barriers may be less than 100 m wide or more than several kilometers in width. Likewise, they range in length from small pocket barriers of a few hundred meters to those along open coasts that extend for more than 100 km. Generally, barriers are wide where the supply of sediment has been abundant and relatively narrow where erosion rates are high or where the source of sediment was scarce during their formation. Barrier length is partly a function of sediment supply but is also strongly influenced by wave versus tidal energy of the region. This relationship is discussed in more detail in a later section.

Barriers consist of many different types of sediment, depending on their geologic setting. Sand, which is the most common constituent of barriers, comes from a variety of sources, including rivers, deltaic and glacial deposits, eroding cliffs, and biogenic material. The major components of land-derived sand are the minerals quartz and feldspar, and rock fragments. These durable grains are a product of physical and chemical decomposition of the bedrock from continents. In northern latitudes where glaciers have shaped the landscape, gravel is a common constituent of barriers, whereas in southern latitudes carbonate material, including shells and coral debris, may comprise a major portion of the barrier sands. Along the southeast coast of Iceland and along portions of the west coast of New Zealand, the barriers are composed of black volcanic sands derived from upland volcanic rocks.

Many environments make up a barrier and their arrangement differs from location to location, reflecting the type of barrier and the physical setting of the region. Generally, most barriers can be divided in three zones: the beach, the barrier interior, and the landward margin (Fig. 8.2).

Beach. Due to continual sediment reworking by wind, waves, and tides, the beach is the most dynamic part of the barrier. Beaches exhibit a wide range of morphologies depending upon a number of factors, including the grain size of the beach, the abundance of sediment, and the influence of storms. Sediment is removed from the beach during storms and returned during more tranquil wave conditions. During the storm and post-storm period, the form of the beach evolves in a predictable fashion. The beach environment is discussed in detail in Chapter 7.

Barrier interior. Along sandy barriers, the beach is backed by a frontal dune ridge (also called the foredune ridge), which may extend almost uninterruptedly along the length of the barrier provided the supply of sediment is adequate and storms have not removed the dune ridge. The frontal dune ridge is the first line of defense in protecting the interior of the barrier from the effects of storms. Landward of this region are the secondary dunes, which have a variety of forms depending on the historical development of the barrier and subsequent modification by wind processes. For example, devegetation of the Provincelands Spit at the tip of Cape Cod in Massachusetts by early settlers led to the formation of large parabolic dunes that are up to a half kilometer long and more than 35 m high. In other locations where barriers have built seaward through time, such as North Beach Peninsula in Washington or Kiawah Island in South Carolina, former shoreline positions are marked by a series of semi-parallel vegetated beach ridges (former foredune ridges) 3–7 m in height. The low areas between individual beach ridges, called swales, commonly extend below the water table and are the sites of fresh and brackish water ponds or salt marshes. In South Carolina, Georgia, and Florida alligators and water moccasins are known to inhabit these ponds. On developed barriers containing golf courses, these ponds often become the sites of water holes.

Landward margin. Along the back side of many barriers the dunes diminish in stature and the low relief of the barrier gradually changes to an intertidal sand or mud flat or a salt marsh. In other instances, the margin of the barrier abuts an open water area associated with a lagoon, bay, or tidal creek. Along coasts where the barrier is migrating onshore, the landward margin may be dominated by aprons of sand that have overwashed the barrier during periods of storms. In time, these sandy deposits may be colonized by salt grasses producing

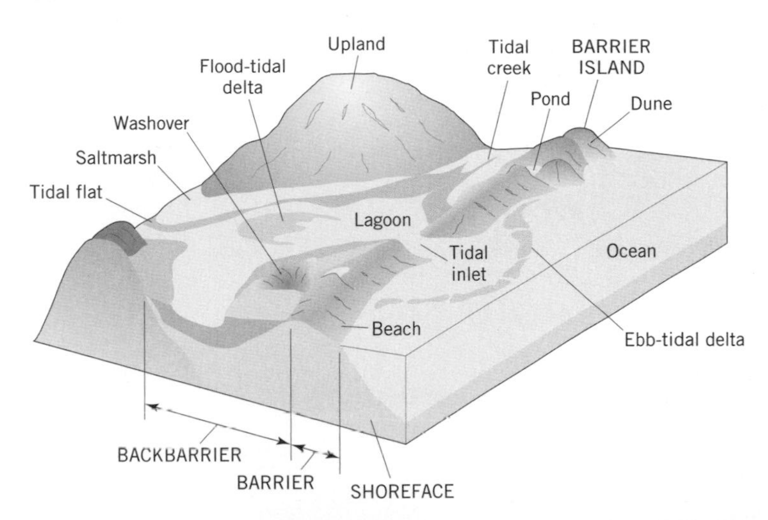

Fig. 8.2 (a) Idealized barrier coast environments. (b) Castle Neck, Massachusetts, can be divided into three zones: (1) **beach** where sand bars are migrating onshore and weld to the lower shore; (2) **barrier interior** consisting of dune and swale topography; (3) **landward margin** where the barrier transitions to tidal flats and salt marsh. In a tropical setting the margin may consist of mangroves.

arcuate-shaped marshes. Overwash deposits may also occur in the interior of the barrier. In still other regions, the back side of a barrier may be fronted by a sandy beach bordering a lagoon or bay. In this setting, wind blown sand from the beach may form a rear-dune ridge outlining the landward margin of the barrier.

Barriers may terminate in an embayment or at a headland. In the case of barrier chains, individual barriers are separated by tidal inlets. These are the waterways that allow for the exchange of water between the ocean and the backbarrier environments. The end of the barrier abutting a tidal inlet is usually the most unstable portion of the barrier due to the effects of inlet migration and associated sediment transport patterns.

8.3 Distribution and coastal setting

8.3.1 General

Barriers comprise approximately 15% of the world's coastlines (Fig. 8.3). They are found along every continent except Antarctica, in every type of geologic setting, and in every kind of climate. Tectonically, they are most common along Amero-trailing edge coasts where low-gradient continental margins

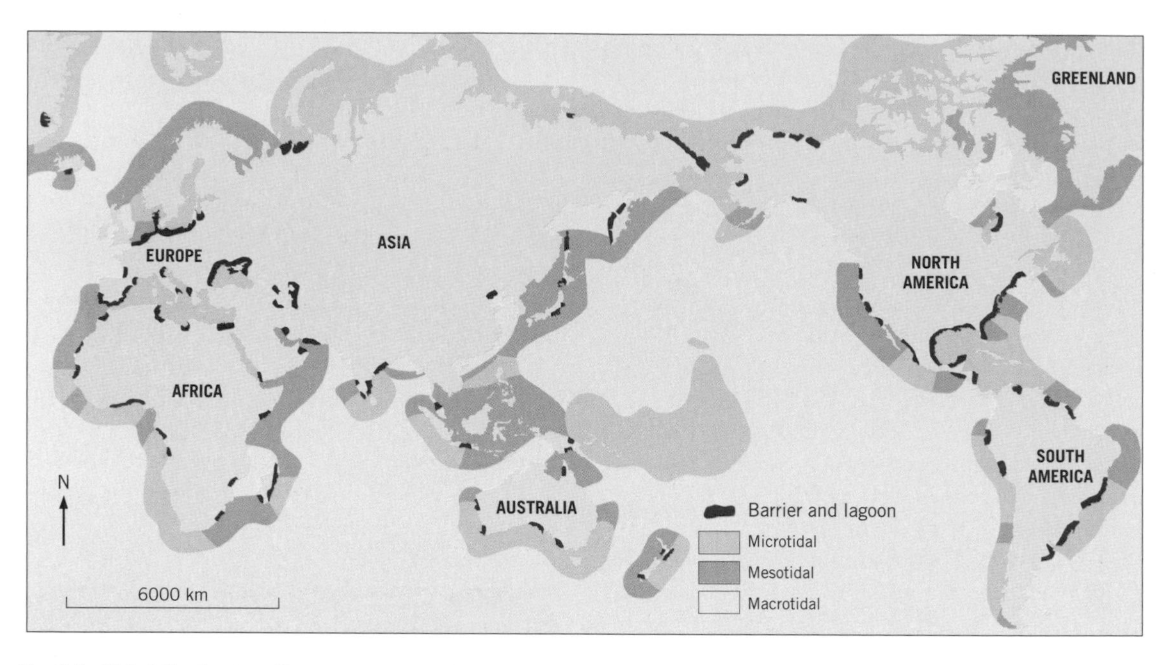

Fig. 8.3 Global distribution of barriers and their tidal range setting. (From M. O. Hayes, 1979, Barrier island morphology as a function of tidal and wave regime. In S. P. Leatherman (ed.), *Barrier Islands*. New York: Academic Press, pp. 1–28, using barrier data from H. G. Gierloff-Emden, 1961, Nehrungen und Lagunen. *Petermanns Geographische Mitteilungen*, **105**, 81–92, 161–76, and tidal range data from D. K. Davies, 1980, *Geographic Variation of Coastal Development*. New York: Longman.)

provide ideal settings for barrier formation. They are also best developed in areas of low to moderate tidal range[1] (microtidal to mesotidal range) and in mid- to lower latitudes. Climatic conditions control the vegetation on the barriers and in backbarrier regions, the type of sediment on beaches, and in some regions, such as the Arctic, the formation and modification of barriers themselves. The disappearance of barriers where tidal energy dominates, such as on the northwest (Big Bend) coast of Florida in the Gulf of Mexico and the German Bight in the North Sea, attests to the requirement for wave energy in the formation of barriers.

8.3.2 Tectonic controls

As discussed in Chapter 2, the tectonic setting of the coast dictates to a large extent the sediment contribution to the coast, the width of the continental shelf, and the general topography of the coast. The tectonic coastal classification can be used to illustrate these relationships and explain the worldwide distribution of barriers. Amero-trailing edge coastlines tend to have abundant sediment supplies due to extensive continental drainage. Their low-relief coastal plains and continental shelves provide a platform upon which barriers can form and migrate landward during periods of eustatic sea-level rise. The longest barrier chains in the world coincide with Amero-trailing edges and include the east coast of the United States (3100 km) and the Gulf of Mexico coast (1600 km). There are also sizable barrier chains along the east coast of South America (960 km), the east coast of India (680 km), the North Sea coast of Europe (560 km), Eastern Siberia (300 km), and the North Slope of Alaska (900 km).

1 Tidal range is the vertical difference between high and low tide. A classification of coasts based on tidal range consists of microtidal coasts (TR < 2.0 m), mesotidal coasts (2.0 ≤ TR ≤ 4.0 m), and macrotidal coasts (TR > 4.0 m).

The unparalleled concentration of barriers along the east coast of the United States is undoubtedly a product of the erosion and denudation of the Appalachian Mountains, which some scientists have speculated may have once rivaled the Himalayas in elevation. The huge volume of sediment derived from the wearing down of these mountains produced a wide, flat coastal plain and continental shelf region, much of which is veneered by layers of unconsolidated sediment or easily eroded sedimentary rocks. It is the reworking and redistribution of these surface sediments by rivers, tides, and waves that are responsible for extensive barrier construction along the east coast. Likewise, the almost continuous chain of barriers in the Gulf of Mexico is related to the wide, flat coastal plain and continental shelf of this margin. It is thought by some scientists that sediment delivery to this coast, particularly along Texas, was much greater in the past when the climate of this region was wetter.

Marginal sea coasts contain a relatively small percentage of world's barriers, even though these regions have some of the largest sediment discharge rivers in the world, particularly along the Asian continent. While it would appear that many of these coasts have ample sediment to build barriers, the irregular topography of these margins leads to much of the sediment filling submarine valleys rather than forming barriers. In addition, the discharge from many Asian rivers has a very high suspended sediment component that may inhibit the concentration of sand-sized material. In fact, many of the marginal sea coastlines containing barriers, such as the chain found along the Nicaraguan and Honduran Caribbean coast, have no large sediment discharge rivers associated with them. The origin of these barriers is probably tied to the reworking of previously deposited shelf sediments, similar to much of the east coast of the United States.

Collision coasts also have few long stretches of barriers, but in this case, it is due to an overall lack of sediment. Most of the rivers discharging along these shores have small drainage areas and contribute little sediment for barrier construction. Additionally, the narrow, steep continental shelves of these margins result in high wave energy and rapid sediment dispersal along the coast. Along parts of the California coast some of the sediment that would ordinarily accumulate along the shore is drained from the beaches during storms through submarine canyons to the deep ocean basins. The proximity of the canyons to the beaches is attributed to the narrow continental shelf. Barrier systems that do occur along the west coast are commonly isolated and related to specific sediment sources, such as nearby rivers or eroding cliffs. Typically, they take the form of barrier spits attached to a bedrock headland. Two of the longest barrier chains along collision coasts are located adjacent to major river mouths, including the southern Washington barriers near the Columbia River and the Gulf of Alaska barrier chain fronting the Copper River delta (Fig. 8.4).

A small percentage of the world's barriers are found along Afro-trailing and Neo-trailing edge coasts due to an overall lack of sediment along these margins. There is little sediment delivery to the neo-trailing edge margins of the Red Sea, Gulf of Aden, and Gulf of California due to the immaturity in the development of river drainage to their coasts and the very low precipitation in these regions. Similarly, much of the northern coast of Africa has little river discharge of sediment due to the arid conditions of the interior region. The extensive barrier system (300 km) along the Ivory, Gold, and Slave coasts in west central Africa is related to the numerous moderately sized rivers of this area.

8.3.3 Climatic controls

By controlling the amount of precipitation and evaporation of a continent, climate exerts a strong influence on the size and number of rivers as well as the overall volume of sediment delivered to the coast. A river's drainage area (defined as the continental area drained by the river) is another important factor governing sediment discharge. Usually, the larger the drainage area (A_d) the greater is the sediment delivery to the coast, but there are exceptions. The Eel River in northern California releases almost twice as much sediment load to the coast as does the much larger Columbia River to the north, despite having a drainage area two orders of magnitude

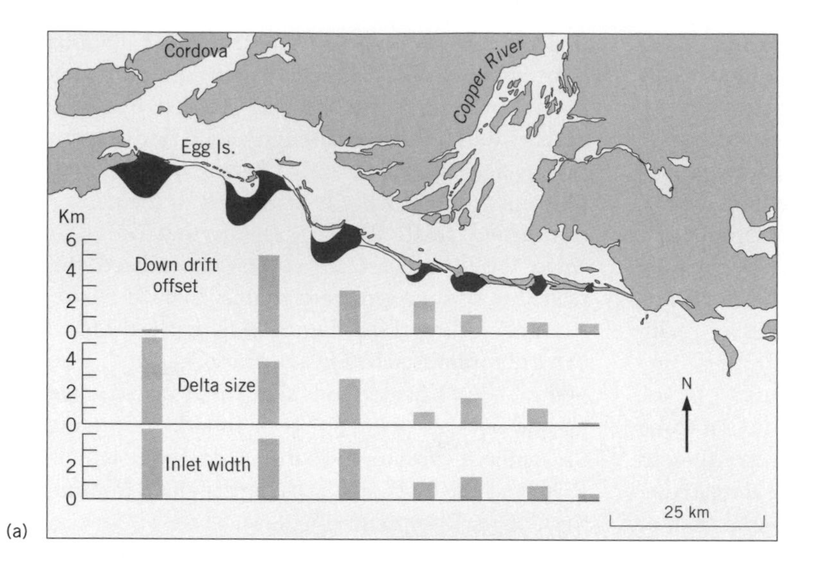

Fig. 8.4 Barrier chains on collision coasts are generally found where a nearby river supplies abundant sediment. Examples include: (a) Copper River delta barrier chain (from Hayes, 1979, details in Fig. 8.3); (b) Long Island–Grays Harbor barrier system along the southern Washington coast north of the Columbia River (from J. R. Dingler & H. E. Clifton, 1994, Barrier systems of California, Oregon, and Washington. In R. A. Davis (ed.), *Geology of Holocene Barrier Island Systems*. Berlin: Springer, pp. 115–66).

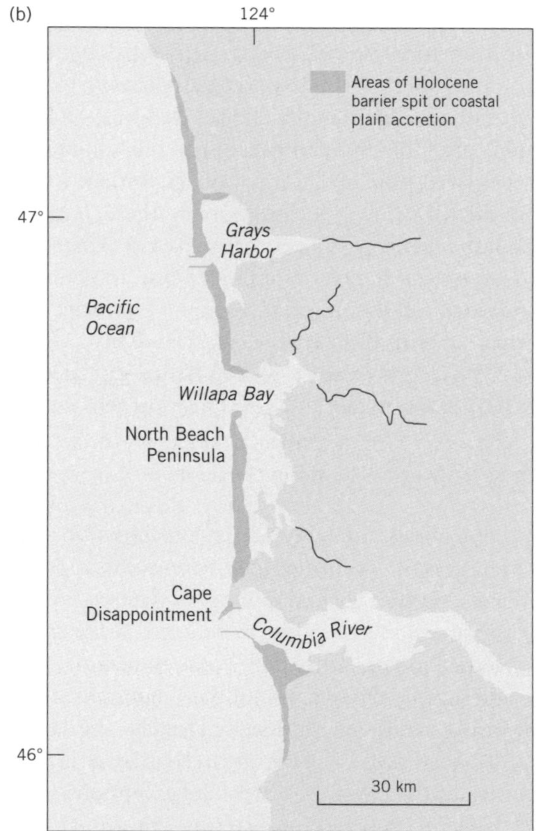

smaller (Eel River A_d = 9000 km^2, Columbia River A_d = 661,000 km^2). The high sediment discharge of the Eel River is a product of the mountainous terrain it drains and a bedrock landscape that weathers easily, producing abundant sediment.

Climate also dictates the kinds of plants, shrubbery, and trees colonizing barriers and the type of vegetation found in the backbarrier region. For example, salt grasses vegetate the backbarrier marshes of mid- to high latitudes, whereas mangroves commonly comprise the backbarrier region in low latitudes. Even the sediment making up the barrier is related to climatic effects and previous climatic conditions, such as the "Ice Age." Generally, the shell and coral content of beaches is relatively high in tropical regions, whereas terrigenous gravel forms a significant component of northern beaches and barriers that have been formed from glacial deposits.

8.3.4 Summary

Barriers are best developed along continental margins where sand-sized sediment is abundant and the coastline is fronted by a moderately wide, gently sloping continental shelf and backed by a coastal plain. Under these conditions, barriers are able to migrate onshore in a regime of slow sea-level rise. Direct contribution of sediment to the coast by rivers

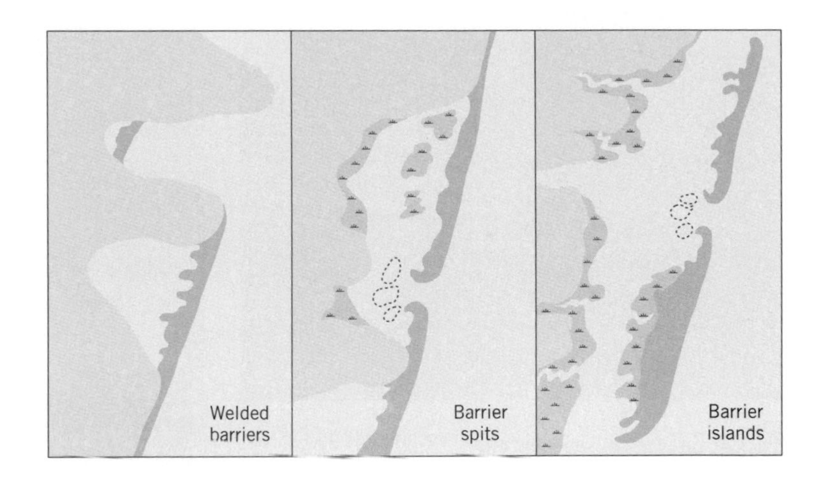

Fig. 8.5 Barriers can be separated into three general mophological classes based on their connection with the mainland: barrier islands, barrier spits, and welded barriers.

is not enough to insure the presence of barriers, as evidenced by the Asian marginal sea coasts where barriers are sparse. The fine-grained sediment discharged from the rivers of this region may overwhelm the ability of waves to concentrate sand and form barriers.

8.4 Barrier types

Although barriers have many different forms, for ease of discussion they are grouped into three major classes based on their connection to the mainland (Fig. 8.5). Barrier spits are attached to the mainland at one end and the opposite end terminates in a bay or the open ocean. Welded barriers are attached to the mainland at both ends. Barrier islands are isolated from the mainland and surrounded by water.

8.4.1 Barrier spits

Barrier spits are most common along irregular coasts where angular wave approach and an abundant supply of sediment result in high rates of longshore sediment transport. These conditions promote spit building across embayments and a general straightening of the coast. Barrier spits are the dominant barrier form along tectonically active coasts. In some instances, spit construction partially closes off a bay, forming a tidal inlet between the spit end and the adjacent headland. In these instances, tidal currents flowing through the inlet prevent spit accretion from sealing off the bay. The Cape Cod shoreline in New England illustrates well this type of coastline evolution. Following deglaciation, Cape Cod appeared very different from how it looks today, consisting of irregularly shaped, sandy glacial and glacio-fluvial deposits (see Chapter 17). This landscape gave rise to large headland areas and broad embayments as eustatic sea level rose in response to melting glaciers and water being returned to the ocean basins. Since 6000 years ago, waves associated with the rising ocean waters attacked the unconsolidated headlands, causing their retreat and the release of large quantities of sediment that were subsequently moved along the shore, leading to the construction of spits across the inundated lowland areas. In this way the shoreline of Cape Cod was smoothed into its present form.

Recurved spits

Many spits have recurved ridges that conform to the general outline of the spit end. Such spits are called recurved spits. Each of the ridges indicates a former shoreline position and collectively they trace the evolution of the spit (Fig. 8.6). Ridge formation is part of the spit extension process whereby sediment is added incrementally to the end of the barrier. Before the subaerial (above the water) portion of the spit can build into a bay or tidal inlet, a platform

Fig. 8.6 Recurved spit building across the mouth of the Santee River in South Carolina. Individual beach ridges represent pulses of sand that lengthen the spit in spurts. The sporadic supply of sand may be related to storm activity.

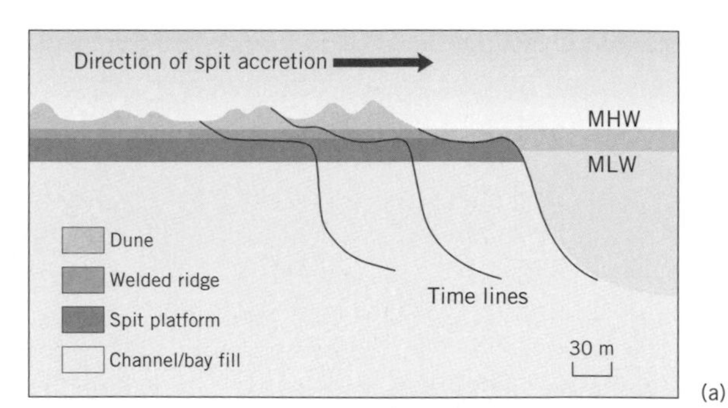

Fig. 8.7 Spit accretion across a bay or into a tidal inlet is preceded by extension of the spit platform. (a) The spit platform shown in cross section (modified from M. O. Hayes & T. Kana, 1978, *Terrigenous Clastic Depositional Environments*. Columbia: Department of Geology, University of South Carolina). (b) Aerial view of Chatham Inlet, Cape Cod, Massachusetts, illustrating swash bars migrating onshore and build ridges, which lengthen the supratidal portion of the spit.

fronting the spit must first be created upon which sediment can accumulate and bars can form. This subaqueous portion of the spit commonly contains much more sand than the part of the spit above water. For example, Sandy Hook, which defines the northernmost end of the New Jersey coast, is accreting into Raritan Bay toward New York Harbor. The spit end of Sandy Hook has an average elevation of 5 m above mean sea level, whereas bay depths immediately offshore of the spit reach 12 m and more. Likewise, the northern spit end of Cape Cod is building into water depths of more than 50 m, contrasting to average land elevations of 10 m.

The spit platform is a shallow-sloping sandy surface that extends from low tide to a depth of several meters below mean low water. The platform is fed with sediment through the longshore movement of sand along the updrift portion of the barrier, enabling the platform to build into deep water (Fig. 8.7). Waves breaking over the spit platform aid in the development of swash bars, subaqueous bars having a length of 100 m or more and a height of one to several meters. These bars, which migrate onshore due to the action of breaking and shoaling waves, tend to wrap around the end of the spit, becoming aligned with the waves that bend around the spit platform. As the bars move onshore, they first gain an intertidal exposure and eventually attach to the upper beach, forming a ridge. Commonly, the welding process is incomplete and a low area (swale) is left between the newly accreted ridge and the dune line that defines the former end of the spit. Gradually, the ridge is colonized by dune grasses and builds vertically by wind processes.

Recurved spits exhibit many different beach ridge patterns characterizing different depositional styles. In some instances, the ridges indicate that there has been a simple extension of the spit, whereas in other cases they demonstrate that the barrier has widened as well as extended along the shore. In both examples, the ridge and swale morphology suggests that spit accretion occurs episodically and each ridge may represent an increase in the rate of sand delivery to the spit platform. In turn, the change in the flux of sand along the coast may be related to storm processes, whereby sediment is excavated from updrift sandy cliffs, river deltas, or beach deposits.

Another common trend along recurved spits is the change in swale environment along the length of the spit, reflecting a gradual filling of swale over time. At the spit end, a newly formed swale may be deep enough to permit tidal inundation from an opening along-ridge on the lagoon side. Sedimentation in the swale resulting from washovers, wind-blown sand, and, to a lesser extent, fine-grained material carried by tidal currents transforms the ponded water region to a marsh environment and eventually to a sandy low interridge area. Thus, the spit end tends to exhibit low ridges with water-filled swales, whereas the updrift, older portion of the spit contains vegetated beach ridges and sandy or grassy swales.

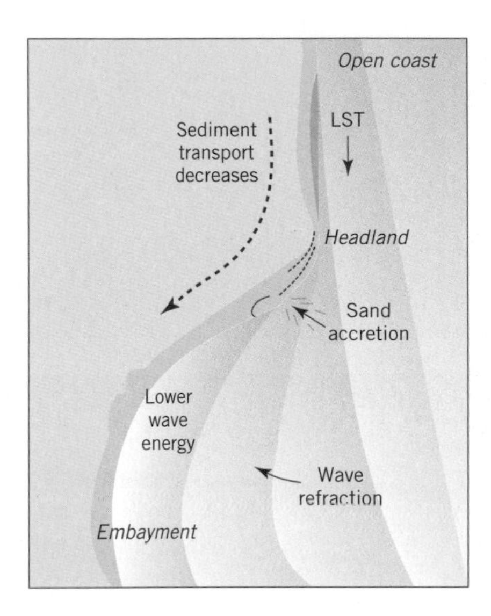

Fig. 8.8 Spit development is due to sand deposition resulting from a decrease in the rate of longshore sediment transport. The rate of sand transport decreases downdrift of the headland due to a reduction in wave energy.

Spit initiation

Spit formation is due to the deflection of sediment that is moving on and along a beach in the surf and breaker zones into a region of deeper water commonly associated with a bay or tidal inlet. This deflection may be initiated at small protrusions of the shoreline or at major headlands. Similarly, the process may occur at the edge of embayments along an otherwise smooth coastline. The explanation of why the longshore movement of sand forms spits rather than following a pathway along the irregular shoreline lies in the fact that wave energy decreases in embayments and thus the capacity of waves to move sand is reduced (Fig. 8.8).

Remember from Chapters 6 and 7 that the movement of sand on and along a coast is a function of wave energy and breaker angle, among other parameters. Consider an idealized case in which a wave crest extends a far distance along the shore such that it can be followed from where it breaks at an angle along the exposed beach to where it breaks within the bay. The section of wave breaking along the exposed beach will be higher and will expend more energy in the nearshore zone than the wave crest that breaks inside the bay. This makes sense because as the wave crest bends into the bay, it travels further than the wave breaking along the exposed coast and thus uses up more energy interacting with the seabed. Because the rate of sand movement is a function of wave height, more sand is transported along the exposed beach than inside the bay. This means that as sediment transport decreases from a relatively high rate along the exposed beach to a lower rate inside the bay, some sediment must be deposited along the way. This sediment accumulates at the edge of the embayment because this is where the change in the longshore transport rate is the greatest. It is this deposition of sediment that initiates spit growth.

Other spit forms

Spits exhibit many different forms in addition to recurved spits, including cuspate spits, flying spits, tombolos, and cuspate forelands. Each of these accretionary landforms develops in specific geologic settings and by particular coastal processes.

Cuspate spits are triangular accumulations of sand that extend from the shoreline into a semi-protected body of water, such as the lagoons and bays found

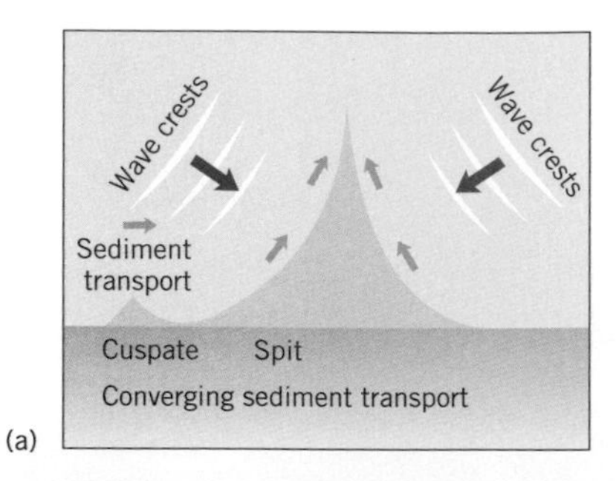

Fig. 8.9 Cuspate spits occur in protected environments such as within bays and lagoons. (a) Model of cuspate spit formation due to bidirectional sand transport (from Hayes & Kana 1978, details in Fig. 8.7). (b) Example of a cuspate spit on the bay side of the Coatue barrier, Nantucket Island, Massachusetts.

behind barrier islands and spits (Fig. 8.9). They range in length from a few tens of meters, such as those in the elongated bays along the southern coast of Martha's Vineyard, to kilometers, including those on the lagoonal side of Santa Rosa Island in the western panhandle of Florida or those in Patos Lagoon along the southern coast of Brazil. Their lengths reflect the size of the bay and the resulting wave energy produced inside the bay. Larger bays and lagoons have longer fetches, which allow larger waves to develop and more sand to be transported along the bay shoreline. The semi-protective nature of the bays means that sand transport along the shoreline is governed by local wind-generated waves and not by ocean swell waves. The multidirectional winds that are common on most coasts produce bidirectional sand transport along these shores. This back and forth movement of sand along bay shorelines is what forms cuspate spits. The tip of the spit may have a narrow intertidal to subtidal tail that changes its orientation as the wave climate changes. Growth and modification of the spit are often revealed in the pattern of beach ridges that are common to cuspate spits.

Flying spits look similar to recurved spits and commonly exhibit cuspate-shaped beach ridges, indicating former shoreline positions. However, unlike recurved spits, they occur along straight to slightly irregular shorelines and extend into deep water at an acute angle to the beach (Fig. 8.10). Like cuspate spits, they are found most commonly along semi-protected shorelines where wave energy is a product of local winds. Flying spits exist in the lagoons bordering the Texas coast, but the most well known and studied flying spit is Presque Isle along the south-central shore of Lake Erie, near Erie, Pennsylvania. Presque Isle is 10 km in length and its spit end is approximately 2 km offshore of the beach. The reason why these spits build into deepwater is not well understood but it may be related to some type of irregularity along the original shoreline or in the nearshore region. Because the trend of flying spits is normally almost parallel to the dominant wave crests of the region, some scientists have argued that spit construction is an attempt of the shoreline to minimize the longshore transport of sediment.

Tombolos are a type of spit that was first identified and named by the Italians because they are common features along their coast (Fig. 8.11). The largest of these is a double tombolo that occurs along the northern Tyrrhenian Sea coast of Italy at Orbetello. In the strictest definition, tombolos are spits of land that build out from the shoreline, eventually providing a sediment bridge from the mainland to an island. At Orbetello, two tombolos, almost 8 km in length, have evolved due to the large size of the island (10 km long) and an abundant sediment supply along the coast.

Tombolos form on the lee side of an island due to the obstruction and redistribution of wave energy caused by the island. The expenditure of wave energy on the island creates a "wave shadow zone" along the landward beach. It is a reduction in the

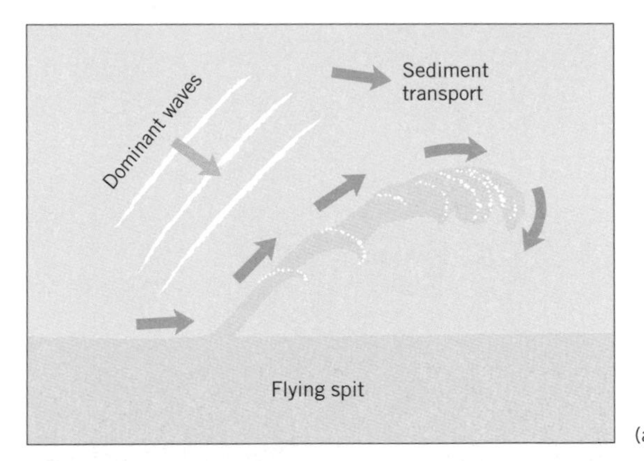

Fig. 8.10 Flying spits are relatively uncommon features found in semi-protected waters. (a) Model of a flying spit (from Hayes & Kana 1978, details in Fig. 8.7). (b) Presque Isle, Pennsylvania (from Ken Winter, US Army Corps of Engineers).

rate of longshore sand transport in this region that produces sediment accumulation and spit growth toward the island. Waves bending around the island also create a reversal in the direction of dominant sand movement along the shore, augmenting the sand trapping processes in the wave sheltered zone. As noted, the double tombolo system at Orbetello is due to the large size of the island. Moderate and small sized islands tend to have a single tombolo.

Cuspate forelands are large triangular-shaped projections of the shoreline that may extend seaward more than 25 km. The apex of the foreland is commonly composed of beach ridges that parallel the two converging shorelines (Fig. 8.12). Alternatively, the ridges parallel a single shoreline and are truncated along the other, indicating a reorientation of the foreland (e.g. Cape Canaveral, Florida). Bays and lagoons or marshy areas generally occupy the region between the foreland and the mainland. Cuspate forelands may be isolated or can occur in a series such as those along the coast of North and South Carolina. Dungeness is a solitary foreland located along the southern coast of Britain where the English Channel is narrowed by the closeness of

Time 1. Wave shadow zone

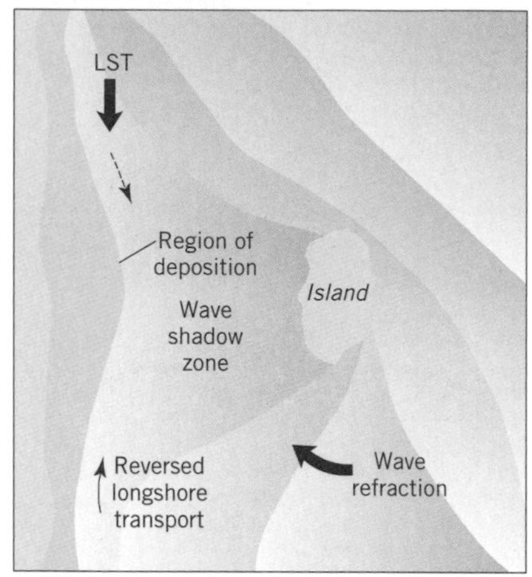

Time 2. Tombolo formation

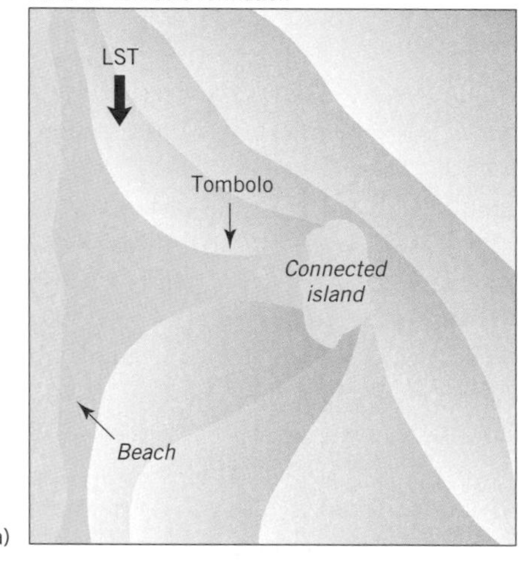

(a)

(b)

(c)

Fig. 8.11 A tombolo is an accretionary landform that connects the mainland to an island. (a) Model of the development of a tombolo. Examples of tombolos in (b) Kodiak Island, Alaska, and (c) Popham Beach, Maine.

France. It is built of gravel ridges and is prograding seaward and to the northeast due to dominant wave energy from the southwest. The Carolina forelands, including Cape Hatteras, Cape Lookout, Cape Fear, and Cape Romain, form an evenly spaced (110–165 km apart) scalloped shoreline with individual capes projecting 25–40 km beyond the adjacent embayments. A similar group of capes with the same approximate spacing (130 km) is found on the North Slope of Alaska along the Chukchi Sea (Icy Cape, Point Franklin, and Point Barrow). Formation of the Carolina capes has been related to giant eddies associated with the northward flowing Gulf Stream, wave refraction around offshore shoals, basement controls, and the reworking of former delta deposits. Although rivers do not presently supply sand to the capes, excepting possibly the Cape Fear River, each of the capes is associated with one or more moderate size rivers that may have delivered sediment to the coast during the Pleistocene, when sea level was lower. These deltaic deposits may have been the source of the cape sediments or may have preceded barrier development and influenced sand transport trends along the coast.

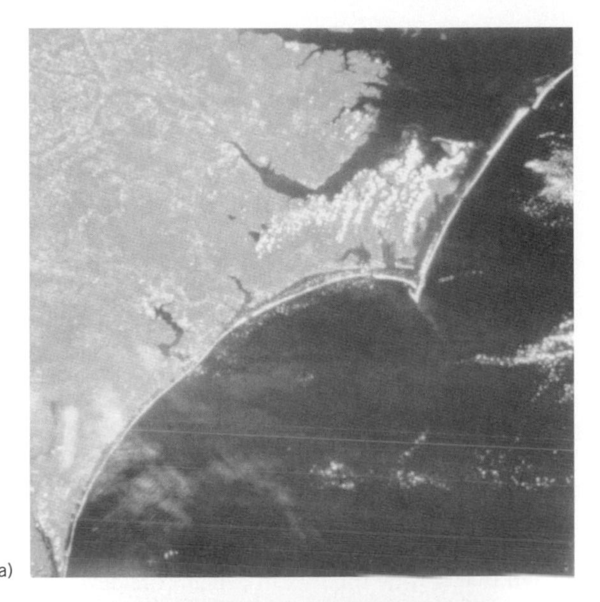

(a)

(b)

Fig. 8.12 Cuspate forelands occur along many sandy coasts, including (a) the North Carolina coast and (b) the Kuwait coast in the Arabian Gulf.

8.4.2 Welded barriers

Welded barriers occur along irregular coasts where the supply of sediment is adequate for barrier construction (Fig. 8.13). They are common features along rocky coasts such as parts of the west coast of the United States and along glaciated coasts includ ing New England, eastern Canada, and Alaska. Most often, these barriers are backed by shallow water bays and lagoons, but freshwater and brackish water marshes also exist. Welded barriers are most common along microtidal coasts where bay areas are diminutive in size. In this type of setting there is insufficient tidal energy to keep a tidal inlet open. If a welded barrier is breached during a storm, the inlet normally closes quickly because the exchange of tidal waters between the bay and ocean cannot remove the sand dumped into the inlet by waves. Thus, environments with moderate to high wave energy promote the development of welded barriers.

Fig. 8.13 Welded barrier fronting a brackish Trustom Pond, Rhode Island. After the tidal inlet closed at the right side of the barrier, tidal exchange ceased and water in the pond became increasingly less saline.

The formation of a welded barrier can be a simple process whereby a spit builds across a bay and attaches to the opposite shoreline. However, in many instances, their evolution represents a long-term response to rising sea level. This process is evident along the southern shore of Martha's Vineyard offshore of Cape Cod, where low welded barriers front elongated bays. Historical evidence shows that some of these barriers were once cut by semi-permanent tidal inlets. Closure of the inlet occurred because the onshore migration of the barrier reduced bay area at a faster rate than rising sea level inundated the bay shoreline.

An interesting phenomenon of welded barriers occurs on the Peninsula coast of Alaska and along other gravelly shorelines of the world. Here, welded barriers that are composed of gravel and sand contain multiple gravel ridges that have been constructed by storm waves. The barriers front ponded water areas and the permeable nature of their sediment permits the seepage of water from the pond through the barrier and into the ocean. Land drainage and the permeability of the barrier sediments control the level of water in the bays. During periods of high water influx, the level of the bay may overtop the crest of the barrier, cutting an ephemeral outlet as water drains into the ocean. After bay water levels have subsided, increased rates of sediment transport commonly accompanying storms and high wave events once again seal off the opening.

8.4.3 Barrier islands

Barrier islands are relatively narrow strips of sand that parallel the mainland coast. They usually occur in chains and, excepting the tidal inlets that separate them, may extend uninterrupted for over a hundred kilometers. For example, Padre Island along the Texas coast is about 200 km long. Barrier chains may consist of a few islands or more than a dozen. The length and width of barriers and overall morphology of barrier coasts are related to several parameters, including tidal range, wave energy, sediment supply, sea-level trends, and basement controls. The fact that they exist along most of the east and Gulf coasts of the United States with different geologic and oceanographic conditions suggests that barriers can form and be maintained in a variety of settings. In the next few sections of this chapter we explore the formation of barrier islands, the different layers comprising barriers, what they look like in cross section, and the morphology of barrier island coasts and individual barrier islands.

Origin of barrier islands

The widespread distribution of barrier islands along the world's coastlines and their occurrence in many different environmental settings has led numerous scientists to speculate on their origin for more than 150 years (Fig. 8.14). Any acceptable theory of their formation must explain the following:

1 Barrier chains are aligned parallel to the coast.
2 Most have formed in a regime of slow eustatic sea-level rise.
3 They are separated from the mainland by shallow lagoons, marshes, and/or tidal flats.
4 Tidal inlets separate individual barriers along a chain.
5 They are composed of sand (some contain gravel).
6 They formed during periods of sand abundance (question: where did the sediment come from?).

The different explanations of barrier island formation can be grouped into three major theories: (i) **offshore bar theory** (de Beaumont, Johnson); (ii) **spit accretion theory** (Gilbert, Fisher); and (iii) **submergence theory** (McGee, Hoyt). It will be shown that no one theory can explain the development of all barriers and, moreover, it is sometimes difficult or impossible to prove how an individual barrier formed, because many barriers have been drastically modified after they were formed due to rising sea level and related processes.

E. de Beaumont One of the earliest ideas of barrier island formation was published in 1845 by a Frenchman, Elie de Beaumont, who studied coastal charts of barriers along the North Sea and Baltic Sea in Europe, and those in the Gulf of Mexico, North Africa, and elsewhere. He believed that waves moving into shallow water churned up sand, which was deposited in the form of a submarine bar when the waves broke and lost much of their energy. As the bars accreted vertically, they gradually built above sea level, forming barrier islands.

G. K. Gilbert Some years later, in 1885, de Beaumont's idea that barriers were formed from offshore sand sources was countered by G. K. Gilbert, who argued that the barrier sediments came from alongshore sources. Gilbert proposed that sediment moving in the breaker zone through agitation by waves would construct spits extending from headlands parallel to the coast. The subsequent breaching of spits by storm waves would form barrier islands. What is truly amazing about Gilbert is that

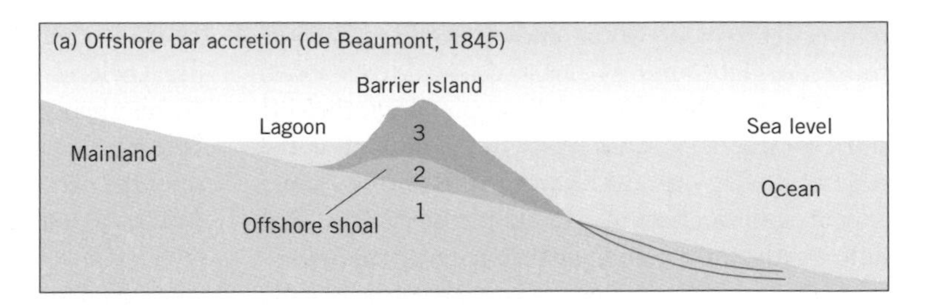

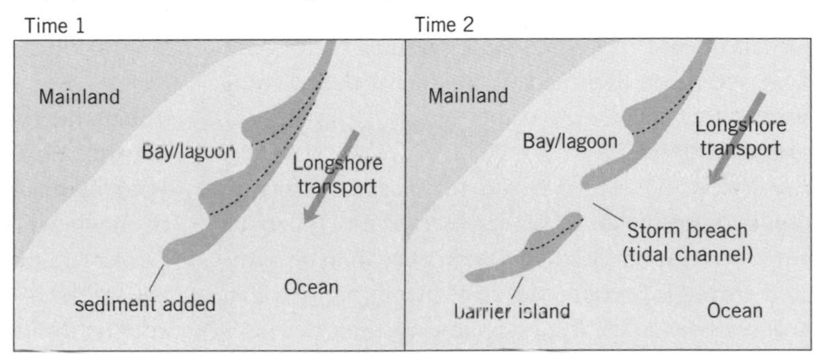

Fig. 8.14 Major models of barrier island formation. (Modified from S. Penland & R. Boyd, 1985, *Transgressive Depositional Environments of the Mississippi River Delta Plain: A Guide to the Barrier Islands, Beaches, and Shoals of Louisiana*. Baton Rouge: Louisiana Geological Survey, p. 233.)

this theory was conceived in western Utah while he was studying former lake deposits associated with the terraces along the Wasatch Mountains. The former lake was called Lake Bonneville and it formed during the Pleistocene when the climate was cooler, reaching a size of 50,000 km^2 or about the size of present-day Lake Michigan. It was along this lake that spits had once developed. As the climate warmed in the Holocene, Lake Bonneville diminished to one-tenth its size to become what is now Great Salt Lake.

W. D. McGee Before the end of the nineteenth century, a third barrier island theory was published by W. D. McGee in 1890. He reasoned that the east and Gulf coasts of the United States were undergoing submergence, as evidenced by the many drowned river valleys that occur along these coasts, including Raritan, Delaware, and Chesapeake Bays. He believed that during submergence coastal ridges were separated from the mainland, forming lagoons behind the ridges. He called the barriers "keys" and used Dolphin, Petit-Bois, Massacre, Horn, Dog, Ship, and Cat islands along the coast of Mississippi as examples where coastal submergence had formed barrier islands.

D. W. Johnson At the turn of the twentieth century there were three viable explanations of barrier island formation, but none of these theories had been tested using laboratory experiments or through the collection of field data. In 1919, D. W. Johnson reinvestigated the various theories and studied the shore-normal profile of barrier coasts. He reasoned that if barriers had formed from spits, then the offshore profile should intersect the mainland at the edge of the lagoon. Additionally, the profile should appear as though sand had simply been deposited onto a uniformly sloping nearshore ramp. However, Johnson discovered that most barrier coasts do not exhibit this type of profile; instead, they seem to indicate that sand had been scooped from the nearshore and moved onshore to build the barrier. Thus, Johnson became a champion of the offshore bar theory.

John Hoyt During the next fifty years the formation of barrier islands was a subject of much debate, but it was not until sediment cores began to be gathered through barriers and in lagoonal regions that progress was made in understanding their development. John Hoyt worked along the coast of Georgia and many of his conclusions concerning barrier island formation were based on coring studies there. In an article published in 1967, he correctly argued that if barriers had developed from offshore bars or through the breaching of spits, then open-ocean conditions would have existed along the mainland prior to barrier formation. Before becoming sheltered, breaking waves and onshore winds would have formed beaches and dune systems at these locations. Hoyt was able to show that for most barrier systems an open-ocean coast never existed along the present mainland coast. In situations where rising sea level and marsh development has led to an encroachment of lagoonal deposits onto the mainland, sediment cores from these regions have shown no evidence of beach and nearshore sediments or the shells of organisms that commonly inhabit the nearshore region.

Hoyt maintained that if coastal dunes or beach ridges were gradually submerged by rising sea level, then the mainland shoreline would never have been exposed to waves and thus nearshore deposits would never have developed. Hoyt went on to show that width of the lagoon was a function of the level of submergence and slope of the land surface. Steeper slopes and lower levels of submergence produced narrow lagoons; whereas flat land gradients and greater levels of submergence led to wider lagoons. In his theory the low areas along the beach ridges became the sites of tidal inlets. He also noted that once the barrier was formed it could then be modified by waves, sediment supply, and sea-level changes.

John Fisher Armed with sedimentologic data, Hoyt had made a compelling case for barrier islands forming by submergence of coastal dunes or beach ridges due to rising sea level. However, in 1968 John Fisher produced a critical review of the submergence theory, pointing out that long, straight, and continuous dune ridges would not occur along a coast being inundated by rising sea level. Moreover, in his own research along the Outer Banks of North Carolina he found that sediments beneath the lagoonal deposits were more like those occurring behind a spit than those in a coastal upland as suggested by Hoyt. In light of these observations, Fisher became a strong proponent of Gilbert's spit accretion theory.

Recent barrier studies

Since the late 1960s there have been many studies of barrier islands aimed at determining the sedimentary layers making up the barrier and deciphering the manner in which these layers were deposited. This research has been aided by the radiometric dating of organic material contained within the barriers' sediments, such as shells, peat, and wood, thereby providing a chronology (timing) of barrier construction. In addition to coring of the barrier sediments, ground-penetrating radar (GPR) is also employed to study barriers. This device sends electromagnetic energy into the ground. Where the electrical conductivity of the sediment changes, such as at the interface of two sediment layers with different grain sizes, mineral compositions, or organic content, some of the energy is reflected back to the surface. These signals are received by an antenna and after processing, provide an X-ray view of the sediment layers comprising the barrier. Sediment cores are taken in conjunction with the GPR surveys to determine the composition of the sediment layers and the environment in which they were deposited. Additional

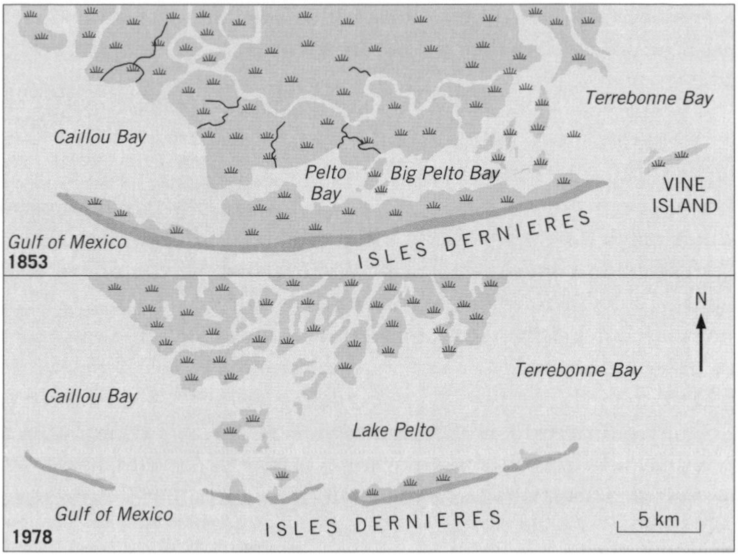

Fig. 8.15 Barrier detachment process illustrated along the Isles Dernieres located on the Mississippi River delta. As the delta plain subsides, marshlands are converted to bays. In a period of only 125 years, semi-protected Pelto and Little Pelto Bays were transformed to large open water environments. (From Penland and Boyd, 1985, details in Fig. 8.14.)

information concerning former barriers and their associated tidal inlets and lagoons has been gathered from the inner continental shelf using high-resolution shallow seismic surveys, a technology similar in principle to GPR. These advancements have provided new insights about the formation of coastal barrier systems.

Scientists now accept the idea that barriers can form by a number of different mechanisms. For example, along the west coast of Florida there are historical documentation and direct observations that barriers have formed from subtidal bars migrating onshore, primarily during storms. Other barrier systems have undoubtedly developed from spits, such as the barriers along the outer coast of Cape Cod or those found along the southern coast of Washington. In both cases, the spits are fed by abundant sand sources derived from eroding glacial cliffs in Cape Cod and the Columbia River in Washington. Along the coast of Louisiana former lobes of the Mississippi River delta have been reworked by wave action, forming beach ridge complexes. Prolonged sinking of the marshes (subsidence) behind the barriers has converted these former vegetated wetlands to open-water areas, leading to barrier detachment from the mainland (Fig. 8.15).

In considering the formation of barrier islands, it is important to recognize that almost all the world's barriers are less than 6500 years old and most are younger than 4000 years old. Most barriers formed in a regime of rising sea level but during a time when the rate of rise began to slow. Sedimentological data from the inner continental shelves off the east coast of the United States, in the North Sea, and in southeast Australia suggest that barriers once existed offshore and have migrated to their present positions. When we core the landward side of barrier islands, which in most cases represents the oldest part of the barrier, we discover that these sediments consist of overwash deposits, often times overlying lagoonal units. This sequence of sediment layers indicates that the barrier was migrating onshore during its initial development. Many barriers eventually stabilized and then prograded seaward when the supply of sediment became more plentiful and the rate of sea-level rise continued to slow.

Final observation

If the initial barriers migrated onshore to their present position and most of the structure of a barrier developed after it stabilized onshore, then it is a bit ironic that so much attention has been given to how barriers formed when little or none of this signature has been preserved in the present barrier system.

8.5 Prograding, retrograding, and aggrading barriers

The overall form of barriers, their stability, and future erosional or depositional trends are related to the supply of sediment, the rate of sea-level rise, storm cycles, and the topography of the mainland. When a barrier builds in a seaward direction it is said to prograde and is called a prograding barrier. The opposite of this condition occurs when a barrier retreats landward, called a retrograding barrier. If a barrier builds vertically and maintains its form as sea level rises, it is labeled an aggrading barrier. These different types are described below.

8.5.1 Prograding barriers

Prograding barriers form in a regime of abundant sand supply during a period of stable or slowly rising sea level (Fig. 8.16). The sand to build these barriers may come from along the shore or from offshore sources. These conditions were met along much of the east coast of the United States and many other regions of the world about 4000–5000 years ago, when the rate of sea-level rise slowed and sand was contributed to the shore from **eroding headlands** (examples: Provinceland Spit in northern Cape Cod, Massachusetts; Lawrencetown barrier along

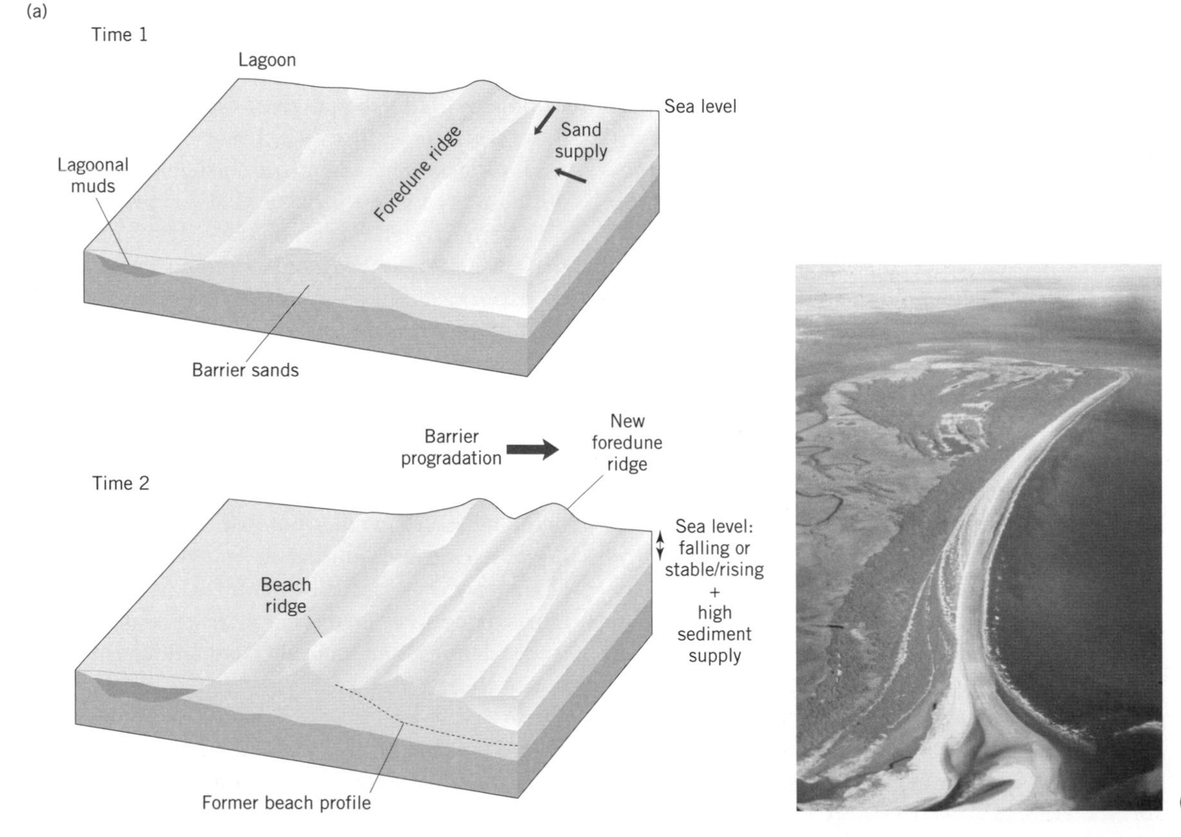

Fig. 8.16 Prograding barriers develop in regions of abundant sediment supply. (a) Model of a prograding barrier. (b) Bull Island is a beach-ridge barrier located north of Charleston, South Carolina.

the Northeast Shore of Nova Scotia), from the **inner continental shelf** (examples: the barrier system along Bogue Banks, North Carolina; Algarve barrier chain, in southern Portugal; Tuncarry barrier in southeast Australia; the East Friesian Islands along the German North Sea coast), and **directly from rivers** (examples: most barriers systems in northern New England; the barrier chain situated north of the Columbia River along the southern coast of Washington).

Prograding barriers commonly build in spurts when the supply of sand is plentiful. Increases in the rate of sediment contribution to the downdrift coast can occur during storms, when unconsolidated cliffs are eroded, releasing large quantities of sand to the littoral system, or during floods, when rivers transport high sediment loads to the coast. In other instances, sediment is moved onshore from the inner shelf during long-term accretionary wave conditions. These processes cause beaches to build in a seaward direction, widening the berm and separating the foredune ridge from the ocean by an enlarging expanse of sand.

A common end product of shoreline progradation is the development of beach ridges. As described earlier, beach ridges are dune systems that usually become vegetated first by grasses, then by shrubbery, and finally by trees. This change in the maturity of the vegetative cover occurs as the ridges are displaced further and further from the shoreline and become more protected from salt spray and the effects of storms. Each beach ridge marks a former shoreline position and their overall pattern indicates how the barrier grew and evolved through time. The interval of time between the formation of successive beach ridges may represent tens to hundreds of years.

Beach ridges form by a number of different mechanisms, involving both erosional and depositional processes (Fig. 8.17):

1 Scarp development. When a beach builds out gradually, the wind molds the upper berm into random hummocks (small incipient dunes). During moderate to large storms, the berm erodes back, producing a continuous scarp along the beach that cuts across these hummocks. This scarp provides a locus against which future wind-blown sand is deposited. In this way a dune ridge is established along the entire length of the barrier. Repetition of this process forms a beach-ridge barrier such as those along the central South Carolina coast.

2 Wrack lines. During the winter and early spring, storms coupled with spring high tides float organic material, largely composed of dead marsh grass, out of the backbarrier, dispersing it along adjacent beaches. The high water levels and large waves accompanying the storms cause the debris to be deposited along the upper portion of the beach. Depending upon the region, these wrack or drift lines, as they are called, may also consist of eel grass, seaweed, or driftwood. The wrack line traps the wind-blown sand, causing dune development. In this way, a dune ridge and future beach ridge may form along the entire length of the barrier.

3 Bar migration. A barrier can prograde through the addition of sand bars that migrate onshore and attach to the upper beach. Landward migrating bars are particularly common in the vicinity of tidal inlets and are discussed in much greater detail in Chapter 12. If one of these bars incompletely welds to the upper berm, then a low area or swale develops between the bar and the landward beach. The bar itself commonly builds vertically from wind-blown sand, forming a dune ridge. When a series of bars migrates onshore and attaches to the beach, the resulting ridge and swale topography often becomes a beach-ridge barrier.

In fact, beach ridges may develop by a combination of two or more of the above processes. A barrier composed of beach ridges is easily identified as one having evolved through progradation, but this morphology does not necessarily indicate that the barrier is still prograding. Conversely, sedimentation conditions along the barrier may have changed such that now it has become erosional.

8.5.2 Retrograding barriers

Retrograding barriers form when the supply of sand is inadequate to keep pace with relative sea-level rise and/or with sand losses. Stated in terms of a sediment budget, a barrier becomes retrogradational when the amount of sand contributed to the barrier

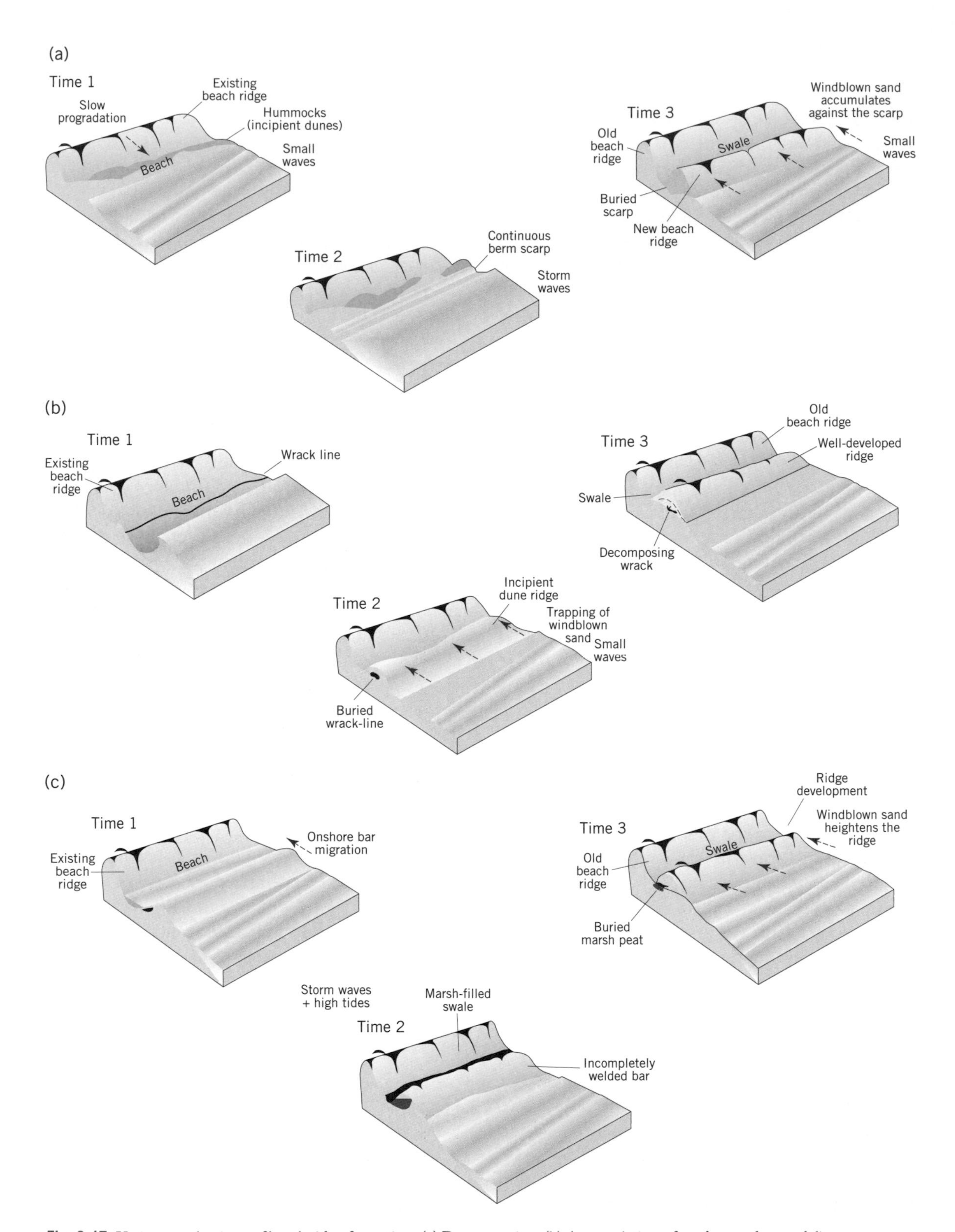

Fig. 8.17 Various mechanisms of beach ridge formation. (a) Dune scarping. (b) Accumulation of sand around a wrack line. (c) Bar welding.

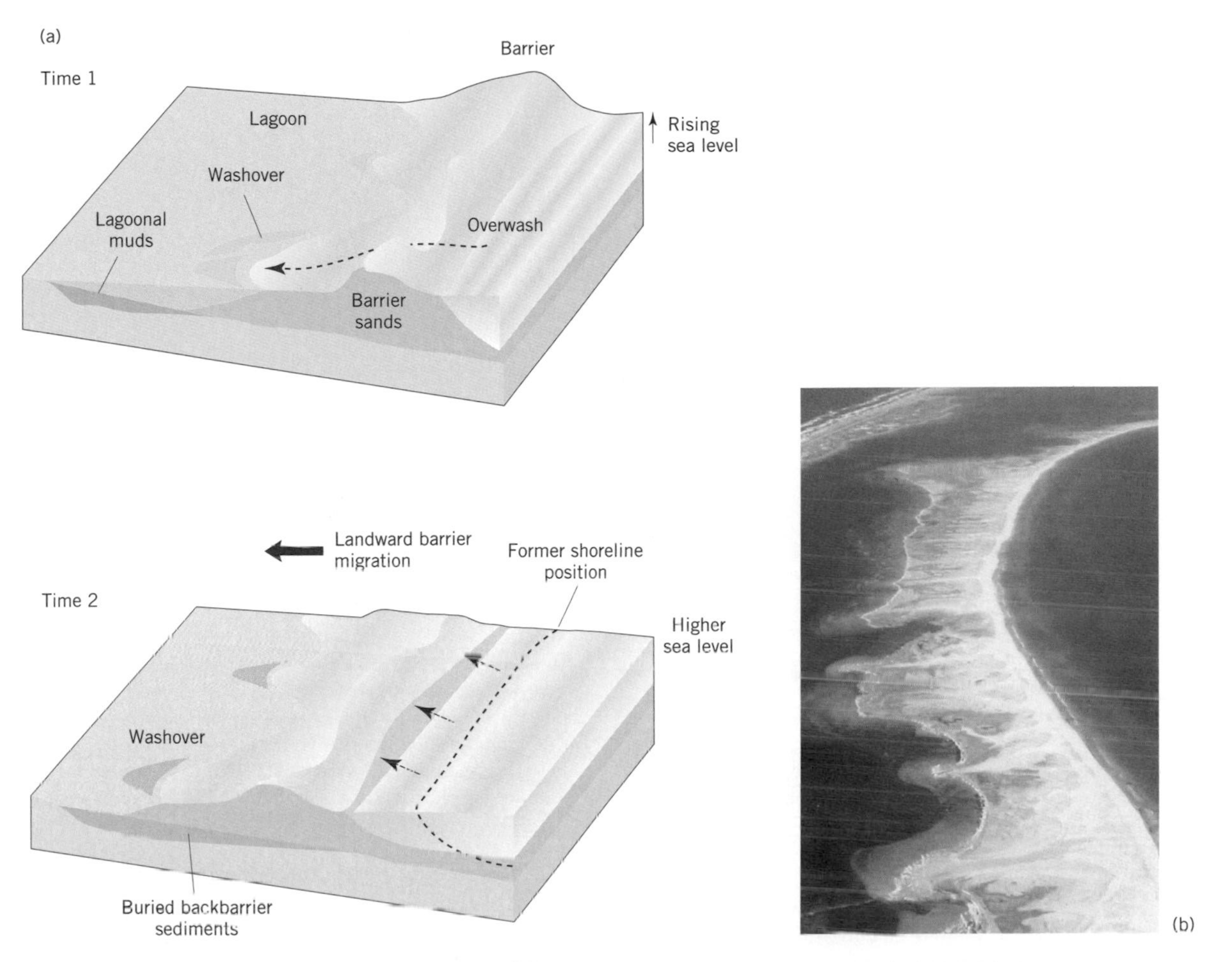

Fig. 8.18 Retrogradational barriers develop in a regime of rising sea level and a depleted sediment supply. (a) Model of a retrogradational barrier. (b) Barrier rollover is occurring along the Magdalen Islands by storm overwash.

is less than the volume transported away from the barrier (Fig. 8.18). The sand may be lost offshore during storms, moved along shore in the littoral system, or transported across the barrier by overwash. The end result is erosion to the front of the barrier, causing a decrease in width of the beach and ultimately destruction of the foredune ridge. Eventually, a narrowing and lowering of the barrier profile produces a retreat of the barrier across the adjacent bay, lagoon, or marsh system. This landward migration of the barrier, termed barrier rollover, is accomplished primarily during storms by overwash. Overwash is a cannibalistic process whereby storm waves transport sand from the beach through the dunes, depositing it along the landward margin of the barrier. In this way the barrier is preserved by retreating landward (Fig. 8.19).

Historical records demonstrate that some barriers, such as the Chandeleur Islands off the Louisiana coast, have migrated onshore from one to several times their widths during the past 100 years. The major cause of barrier retreat in Louisiana and much of the Gulf coast region is attributed to relative sea-level rise and exhausted sediment supplies. Retreat of the Chandeleurs is coincident with storm surges and large waves associated with the passage of

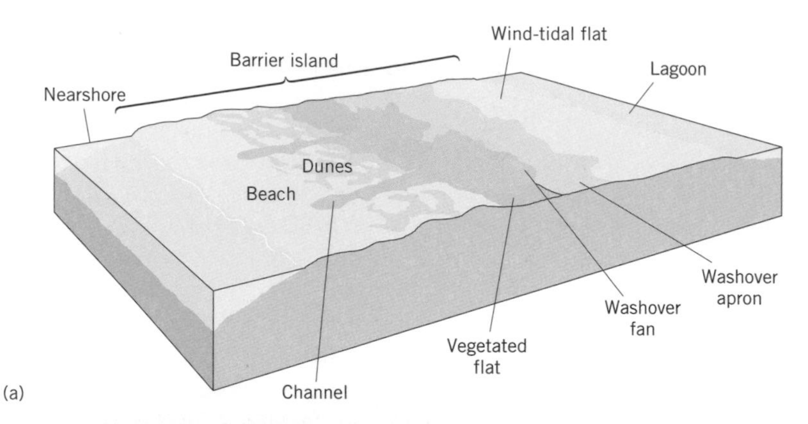

(a)

(b)

Fig. 8.19 (a) Washovers are the primary mechanism whereby a barrier migrates onshore (from R. A. Davis Jr (ed.), 1994, *Geology of Holocene Barrier Systems*. New York: Springer). (b) In some instances a washover is a precurser to barrier breaching and tidal inlet formation, such as at Monomoy Island along the southeast coast of Cape Cod, Massachusetts.

hurricanes. A somewhat different scenario of barrier retreat has occurred along the central Maryland coast. Here in 1933 a severe hurricane breached the northern end of Assateague Island, creating a tidal inlet (Ocean City Inlet). In 1935, jetties built to stabilize the entrance to the inlet blocked the southerly transport of sand that had once nourished Assateague Island. In addition, strong ebb tidal currents produced in the jettied channel transported large quantities of sediment offshore, forming a massive sand shoal known as an ebb-tidal delta. Due to these sand-trapping mechanisms, the sediment-starved shoreline immediately downdrift of the inlet began to erode. Eventually, erosion along northern Assateague Island reached a critical width (~200 m) and low overall elevation such that overwash activity produced barrier rollover. In less than 50 years after the island was breached, the barrier south of the inlet had retreated one island width across the adjacent lagoon.

Retrograding barriers are identified by their overall narrow width, single or nonexistent foredune ridge, and washover aprons. Because these barriers have been migrating over various types of back-barrier settings, including lagoons and marshes, the sedimentary components comprising these environments are commonly exposed along the front side of the barrier, usually in the intertidal zone. This explains the appearance of stiff muds and the remnants of peat deposits along the lower beach of many retrograding barriers. In some instances, tree stumps may extend through the sands of the intertidal zone, indicating that the barrier has migrated onshore a distance great enough for trees once growing on the mainland to have resurfaced as stumps on the seaward side of the barrier.

8.5.3 Aggrading barriers

If a barrier has built vertically through time in a

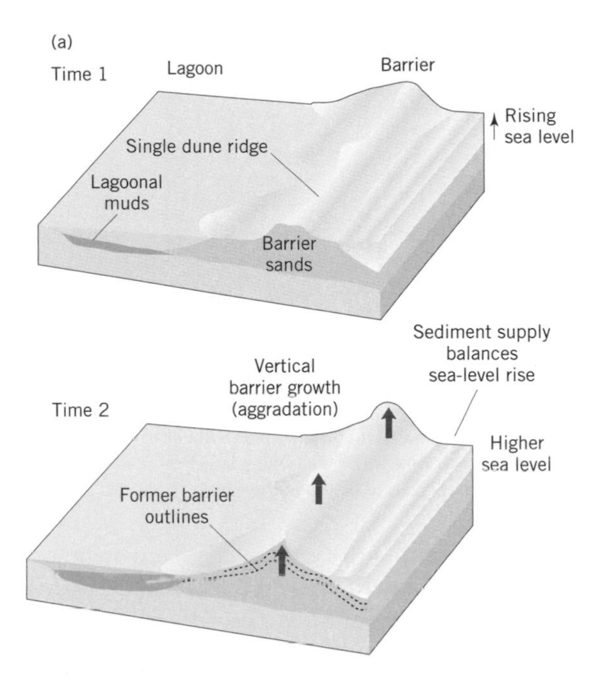

Fig. 8.20 Aggrading barriers build vertically in a regime of slow sea-level rise. (a) Model of an aggrading barrier. (b) Good Harbor Beach is a pocket barrier in northen New England. Coring studies have revealed that the barrier has accreted vertically during the past 2500 years and is more than 7 m in thickness.

regime of rising sea level and occupies approximately the same foot-print as it did when it first formed or stabilized, it is termed an aggrading barrier (Fig. 8.20). These barriers are rare because sediment must be supplied (to the barrier) at a rate that exactly compensates for rising sea level. Too little sand and the barrier migrates onshore (retrograding), whereas too large a supply and the barrier builds seaward. Part of Padre Island along the Texas coast is an example of an aggrading barrier. Without subsurface information they are difficult to recognize, because morphologically they may appear similar to a non-beach ridge, a prograding barrier, or even a retrograding barrier that has stopped moving onshore.

8.6 Barrier stratigraphy

Barriers exhibit a variety of architectures consisting of many different types of sedimentary deposits, depending upon their evolutionary development (Fig. 8.21). The sequence and composition of the layers making up the barrier, termed its stratigraphy, are defined by a set of grain size, mineralogic, and other characteristics of the layers. Factors such as sediment supply, rate of sea-level rise, wave and tidal energy, climate, and topography of the land dictate how a barrier develops and its resulting stratigraphy. For example, barriers that formed in the vicinity of the Mississippi River delta consist of fine to very fine sand because this is the most abundant coarse sediment size delivered by the river. In contrast, barriers along glaciated coasts tend to contain coarser-grained sediment, including gravel, due to the coarse-grained, often heterogeneous, nature of glacial deposits found along these coasts. Even the rivers of these regions tend to deliver fine to coarse sand to the coast. The gravelly sand barriers along the northeast shore of Nova Scotia examplify this condition, having been formed from the erosion of glacial features called drumlins (see Chapter 17). In some regions the sediment comprising the barriers may have come from more than one source. For example, on the Gulf coast of Florida, barrier sediments are composed of two populations: (i) a carbonate sand derived from shells and other carbonate material; and (ii) a terrigenous sand originally brought to the peninsula from the Appalachain Mountains.

Barriers exhibit highly variable thicknesses from the thin (~2 m) Chandeleur Islands off the Louisiana coast or (1.2 m) Coke Island in North Carolina

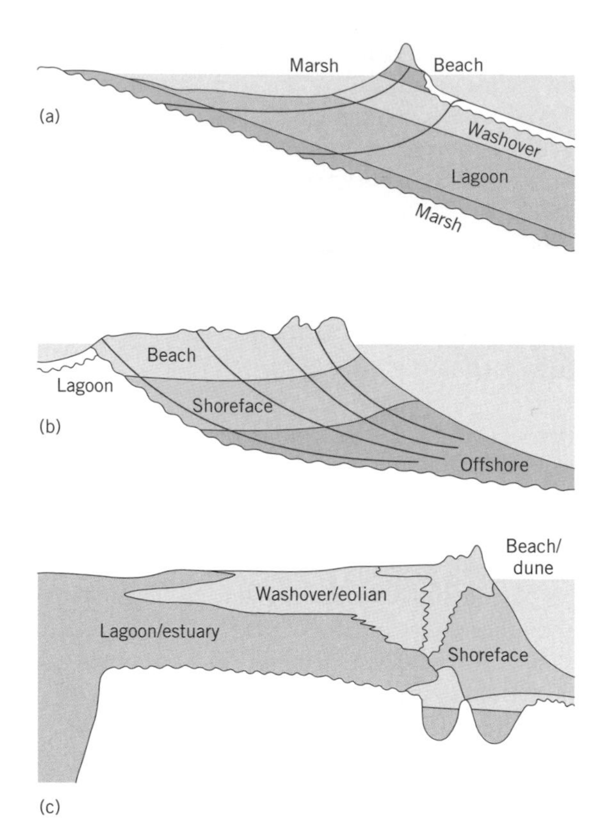

Fig. 8.21 Stratigraphic models of different types of barrier sequences (from W. E. Galloway & D. K. Hobday, 1980, *Terrigenous Clastic Depositional Systems*. New York: Springer). (a) Transgressive barriers are underlain by washover and lagoonal facies. (b) Regressive barriers commonly contain prograding beach ridges. (c) Aggradational barriers represent steady state conditions.

Fig. 8.22 Coke Island in southern North Carolina is a relatively thin retrogradational barrier consisting largely of coalescing overwash fans deposited on top of the marsh.

(Fig. 8.22) to Tuncurry barrier in southeast Australia, where barrier sands extend to depths of more than 20 m, or Plum Island in northern Massachusetts, which ranges from 15 to 20 m in thickness. Along the west coast of Florida there is little new sand being added to this coast due to the lack of any riverine supply and the dearth of sediment on the inner continental shelf. Thus, the barrier deposits here are mostly less than 5 m thick. Generally, there is a direct correspondence between barrier thickness and sediment abundance. Large sand supplies cause barriers to build seaward and aid in dune construction. Both processes contribute to thick barrier sequences. Another important factor affecting barrier thickness is accommodation space, which defines how much room is available for the accumulation of barrier sands. Steeper gradient coasts produce more accommodation space as a barrier progrades than do flat-lying coasts. Likewise, as a spit builds into a deep bay, the thickness of the barrier sands will increase as the accomodation space increases. Rising sea level can produce a similar effect. For example, in Cape Cod Bay the Sandy Neck barrier has been building across Barnstable Bay for the past 3500 years. During that time, sea level has risen approximately 3 m and thus if the barrier maintains nearly the same elevation above mean high water through time, then the spit will be about 3 m thicker at its end than where spit growth was initiated.

Barrier sequences often contain tidal inlet deposits, especially along barrier coasts where tidal inlets open and close and/or where tidal inlet migration is an active process. A tidal inlet migrates by eroding the downdrift side of its channel while at the same time sand is added to the updrift side of its channel. In this way, the updrift barrier elongates, the downdrift barrier becomes shorter, and the migrating inlet leaves behind channel fill deposits underlying the updrift barrier (Fig. 8.23). Independent studies along New Jersey and the Delmarva Peninsula, North Carolina, and South Carolina indicate that 20–40% of these barrier coasts are underlain by tidal inlet fill deposits. Long-term tidal inlet migration along Shackleford Banks in North Carolina

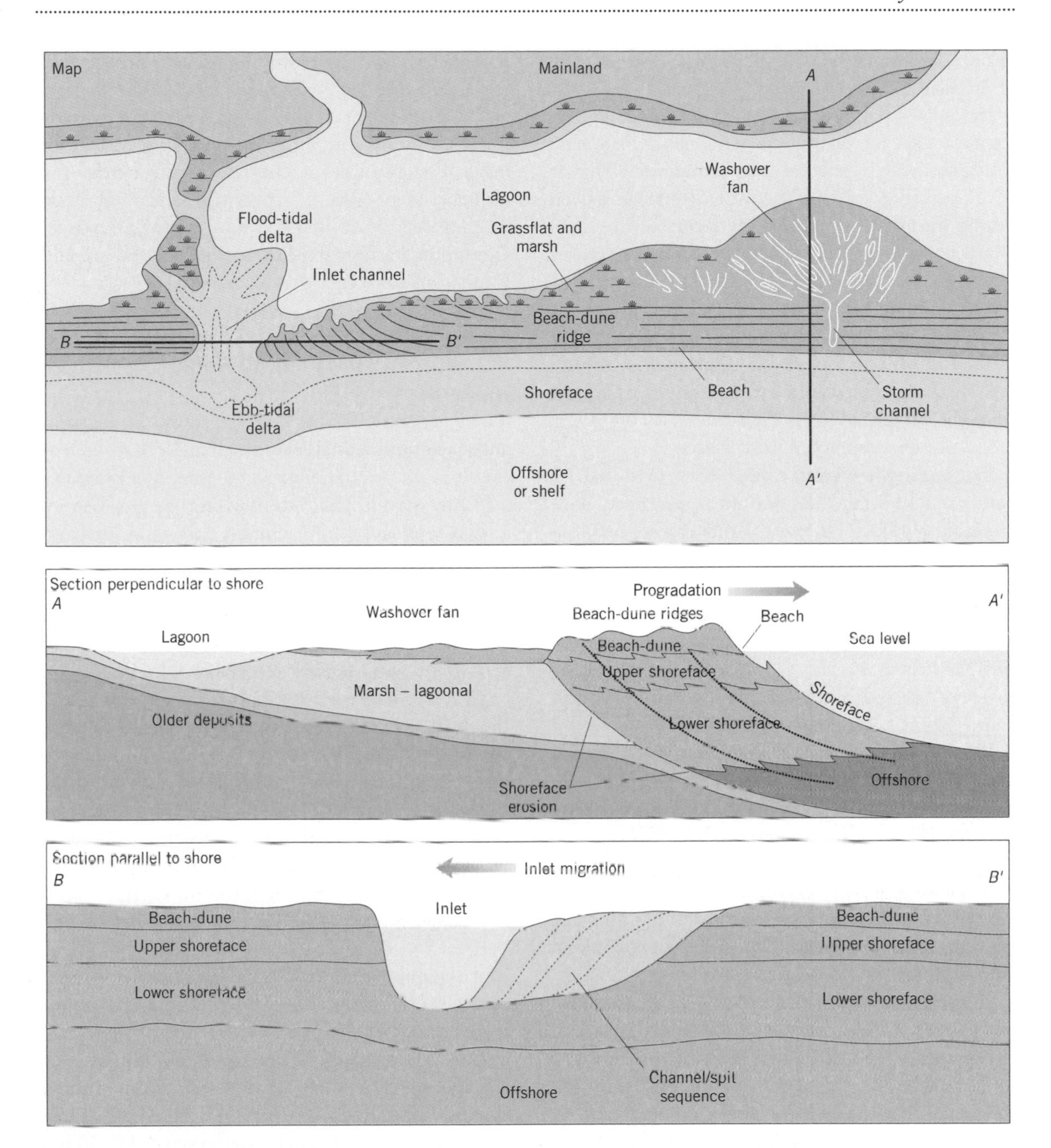

Fig. 8.23 The stratigraphy of a barrier is dependent of its evolutionary history (from Donald G. McGubbin 1982, Barrier-island and strand-plain facies. In P. A. Scholle & D. Spearing (eds), *Sandstone Depositional Environments*, p. 248, Fig. 1). It may contain elements of storm washovers (section *A–A'*), beach ridge progradation (section *A–A'*), or tidal inlet migration and spit accretion (section *B–B'*).

has produced inlet fills 10–20 m thick beneath 90% of the island.

In terms of prograding, retrograding, and aggrading barriers, each of these systems has a diagnostic stratigraphy that reflects the manner in which it developed (Fig. 8.21). If we were able to cut a deep trench through the barrier, the layers of the sediments comprising the barrier would be revealed. In actuality, the stratigraphy of barriers is determined from numerous sediment cores drilled throughout the barrier, and in some instances augmented with GPR and other geophysical information. In each case discussed below, the stratigraphy is described from the Holocene (post Pleistocene, 10,000 years ago to present) contact to the surface.

1 Prograding barrier. Because this type of barrier builds in a seaward direction, the barrier sequence is commonly thick (10–20 m) and overlies offshore deposits, usually composed of fine-grained sands and silts. The barrier sequence consists of nearshore sand, overlain by beach deposits and topped by dune sand. The contacts between the units are gradational and for the most part the sedimentary sequence coarsens upward except for the uppermost fine-grained dune sands.

2 Retrograding barrier. This barrier type migrates in a landward direction over the marsh and lagoon by overwash processes. The Holocene sequence typically bottoms in lagoonal muds, but if the barrier has retreated far enough landward, mainland deposits may be preserved, forming the base of the sequence. In these instances, we may find tree stumps, soils, and other deposits. The mainland units are overlain by backbarrier sediments, including a variety of units such as lagoonal silt and clay and marsh peat that had formed in intertidal areas. In the vicinity of tidal inlets, backbarrier deposits consist of channel sand and large sand shoals called flood-tidal deltas (see Chapter 9 on tidal inlets). Overlying the backbarrier deposits is the thin barrier sequence (<3–4 m) consisting of overwash, beach deposits, and dune sediments if they are present.

3 Aggrading barrier. These barriers build upward in a regime of rising sea level and, in an ideal case, the deposits from the same environmental setting are stacked vertically. In most cases, however, the barrier has shifted slightly landward and seaward through time due to changes in sediment supply and rates of sea-level rise. Therefore, most aggrading barriers exhibit some interstacking of various units. For example, in the rear of the barrier the sequence may consist of overwash and dune units interlayered with marsh and lagoonal deposits. Aggrading barriers tend to be thick (10–20 m) and for reasons stated above are uncommon.

8.7 Barrier coast morphology

There are many factors that determine the location and size of individual features along a coast, such as the slope of the land dictating the size of a lagoon or a former stream valley controlling the position of a tidal inlet. Despite the many factors influencing coastal morphology in coastal plain settings, the overall distribution of barriers, tidal inlets, and various backbarrier environments is primarily related to the relative magnitude of wave and tidal energy. In a simplification of their respective roles, waves are responsible for the transport of sediment along the shore, which tends to elongate barriers. The rise and fall of the tides cause a filling and emptying of backbarrier areas. Tidal inlets through which this exchange of water occurs are the sites of strong tidal flow. These currents transport sand in onshore and offshore directions.

8.7.1 Hayes models

In a scheme conceived by Miles Hayes and later modified by him and others, depositional coastlines are separated into three classes based on the wave height and tidal range of the region. The three major divisions include wave-dominated, mixed energy, and tide-dominated settings (Fig. 8.24). Barriers are found almost exclusively in the wave-dominated and mixed energy environments. It should be noted that it is not the absolute value of either wave height or tidal range that is significant; it is the ratio of the two parameters that dictates the presence and distribution of barriers. For example, in the Ten Thousand Island region along the southern Florida

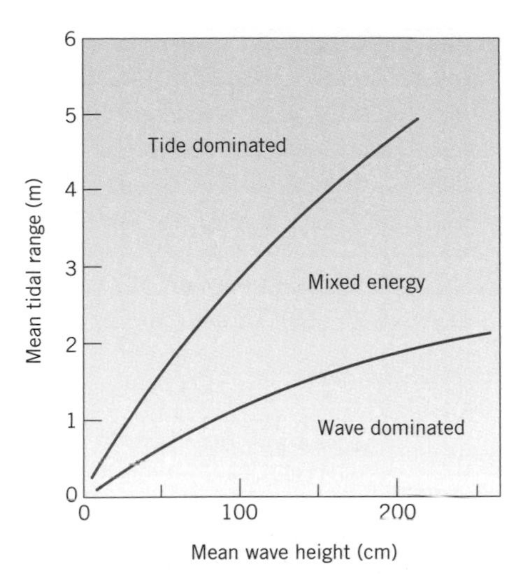

Fig. 8.24 Depositional coastlines can be separated into three major types based on their wave and tidal energy. Barrier coasts occur almost exclusively in the wave-dominated and mixed energy environments. Tide-dominated coasts are generally funnel-shaped and associated with a river. (From R. A. Davis Jr & M. O. Hayes, 1984, What is a wave-dominated coast? *Marine Geology*, **60**, 313–29, after M. O. Hayes, 1979, Barrier island morphology as a function of tidal and wave regime. In S. P. Leatherman (ed.), *Barrier Islands*. New York: Academic Press.)

Gulf coast the tidal range is only about 1 m, but this section of coast, consisting primarily of mangrove islands, is clearly tide-dominated. The lack of barriers along this coast is the result of very low wave energy. Remember that barriers are wave-built accumulations of sand, and where wave energy is insufficient, barriers will not form. Similarly, in the German Bight of the North Sea extensive barrier development disappears toward the apex of the bight at the entrance to the Elbe River, where the tidal range increases to almost 3.5 m. The expansive tidal flats and large tidal range of this region diminsh wave energy, precluding the formation of barriers. Moreover, the importance of wave energy is illustrated along the west coast of the United States, where barriers exist despite spring tidal ranges approaching 4 m. Here an abundant sand supply and strong wave energy overcome the effects of tides to produce some very long retrogradational barriers.

Given the same ratios of wave height and tidal range, depositional coasts throughout the world exhibit similar morphologies, as described below.

Wave-dominated coast

These coasts are dominated by wave-generated longshore sediment transport; tides play a secondary role. They are characterized by long, linear barrier islands, and few tidal inlets (Fig. 8.25). The backbarriers are composed of mostly open-water lagoons or bays. Marshes occur along the backside of the barriers, commonly on former overwash deposits, and at the edge of the mainland. Large sand shoals are usually found on the landward side of the tidal inlets.

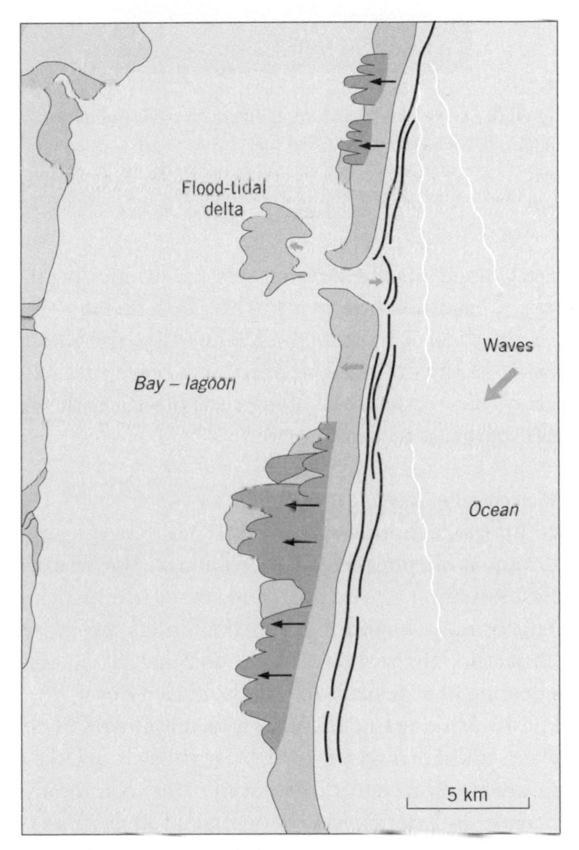

Fig. 8.25 (a) Model of a wave-dominated barrier coast (from M. O. Hayes, 1975, Morphology of sand accumulations in estuaries. In L. E. Cronin (ed.), *Estuarine Research, 2*. New York: Academic Press, pp. 3–22; Hayes, 1979, details in Fig. 8.3).

Fig. 8.25 (*cont'd*) (b) The Outer Banks of North Carolina are an example of a wave-dominated coast. The cuspate forelands of Cape Hatteras (north) and Cape Lookout (south) are joined by long linear barrier islands that are interrupted by few tidal inlets.

Sand shoals on the seaward side are diminutive in size. Coastlines fitting into this class include the coasts of Texas, the panhandle of Florida, the Outer Banks of North Carolina, Maryland, northern New Jersey, the barrier coast along the Nile River delta, and southeast Iceland.

Mixed energy coast

In this model both wave and tidal processes are important in shaping coastal morphology. Barriers on these coasts tend to be short and stubby (drumstick-shaped; see Chapter 12) and tidal inlets are more numerous than along wave-dominated coasts, reflecting the greater role of the tides (Fig. 8.26). The backbarriers of these regions are mostly filled with sediment and covered by expansive marshes incised by tidal creeks. Open water areas commonly increase in extent near the inlets. In at least two locations of the world, the backbarriers of this coastal type (East Friesian Islands along Germany; Copper River delta barriers in Alaska) consist of extensive tidal flats. It is likely that the change from open water lagoons of wave-dominated coasts to the intertidal environments of mixed energy coasts is the result of more sediment being transported into the backbarrier by tidal currents. In addition, an increase in tidal range produces larger interidal areas, which

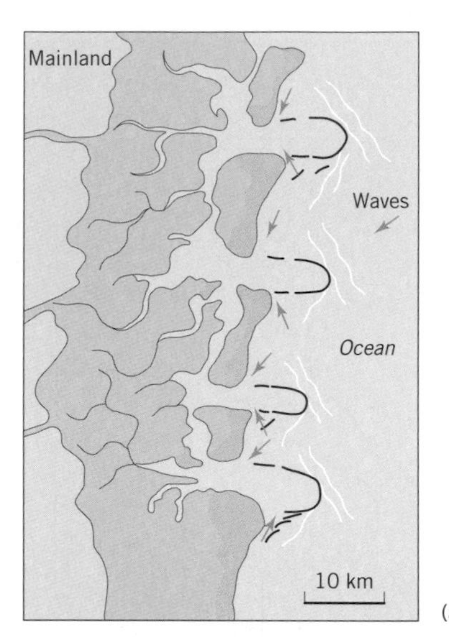

Fig. 8.26 (a) Model of a mixed energy barrier coast (from Hayes, 1975, details in Fig. 8.25; 1979, details in Fig. 8.3). (b) The beach ridge barriers in Georgia are an example of a mixed energy coast. Many of the barriers along the Georgia Bight have formed around pre-existing Pleistocene-age barrier islands.

promote the development of marshes and the stabilization of sediment. Mixed energy coasts include the barrier coasts of northern New England, southern New Jersey, Virginia, South Carolina, Georgia, much of the west coast of Florida, the Friesian Islands in the North Sea (see Box 8.1), and the Copper River delta barrier system in Alaska.

Tide-dominated coasts

Many tide-dominated coasts coincide with very large funnel-shaped embayments (>100 km across). This coastline configuration enhances the tidal wave, which produces large tidal ranges, strong tidal currents, and a sedimentation regime that is dominated by onshore–offshore transport. Due to the low wave energy along these coasts, barriers and tidal inlets are absent. Rivers commonly discharge sediment at the heads of these embayments. The coarse-grained sediment is transported offshore and deposited on large subtidal sand ridges that parallel the tidal flow. The fine-grained sediment tends to accumulate onshore, forming expansive tidal flats. Landward of

Box 8.1 Three-hundred year history of the East Friesian Islands

The East Friesian Islands provide a unique opportunity to assess the long-term historical development of a barrier island chain. Hahns Homeier and Gunter Luck of the Forschungsstelle Nordorney, Germany, assembled a series of detailed maps that depict shoreline and bathymetric changes of this barrier system that extend back to 1650. There are few barrier coasts in the world for which the historical database goes back this far. The sequential maps demonstrate patterns of inlet migration, growth of individual barriers, reduction in size of the backbarrier drainage areas, and decrease in size of the ebb-tidal deltas. Interestingly, some of the historical changes are a product of a strong wind regime, but others, including some large-scale morphological changes, are solely related to actions by humans.

The East Friesian Islands are composed of seven barriers and six tidal inlets spanning the southeast coast of the North Sea between Ems River to the west and Weser River to the east (Fig. B8.1). The barrier chain is 90 km long and separated by from the mainland by a 4–12 km wide tidal flat, which is incised by a network of tidal channels. The winds in this region blow from the westerly quadrant during the entire year, producing a net easterly longshore transport rate of

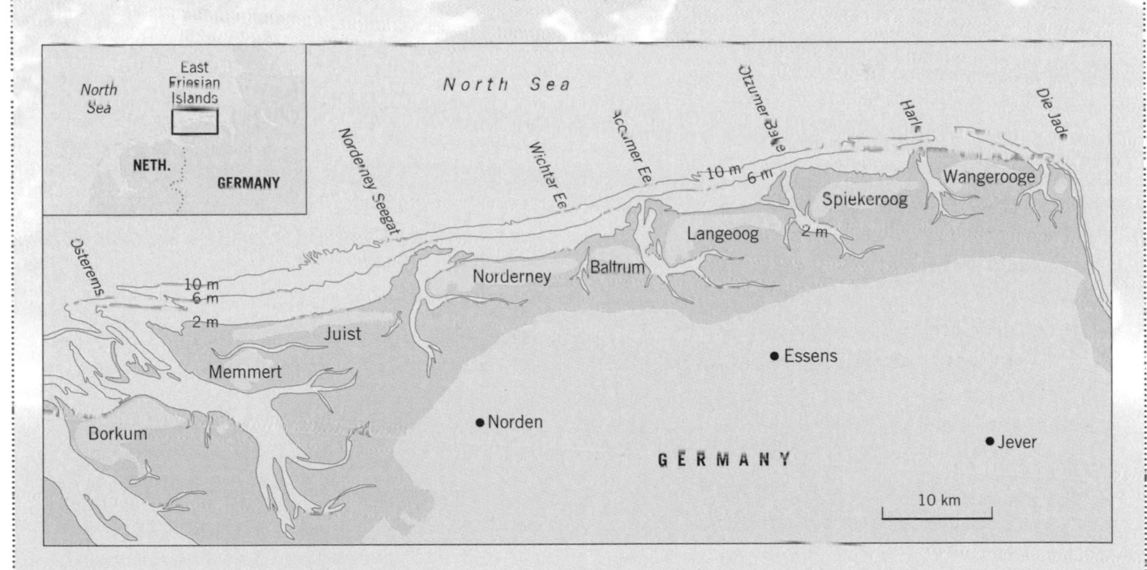

Fig. B8.1 The East Friesian Islands are located along the coast of Germany on the North Sea.

Box 8.1 (*cont'd*)

270,000 m^2 yr^{-1} of sand. Spring tidal ranges vary systematically from 2.5 m at the western end of the chain to 2.9 m at the eastern end.

Much of the sand moving eastward in the littoral system is eventually delivered to the seegats (tidal inlets), where it bypasses the inlets through a complex pattern of wave and tidal processes. Sand enters the inlets from the west and is transported across the ebb-tidal delta, ultimately reaching the downdrift inlet shoreline in the form of large landward migrating swash bars. These bars are elongate parallel to shore and are 1–1.5 km in length, containing more than 100,000 m^2 of sand. As discussed in greater detail in Chapter 12, the shape of the barriers is dictated by where the swash bars weld to the beach. Drumstick-shaped barriers develop where the bars attach to the updrift end of the island, hump-backed barriers form when bars weld to the middle of the island, and a downdrift bulbous barrier shape is a product of bars moving onshore at the distal eastern end of the island.

The East Friesian Islands have evolved in a regime of rising sea level, like most other barrier sytems in the world. The fact that the Friesian barriers have grown in size during the past 300 years is evidence that the supply of sand to the island chain has more than compensated for land loss due to continued sea-level rise. The increase in dimensions of the barriers is a product of new sand being added to islands as well as due to human modification of the backside of the barriers through poldering (Fig. B8.2). Poldering is the process whereby land is reclaimed from the sea. The practice involves building dikes across tidal flats and allowing sediment-laden tidal waters to enter regions that have been diked. After the suspended sediment from the seawater is deposited, the clear water is discharged from the dikes at low tide and new muddy seawater is allowed to flood through the dikes during the next high tide. After this procedure is repeated many hundreds of times, the tidal flat region accretes to an elevation where it can become productive farmland. Poldering explains how the back sides of the barrier have grown in size since 1650.

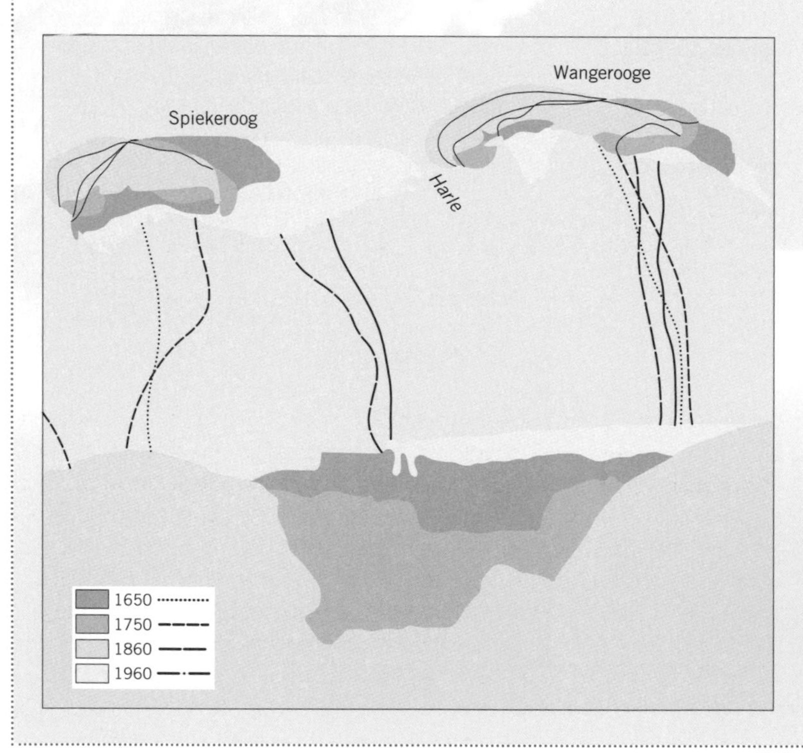

Fig. B8.2 Changes in areal extent of the drainage system of Harle Inlet. Land reclamation of tidal flats has decreased the size of the backbarrier, resulting in a smaller inlet tidal prism.

Box 8.1 (*cont'd*)

During the same period of time, individual barriers along the East Friesian chain have lengthened and some have prograded seaward. Between 1650 and 1960 the barriers increased in aerial extent by almost 42 km^2, an increase of 80%. Of this new land, 56% was attributed to poldering and 44% was the result of barrier accretion. The source of new sand was puzzling to scientists until a historical analysis was made of the backbarrier region. It was found that from 1650 to 1960 the area drained by the tidal inlets (drainage area) decreased by 30%, amounting to 149 km^2 loss in tidal flats and open water areas. The decrease in drainage area was due in part to spit accretion at the eastern end of the barriers, but was primarily a result of poldering, not only along the landward side of the barriers but also along the mainland shore. As the size of the backbarrier areas decreased, so too did the tidal prism of the inlets. Because inlet tidal prism controls the volume of sand contained in the ebb-tidal deltas, extensive poldering in the backbarrier during the 1650–1960 period ultimately led to smaller equilibrium-sized ebb deltas. As the ebb deltas reduced in volume, wave energy transported the deltaic sand back onshore, increasing the sand supply to the adjacent beaches.

The end result of backbarrier poldering has been highly beneficial to the stability and long-term depositional trend along most of East Friesian Island chain. The striking growth of the barriers between 1650 and 1960 is reflected in an increase in the total length of the barriers by more than 14 km. Most of this increase has been at the expense of the tidal inlets, which collectively have narrowed by over 10 km (Fig. B8.3). Thus, the history of the East Friesian Island chain demonstrates that changes to the backbarrier can affect the sediment supply to the fronting barriers and, more importantly, human alterations can have a pronounced impact on the erosional–depositional trends of a barrier coast.

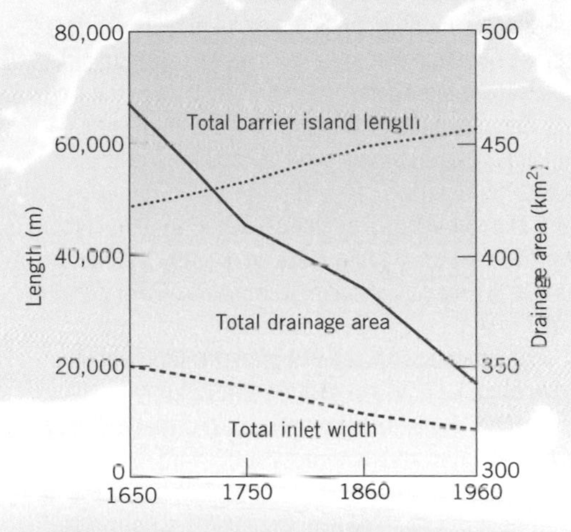

Fig. B8.3 Graph showing morphological changes of the East Friesian Inlets between 1650 and 1960. During this 310-year period the drainage areas decreased in size, resulting in smaller tidal inlets and longer barrier islands. (From D. M. FitzGerald, 1988, Shoreline erosional–depositional processes associated with tidal inlets. In D. G. Aubrey & L. Weishar (eds), *Hydrodynamics and Sediment Dynamics of Tidal Inlets*. Berlin: Springer, pp. 186–225.)

the tidal flats are wide marshes. In equitorial regions marshes are replace by mangroves. The Bay of Fundy in Nova Scotia, Gulf of Cambay in India, head of the Bay of Bengal in Bangladesh, mouth of the Amazon River in Brazil, and Bristol Bay in Alaska are examples of tide-dominated coasts.

8.7.2 Georgia Bight

A particularly interesting section of coast, demonstrating the relative influence of wave versus tidal energy, is seen in the Georgia Bight, encompassing the region from North Carolina to Florida. The arcuate shape of this coast produces relatively narrow and steep continental shelves along the flanks of the bight and a wide, shallow shelf at the apex to the bight. From the Outer Banks of North Carolina to the Georgia coast the widening of the continental shelf increases the amplification of the tidal wave as it moves across the shelf, resulting in larger tidal ranges. Conversely, as the inner continental shelf flattens along this same stretch of coast, a

Box 8.2 Rapid changes at Anclote Key, Florida

We generally thnk that major changes to barrier islands either take a long time to eventuate or are the result of hurricanes or other intense storms. Major changes in the length and morphology of Anclote Key on the Gulf coast of the Florida peninsula have taken place over only a few decades without the benefit of any significant storm. This island, which is the northernmost barrier along this section of coast, is separated from the mainland by a 5 km wide shallow expanse of water. Anclote Key is a wave-dominated linear barrier exhibiting recurved beach ridges at both ends. Despite having no nearby adjacent barriers, fairly deep, inlet-like channels exist at both ends of the island. Radiocarbon dates of shells and organic material obtained during extensive coring of the island have shown it to be about 1500 years old. The barrier sand rests on a pavement of Miocene limestone that is from 3 to 5 m below mean sea level.

Historical maps and charts of the region indicate that Anclote Key did not change significantly from 1881 to the 1960s. Most of the changes that did occur were associated with a modest southward extension of the island. During this time the landward migration and attachment of swash bars to the southern shoreline relegated an old coastguard pier to an upland position. The northern end of the island displayed no noticeable change until the early 1960s. Over the next 20 years, spit accretion extended the northern end of Anclote Key by more than a kilometer. Historical records reveal that the spit system has consistently terminated at the site of a tidal channel. Strong flood-tidal currents in this channel (landward directed currents), along with wave action and a large sediment supply, have produced a long spit that has widened the northern end of the island by about a kilometer (Fig. B8.4).

The growth of a barrier over a 20-year period is in itself remarkable, particularly along a segment of coast that is sediment-starved. Where did all the sand come from? The answer to this question was found in an analysis of aerial photographs taken along this coast during the 1950s (1951 and 1957). Both of these sets of photographs show that sea grass once covered an expansive area of the inner shoreface to within 30 m or so of the shoreline. A 1963 aerial photograph of the same region shows that the vegetation had disappeared. Some process or condition, as yet undetermined, caused the demise of the sea grass, which had been stabilizing the sand substrate. Once the stabilizing effect of the grass was removed, wave action transported the nearshore sand onto the adjacent beaches of Anclote Key. Much of the newly accreted sand was then carried by longshore currents to the northern end of the island, forming recurved spits. This same phenomenon also occurred a few kilometers south of Anclote Key and was responsible for the construction of a long spit on Honeymoon Island, as well as the formation of a new barrier island called Three-Rooker Bar.

Fig. B8.4 Aerial photograph of northern Anclote Key.

greater proportion of the deepwater wave energy is attenuated. Thus, all other factors being about equal, tidal energy increases and wave energy decreases from North Carolina to Georgia. The coastal response to this change in the physical setting is dramatic (Fig. 8.27). The wave-dominated Outer Banks consist of long linear barriers interrupted by few tidal inlets and separated from the mainland by broad shallow bays. Contrastingly, the mixed energy Georgia coast is composed of relatively short, beach ridge barriers separated by large tidal inlets with well developed sand shoals. The backbarrier consists of marsh and tidal creeks. Whereas other factors control the width of the backbarrier, location

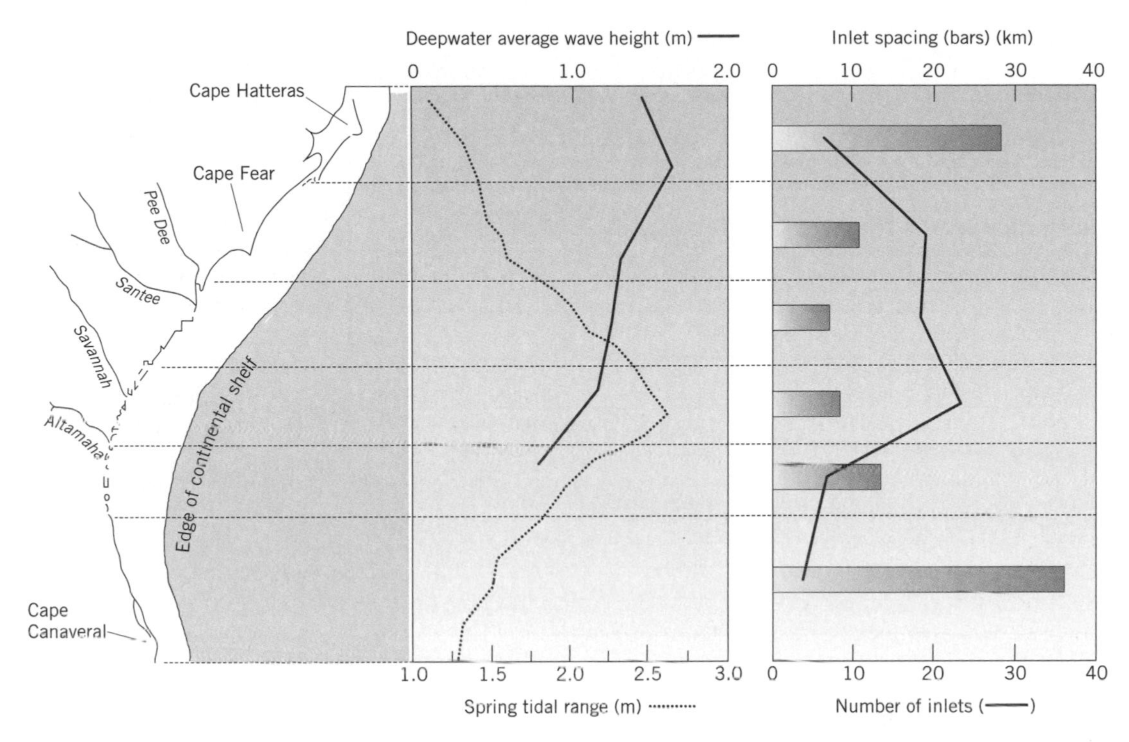

Fig. 8.27 The Georgia Bight exhibits a systematic alteration in coastal morphology that is largely a product of change in wave versus tidal energy, which in turn is a function of continental shelf width. (From D. M. FitzGerald, 1996, using data from D. Nummedal, G. Oertel, D. K. Hubbard & A. Hine, 1977, Tidal inlet variability – Cape Hatteras to Cape Canaveral. In *Proceedings of Coastal Sediments '77*. Charleston, SC: ASCE, pp. 543–62; Hayes, 1979, details in Fig. 8.3.)

of inlets, sediment supply, and longshore transport directions, the overall morphology of the coast is a function of wave versus tidal energy, which, in turn, is strongly influenced by continental shelf width.

8.8 Barrier coasts: morphology and evolution

In some areas of the world, the historical records are long enough to record the physical processes that produce major changes to the coast, thereby revealing how these barriers evolve. In other locations detailed stratigraphic studies have led to a good understanding of barrier development. Using these studies, examples of different barrier settings, their morphology, and their evolution are presented below.

8.8.1 Eastern shore of Nova Scotia

Along glaciated coasts local glacial deposits are commonly the major source of sediment to form barriers. These deposits are usually composed of till (poorly sorted mixture of gravel, sand, silt, and clay) or somewhat better sorted outwash sediments consisting of sand and fine gravel layers laid down by glacial melt water streams. Due to the mixed nature of these sediments, the barriers that they produce are composed of sand and gravel. In 1987, Ron Boyd and colleagues at Dalhousie University proposed a six-stage model to explain the evolution of the eastern shore of Nova Scotia (Fig. 8.28). According to their model, when the last continental glaciers retreated from a position on the continental shelf a variety of glacial deposits was left behind (Stage 1). As sea level rose, the large valleys excav-

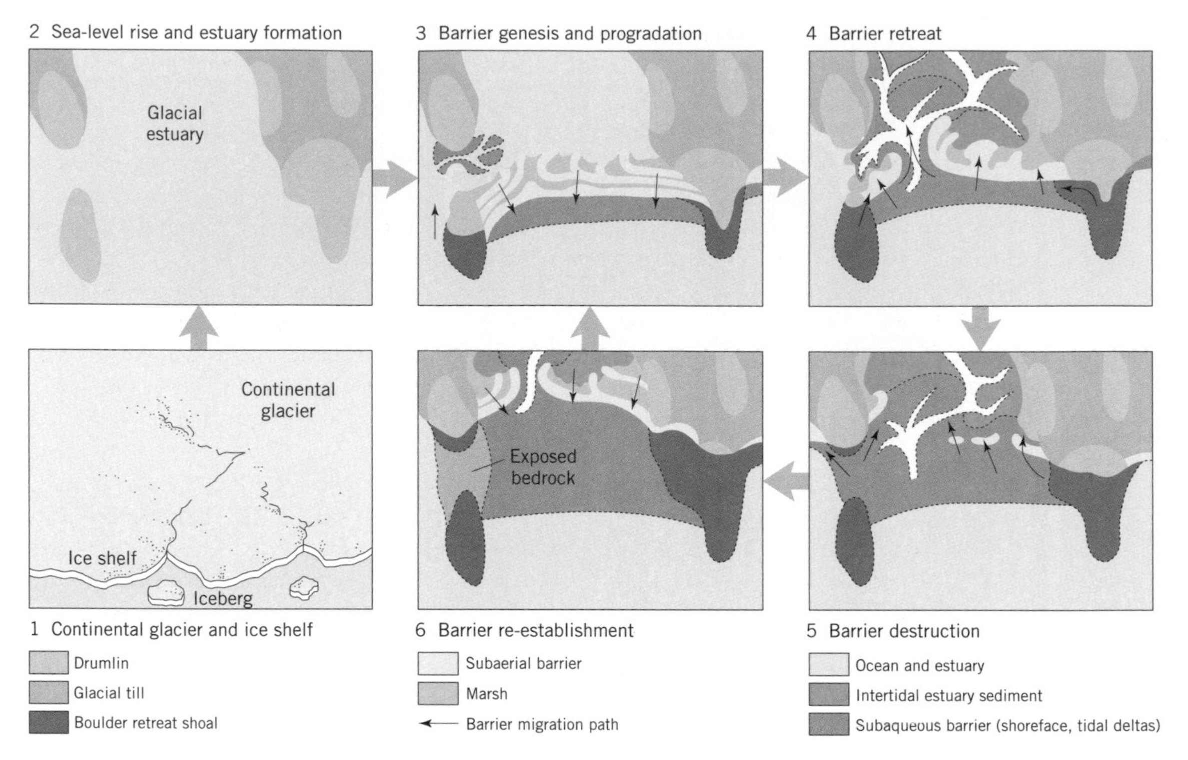

Fig. 8.28 Model of shore evolution for the drumlin coast of the eastern shore of Nova Scotia. The pathway of barrier evolution (progradational versus retrogradational) is controlled by sediment supply and the rate of sea-level rise. (From R. Boyd, A. J. Bowen & R. K. Hall, 1987, An evolutionary model for transgressive sedimentation on the eastern shore of Nova Scotia. In D. M. FitzGerald & P. S. Rosen (eds), *Glaciated Coasts*. New York: Academic Press, pp. 87–114.)

ated by the glaciers were drowned and transformed into embayments bordered by glacial headlands (Stage 2). As these headlands eroded due to wave attack, sand and gravel were released, building spits into the adjacent bays. Where two spit systems growing from opposite sides of the embayment joined, a baymouth barrier (type of welded barrier) was formed (Stage 3) (Fig. 8.29). While sediment supplies are adequate, the barriers may prograde, resulting in a number of beach ridges and swales. As the headlands continue to erode, boulder retreat shoals define their former extent. Eventually, continuing sea-level rise and diminishing sediment supplies lead to erosion and breaching of the barriers (Stage 4). By Stage 5 the barriers are mostly destroyed and some of the sediment once contained in them is transported landward in the form of intertidal and subtidal shoals by flood tidal currents

Fig. 8.29 The welded barrier and spits of central Nova Scotia portray Stage 3 of the drumlin coast model of Boyd *et al.* (1987, details in Fig. 8.28).

and wave action, especially during storms. With a further rise in sea level, portions of the subaqueous shoals migrate onshore and attach to new headland areas. Sediment contributed from the shoals and from the erosion of the new headlands helps to re-establish the barriers at the landward sites (Stage 6). At this point, the growth of the new barrier proceeds to Stage 3. With time, there may be several cycles of barrier growth, retreat, destruction, and re-establishment. Along glaciated coasts, such as the eastern shore of Nova Scotia, it is possible to find examples of different stages of this evolutionary model.

8.8.2 Mississippi River delta barriers

The Mississippi River delta in the Gulf of Mexico is composed of seven major delta lobes that were deposited approximately during the past 7000 years. Each lobe represents a period when the Mississippi River was debouching its sediment load essentially in one general location. Thus, deltaic sedimentation has been characterized by periods of delta progradation followed by abandonment and delta building at a different nearby site. This process of delta lobe switching has produced five different barrier shorelines along the Holocene delta plain.

A model of barrier evolution for the Mississippi River delta was presented by Shea Penland, Ron Boyd, and John Suter in 1985 (Fig. 8.30). In the first stage of their model, abandonment of the active delta leads to a subsidence of the delta plain and a high rate of relative sea-level rise. The active delta is transformed into an eroding headland with retreat of the shoreface. Sediment eroded from the headland forms flanking barrier spits, which are subsequently breached during storms. The flanking barrier spit stage is replaced by a barrier island arc as the headland region continues to sink and sea level continues to rise rapidly (Stage 2). In time, the barrier island arc is separated from the mainland by a large expanse of shallow open water. The barrier island arc migrates

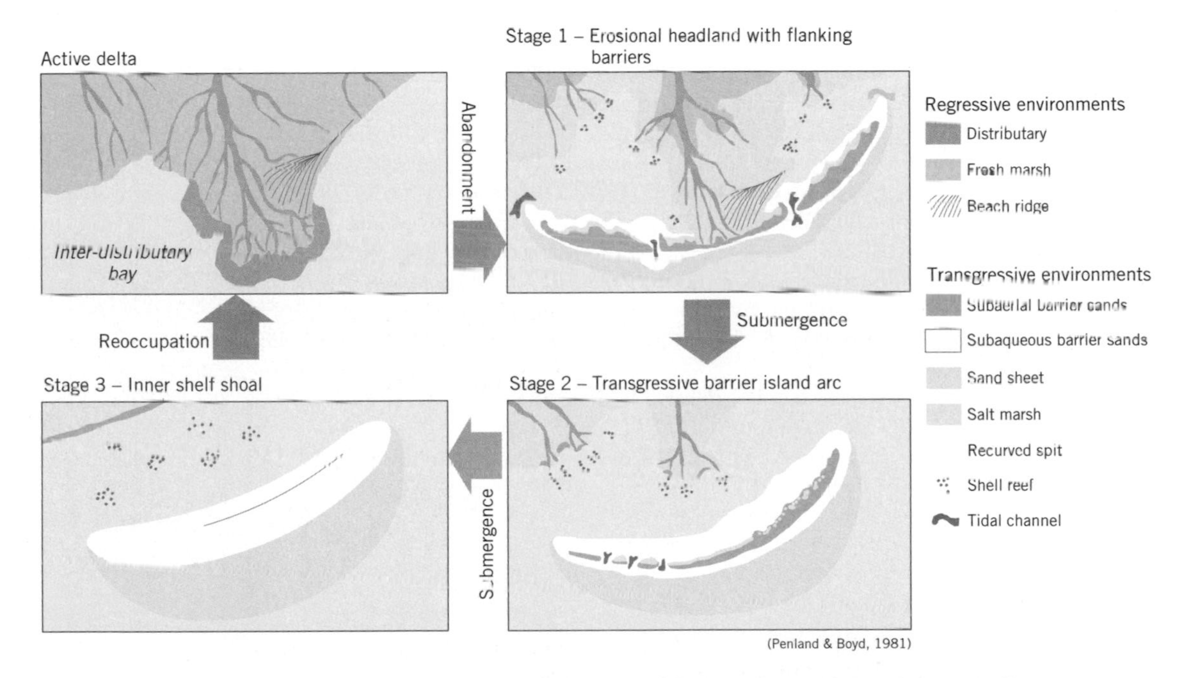

Fig. 8.30 Model of barrier evolution for the Mississippi River delta (from S. Penland, R. Boyd & J. R. Suter, 1988, Transgressive depositional systems of the Mississippi Delta plain: model for barrier shoreline and shelf sand development. *Journal of Sediment Petrology*, **58**, 932–49). The driving forces behind this model are the subsiding delta plain and rising sea level, and the changes in the locus of sediment discharge by the Mississippi River.

onshore through overwash activity and the construction of flood shoals on the landward side of the inlets. For a time, the arc maintains a position above sea level by migrating on top of a stacked sequence of flood shoals and washover deposits. Eventually, the arc is transformed into a subtidal shoal as sea-level rise outpaces the ability of the barrier to build vertically (Stage 3). At this point, the active delta may reoccupy this region and the process can begin anew. Each stage in the evolution of the barrier coast can be seen today along the Mississippi River delta.

8.9 Summary

Barriers occur on a worldwide basis but are predominantly found along Amero-trailing edge continental margins in coastal plain settings where the supply of sand is abundant. The coastal plain setting appears to be an important requirement for the wide distribution of barriers, in that it provides a sediment source and a platform upon which sediment can accumulate. Barriers have many different forms but can be grouped into three major classes depending on their connection of the mainland (barrier islands, barrier spits, and welded barriers). Barriers have formed by different mechanisms and many have migrated onshore to their present position. Retrograding barriers are continuing to move onshore through rollover processes, whereas prograding barriers are building seaward due to abundant sediment supplies and/or stable sea levels or slow rates of sea-level rise. Aggrading barriers are accreting vertically and keeping pace with rising sea level. The stratigraphy of barriers is dependent on their evolution and factors such as sediment supply, rate of sea-level rise, wave and tidal energy, climate, and topography of the land.

Suggested reading

Carter, R. W. G. & Woodroffe, C. D. (eds) (1994) *Coastal Evolution*. New York, Cambridge University.

Coates, R. (ed.) (1973) *Coastal Geomorphology*. Binghamton: State Univeristy of New York.

Cronin, L. E. (ed.) (1978) *Estuarine Research, volume 2*. New York: Academic Press.

Davis, R. A. (ed.) (1994) *Geology of Holocene Barrier Island Systems*. Berlin: Springer-Verlag.

Fletcher, C. H. & Wehmiller, J. F. (eds) (1992) *Quaternary Coasts of the United States: Marine and Lacustrine Systems*. SEPM Spec. Pub. 48.

Hayes, M. O. & Kana, T. (1978) *Terrigenous Clastic Depositional Environments*. Columbia: Department of Geology, University of South Carolina.

King, C. A. M. (1972) *Beaches and Coasts*. New York: St Martin's Press.

Leatherman, S. P. (1983) *Barrier Island: From the Gulf of St Lawrence to the Gulf of Mexico*. NewYork: Academic Press.

Oertel, G. F. & Leatherman, S. P. (1985) *Marine Geology*, **63**, special issue on barrier islands.

Nummedal, D., Pilkey, O. H. & Howard, J. D. (eds) (1987) *Sea Level Fluctuation and Coastal Evolution*. SEPM Spec. Pub. 41.

Schwartz, M. L. (ed.) (1973) *Barrier Islands*. Stroudsburg, PA: Dowden, Hutchinson, and Ross.

9 Coastal dunes

Fig. 9.1 Large vegetated dunes developed along the panhandle coast of Florida. Onshore wind and an abundance of sediments are the main factors in coastal dune development.

Fig. 9.2 Huge dunes that have developed along the mainland coast of Namibia. This coastal desert has what is probably the largest and most extensive coastal dune complex in the world. (Courtesy of N. Lancaster.)

9.1 Introduction

Sand dunes are an important part of many, but not all, coastal areas. These dunes are large piles of sand that accumulate as a result of similar processes and in generally similar shapes and patterns to dunes on inland deserts. The fundamental prerequisites in both cases are an abundant sediment supply and the wind to transport it. In most coastal areas, the wind is typically not a limiting factor but the sediment supply may be. Nevertheless, we have coasts where dunes may be several meters in elevation (Fig. 9.1).

Coastal dunes are not restricted to barrier islands, although nearly all barriers have at least small dunes. Some coasts without barriers have tremendous dune fields. Particularly good examples are the southern coast of Oregon, where the dunes extend a few kilometers inland from the coast, and the southeastern part of Lake Michigan, where dunes nearly 100 m high have developed. Among the largest dunes in the world are those on the coast of Namibia (Fig. 9.2) in southwest Africa, where barriers are absent.

9.2 Types of coastal dunes and their distribution

Any coast where sand accumulates in significant quantities has the potential for the development of dunes. Dunes are about the best protection we have against severe storms and their related large surges. Prevailing winds or diurnal sea breezes provide the typical transport mechanism for sand-sized sediment on most coasts. The prevailing winds typically have some onshore or shore-parallel component, and the sea breeze may be a major factor in many areas. It is, for example, the dominant wind along the southwest coast of Australia near Perth. Any dry part of the beach or other coastal environment that is sparsely or not vegetated is subject to eolian (wind) transport.

Dunes typically form in a linear fashion just landward of the beach, but they may also develop on the inland portion of a barrier island or the mainland. The linear dunes landward of the beach are called foredunes (Fig. 9.3). Some coasts have several lines of dunes behind the beach (Fig. 9.4), demonstrating their progradation due to abundant sediment and appropriate wind conditions. Each of these ridges was a foredune ridge at the time that it formed but was then fronted by a new one. Some barrier islands contain a complicated assortment of dune ridge arrangements that show sets of ridges at acute angles to one another (Fig. 9.5). This condition indicates periods of erosion that separated periods of dune accumulation and barrier progradation. Some barriers have only a single foredune ridge and it may be a high one. One of the best examples

Fig. 9.3 Foredunes along the central Texas coast. Here the abundance of sediment and the wide beach in the presence of sustained wind produce prograding accumulations of dunes. The seawardmost linear dune ridges are called foredunes.

Fig. 9.4 Barrier island along the southeastern coast of Australia displaying multiple dune ridges, indicating the presence of a large sediment supply. The numerous parallel ridges indicate that conditions for the formation of these dune ridges have prevailed for at least several centuries.

of this is the barrier at Ninety-Mile Beach in southeastern Australia. Here the single dune ridge is 10–15 m high along most of the barrier.

Dunes that form on the interior of barriers or on the coastal zone mainland develop as the result of lack of vegetation to stabilize the substrate. This is generally due to an arid climate that gives rise to desert conditions or to removal of vegetation by grazing or other human-related activities. A good example of how a combination of these factors has

Fig. 9.5 Barrier island along the Gulf coast of Florida showing sets of low-lying dune ridges that are at angled to each other. This condition shows that the shoreline orientation of the barrier has changed many times and that it has remained in a particular orientation for a significant period of time.

contributed to an extensive, active dune complex is on Padre Island on the Texas coast. Generally subarid conditions along with abundant sand along the beaches have resulted in an extensive active dune complex in the Padre Island National Seashore on the central part of the island (Fig. 9.6). An additional assist was provided by extensive cattle grazing on the island during the late nineteenth and early twentieth centuries. This portion of the Texas coast is one of considerable sediment accumulation and persistent onshore winds. As a consequence, the island is extremely wide and the mainland is dominated by an extensive dune complex. The barrier itself contains extensive active dunes that range from being only a meter or so high on the landward side near Laguna Madre, to several meters high in the central island. The limited development of the dunes on the landward part of the island is due to their destruction as the result of storm surge associated with hurricanes. These small dunes are on the wind tidal flats where storm surges of a meter or so are fairly common and can flood the dunes, destroying them by a combination of waves and currents. After the surge subsides, the wind and available sand must start to construct the dunes again.

A different situation is exemplified by the coastal dune complex along the southern Oregon mainland. Here the strong winds off the Pacific, along

(a)

(b)

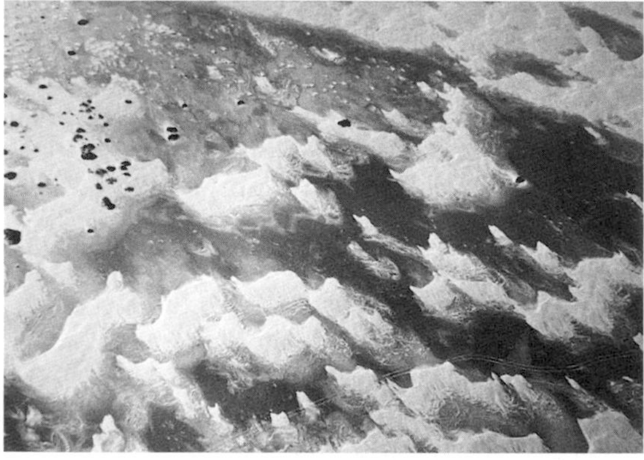

Fig. 9.6 (a) Ground and (b) aerial views of well developed dunes on Padre Island that mimic an inland desert. These rippled dunes have developed from an absence of vegetation due to cattle grazing.

with great amounts of sand, have produced huge mobile dunes that extend 3–4 km inland. These coastal dunes have inundated coniferous forests of mature trees as they migrate in a southerly direction perpendicular to the coast.

9.3 Dune formation

Dry sand and proper wind conditions are common along the backbeach environment because it is rarely wet and generally devoid of vegetation. Wet sand has too much cohesion to respond to wind. The dry backbeach shows various signs of wind transport, including ripples, sand shadows (Fig. 9.7), and heavy minerals or gravel lag concentrates of shells and shell debris. The sand shadows indicate a recent wind direction and may show scour around a shell or pebble. The gravel or shell lag deposit results from wind blowing the fine sand from the beach and leaving the larger particles that cannot be transported. After a while the large particles become concentrated and actually form almost a pavement. Such a pavement inhibits further wind erosion and is called a desert armor because of its importance in limiting wind erosion. Once the concentration becomes nearly continuous across an area, wind cannot access the smaller particles that it can transport.

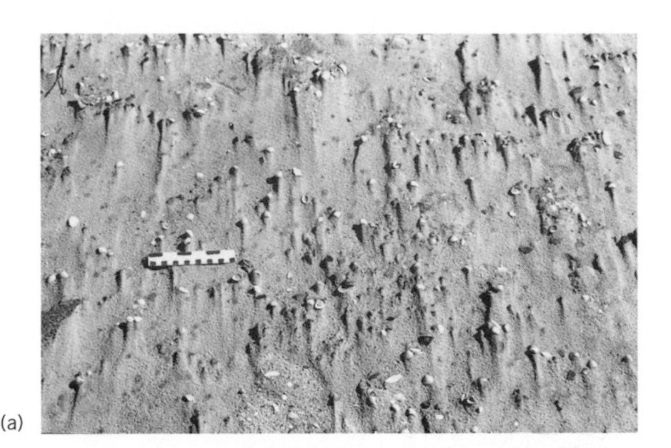
(a)

(b)

Fig. 9.7 The backbeach includes various surface features, such as (a) sand shadows and (b) ripples, both of which indicate wind direction by their orientation. Sand shadows parallel wind and ripples are perpendicular to it.

Fig. 9.8 Opportunistic plants are effective stabilizers of the backbeach and help to begin the formation of dunes. The small tufts of plant serve as an anchor for sand blown across the backbeach. The sand accumulates in the shadow of the plant much like the sand shadows in Fig. 9.7a.

Much of the wind-blown beach sand tends to accumulate just landward of the active backbeach. It is stopped from further transport by any type of obstruction that may be present, including bedrock cliffs, vegetation, existing dunes, or even human construction, such as buildings or sea walls. Once the initiation of eolian sediment accumulation begins, it continues unless conditions change, such as loss of sediment supply, the destruction of the stabilizing factor, or wave-induced erosion.

One of the best and most widespread aids in dune development is vegetation. Any type of plant serves as a focus for anchoring wind-blown sediment. Typically the relatively inactive backbeach is covered with opportunistic plants (Fig. 9.8), such as the beach morning glory, beach *Spinifex*, and marram grass. One of the most effective dune stabilizers on southern dunes up to the latitude of Virginia is sea oats (*Uniola*), while the American beach grass (*Ammophila*) extends from Virginia up to Nova Scotia in Canada. It is quite common to see small piles of sand around isolated plants on the backbeach area (Fig. 9.9). After only months these piles increase in size and the plants spread, thus increasing their effectiveness. In fact, even pieces of wood or any other sizable obstacle can act as a seed for dune development. Eventually small dunes or coppice mounds (Fig. 9.10) appear. As the sand becomes trapped by the vegetation this provides an enlarged area of stability. More sand becomes trapped and eventually a small dune develops. These small sand accumulations are typically a meter or two in diameter and about half that in height. Under the proper conditions they will eventually become larger and coalesce into a continuous foredune ridge.

Such small incipient dunes are quite vulnerable: even a modest storm can destroy them, requiring the building process to begin again. This is the reason why so much attention is paid to preserving vegetation on the backbeach and at the foot of the dunes. The absence of intense storms, along with an abundant supply of sand and a regular mechanism for delivery of the sand, eventually produces a dune. Dune size is dependent largely on the supply of sand-sized sediment.

Fig. 9.9 Small mounds of sediment captured by plants on the backbeach environment. As time passes the plants grow larger and are able to trap more sediment.

Fig. 9.10 The small dunes that begin to form on the landward portion of the backbeach are called coppice mounds. They form the third stage in the continuum of Figs 9.8 to 9.10.

A major geomorphic or structural feature along the coast that presents some vertical component is also an effective trap for wind-blown sand. The base of a rocky cliff or sea wall can accumulate sediment and become vegetated. Assuming that sufficient sediment supply is available, this type of accumulation can eventually become a dune.

9.4 Dune dynamics

The existence of dunes is testimony to the mobility of sand through wind transport on the coast, and attack by waves is an obvious factor in dune stability.

Fig. 9.11 Storm erosion causes foredunes to be scarped and vegetation to be exposed. Although vegetation is a good stabilizer of dunes, it is not sufficient to prevent erosion by waves crashing at the base of the dunes during a storm.

Although vegetation is an effective stabilizer of these accumulations, there are conditions when even vegetated dunes may become mobile or may be eroded.

The first and most obvious factor is attack by waves. Even though dunes are out of the regular influence of waves, they are quite vulnerable to only modest surges produced by storms. In areas of generally erosive beach conditions dune retreat is especially a problem because there is no backbeach to protect them. Elevated water level with superimposed storm waves produces swash and, in some cases, direct wave attack at the toe of the dune. The sand is easily washed away and carried both offshore and alongshore. Even though a dense dune grass cover is present, the sand is easily removed, commonly leaving a dense root system hanging over the scarp in the dune (Fig. 9.11). Post-storm recovery may occur and return some, or even all, of the sand to the beach. Proper conditions can start the rebuilding process of the dune but it can take many years to restore the loss of just a single storm. It is generally rather easy to recognize dunes that have been eroded and then rebuilt by the change in profile and perhaps even by the type of vegetation. Rising sea level presents another scenario for dune erosion, by providing a continual increase in the accessibility of the dunes to wave attack.

The other major aspect of dune dynamics is concerned with the migration of part or all of the dune

Box 9.1 Provincetown dune field on Cape Cod

The Cape Cod peninsula of Massachusetts owes its existence to glaciers that covered most of the northeastern part of the United States until a few tens of thousands of years ago. Most of the north–south trending outer part of the Cape is outwash associated with the glacial moraine that trends east–west. This outwash is mostly sand and gravel and has contributed sediment for the beaches and the islands to the south of the village of Chatham (Fig. B9.1). On the northern portion of this feature is a large recurved spit, on which is located the community of Provincetown, one of the largest settlements on Cape Cod. This spit has accumulated as the result of northward-moving sand carried by longshore currents from the eroding moraine to the south.

The combination of abundant sand supply and the onshore wind in this area has produced an extensive dune field, with many dunes more than 10 m high. There are numerous dune ridges in this recurved spit complex and it continues to grow as the Cape Cod bluffs continue to erode. The dune sand has become a bit of a problem for the local community and adjacent areas. Commonly it covers the road in the same fashion that snow drifts across the road in the northern latitude winters. Unlike snow, the sand doesn't melt, so it becomes an expensive task to keep removing it. Sand has also invaded homes and other buildings as it blows inland from the beach area. The rapid mobility of the sand makes it impossible to develop natural vegetation cover to stabilize it. Attempts have been made to artificially stabilize the blowing sand, but with limited success.

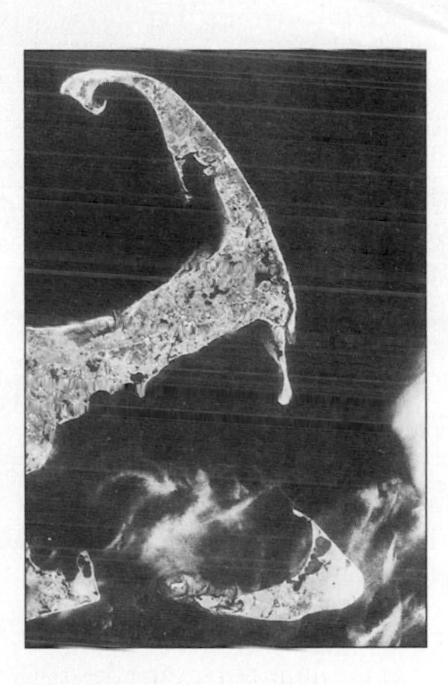

Fig. B9.1 Aerial photo of Cape Cod, Massachusetts. The elongate area to the north (top) includes the Provincetown Dune field.

Fig. B9.2 Large, mobile dunes in the Provincetown area of Cape Cod. The sediment that forms these dunes was originally eroded from the area several kilometers to the south and transported northward by longshore currents.

through eolian processes. The same mechanism that forms the dune also can cause it to move, sometimes great distances. Generally, dune mobility is associated with an absence of vegetation. Climatic conditions may reduce or eliminate vegetative cover, or overgrazing may remove much of the vegetation. Regardless of the reason the result is the same: sediment begins to move.

Fig. 9.12 The steep landward slope of a migrating dune. Onshore wind blows unstabilized dunes landward, and they move over anything that is in their path.

Fig. 9.13 Blowover migration of coastal dunes may cover forests (Fig. 9.12) or buildings. Buildings such as this may be buried and exposed in only decades as the large dunes migrate.

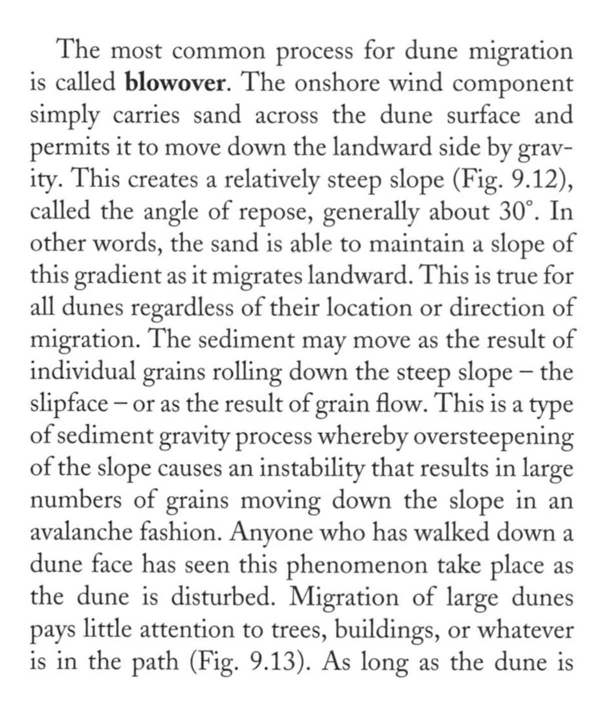

The most common process for dune migration is called **blowover**. The onshore wind component simply carries sand across the dune surface and permits it to move down the landward side by gravity. This creates a relatively steep slope (Fig. 9.12), called the angle of repose, generally about 30°. In other words, the sand is able to maintain a slope of this gradient as it migrates landward. This is true for all dunes regardless of their location or direction of migration. The sediment may move as the result of individual grains rolling down the steep slope – the slipface – or as the result of grain flow. This is a type of sediment gravity process whereby oversteepening of the slope causes an instability that results in large numbers of grains moving down the slope in an avalanche fashion. Anyone who has walked down a dune face has seen this phenomenon take place as the dune is disturbed. Migration of large dunes pays little attention to trees, buildings, or whatever is in the path (Fig. 9.13). As long as the dune is larger than the obstruction it will move over it. Houses have been buried and then many years later exumed as a result of migrating dunes.

9.5 Summary

Dunes serve as the best protection for depositional coasts. Granted, they are not as good a form of protection from waves and erosion as are cliffs or other bedrock, but they provide a barrier from wave attack during storms. The addition of vegetation, especially mature trees, results in what is almost like a dam from high water.

In order for dune development to proceed to a level where this protection is in place, it is necessary to have a large supply of sediment and a well developed beach. The dry beach is the immediate source of sediment to form the coppice mounds and then the foredunes.

Suggested reading

Goldsmith, V. (1987) Coastal dunes. In R. A. Davis (ed.), *Coastal Sedimentary Environments*. Heidelberg: Springer-Verlag.

Hesp, P. (1999) The beach backshore and beyond. In A. D. Short (ed.), *Handbook of Beach and Nearshore Morphodynamics*. Sydney: John Wiley & Sons.

Carter, R. W. G. (1988) *Coastal Environments*. London: Academic Press.

10 Coastal lagoons

10.1 Introduction

Lagoons are very restricted coastal bays. They occur as the result of specific climatic, geologic, and hydrographic situations. Lagoons have specific hydrologic characteristics, they host their own special fauna and flora, and they have their own sediment signature. These special coastal water bodies are not particularly common globally but they are important because they provide a special environment with unique characteristics. In general, coastal lagoons represent what is considered to be a stressed environment because of the extreme conditions that prevail.

This chapter considers the various characteristics of coastal lagoons and what distinguishes them from other coastal water bodies. Some important examples are discussed and compared.

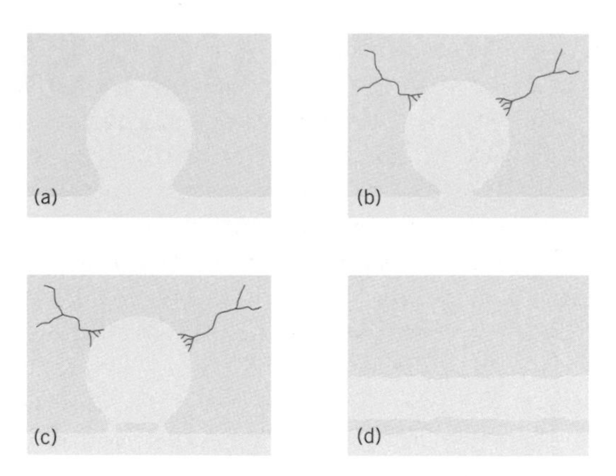

Fig. 10.1 Various settings of different types of coastal bays, including: (a) an open embayment of the sea; (b) an open estuary; (c) an estuary with a barrier; and (d) a lagoon that lacks streams and tidal flux.

10.2 Definition

Most definitions of the term lagoon are nonspecific. Many authors apply the name lagoon to any water body that is landward of a barrier. Typically, lagoons are indicated as being parallel to the coast and separated from the open marine water by a natural barrier. The major problem with this loose use of the term lagoon is that there is no restriction or limitation on the nature of the circulation and water characteristics within the water body.

Some coastal bays are simply open embayments of the sea (Fig. 10.1a), with similar tides, salinities, and other characteristics to the open marine environment, but with lower wave energy. Other coastal bays have important freshwater contributions from streams, and thus lower salinities. These may be open embayments or they may have various forms of barriers that separate them from the open sea (Fig. 10.1b, c). They are properly called estuaries and are discussed at length in Chapter 15.

The other end of the coastal bay spectrum is the lagoon, where there is an absence of significant freshwater influx and where there is no significant tidal flux because of the presence of an efficient barrier blocking interaction between the bay and the open marine environment (Fig. 10.1d). Lagoons, therefore, will be expected to have an elevated salinity due to a general excess of evaporation over precipitation, which is the primary mode of introduction of water to the lagoons. Storms may wash over low barriers, providing another method for introducing water to the lagoon. Such coastal water bodies with high salinities have very different characteristics than the open bays and the estuaries. For these reasons it is important to maintain strict definitions for the various types of coastal bays.

10.3 Morphology and setting

Lagoons display various shapes, but most are elongate parallel to the coastline and virtually all are separated from the open marine environment by a barrier island, or in some places by a reef, such as in the Persian Gulf. Because the wave-dominated nature of the coast produces long barrier islands and coral reefs, many lagoons extend for tens of kilometers along the coast. These water bodies may form in a variety of geologic settings as long as there is some embayment from the open marine environment and a mechanism for isolating it, such as the development of a barrier or spit. Lagoons may develop along high-relief areas such as the Pacific coast of Mexico, where high wave energy and pronounced longshore drift develop barrier spits that are wave-dominated and become closed to significant tidal flux. Most lagoons are developed along coastal plains where barriers have efficiently separated a coastal bay from the open ocean.

Climate is another factor that plays a role in most coastal plain or prograding strandplain lagoons. In order to eliminate freshwater runoff into the coastal bay it is necessary for rainfall to be significantly limited. Typically this occurs along coasts where near-desert conditions prevail. In fact, many of the world's lagoons have developed along such coasts. These include the coasts of southern Australia, northern Africa, the Persian Gulf, south Texas, much of Mexico, southern Brazil, and southeast Africa.

The nature of the coastal lagoons makes them typically quite shallow; a meter or less is common. Some are actually emphemeral, with dry beds during the dry season and with standing water in the lagoonal basin only during the wet season.

Fig. 10.2 Vertical aerial photo of Lake Reeve, Victoria, Australia, showing some dry portions. This coastal lagoon is partly dry during the summer and fills with water during the wet season in the winter. The lagoon shown here is about a kilometer wide, from top to bottom of the photograph.

10.4 General characteristics

Although there are significant differences among various coastal lagoons, there are many similarities. We can characterize them by their salinity, organisms, processes, and sediments.

10.4.1 Salinity

Lagoons are commonly **schizohaline** or **hypersaline**, at least in the broad senses of these terms. Schizohaline water bodies are those which display great change in salinity from brackish to hypersaline, generally in response to seasonal rainfall or some cyclic phenomena. Some shallow coastal lagoons fall into this category. A rather extreme example is Lake Reeve in Victoria, Australia, which is a long, shallow lagoon (Fig. 10.2). During the wet season it is almost fresh, whereas in the dry season it dries up locally and becomes quite saline in those areas where water remiains. This type of salinity pattern would characterize lagoons of the mid-latitudes, with strong seasonality in precipitation and with high evaporation to precipitation ratios.

Hypersaline lagoons are those in which salinities are continually above normal marine concentrations. They characterize the semi-arid and arid coastal areas where little or no freshwater influx occurs. Salinity commonly increases away from connections with the open sea, if any are present. Laguna Madre in Texas and the Coorong in South Australia are examples and are discussed in detail below.

10.4.2 Organisms

Hypersaline or schizohaline conditions cause serious problems for organisms. Typical marine or estuarine species cannot tolerate either of these salinity situations, so we find special communities present in lagoons. The usual situation for such extreme environments is to have very few species because of the special adaptations that are required for such severe salinity conditions. Generally, however, the numbers of individuals within these specially adapted species is very high.

A good example of an organism that can tolerate major fluctuations in salinity is the killifish, *Funulus*, which is common in parts of Laguna Madre, Texas. This small fish inhabits many of the isolated or nearly isolated ponds and embayments of this lagoon, where salinity may change from nearly fresh water (<5 parts per thousand, p.p.t.) levels to as much as 200 p.p.t. This is almost six times higher than normal marine conditions, which are 35 p.p.t. During summer, evaporation will raise salinity to near the high end of this range and one rainfall can lower it to brackish levels of 10–15 p.p.t. *Funulus* can tolerate such extreme and rapid changes in salinity through its ability to rapidly osmoregulate it body fluids to match those in its aquatic environment.

Another excellent example of an organism that can live under severe salinity conditions is found in both of the southeastern Australian lagoons: the Coorong and Lake Reeve. Here a single species of cerithid gastropod, *Rosiella*, appears in huge numbers along the shallow and exposed wet margins of the lagoons. These detritus feeders graze while moving slowly over the surface. They also contribute large numbers of small pellets to lagoonal sediment. Their numbers reach such high concentrations that is is common for the small beaches along the lagoon shores to be made exclusively of the shells from this single species (Fig. 10.3).

Fig. 10.3 Cerithid snails, which abound in many schizohaline lagoons. The extreme conditions in these lagoons give rise to a fauna that tends to contain one or only a few species. In this case it is only one species of snails.

10.4.3 Chemical precipitates

High concentrations of dissolved elements and ions in lagoons may result in actual precipitation of minerals. There is a generally predictable order of mineral precipitation depending on concentration. The typical salinity of open lagoons ranges from about 40 through 70 p.p.t. most of the time. These levels of salinity do not result in the formation of any of the typical evaporite minerals, but carbonate minerals may precipitate. These carbonate minerals (aragonite and calcite) are different in that they do precipitate from solution in elevated salinities, but they are also influenced by photosynthesis, pH, and other factors.

The true evaporite minerals that may appear in lagoons do not precipitate until a salinity of about 200 p.p.t. is reached and gypsum (calcium sulfate) forms. The next common mineral to precipitate is halite, common salt, which appears at levels of 300 p.p.t. In very extreme situations, other evaporite minerals may form, but they are typically restricted to saline lakes and intermittent lakes in inland environments. Even gypsum and halite are uncommon in coastal lagoons, and are generally restricted to local ponds and embayments of the major lagoonal water body, where salinities can reach extreme levels.

10.4.4 Physical processes

Typical coastal processes, such as tides and waves, are not prevalent in lagoons. The definition of the term and the nature of lagoons limits any tidal influence. Tidal flux is absent or very local adjacent to typically small inlets. Thus, there is no tidal mechanism for transporting sediments into or out of coastal lagoons. Waves are limited in their influence by the short fetch that is typical of the normally elongate lagoons. The width of these coastal bays is generally only a kilometer or so, which would produce short and steep waves if strong winds were present. These waves do not have much influence on the lagoon floor but they do develop narrow beaches and small beach ridges (Fig. 10.4). These choppy waves can also cause coastal erosion along lagoonal shorelines.

Fig. 10.4 Small beach and related beach ridges in Lake Reeve, Australia, both wave-formed features of this lagoon. Although the lagoon is only a kilometer or so across and is shallow at best, small beach ridges form by waves generated during strong winds.

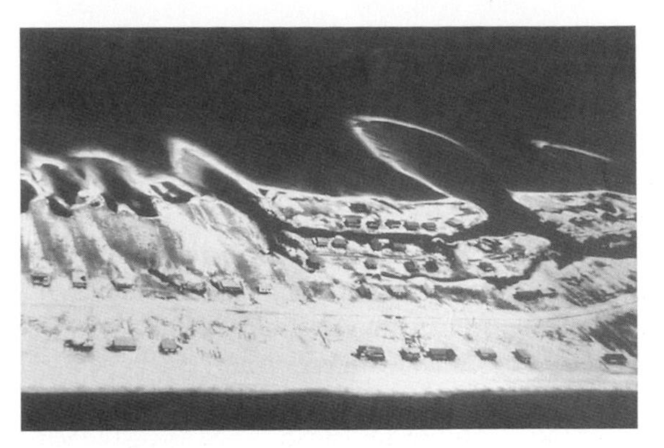

Fig. 10.5 Washover fan extending into a coastal lagoon as the result of a severe storm that overtopped this barrier island. This is a common result on low-lying islands during severe storms with a large surge. These were formed on Dauphin Island, Alabama, as the result of Hurricane Frederick in 1979. (Photograph by D. Nummedal.)

There are other physical processes that have a major influence on coastal lagoons. All are derived from some influence of wind. This can be the result of storm activity or ambient onshore winds that prevail along most coasts. Wind can produce circulation in lagoons that homogenizes any stratification of salinity (density) that might develop from very low energy conditions.

Storms, especially hurricanes, produce large waves and elevated water level or storm surge. It is common for this combination to breach dunes of barrier islands and transport both water and sediment into the lagoon. Storm surges of over 3 m are fairly common in severe hurricanes and, when combined with large waves, can develop large washover fans that extend well into the lagoon (Fig. 10.5). This phenomenon transports large volumes of sediment into the lagoon in a very short time. The same is true for water. Such high-intensity hurricanes are most common in fairly low latitudes. Many coastal lagoons do not receive this input of either water or sediment. This is especially true if the barriers contain dunes of several meters' elevation.

The other storm-related process that affects lagoons is the wind tide that results from setup in the water body itself. When wind blows over these shallow water bodies, water is pushed toward the downwind side of the lagoon. Because lagoonal shorelines generally have a gentle gradient, there may be significant flooding by this elevated, wind-blown water; causing a wind tide. When this happens, it is common for sediment to be transported onto these wind tidal flats (Fig. 10.6). On the landward side of Padre Island, Texas, which borders Laguna Madre, and along the mainland shoreline, it is common for fine-grained sediment, suspended by the turbulence of the storm, to accumulate in a thin veneer on these normally sandy wind-tidal flats.

Fig. 10.6 Extensive wind tidal flats along the barrier island side of Laguna Madre, Texas. These extensive flats are just above high tide elevation and are covered with cyanobacteria (blue-green algae).

Another important wind-related phenomenon that influences lagoonal environments is the prevailing winds. These winds transport dry and unstabilized sediment into the lagoon. This dry sediment might be from the backbeach environment but is most typically from dunes. Its transport might also be associated with the daily sea breeze cycle. As a result, large amounts of sediment are transported into lagoons. This transport can take place in two ways: (i) by individual grains being picked up and transported in suspension by the wind and (ii) by large numbers of grains being carried along the sediment surface as the dunes migrate into the lagoon.

Most people who have been to the beach have experienced the situation where wind blows sand into your eyes, onto your blanket or into your sandwich. This is the same phenomenon that transports sand grains from the front part of the barrier island to the backbarrier and into the lagoon. This condition results in a persistent but slow rate of sediment transport to the lagoon.

Unvegetated dunes are vulnerable to wind transport because of the absence of any stabilizing plants. The absence of vegetation may be the result of arid conditions, loss due to overgrazing of the barrier islands, deforestation by human activity, or the catastrophic loss of vegetation due to storms or disease. As the wind blows with an onshore component, there is essentially a mass transport of sediment across the dune surface and down the leading edge or slipface of the dune. This is how dunes migrate, and without any stabilization there can be an almost catastrophic rate of sediment transport. In many locations there are very high and steep slipfaces on dunes that are migrating landward (Fig. 10.7), including some that are actually migrating over forests. In several situations this migration of dunes has transported huge amounts of sediment into coastal lagoons. The best example is probably Laguna Madre in Texas, where windblown sand has filled the entire width of the lagoon and dredging is necessary to maintain a navigable channel for the Intracoastal Waterway (Fig. 10.8).

In summary, although there is no significant sediment introduction to the lagoon from either river discharge or tidal flux, a significant amount of sediment is delivered to these special coastal water bodies. The rates of sediment introduction are low in many situations and the methods of delivery are dominantly the result of wind-generated processes. The result is that most lagoonal sediment is derived from the seaward direction and is carried to the lagoon throughout its length; no point sources of sediment are available.

Fig. 10.7 The steep slipface of a barrier dune as it migrates landward to encroach on the adjacent lagoon. Onshore winds with abundant sediment and an absence of vegetation produce the conditions for this to happen.

10.4.5 Lagoonal sediments

The nature of lagoonal sediments is quite diverse because there are diverse sources and mechanisms for their presence. There are three types of sediment that accumulate in lagoons: (i) chemical precipitates; (ii) sediment particles carried in by various wind-related processes; and (iii) skeletal material from organisms that live in the lagoon.

Chemical precipitation of evaporite minerals is typically limited to local sites where high evaporation rates are present and where salinities are extreme. The most common situation for this type of sediment accumulation is in small ponds and lakes that have been separated from the main lagoon. This might occur during the dry season or in situations where a longer isolation of the pond or lake takes place.

The more common chemical precipitate is various species of calcium carbonate. This usually occurs as

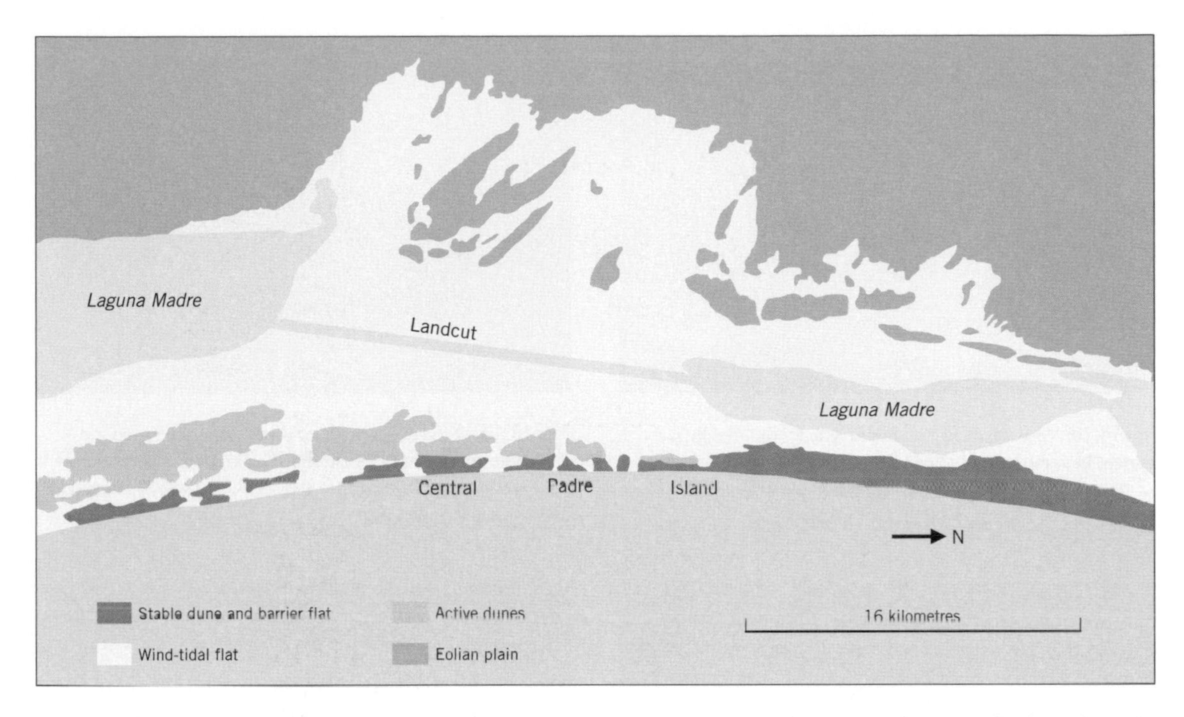

Fig. 10.8 Area called the "Landcut" in Laguna Madre, Texas, where wind-blown sand closes the entire lagoon, and a channel must be dredged to keep the Intracoastal Waterway open because this coastal thoroughfare is an important commercial waterway.

clay-size particles of aragonite or calcite, both types of $CaCO_3$. The precipitation of these minerals is commonly the result of photosynthesis, which in turn alters the pH of the shallow aquatic environment of the lagoon. The carbonate muds, commonly called micrite, generally occur in thin and discontinuous layers in the open lagoons (Fig. 10.9), but also may be present as thick and extensive layers (Fig. 10.10). In many situations the carbonate muds accumulate in association with filamentous blue-green algae, also called cyanobacteria. These microorganisms typically develop so-called algal mats (Fig. 10.11) along the periphery of coastal lagoons. It is the photosynthesis of these cyanobacteria that assists in the precipitation of the carbonate muds.

The sediment that is introduced by wind activity is typically well sorted and fine to medium sand. Although it is generally quartz in composition, virtually any composition is possible depending on the nature of the overall barrier sediment. This sediment appears in a variety of forms. It may be

Fig. 10.9 Thin and discontinuous lens of calcium carbonate mud (micrite) in Lake Reeve, Australia. Conditions of seasonal evaporation increase salinity, which, when combined with photosynthesis, results in the precipitation of calcium carbonate.

Fig. 10.10 Very extensive carbonate mud accumulation in a lagoon. In some schizohaline lagoons the calcium carbonate may be more than 30 cm thick and cover large areas of the lagoon, such as here in the Coorong of South Australia.

Fig. 10.12 Dune migrating landward, encroaching on the lagoon at the Coorong, Australia. This coastal area is a desert with abundant sediment from the Southern Ocean, much of it biogenic shell material.

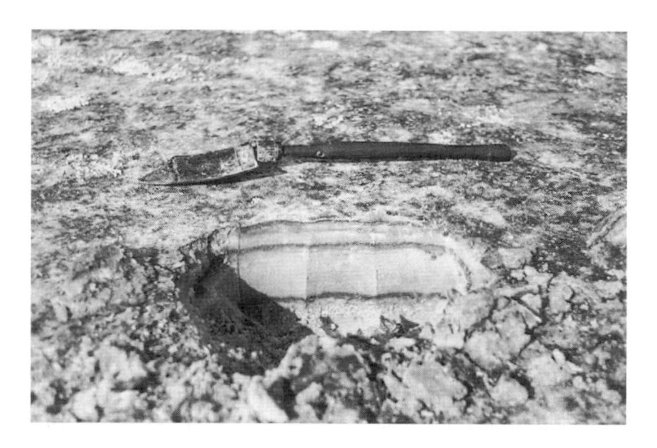

Fig. 10.11 Algal mats (dark layers and surface) that are formed by large populations of cyanobacteria (blue-green algae). These mats are typically from several millimeters to a centimeter or so thick. They may be buried by sediment and then established on the current surface, such as shown in this photograph.

Fig. 10.13 Flakes of a thin mud that has been desiccated due to its position on a wind tidal flat (Baffin Bay, Texas). Flooding produced a thin layer of mud that was dried, forming these flakes of mud.

isolated sand grains in an otherwise muddy lagoonal sediment. Extreme storms that generate washover processes can transport thin layers of sand into the lagoon, with thickness up to tens of centimeters. In the case of migrating sand dunes, the entire dune may become part of the lagoon (Fig. 10.12).

Fine sediment that settles out of suspension from the water column during and just after storms is most recognizable along the margins of the lagoon, where it accumulates in thin layers only a few millimeters thick. After high water conditions cease, the thin mud layer is exposed and dries quickly. Shrinkage takes place, causing the layer to become a series of curled up flakes (Fig. 10.13) that can easily be removed by wind or the next period of high water.

Although the variety of mollusks, ostracods, and other skeleton-bearing organisms is not great, each species present is usually very abundant, thereby producing a significant amount of gravel and sand-sized sediment in the lagoon. This skeletal sediment component may be scattered throughout the lagoonal sediment or it may be concentrated along the shoreline by wave action.

10.5 Example lagoons

There are many good examples of lagoons throughout the world. Some are large and others limited in their extent. Because of the nature of the lagoon and its high evaporative environment, the examples included here represent a spectrum of increasing salinity and, therefore, increasing amount of chemical precipitate. The three lagoons considered, in order of increasing salinity, are the Laguna Madre/Baffin Bay complex in south Texas, Lake Reeve in Victoria, Australia, and the Coorong in South Australia. All are very long and have only one inlet at one end of the lagoon.

10.5.1 Laguna Madre

This lagoon extends from near Corpus Christi along the entire south Texas coast to near Brownsville at the Texas–Mexico border (Fig. 10.14). It has some circulation at its north end and is connected to the open Gulf via an artificial inlet called Mansfield Pass. There is no significant perennial stream that enters Laguna Madre or the adjacent Baffin Bay. The lagoon is pristine, with significant development only at each end. Most of the adjacent barrier is a National Seashore and the adjacent mainland is an area of cattle grazing and cotton fields. Oil exploration has taken place in various locations throughout the lagoon.

The salinity ranges from about the low 40s to near 90 p.p.t. from north to south. There are seasonal variations and local places of higher concentrations. Evaporite precipitation is not present in surface waters of the open lagoon but does occur in the pore waters of the sediment along the coast and probably beneath it as well. Carbonate mud occurs only in small patches in the southern part of the lagoon.

Laguna Madre has extensive wind-tidal flats, most of which are covered by algal mats. In some places it is possible to see multiple layers of these mats interspersed between sandy layers produced by washover events. Back island dunes are widespread and are mobile due to the absence of vegetation caused by overgrazing (Fig. 10.15).

Landward of Laguna Madre and connected with it is a separate lagoonal basin, Baffin Bay. This irregularly shaped lagoon is a relict estuary that has lost its freshwater influx due to a change in climate. No significant streams are present in adjacent south Texas. This shallow, hypersaline water body has an additional element of interest. These are small but numerous serpulid worm reefs. These reefs rise nearly a meter above the surrounding waters (Fig. 10.16) and contribute considerable skeletal material to the sediment of the lagoon floor. The reefs do display some asymmetry that indicates the direction of high energy, in this case waves.

10.5.2 Lake Reeve

The southeastern coast of Victoria, Australia, is characterized by a continuous barrier island known as Ninety-Mile Beach. This barrier has one small opening on the northeast end at Lakes Entrance (Fig. 10.17). The backbarrier aquatic environment is Lake Reeve, a shallow, coast-parallel lagoon that is schizohaline in its character. This lagoon displays a wide range of salinities, from nearly fresh at its northeast end to near 100 p.p.t. in various of the small and sometimes isolated basins within the lagoon. Salinity also varies greatly with the season; the winter is wet and low salinity and the summer is dry and high salinity.

Because the dunes on Ninety-Mile Beach rise 10 m or more above sea level, there is no washover of the barrier and there has not been for most of its history. Additionally, the dunes are well stabilized by vegetation. This means that there has been virtually no introduction of sediment from the seaward direction. The floor of most of the lagoon is comprised of shelly sand with very little mud. Radiocarbon dating of the shells shows that they are more than 4000 years old, testimony to the absence of sediment introduction.

Lake Reeve is accumulating two primary types of sediment at the present time. One is small pellets of mud produced by small cerithid snails that

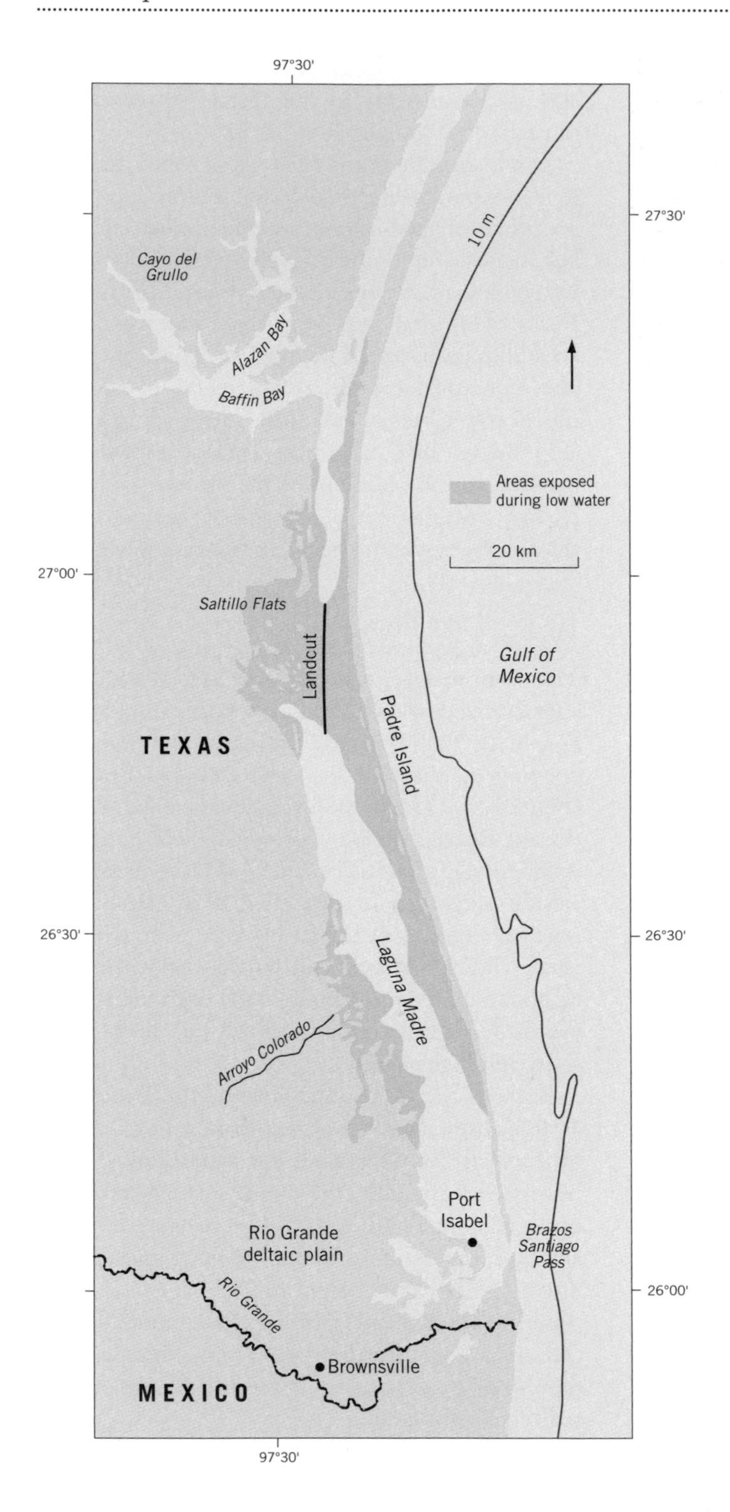

Fig. 10.14 Map of Laguna Madre, Texas, the longest coastal lagoon in North America. This lagoon has essentially no connection to the open Gulf of Mexico. Salinity increases from top (north) to bottom (south).

Fig. 10.15 Unstable back island dunes adjacent to Laguna Madre. These serve as a major supplier of sediment for the lagoon. These dunes are developed on the wind tidal flats.

inhabit the lagoon in huge numbers. These snails graze over the typically algal mat covered margin or dried portion of the lagoon, where they feed on the cyanobacteria and produce large numbers of these pellets. The other type of sediment is calcite, which precipitates directly from the lagoonal waters as the result of high salinities and photosynthesis. The

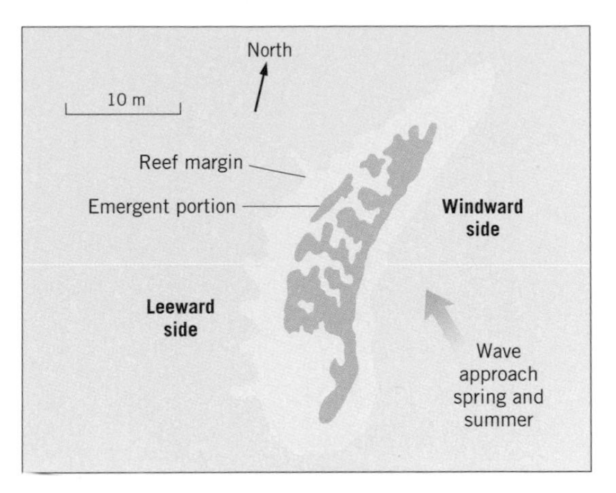

Fig. 10.16 Map showing serpulids in Baffin Bay. The smooth side faces the prevailing winds and the irregular side is on the leeward side.

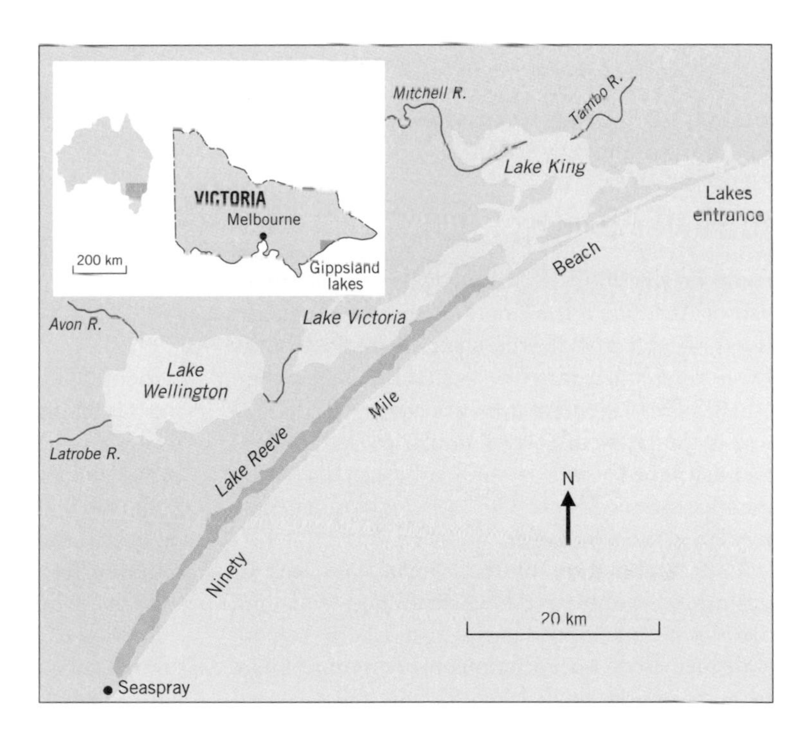

Fig. 10.17 Map of Lake Reeve, Australia, showing the only opening to the open ocean at the northeast end, Lakes Entrance. Such a condition results in virtually no circulation with adjacent estuaries or the open ocean.

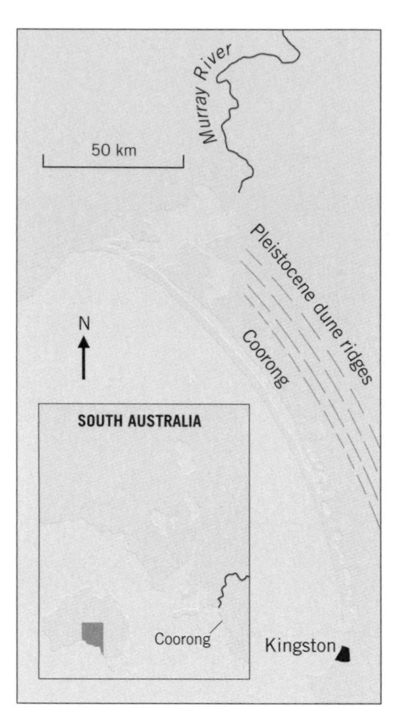

Fig. 10.18 Map of the Coorong in South Australia with the opening at the north end where the Murray River discharges. This is a similar condition to that of Lake Reeve, shown in Fig. 10.17.

mud may be in thin lenses or it may extend over large areas of the lagoon floor.

10.5.3 The Coorong

Another very long coastal lagoon in the southern part of Australia is the Coorong, south of Adelaide. It has an inlet at its northern end where the Murray River discharges into the Southern Ocean (Fig. 10.18). This lagoon and its associated barrier represent the latest of several similar coastal systems that developed as this part of South Australia during the Pleistocene Epoch. The barrier is dominated by very large, mobile dunes.

The combination of the climate, the lack of freshwater except at the northernmost end, and the marine influx produces a high-salinity lagoon, with increased concentrations to the south. There are also several small lakes that are relict coastal lagoons. Unlike Lake Reeve, the Coorong is receiving significant sediment from the barrier and it is precipitating a variety of minerals throughout much of the lagoon and the related small lakes.

Fig. 10.19 Extensive and thick carbonate mud that has precipitated in one of the many small isolated ponds in the Coorong complex. These small lakes were originally part of the main lagoon and have become isolated. They are now evaporitic and high in dissolved calcium and magnesium.

Fig. 10.20 Halite deposits that have formed as the result of evaporite precipitation. When evaporation brings salinity up to about 200 p.p.t. (20%) salt will form. These deposits in the Coorong area were mined during the nineteenth century.

Onshore winds from the Southern Ocean can be quite strong and cause individual grains to be removed from the unvegetated dunes and carried into the lagoon. There is also extensive landward migration of the dunes into the Cooring. Chemical precipitates include thick carbonate muds (Fig. 10.19), and both gypsum and halite (Fig. 10.20). The halite or common salt was mined in the nineteenth century by Chinese immigrants who accidently discovered these deposits on their way to the gold mines further inland in Australia.

10.6 Summary

Lagoons are quite special coastal bays. Their typically hypersaline conditions result from a lack of significant freshwater influx and of tidal flux. Because climate tends to be a factor in lagoonal development, most are associated with arid coastal conditions. These characteristics result in unusual biota, with few species but abundant numbers. Limited methods of sediment introduction result in slow rates of influx except in the case of major storms. Additional accumulation of material in lagoons is produced by chemical precipitation of carbonate and evaporite minerals.

Suggested reading

Eisma, D. (1998) *Intertidal Deposits: River-mouths, Tidal Flats and Coastal Lagoons.* Boca Raton, FL: CRC Press.

Isla, F. I. (1995) Coastal lagoons. In G. M. E. Perillo (ed.), *Geomorphology and Sedimentology of Estuaries.* Developments in Sedimentology No. 53. Amsterdam: Elsevier.

11 Tides of the ocean

11.1 Introduction

Anyone who has spent time along the seashore knows that the ocean level changes on an hourly basis. Fishermen plan their activities around high and low tide, such as those who dig clams when tidal flats are uncovered or gather mussels at low tide. Boaters are aware that some channels are only navigable at high tide when waters are deep enough for their boats to pass. Large vessels sailing to port normally enter at slack water or during an ebbing tide. Moving against the tidal flow provides greater steerage for ships than traveling with the currents.

The rise and fall of the tides is one of the major rhythms of Planet Earth. It has a dramatic effect on shoreline processes and coastal landforms. Were it not for the tides there would be no tidal inlets and few natural harbors along the east and Gulf coasts of the United States or along many other barrier coasts of the world. Those familiar with the coast recognize that times of high and low tide occur approximately an hour later each day. More astute observers know that daily tides gradually change in magnitude over the course of a month and that these variations are closely related to phases of the Moon.

Tides are a manifestation of the Moon's and the Sun's force of the gravity acting on the Earth's hydrosphere, as well as the relative orbits of these celestial bodies. Having wavelengths measuring thousands of kilometers, tides are actually shallow-water waves affecting the world's oceans from top to bottom. The surface expression of tides is most dramatic in funnel-shaped embayments, where vertical excursions of the water surface can reach more than 10 m in such areas as the Gulf of Saint-Malo, France (Fig. 11.1) or the Gulf of San Matias, Argentina, and even as high as 15 m in the Bay of Fundy, Canada. Pytheas, a Greek navigator, recorded the Moon's control of the tides in the fourth century BC. However, it was not until Sir Isaac Newton (1642–1727) published his *Philosophiae naturalis principia mathematica* (Philosophy of natural mathematical principles) in 1686 that we finally had a scientific basis for understanding the tides.

Fig. 11.1 Mont Saint Michel in the Gulf of Saint Malo is surrounded by water at high tide due to a 12 m tide that inundates the tidal flats and even floods some of the parking lots.

In this chapter we discuss the origin of the tides and how their magnitude is governed by the relative position of the Earth, Moon, and Sun. It will be shown that the complexity of the tides is a function of many factors, including the elliptical orbits of the Earth and the Moon, the declination of the Earth and the Moon, and the presence of continents that partition the oceans into numerous large and small basins. The phenomena of tidal currents and tidal bores are also discussed.

11.2 Tide-generating forces

11.2.1 Gravitational force

As a basis for understanding the Earth's tides, we begin with Newton's universal **law of gravitation**, which states that every particle of mass in the universe is attracted to every other particle of mass. This force of attraction is directly related to the masses of the two bodies and inversely proportional to the square of the distance between them. The law can be stated mathematically as:

$$F = G\frac{M_1M_2}{R^2}$$

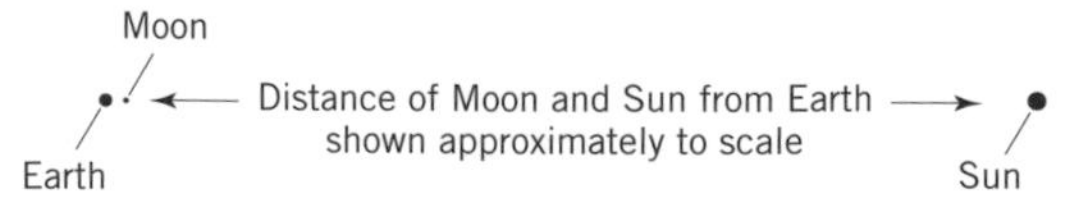

Moon

Earth ← Distance of Moon and Sun from Earth shown approximately to scale → Sun

	Average distance from Earth	Mass
Moon	385,000 km	7.3×10^{19} t
Sun	149,800,000 km	2×10^{27} t
Sun compared to Moon	390 times further away	27 million times more massive

Newton's Law of Gravitational Attraction (F_g)

$$F_g \sim \frac{\text{mass}}{\text{distance}^2} = \frac{\text{Sun 27 million times more massive}}{(\text{Sun 390 times further away})^2} = 180$$

Tide-producing force (F_t)

$$F_t \sim \frac{\text{mass}}{\text{distance}^3} = \frac{\text{Sun 27 million times more massive}}{(\text{Sun 390 times further away})^3} = 0.46$$

Therefore, the Sun has 46% of the control on tides compared to the Moon

Fig. 11.2 The Earth's tides are primarily controlled by the Moon because in the equation governing the tide-generating force, distance between the masses is cubed. Thus, even though the Sun is much more massive than the Moon, it is also much further away from the Earth than is the Moon.

where F is the force of gravity, G is the gravitational constant, M_1 and M_2 are the masses of the two objects, and R is the distance between the two masses. From the equation it is seen that the force of gravity increases as the mass of the objects increases and as the objects move closer together. Distance is particularly critical because this factor is squared (R^2) in the equation. Thus, the celestial bodies producing the Earth's tides are the Moon, due to its proximity, and the Sun, because of its tremendous mass. The other planets in the Solar System have essentially no effect on the tides due to their relatively small mass as compared to the Sun and their far distance from Earth as compared to the Moon. Although the attractive forces of the Moon and Sun produce slight tides within the solid Earth and large oscillations in the atmosphere, it is the easily deformed liquid Earth (hydrosphere) where tides are most clearly visible.

As illustrated in Fig. 11.2, the Sun is 27 million times more massive than the Moon, but at the same time it is 390 times further away. After substituting the respective mass and distance values for the Moon and Sun into Newton's gravitation equation, it can be seen that the attractive force of the Sun is approximately 180 times greater than that of the Moon. However, we still know that the Moon has a greater control on the Earth's tides than does the Sun.

11.2.2 Centrifugal force

To understand how the gravitational force actually produces the tides it is necessary to learn more about orbiting celestial bodies, including the Moon–Earth system and the Earth–Sun system. First, it is important to recognize that counteracting the gravitational attraction between the Moon and Earth is the centrifugal force (Fig. 11.3). Centrifugal force is a force that is exerted on all objects moving in curved paths, such as a car moving through a sharp right bend in the road. The centrifugal force is directed outward and can be felt by the car's driver as he or she is pressed against the car door through the turn. Likewise, the centrifugal force balances the attractive force between the Earth and the Moon. If the Moon were stopped in its orbit, the centrifugal force would disappear and gravitational force would cause the Earth and the Moon to collide. Conversely, if the gravitational force ceased between the two bodies, the Moon would career into space.

Thus far, we have been careful not to say that the Moon orbits the Earth. In fact, the Earth and the Moon form a single system in which the two bodies

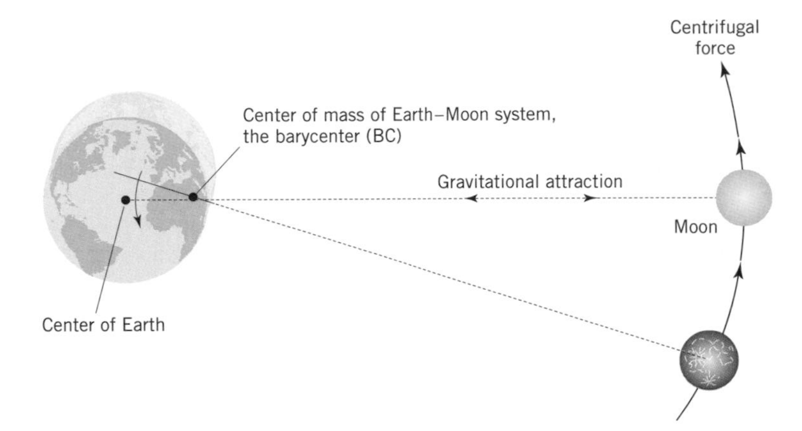

Fig. 11.3 The Earth–Moon system rotates around a common center of mass called the barycenter, which is within the Earth. The gravitational attraction between the Earth and the Moon is balanced by the centrifugal force.

revolve around a single center of mass. Because the Earth contains approximately 81.5 times more mass than the Moon, the center of mass of the system, called the barycenter, is within the Earth. The barycenter can be determined from knowledge that the average distance between the center of the Earth and the center of the Moon is 385,000 km. The center of mass must be 81.5 times closer to the Earth's center than to the Moon's. By dividing 385,000 by 81.5 we calculate that the center of mass is 4724 km from the Earth's center. The Earth's radius is 6380 km and therefore the center of the Earth–Moon system is located 1656 km (6380 − 4724) beneath the surface of the Earth. The Earth–Moon system can be visualized by considering a dumbbell with a much larger ball at one end (81.5 times more massive) than the other. If this dumbbell were thrown end over end, it would appear as though the large ball (the Earth) wobbled and the small ball (the Moon) orbited the large ball.

It should be understood that because the entire Earth is revolving around the center of the Earth–Moon system, every unit mass on the surface of the Earth is moving through an orbit with the same dimensions. The average radius of the orbits is 4724 km. The movement of the Earth around the Earth–Moon center of mass should not be confused with the Earth spinning on its axis, which is a separate phenomenon. Thus, if every unit mass on the surface of the Earth has the same size orbit, then it follows that the centrifugal force of the unit masses must also be equal.

11.2.3 Tide-producing force

Ocean tides exist because gravitational and centrifugal forces are unequal on the Earth's surface (hydrosphere) (Fig. 11.4). In fact, the gravitational attraction between the Moon and Earth only equals the centrifugal force of the Earth at the center of their two masses, which is determined to be 1656 km inside the Earth's surface. Thus, if we consider a unit mass at the surface of the Earth at a site facing the Moon, this mass experiences a force of attraction by the Moon that is greater than the counterbalancing centrifugal force. The larger gravitational force

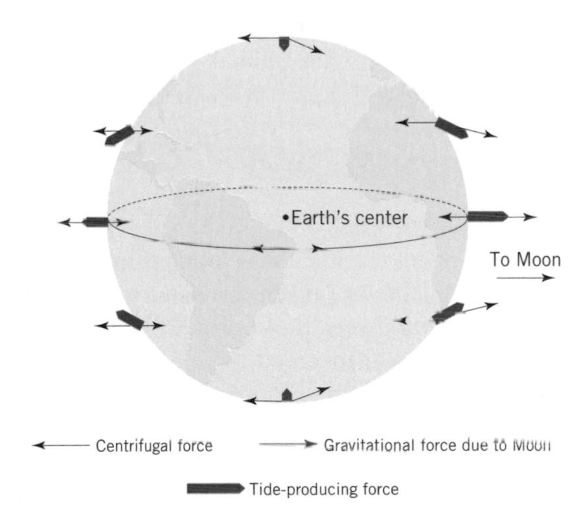

Fig. 11.4 Tides are generated in the Earth's hydrosphere due to differences between the centrifugal force and the Moon's attractive force.

is explained by the fact that at this location the distance to the Moon is less than the distance between the center of masses and the Moon, where the two forces are equal. Remember in Newton's gravitation equation that as distance decreases, the force of gravity increases. Conversely, for a unit mass on the opposite side of the Earth the centrifugal force exceeds the gravitational attraction exerted by the Moon because this site is farther away from the Moon than the center of mass, where the forces balance. Thus, the unequal forces on either side of the Earth cause the hydrosphere to be drawn toward the Moon on the near side of the Earth and to be directed away from the Moon on the opposite side. This produces two tidal bulges of equal size that are oriented toward and away from the Moon. These forces also result in depressions in the hydrosphere that are located halfway between the two bulges on either side of the Earth. If we disregard the curvature of the Earth, the tides can be thought of as a long wave, with the crest being the bulge and the depression being the trough. This waveform is called the tidal wave, which should not be confused with a tsunami, which sometimes is inappropriately referred to as a tidal wave.

The above description reveals that forces generating the Earth's tides are very sensitive to distance. The tide-generating force is derived by calculating the difference between the gravitational force and the centrifugal force. A simplified form of the relationship is given by:

$$F \approx \frac{M_1 M_2}{R^3}$$

The tide-generating force F is proportional to the masses M_1 and M_2 and inversely related to the cube of the distance between the bodies R^3. When these computations are performed for unit masses over the surface of the Earth, it is seen that the resulting vectors are oriented toward and away from the Moon (Fig. 11.4). Note also that **distance** is cubed in the equation, which explains why the Moon exerts a greater control on the Earth's tides than does the Sun. As illustrated in Fig. 11.2, after substituting the respective mass and distance values into the above equation, it is calculated that the tide-generating force of the Sun is only 46% of that of the Moon.

11.3 Equilibrium tide

The equilibrium tide is a simplified model of how tides behave over the surface of the Earth given the following assumptions:

1 The Earth's surface is completely enveloped with water, with no intervening continents of other landmasses.

2 The oceans are extremely deep and uniform in depth, such that the seafloor offers no frictional resistance to movement of the overlying ocean water.

3 There are two tidal bulges that remain fixed toward and away from the Moon.

11.3.1 Tidal cycle

In our initial discussion of the equilibrium tide model we will neglect the effects of the Sun; they are treated below. If we consider a stationary Moon, then the Earth passes under the two tidal bulges each time it completes a rotation around its axis (Fig. 11.5). In this idealized case the wavelength of the tidal wave, which is the distance between the tidal bulges, would be half the circumference of the Earth. High tide coincides with the Earth's position

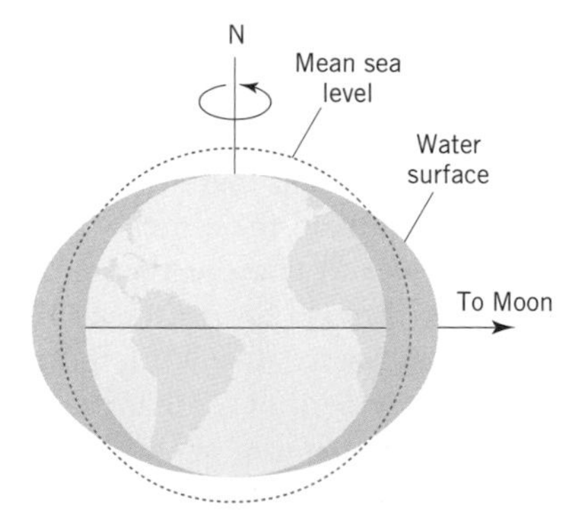

Fig. 11.5 Under idealized equilibrium tide conditions the Earth passes beneath two equal and opposite tidal bulges every 24 hours. This situation assumes an absence of continents, a uniform depth ocean, and the moon aligned with the equator.

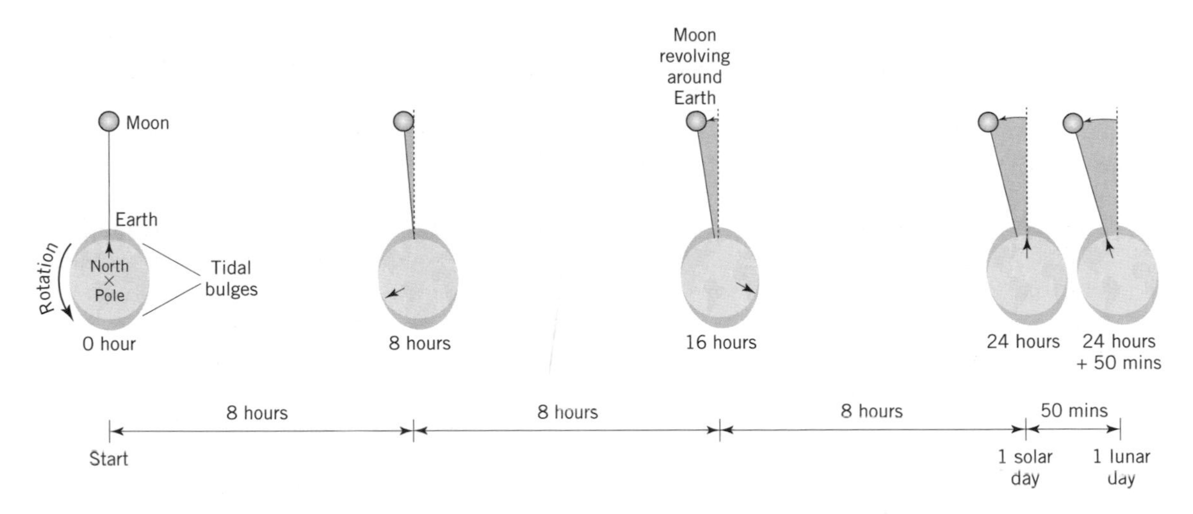

Fig. 11.6 Cartoon illustrating why times of high and low tide occur 50 minutes later each successive day.

under the bulges and low tide corresponds to the troughs located midway between the bulges. The tidal period would be 12 hours (interval between successive tidal bulges). The rhythm of tidal changes referred to as the tidal cycle can be better conceptualized if we choose a position on the equator directly facing the Moon and record how the water level fluctuates at this site through time as the Earth spins on its axis. At 12:00 midnight the ocean is at high tide because the Earth is directly under the maximum extent of the tidal bulge. After high tide the water level gradually drops, reaching low tide six hours later at 6:00 a.m. During the next six hours the tide rises, attaining a second high tide at 12:00 noon. The cycle repeats itself over the next 12 hours.

11.3.2 Orbiting moon

In the real world the Moon is not stationary but it completes an orbit around the center of the Earth–Moon system in a period of 27.3 days. The Moon moves in the same direction as the Earth spins on its axis and therefore after the Earth completes a full 24 hour rotation the Moon has traveled 13.2° of its orbit. For the Earth to "catch up" to the Moon, it must continue to rotate for an additional 50 minutes. Thus the Moon makes successive transits above a given location on the Earth in a period of 24 hours and 50 minutes, which is called the lunar day (Fig. 11.6). Because the Moon is moving, high and low tides do not take place every 12 hours as discussed in the simple model above, but every 12 hours and 25 minutes. The time interval between high and low tide is about 6 hours and 13 minutes. When there are two cycles in a day (actually 24 hours and 50 minutes) they are called semi-diurnal tides.

11.3.3 The Moon's declination

So far in our discussion of the tides, we have simplified matters by envisioning a Moon that is always directly overhead of the equator. However, the Moon's orbit is actually inclined to the plane containing the equator. Over a period of a month the Moon migrates from a maximum position 28.5° north of the equator to a position 28.5° south of the equator and back again. When the Moon is directly overhead of the tropics the tides are called **tropic tides** and when it is over the equator they are called **equatorial tides**. Because the Moon is the dominant controller of the tides it follows that when the Moon is positioned far north or south of the equator the tidal bulges will also be centered in the tropics. This arrangement of the tidal bulges leads to a semi-diurnal tidal inequality, meaning that successive tides have very different tidal ranges (tidal range is

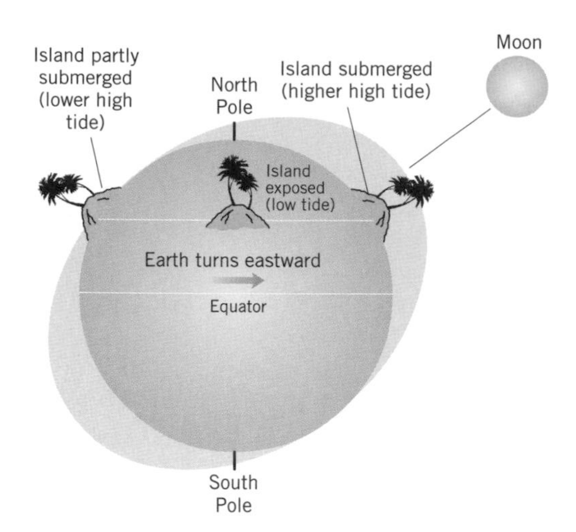

Fig. 11.7 In its orbit around the Earth, the Moon moves above and below the equator. When the Moon is not aligned with the equator, successive tidal bulges and the corresponding tidal ranges are unequal. This phenomenon is called a semi-diurnal inequality.

the vertical difference in elevation between low and high tide). For example, if you are along the east coast of Florida at tropic tide conditions, during one tidal cycle the tide will come up very high and then go out very far, generating a relatively large tidal range. During the next tidal cycle, a low high tide is followed by a high low tide, producing a small tidal range. As seen in Fig. 11.7, the diurnal inequality is explained by the fact that when the tidal bulges are asymmetrically distributed about the equator, the Earth will rotate under very different sized tidal bulges. This translates to unequal successive high and low tides. It should be noted that during equatorial tide conditions there is little to no inequality of the semi-diurnal tides, whereas they reach a maximum during tropic tides.

11.4 Interaction of the Sun and the Moon

It was shown above that the Sun's tide-generating force is a little less than half that of the Moon (46%). It is important to note that, just like the Moon, the Sun produces bulges and depressions in the Earth's hydrosphere. These are called solar tides and they have a period of 12 hours, unlike the 12 hours and 25 minutes period of the lunar tides. The period is 12 hours because the Earth passes through two solar bulges every day (24 hours). One way of explaining the interaction of the Moon and Sun is to show how the Sun enhances or retards the Moon's tide-generating force. In order to do this we must first understand how the phases of the Moon correlate with the position of the Earth, the Moon, and the Sun.

New Moons and full Moons result when the Earth, the Moon, and the Sun are aligned, a condition referred to as **syzygy** (a great Scrabble word worth many points). A new Moon occurs when the Moon is positioned between the Earth and the Sun, whereas a full Moon results when the Moon and the Sun are on the opposite sides of the Earth. When the Moon forms a right angle with the Earth and the Sun (**quadratic** position), only half of the Moon's hemisphere is illuminated. This phase occurs during the Moon's first and third quarters. The Moon cycles through these different phases over a period of 29.5 days.

During new and full Moons when the Earth, the Moon, and the Sun are all aligned, the tide-generating forces of the Moon and the Sun act in the same direction and the forces are additive (Fig. 11.8). Conceptually, one can envision the Sun's bulge sitting on top of the Moon's bulge. Between the bulges the Sun's trough further depresses the Moon's trough. The double bulges and double troughs lead to very high high tides and very low low tides. This condition is called a spring tide and is characterized by maximum tidal range (Fig. 11.9).

During quadratic conditions the Sun's effects are subtractive from the Moon's tide-generating force. Because the Moon is positioned at 90° to the Earth and the Sun, its bulge coincides with the Sun's trough and its trough is positioned at the Sun's bulge. The superposition of bulges and troughs causes destructive interference and produces low high tides and high low tides. This condition is called a neap tide and is characterized by minimum tidal ranges (Fig. 11.9). When the Earth, Moon, and Sun are arranged in positions between syzygy and quadratic we experience mean tides with average tidal ranges.

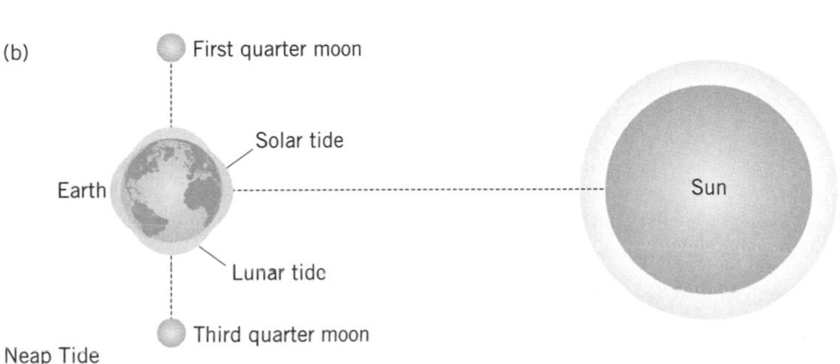

Fig. 11.8 The alignment of the Earth, Moon, and Sun determines the size of the tidal bulges and the magnitude of the tidal range. (a) During periods of full and new moons the tidal bulges are additive, producing relatively large tidal ranges. (b) When the Moon, Earth, and Sun are at right angles during half moon conditions the tidal forces are subtractive and tidal ranges are relatively small.

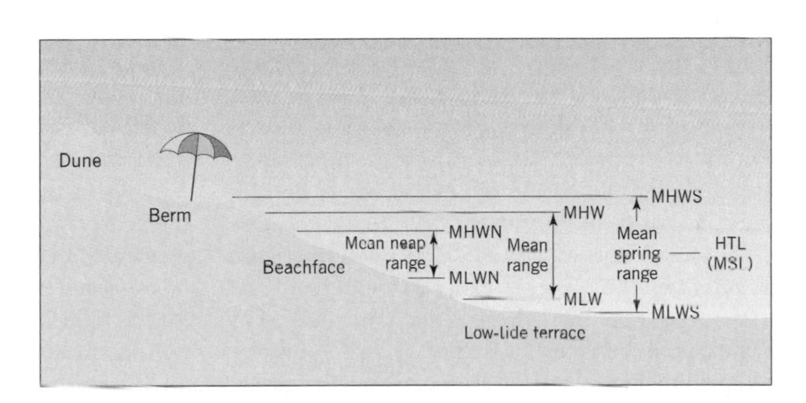

Fig. 11.9 Tidal ranges and the elevation of high and low tides are a function of the position of the Earth, Moon, and Sun.

Spring and neap tides occur approximately every 14 days, whereas mean tides occur every seven days.

11.5 Effects of orbital geometry

Remembering that the tide-producing force is particularly sensitive to distance, it is understandable that the geometry of both the Earth's and the Moon's orbits affects the tides. The Earth revolves around the Sun in an elliptical orbit and the Sun is situated at one of the foci of the ellipse. In early January the Earth is nearest to the Sun, at a position called perihelion (Fig. 11.10). Six months later (July) the Earth is at aphelion, furthest from the Sun. The difference in distances is approximately 4%. The Moon's orbit around the center of the Earth–Moon system is also elliptical. When the Moon is close to the Earth the position is referred to as perigee and when it is most distant it is called apogee. There is a 13% difference between perigee and apogee.

If we consider all the various factors that influence the magnitude of the tides, we begin to understand why tidal ranges and high and low tidal elevations change on a daily basis. Tide levels are especially important during storms. In early February 1978 a major northeast storm, named the Blizzard of 1978, wreaked havoc in New England, dropping over two feet of snow and completely immobilizing the residents for several days. The Blizzard of 1978 was

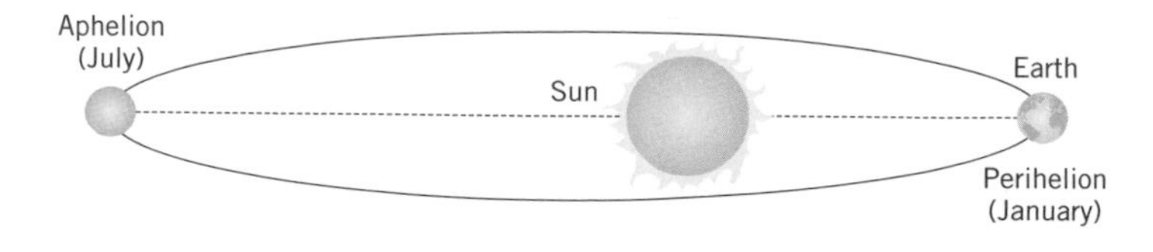

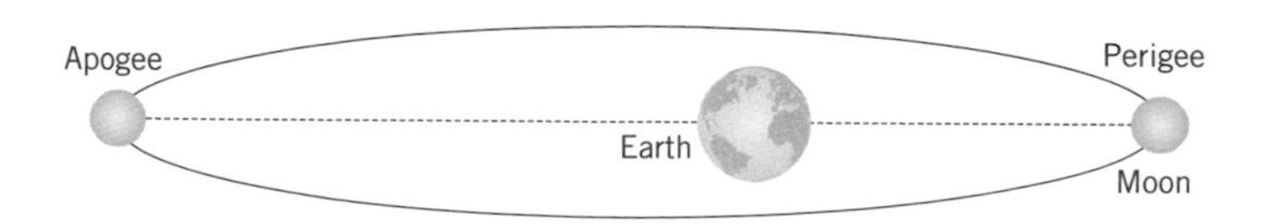

Fig. 11.10 The elliptical orbit of Earth affects its distance from the Sun and therefore the magnitude of the tide-producing force during the year.

a particularly menacing storm causing widespread beach erosion, the destruction of hundreds of coastal dwellings, and hundreds of millions of dollars worth of damage to roadways and other infrastructure. One of the reasons why this storm was so severe was the astronomic conditions that existed at the time of the storm. The Earth, Moon, and Sun were in syzygy so the storm hit during spring tide conditions. At the same time the Moon was at perigee and the Earth and Sun were close to a perihelion position. Syzygy, perigee, and perihelion combined to raise high tide levels 0.55 m above normal. The extreme astronomic tides coupled with the 1.4 m storm surge caused extensive flooding. Storm waves elevated by high water levels broke directly against foredune ridges, across barriers, and over seawalls. Had the storm hit during quadratic, apogean, and aphelion conditions, high tide waters would have been 1.1 m below the February Blizzard levels and damage would have been an order of magnitude less (Fig. 11.11).

11.6 Effects of partitioning oceans

We have been treating tides as if the Earth were completely enveloped by a uniformly deep ocean. However, we know that continents and island archipelagoes have partitioned the hydrosphere into several interconnected large and small ocean basins whose margins are generally irregular and quite shallow. Because oceans do not cover the surface of the Earth, the tidal bulges do not behave as simplistically as has been presented thus far. In addition to the complexities imparted by the presence of landmasses, the equilibrium tide concept is further complicated by the fact that the Earth spins faster in lower latitudes and slower in higher latitudes than the tidal wave. Thus, the oceans do not have time to establish a true equilibrium tide. Finally, the ocean tides are affected by the Coriolis effect, which is generated by the Earth's rotation. This causes moving objects, including water masses, to be deflected to the right in the northern hemisphere and to left in the southern hemisphere. For example, the Gulf

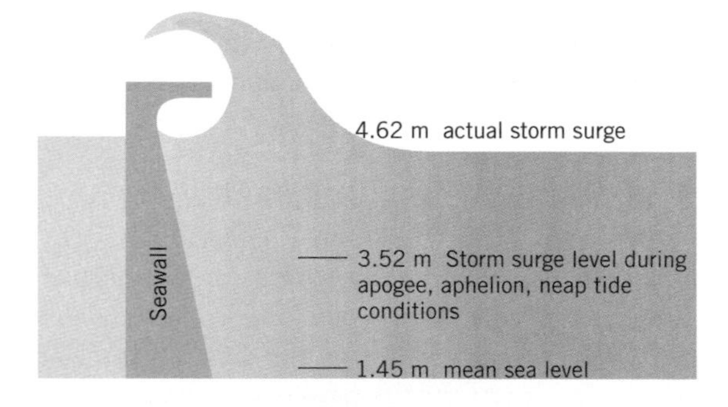

Fig. 11.11 Cartoon of seawall along Winthrop Beach, Massachusetts that was overtopped by storm waves during the blizzard of 1978. Note that if the storm had occurred during low astronomic tidal range conditions, fewer waves would have broken over the seawall and the overall damage to the New England shoreline would have been far less severe.

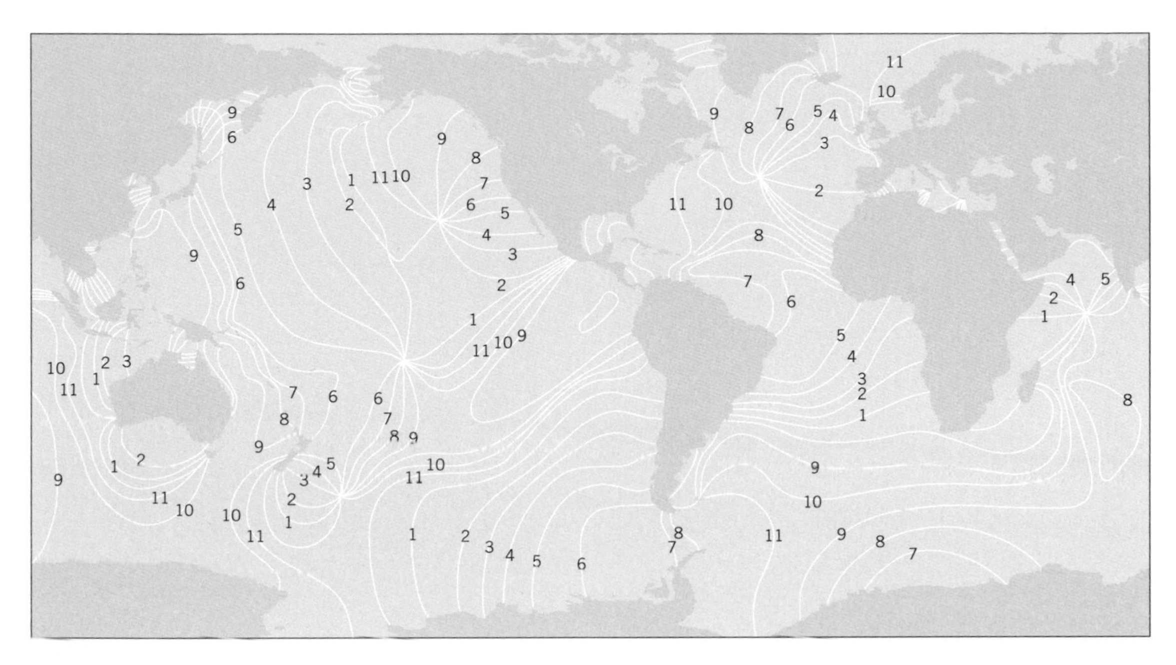

Fig. 11.12 Amphidromic systems throughout the world's oceans. The tidal wave rotates in a counterclockwise direction around amphidromic points in the northern hemisphere and in a clockwise direction in the southern hemisphere. The lines radiating from the amphidromic points are co-tidal lines. They indicate hypothetical times in which the crest of the tidal wave passes through the ocean basins.

Stream that flows northward along the margin of North America is deflected northeastward toward Europe due to the Coriolis effect.

In the dynamic model of ocean tides we no longer envision static ocean bulges that remain fixed toward the Moon and under which the Earth spins; instead, the tidal bulges rotate around numerous centers throughout the world's oceans (Fig. 11.12). An individual cell is called an amphidromic system and the center of the cell around which the tidal wave rotates is known as the amphidromic point or nodal point. The rotation of the tidal wave is due to the Coriolis effect and is counterclockwise in the northern hemisphere and clockwise in the southern hemisphere.

To understand the behavior of an amphidromic system let us first begin with a hypothetical square-shaped ocean basin that responds to the Moon's tide-generating forces (Fig. 11.13). The tidal wave that develops under these conditions exhibits elements of both a standing wave and a progressive wave. As the Earth spins and the Moon travels from east to west over the hypothetical basin, the tidal bulge sloshes against the western side of the basin in an attempt to keep abreast of the passing Moon. As the Earth continues to rotate, the bulge of water begins to flow eastward back toward the low center of the basin. However, the Coriolis effect deflects this water mass to the southern margin of the basin, and water piles up there. This in turn creates a water surface that slopes northward and the process is repeated. The end result is a tidal wave that rotates in a counterclockwise direction around the basin with a period of 12 hours and 25 minutes. High tide is coincident with the tidal bulge and low tide occurs when the bulge is along the opposite side of the basin.

When a line is drawn along the crest of the tidal wave every hour for a complete rotation, the resulting diagram looks like a wheel with spokes. It depicts how the tidal wave rotates within the hypothetical basin and its center is the amphidromic point. The spokes are called co-tidal lines and they define points within the basin where high tide (and low

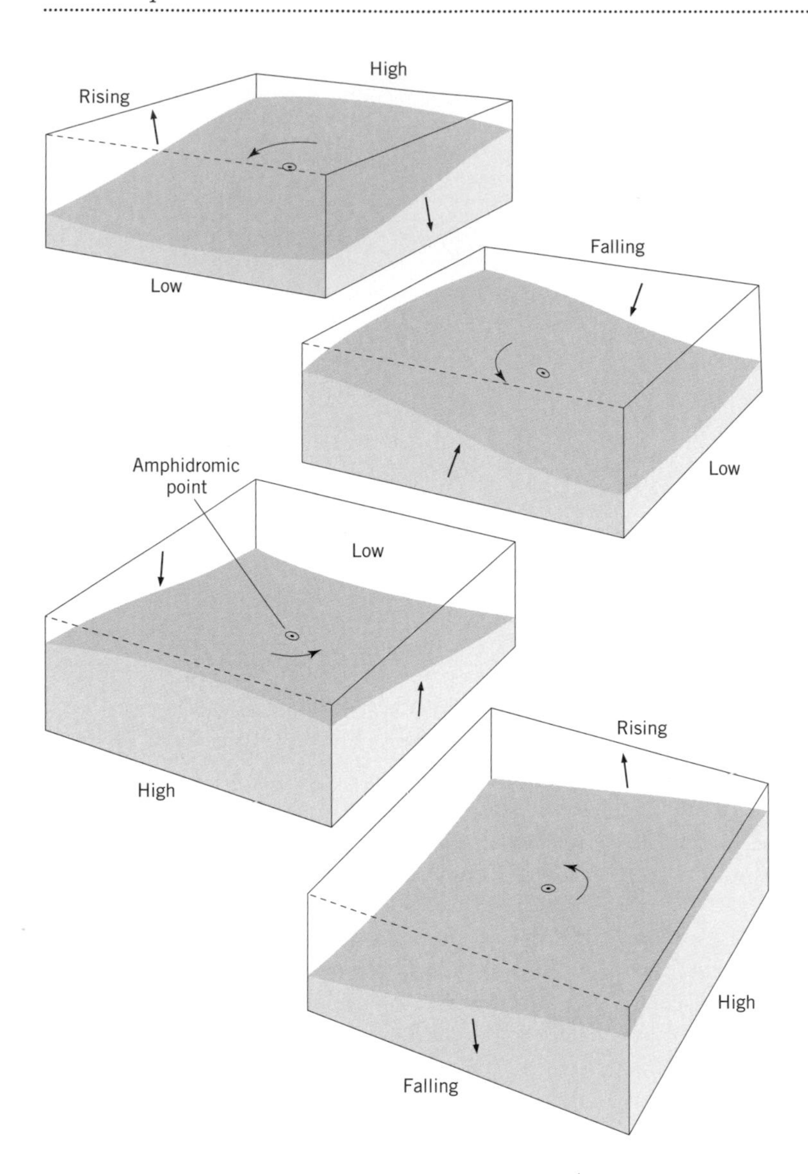

Fig. 11.13 Hypothetical tidal wave rotating around an amphidromic point in a square-shaped basin. Note that water elevation changes (tidal range) increase outward from the amphidromic point. The Coriolis effect causes rotation of the tidal wave.

tide) occurs at the same time (Fig. 11.12). If points of equal tidal range are contoured within the basin a series of semi-concentric circles are formed around the amphidromic point. These contours with equal tidal range are referred to as co-range lines. Ideally, tidal range is zero at the amphidromic point and gradually reaches a maximum toward the edge of the basin. Due to the land barriers and other factors the world's oceans are divided into approximately 15 amphidromic systems. This does not include smaller seas that have their own amphidromic cells, such as the Gulf of Mexico (one system), Gulf of Saint Lawrence (one system), and North Sea (three systems).

11.7 Tidal signatures

In the ideal case, we expect two tidal cycles daily (actually 24 hours and 50 minutes). However, the highly variable basinal geometries of the world's ocean and modifications of the tidal wave as it

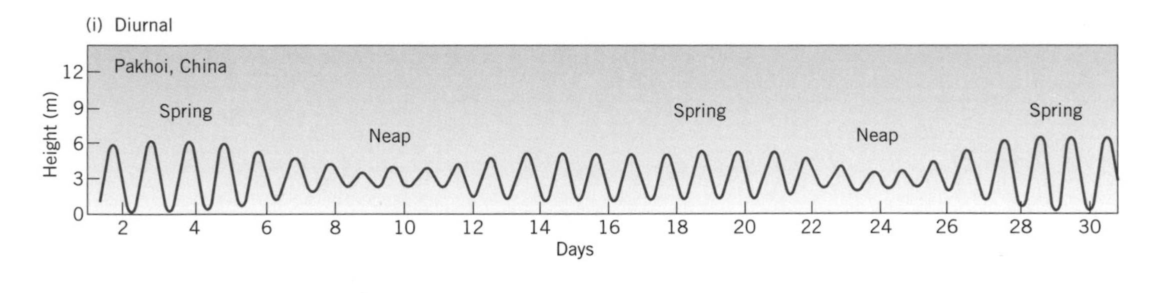

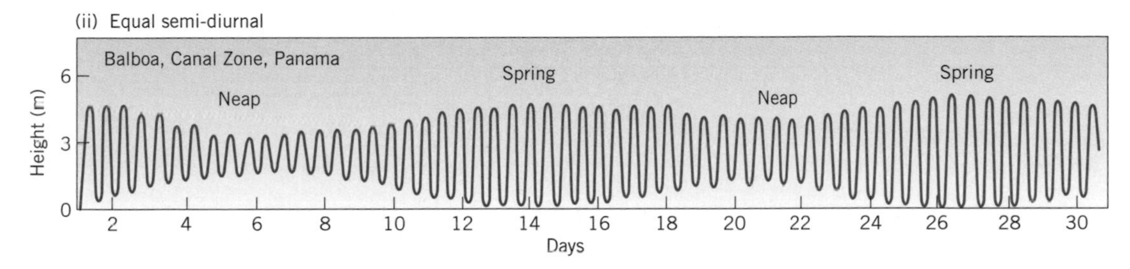

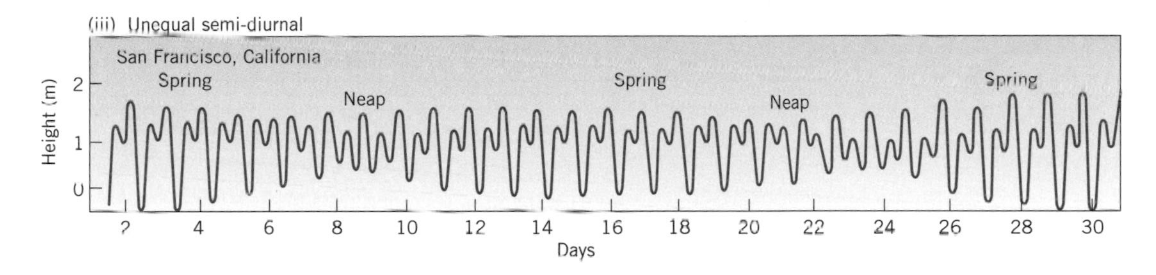

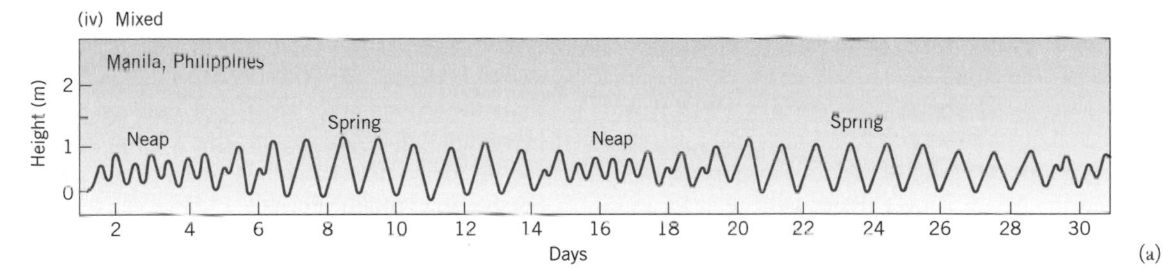

(a)

Fig. 11.14 Coastlines throughout the world experience a variety of tidal signatures. (a) The major types include: (I) diurnal tides, one tide daily; (II) semi-diurnal tides, two tides daily; (III) semi-diurnal tides with strong inequality; (IV) mixed tides, combination of diurnal and semi-diurnal tides. (From R. A. Davis, 1977, *Principles of Oceanography*, 2nd edn. Reading, MA: Addison-Wesley, p. 155, Fig. 8.9.)

shoals across the continental shelf, as well as other factors, have combined to produce a variety of tidal signatures throughout the world's coastlines. There are three major types of tides (Fig. 11.14):

1 Diurnal tides. Coasts with diurnal tides experience one tidal cycle daily, with a single high and low tide. They have a period of 24 hours and 50 minutes. This type of tide is rare and commonly associated with restricted ocean basins, including certain areas within the Gulf of Mexico, the Gulf of Tonkin along Southeast Asia, and the Bering Sea. In these areas distortions of the tidal wave produce

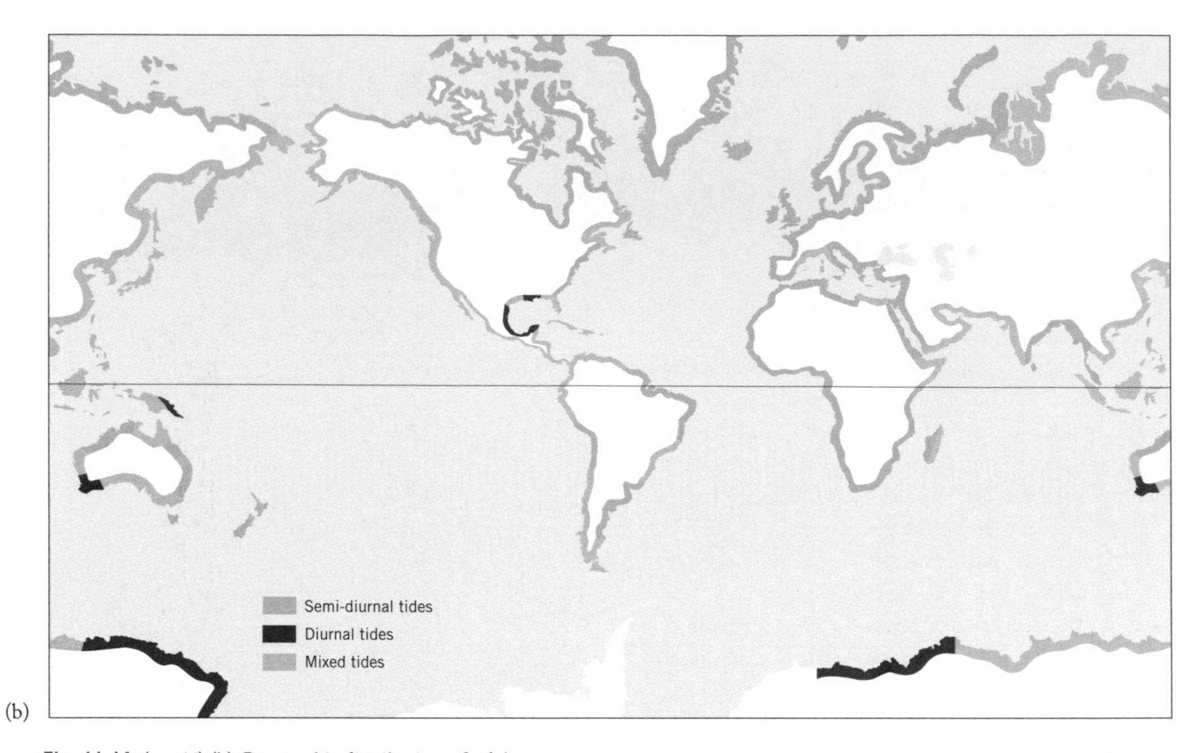

Fig. 11.14 (*cont'd*) (b) Geographic distribution of tidal types.

a natural tidal oscillation coinciding with 24 hours and 50 minutes. The open southwest coast of Australia is an exception to this general trend.

2 Semi-diurnal tides. This is the most common type of tide along the world's coast. Semi-diurnal tides are characterized by two tidal cycles daily with a period of 12 hours and 25 minutes. Seldom, however, are the two tides of the same magnitude except when the Moon is over the equator or at locations near the equator. As discussed above, when the Moon is over the tropics successive tidal bulges (the tidal wave) have unequal magnitudes, producing different tidal ranges with high and low tides reaching different elevations. This condition of unequal tides is called a semi-diurnal inequality.

3 Mixed tides. This type of tide occurs extensively throughout the world. As the name implies mixed tides have elements of both diurnal and semi-diurnal tides. The signature varies during a lunar cycle from a dominant semi-diurnal tide with a small inequality to one that exhibits a very pronounced inequality. At some sites, including San Francisco, Seattle, and Port Adelaide in Australia, during part of the lunar month one of the two daily tides manifests itself as a very small vertical excursion measuring no more than 0.1–0.3 m. These tides have a distinct diurnal signature. Along other coasts, such as Los Angeles, Honolulu, and Manila in the Philippines, the second daily tide essentially disappears and the tide becomes totally diurnal.

Thus, the complexities that produce and modify the Earth's tides are revealed by the variability in their tidal signature throughout the world and even temporally as viewed during a lunar cycle.

11.8 Tides in shallow water

11.8.1 Continental shelf effects

In the middle of the ocean the tidal wave travels with a speed of 700 km h^{-1}. In these regions the tidal range is only about 0.5 m. The tidal wave that reaches the coast travels from the deep open ocean across the continental margin to the shallow inner continental shelf. Similar to wind-generated waves, shoaling of the tidal wave along this pathway causes it to slow down. The tidal wave that traverses the entire continental margin is reduced in speed to about 10–20 km h^{-1}. Like wind waves, the tidal waves also steepen, which is reflected in an increase in tidal range. For example, the tidal wave in the north Atlantic is estimated to be 0.8 m in height (tidal range) at the edge of the continental shelf. The wave steepens as it propagates through the Gulf of Maine, producing a tidal range of 2.7 m along the coast of Maine. On a worldwide basis, using an average shelf width of 75 km, it is estimated that the tidal wave will heighten from 0.5 m in the deep ocean to about 2.4 m along the coast after it traverses the continental shelf. Variations from this value are due to differences in shelf width and slope, and variability in the configuration of the coast. It is of interest to note that along the east coast of the United States, the continental shelf is relatively wide off the Georgia coast, where tidal ranges reach 2.6 m. North and south of this region the shelf narrows and tidal ranges correspondingly reduce to 1.1 m at Cape Hatteras and less than a meter along central Florida (Fig. 11.15).

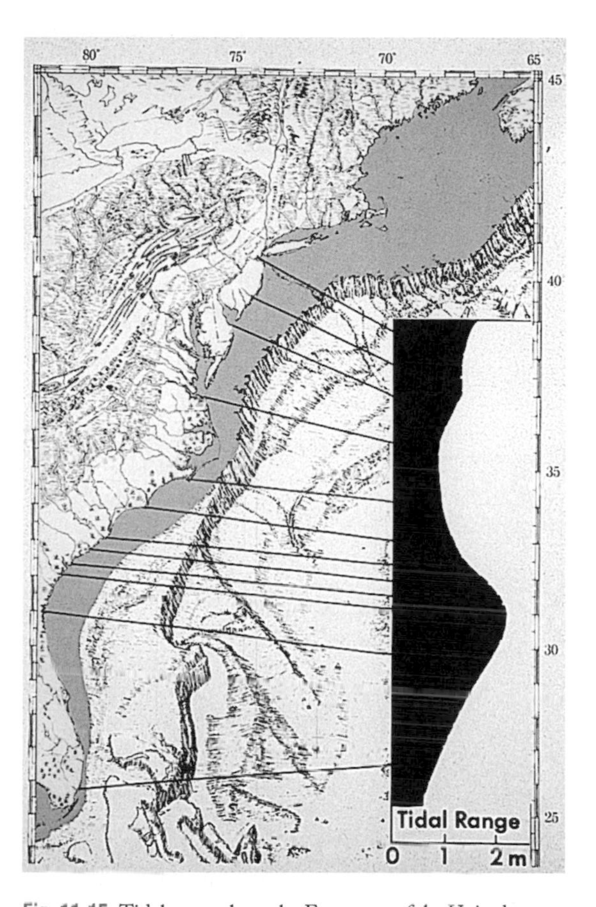

Fig. 11.15 Tidal range along the East coast of the United States is controlled in part by the width of the continental shelf. As the tidal wave travels from the deep ocean across the shallow continental shelf, its speed slows and its crest steepens. The amount of wave steepening, which increases the tidal range, is proportional to the length and gradient of the shelf. Thus, the wide, gentle shelf off the Georgia coast produces a tidal range greater than 2 m, whereas the steep narrow shelves bordering south Florida and Cape Hatteras in North Carolina generate tidal ranges less than 1 m. (From D. Nummedal, G. Oertel, D. K. Hubbard & A. Hine, 1977, Tidal inlet variability – Cape Hatteras to Cape Canaveral. In *Proceedings of Coastal Sediments '77*. Charleston, SC: ASCE, pp. 543–62.)

11.8.2 Coriolis effect

Just as the Coriolis effect produces a counterclockwise rotation of the tidal wave in large amphidromic systems in the northern hemisphere, it also influences the propagation of the tidal wave into gulfs and seas from the open ocean. This phenomenon is demonstrated well in the North Sea, where the Coriolis effect dramatically modifies tidal ranges. The North Sea is a shallow (<200 m), rectangular-shaped basin approximately 850 km long from the Shetland Islands southeastward to the German Friesian Islands and 600 km wide from Great Britain eastward to Jutland in Denmark. Tides in the North Sea are forced by the North Atlantic amphidromic system (Fig. 11.12). The tidal wave approaches through the northern open boundary of the North Sea and propagates southward. Ultimately, the wave partially reflects off the southern margin of the basin and interacts with the next incoming wave. The resulting oscillations combined with the Coriolis effect produce three amphidromic systems, two of which are displaced toward the coasts of Norway

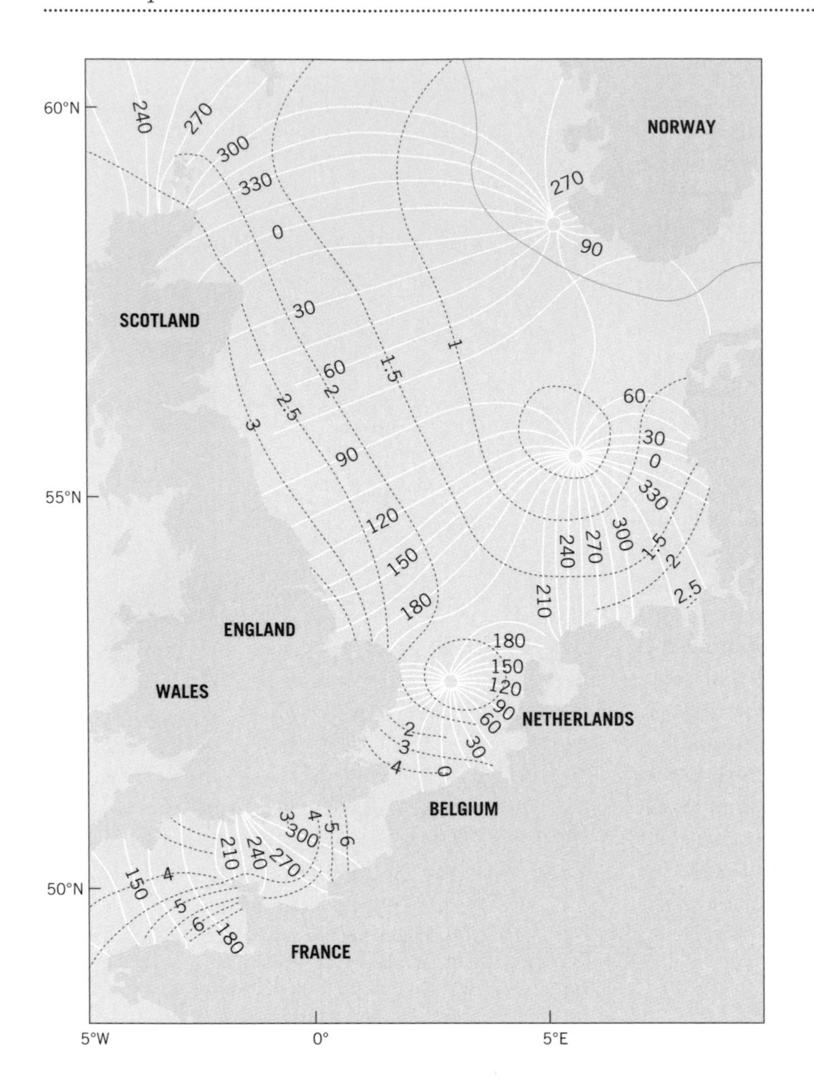

Fig. 11.16 The distribution of tidal ranges in the North Sea illustrates how the Coriolis effect affects tidal wave propagation into a shallow sea. (From D. A. Huntley, 1980, Tides on the north-west European continental shelf. In F. T. Banner, M. B. Collins & K. S. Massie (eds), *The North-West European Shelf Seas: The Sea Bed and the Sea in Motion. II Physical and Chemical Oceanography and Physical Resources.* Amsterdam: Elsevier.)

and Denmark. As illustrated by the co-range lines in Fig. 11.16, tidal ranges are much higher along the east coasts of England and Scotland (~4 m) than on the Norwegian and Danish coasts (<1 m). This disparity is caused by the tidal wave being deflected to the right as it moves into the North Sea. Water is piled up on the western side of the basin and diminishes the tidal ranges along the eastern side.

A similar situation occurs in the English Channel. Here the tidal wave approaches from the southwest and propagates eastward through the channel. The Coriolis effect deflects water away from the English side of the channel and toward the coast of France. Tidal ranges along the coasts of Brittany and Normandy are greater than 4–5 m, whereas along the southwest English coast they are less than 3 m.

11.8.3 Funnel-shaped embayments

The configuration of the coast can also have a pronounced influence on tidal ranges. Rotary tidal waves do not exist in funnel-shaped embayments due to their narrowness. Instead, the tidal wave propagates into and out of funnel-shaped bays. The wave is constricted by the seabed, which shallows in a landward direction, and by the increasingly narrow con-

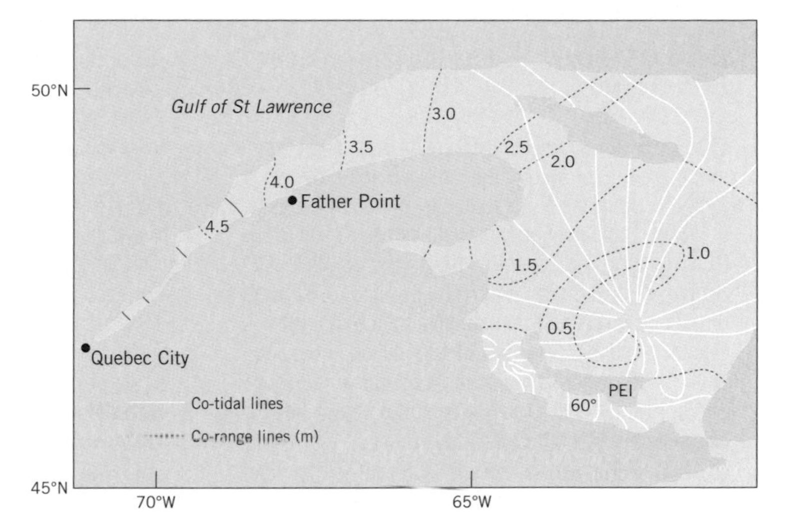

Fig. 11.17 The Gulf of St Lawrence narrows from a width of 150 km at the entrance to the St Lawrence River to less than 15 km wide just downstream of Quebec City. Gradual constriction of the tidal wave in this funnel-shaped embayment increases spring tidal ranges from 1.0 m at the entrance to over 5.0 m at Grosse Isle near Quebec City.

fines of the embayment. Although frictional elements serve to decrease the energy of the propagating wave, the overall steepening of the tidal wave causes an amplification of the tidal range (Fig. 11.17). For example, the 2.0 m tidal range at the entrance to the Saint Lawrence estuary increases to over 5 m during spring tidal conditions at Quebec City, some 600 km upstream. Funnel-shaped embayments are found all over the world, including the Bay of Fundy in Canada (see Box 11.1), the Gironde and Seine estuaries in France, the Wash and the Severn Estuary in the United Kingdom, Cambridge Gulf in Australia, Cook Inlet and Bristol Bay in Alaska, the Gulf of Cambay in India, the head of the Gulf of California, and the Rio de Plata in South America.

11.8.4 Tidal bore

In some estuaries large tidal ranges lead to the formation of tidal bores (Fig. 11.18). A tidal bore is a steep-crested wave or breaking wave that moves upstream with the rising tide. Their occurrence coincides with large funnel-shaped estuaries that have tidal ranges exceeding 5 m and a channel that progressively shallows upstream. The height of most bores is less than 0.4 m but there are some spectacular bores that adventurers surf on as the wave advances upriver. A tidal bore is formed when the propagating tidal wave oversteepens and breaks due to a constriction of the channel and retarding effects of the river's discharge. Tidal bores are best developed during spring tide conditions when tidal ranges are near maximum. Bores are found in the Severn and Trent estuaries in the United Kingdom, the Seine in France, the Truro and Petitcodiac rivers that discharge into the Bay of Fundy, the Ganges in Bangladesh, and several rivers along the coast of China. Some of the largest tidal bores in the world occur in China-tang River in northern China and in the Pororoca River, a branch of the Amazon. Their heights have been reported to approach 5 m and they travel at speeds close to 20 km h^{-1}. Fisherman and shippers will often travel upstream by riding the ensuing strong currents that follow the passage of a tidal bore.

11.8.5 Tidal currents

Tidal currents are most readily observed in coastal regions where the tidal wave becomes constricted. As the tidal wave approaches the entrance to harbors, tidal inlets, and rocky straits, the tide rises at a faster rate in the ocean than it does inside the harbor or bay. This produces a slope of the water surface and, just like a river system, the water flows downhill, producing a tidal current (Fig. 11.19). The water moving through a tidal inlet and flooding a bay is called a **flood-tidal current**. The water emp-

Box 11.1 Bay of Fundy: the largest tides in the world

The Bay of Fundy in eastern Canada is perhaps the most famous funnel-shaped embayment in the world. It has record tides equal to the height of a five-story building. The bay connects to the northern end of the Gulf of Maine and separates the provinces of New Brunswick and Nova Scotia (Fig. B11.1). The formation of the bay is related to the opening of the Atlantic Ocean, which occurred about 180 million years ago when North America began separating from Europe and northern Africa. The bay is part of a rift valley that developed within a broad sandy arid plain. Remnants of basaltic eruptions associated with early rifting can still be found at several locations along the bay's margin. During repeated episodes of Pleistocene glaciation, ice sheets scoured and deepened the basin and then deposited a thick carpet of glacial sediment. Following deglaciation, isostatic rebound (see Chapter 17) caused the sea to retreat from the Bay of Fundy, exposing the basin to riverine reworking. Approximately 6000 years ago, rising eustatic sea level inundated the bay and allowed waves to erode the soft sedimentary rocks that surround most of the Bay of Fundy shoreline. Sandstone cliffs, intertidal marine platforms, and vast sand shoals are a product of this erosion.

The Bay of Fundy is 260 km long and 50 km wide at its opening, gradually narrowing into two separate bays: Chignecto Bay to the northeast and the Minas Basin to the east. It is deepest at its mouth and progressively shoals toward its eastern end, with an average depth of about 32 m along its length. The immense tidal range in the Minas Basin (Fig. B11.2) leads to more than 14 km^3 of water flushing the bay twice daily (actually every 12 hours and 25 minutes). Tidal currents at the entrance to the Minas Basin exceed 4 m s^{-1}. These strong currents mold the sandy bottom into giant sand ripples called sandwaves, which have heights of more than 10 m. Flow in the 5 km wide tidal channel leading into the Minas Basin is equal to the combined discharge of all the streams and rivers on Earth. The weight of the huge volume of water entering the Minas Basin actually causes a slight tilting of western Nova Scotia.

Although the funnel shape of this coast contributes to very large tides, it is **tidal resonance** that generates the world's largest tidal ranges. The length and depth of the Bay of Fundy promote the development of a standing wave (similar to the sloshing back and forth of water in a bathtub or a coffee cup) that is

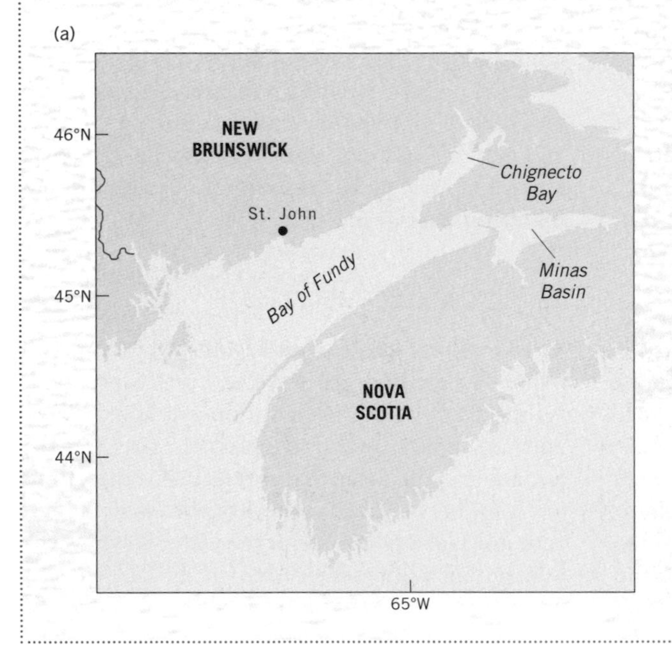

Fig. B11.1 Location map of the Bay of Fundy.

Box 11.1 (*cont'd*)

constructively perturbed by the tide-generating forces in the Atlantic Ocean. Resonance occurs in elongated embayments when the advancing tidal wave reflects off the head of the bay back toward the bay's entrance. A standing wave is produced when the geometry of the bay is of the correct dimensions, such that the reflected wave arrives at the bay entrance at the same time as the next incoming tidal wave. Each incoming wave amplifies the standing wave until the energy that is added balances the energy lost due to friction. Tidal ranges at the mouth to the Bay of Fundy are a modest 3 m but resonance and funneling effects gradually increase the range to an amazing 16 m near Wolfville at the eastern end of the Minas Basin (Fig. B11.3).

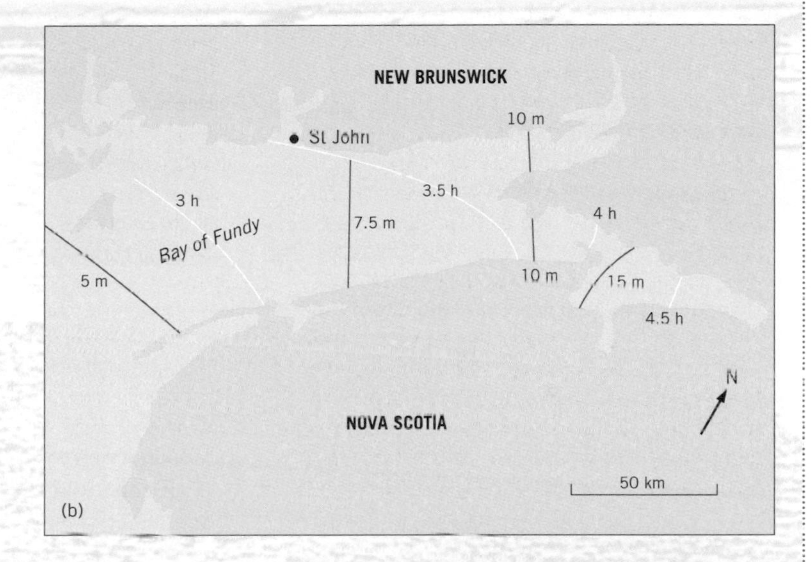

Fig. B11.2 Co-range and co-tidal lines for the Bay of Fundy.

Fig. B11.3 Pictures illustrating large tidal ranges, showing high and low tides.

Fig. 11.18 Tidal bores occur in funnel-shaped estuaries with tidal ranges greater than 5 m. (a) They are produced when shoaling and constriction of the landward moving tidal wave oversteepens and may begin to break. (b) View of tidal bore in the Salmon River in the Minas Basin along the Bay of Fundy, Nova Scotia. (c) Close-up view of tidal bore, which is approximately 30 cm in height.

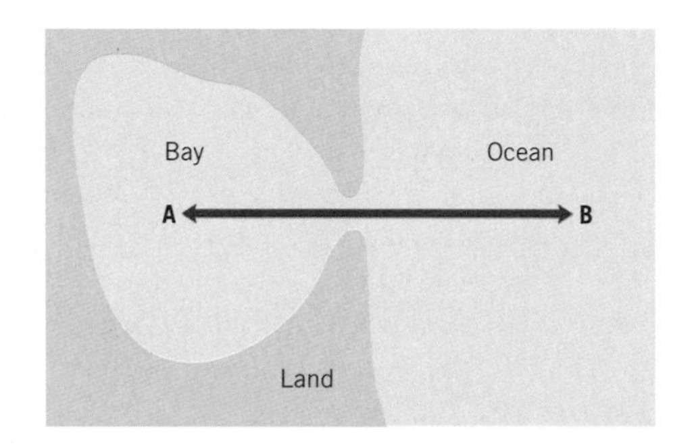

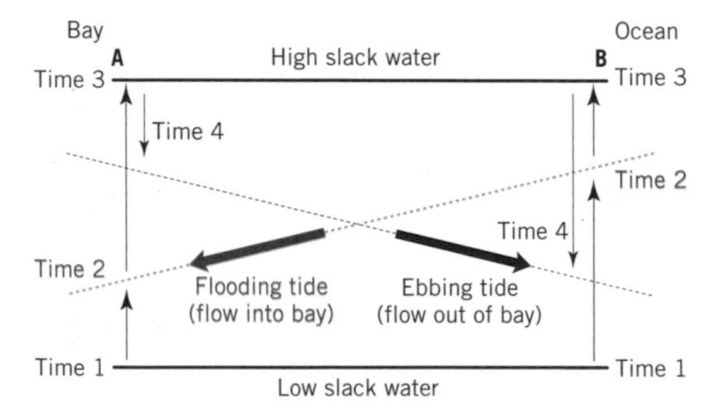

Fig. 11.19 Tidal currents occur at the entrance to bays, harbors, and tidal inlets and are due to a constriction of the tidal wave.

tying out of a bay and moving seaward is referred to as an **ebb-tidal current**. In a slight oversimplification, when the tidal waters in the ocean and bay are at the same elevation, there is **slack water** at the tidal inlet. This condition usually occurs at high tide and low tide. Likewise, the strongest current velocities are produced when the water surface through the inlet achieves the steepest slope, which commonly is near midtide but may also occur closer to high or low tide. During spring tide conditions when the maximum volume of water is exchanged between the ocean and bay, tidal currents can reach velocities of 3 m s^{-1}.

Along non-sandy shorelines, tidal currents can achieve strong current velocities, particularly in regions with large tidal ranges, large bay areas, and narrow constrictions. One such location is along the Norwegian coast north of Bodø where Vestfjord connects to the Norwegian Sea. The fierce tidal currents that flow through the straits reach speeds greater than 4.0 m s^{-1}. This creates strong whirlpools that make travel through the strait extremely dangerous during peak current flow. The Norwegians call these whirlpools the **maelstrom** and fishermen time their passage to avoid these perilous eddies.

The strong tidal currents that are generated in funnel-shaped estuaries and elsewhere along the world's coastlines can be harnessed to provide a source of energy. For example, in the Gulf of Saint Malo along the Brittany coast of France the tidal range in the Rance Estuary can exceed 12 m. This exceptional large tidal fluctuation produces very strong tidal currents. A 750 m wide barricade has been constructed across the river and houses 24 hydroelectric power stations. The reversing tidal currents of the Rance have been providing electricity since 1966.

11.9 Summary

The Moon's and the Sun's force of attraction exerted on the Earth's hydrosphere causes ocean tides. The Moon's tide-generating force is about twice that of the Sun because it is much closer to the Earth. The Moon and Earth revolve around a common center of mass inside the Earth, which produces a centrifugal force that balances the forces of attraction. In the equilibrium tide model, two tidal bulges are developed because masses on the Earth's surface are acted on unequally by gravitational and centrifugal forces. One bulge faces the Moon and the other is directed away from the Moon. The tidal period is 12 hours and 25 minutes rather than 12 hours (half of the Earth's rotation) because it takes the Earth an additional 50 minutes each day to catch up with the Moon in its orbit. As the Moon revolves around its common center of mass with the Earth its orbit makes excursions north and south of the equator. When the Moon is over the tropics, the Earth passes through unequal successive tidal bulges, producing different elevations in successive high and low tides and unequal tidal ranges. This tidal condition is called a semi-diurnal inequality.

The effects of the Sun can enhance or retard the Moon's tide-generating force. When the Moon, Earth, and Sun are aligned (a position called syzygy), the Sun's effects are additive and we experience spring tides and large tidal ranges. When the Moon, Earth, and Sun are at right angles (quadratic position), the Sun's effects diminish the Moon's tide-generating forces and we have neap tides and relatively small tidal ranges. Mean tides and average tidal ranges occur between syzygy and quadratic positions. Due to the elliptical orbits of the Moon and Earth, the height and range of the tides increases when the Moon is proximate to the Earth (perigean tides) and the Earth is close to the Sun (perihelion tides).

The continents and island archipelagos partition the Earth's hydrosphere into several interconnected large and small ocean basins. Based on their dimensions, Coriolis and tide-generating forces cause the tidal wave to rotate around one or more amphidromic points within these basins, counterclockwise in the northern hemisphere and clockwise in the southern hemisphere. Tidal range increases with distance from the amphidromic point but ranges in the open ocean are generally quite low (<0.6 m). When the tidal wave propagates across the continental margin the wave slows down and the crest steepens, resulting in an increase in the tidal range (1.0–2.0 m at the coast). The tidal signature along the coast reflects the geometry of the basin and shoaling behavior of the tidal wave. Most open ocean coasts experience semi-diurnal tides (two tides daily) or mixed tides, which is a tidal signature that exhibits periods of semi-diurnal tides, and distinctly diurnal (one tide daily) tides during other portions of the lunar month. Diurnal tides are most common in restricted basins where the tidal wave resonates with a period close to 24 hours and 50 minutes.

The tidal wave can undergo dramatic distortions as it moves into restricted ocean basins and through straits due to Coriolis effects and shoaling effects. Deformation of the tidal wave can result in dramatic differences in tidal range over distances less than 50 km. This is particularly apparent in funnel-shaped embayments, where steepening of the advancing tidal wave can increase the tidal range by 2–4 m at the head of the bay. Even greater tidal ranges can result if a standing wave is produced in the bay that is constructively interfered by the incoming tidal wave. This phenomenon is best developed in the Bay of Fundy, where tidal ranges (up to 16 m in the

Minas Basin) are the largest in the world. In some estuaries with very large tidal ranges, the advancing tidal wave steepens, forming a steep-crested wave or breaking wave called a tidal bore, which moves upstream with the rising tide. Tidal currents are produced in coastal settings when the tidal wave becomes constricted, such as at the entrance to a bay or tidal inlet.

Suggested reading

Defant, A. (1958) *Ebb and Flow: The Tides of Earth, Air, and Water*. Ann Arbor: University of Michigan Press.

Fischer, A. (1989) The model makers. *Oceanus*, **32**, 16–21.

Greenberg, D. A. (1987) Modeling tidal power. *Scientific American*, **247**, 128–131.

Lynch, D. K. (1982) Tidal bores. *Scientific American*, **247**, 146–57.

Open University (1989) *Waves, Tides, and Shallow-water Processes*. Oxford: Pergamon Press.

Redfield, A. C. (1980) *Introduction to Tides*. Woods Hole, MA: Marine Science International.

Sobey, J. C. (1982) What is sea level? *Sea Frontiers*, **28**, 136–42.

von Arx, W. S. (1962) *An Introduction to Physical Ocenaography*. Reading, MA: Addison-Wesley.

12 Tidal inlets

12.1 Introduction

Tidal inlets are found along barrier coastlines throughout the world. They provide a passageway for ships and small boats to travel from the open ocean to sheltered waters (Fig. 12.1). Along many coasts of the world, including much of the east and Gulf coasts of the United States, the only safe harborages, including some major ports, are found behind barrier islands. The importance of inlets in providing navigation routes to these harbors is demonstrated by the large number of improvements that are performed at the entrance to inlets, such as stabilization by the construction of jetties and breakwaters, dredging of channels, and the operation of sand bypassing facilities.

Tidal inlets are also conduits through which nutrients are exchanged between backbarrier lagoons and estuaries and the open coastal waters. Numerous species of finfish and shellfish rely upon tidal inlets for access to backbarrier regions for feeding, breeding, and nursery grounds for their young. The fact that many fish travel through inlets in search of food makes tidal inlets prize locations for saltwater sportfishing. In many lagoons, tidal inlets maintain the salinities, temperatures, and nutrient levels that are necessary for the reproduction and growth of valuable shellfish. For example, along the coast of Massachusetts, there are several sites in which the saltwater passageways (tidal inlets) to the bay or lagoon periodically close. When this occurs, the freshwater influx gradually reduces the salinity in the lagoon, making the environment inhospitable to many saltwater shellfish. The spits fronting these lagoons are then artificially opened to maintain the proper habitat for various types of clams that are harvested by local fishermen.

Fig. 12.1 Tidal inlets serve as passageways to harbors and conduits through which nutrients are exported to coastal waters.

Fig. 12.2 The apartment building at the northeastern end of Wrightsville Beach, North Carolina, is endangered by the southerly migration of Masons Inlet.

An understanding of tidal inlet processes is important not only for the maintenance of navigable waterways, but also for the management of adjacent barrier shorelines. Tidal inlets interrupt the longshore transport of sediment, affecting both the supply of sand to the downdrift beaches and erosional–depositional processes along the inlet shoreline. As will be shown, the greatest magnitude of shoreline changes along barriers occurs in the vicinity of inlets, and these are a direct consequence of tidal inlet processes. The changes may be due to inlet migration, concentrated wave energy, large bars migrating onshore, sand losses to the backbarrier, and other processes that are treated in this chapter. This information may become particularly important when one is considering the purchase of real estate on barrier islands in the vicinity of inlets (Fig. 12.2).

12.2 What is a tidal inlet?

A tidal inlet is defined as an opening in the shoreline through which water penetrates the land, thereby

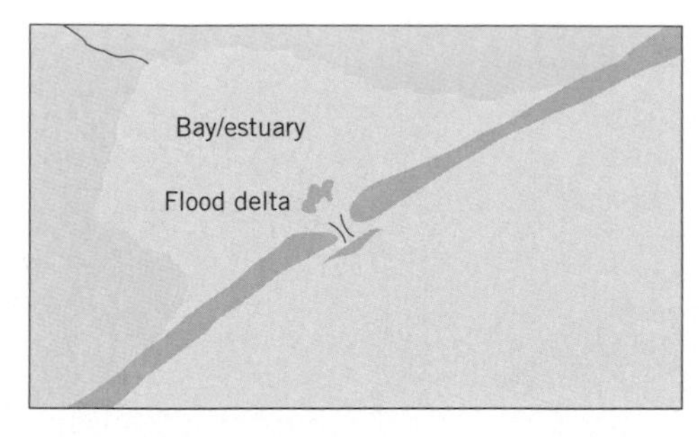

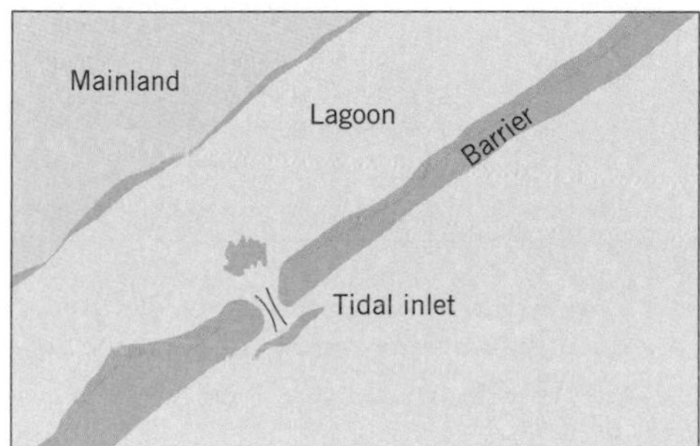

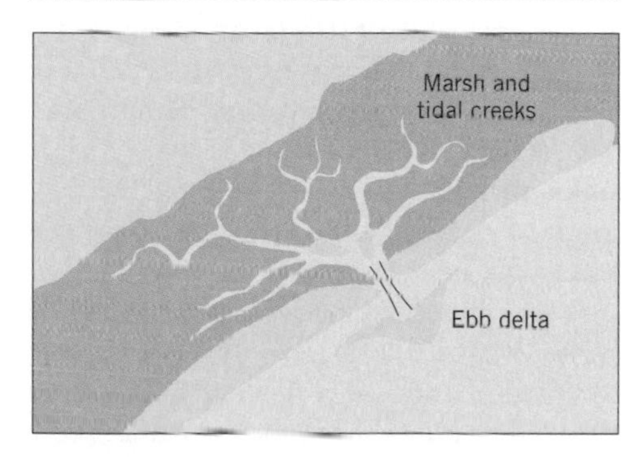

Fig. 12.3 Tidal inlets are the openings along barrier shorelines. They allow the exchange of tidal waters between the ocean and backbarrier, which consists of bays, lagoons, and marsh and tidal creeks. They are fronted by ebb-tidal deltas and backed by flood-tidal deltas.

providing a connection between the ocean and bays, lagoons, and marsh and tidal creek systems. The main channel of a tidal inlet is maintained by tidal currents (Fig. 12.3).

The second half of this definition distinguishes tidal inlets from large, open embayments or passageways along rocky coasts. Tidal currents at inlets are responsible for the continual removal of sediment dumped into the main channel by wave action. Thus, according to this definition tidal inlets occur

Fig. 12.4 Entrance to York Harbor, Maine. This is a bedrock passageway. It is not a tidal inlet because there is little sediment deposited in the channel by wave action and tidal currents are not needed to keep the channel open.

along sandy or sand and gravel barrier coastlines, although one side may abut a bedrock headland. For example, along the coast of Maine the entrance to York Harbor is bordered on both sides by bedrock and there is very little mobile sediment found in the channel or seaward of the harbor. In this case, the tidal currents generated by the 2.7 m tidal range remove little or no sediment from the entrance channel. Because tidal currents are not required to sustain the dimensions of the channel, the entrance to York Harbor is not a tidal inlet (Fig. 12.4).

Some tidal inlets coincide with the mouths of rivers (estuaries), but in these cases inlet dimensions and sediment transport trends are still governed, to a large extent, by the volume of water exchanged at the inlet mouth and the reversing tidal currents, respectively.

Tidal currents are produced at inlets due to the rise and fall of the tides (Fig. 12.5). During the rising tide, the water level of the ocean rises at a faster rate than that inside the inlet. The water surface slope created by this condition causes the sea to flow into the inlet. This landward flowing water is called a flood-tidal current. During the falling tide, the water level of the ocean drops ahead of that of the bay inside the inlet. The seaward sloping water surface produces ebb-tidal currents.

At most inlets over the long term, the volume of water entering the inlet during the flooding tide

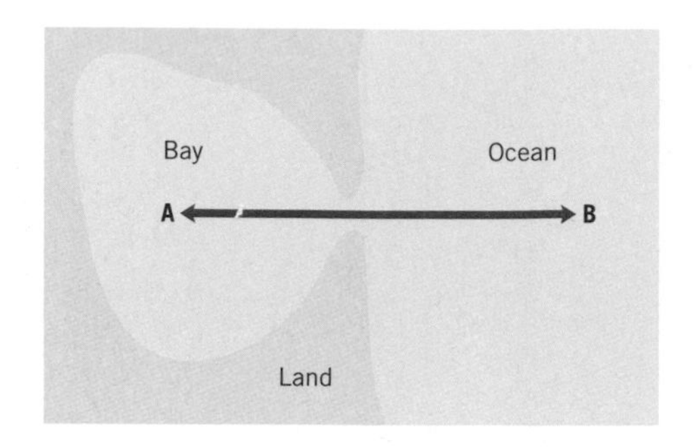

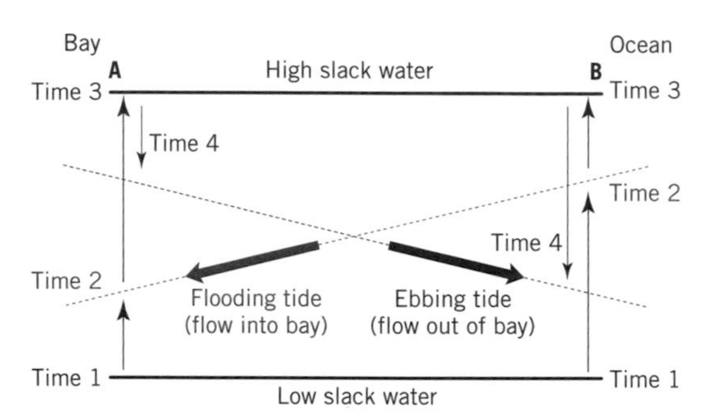

Fig. 12.5 Tidal currents at inlets commonly reach velocities of 1–2 m. They are produced by a constriction of the tidal wave whereby the changing water level in the ocean precedes the tide level inside the inlet.

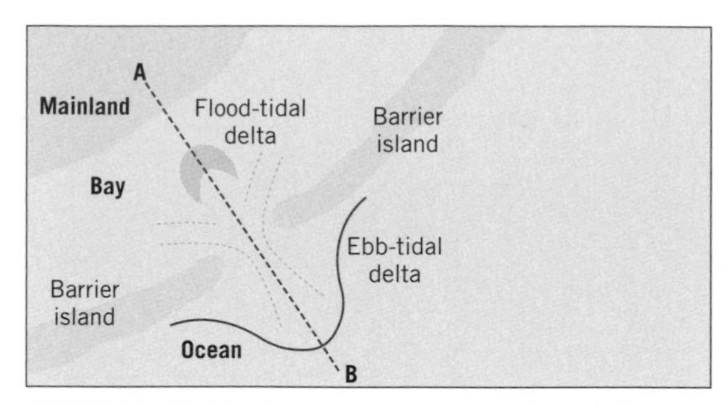

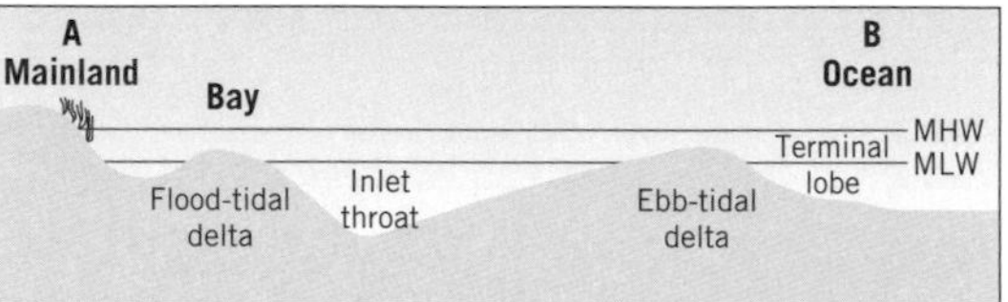

Fig. 12.6 Longitudinal and cross sections of a tidal inlet. Note that the inlet throat is the narrowest and deepest region of the tidal inlet.

equals the volume of water leaving the inlet during the ebbing cycle. This volume is referred to as the **tidal prism**. The tidal prism is a function of the open water area in the backbarrier and the tidal range. For example, a rough estimate of the tidal prism going into and out of Mobile Bay, Alabama, is determined by multiplying the area of Mobile Bay by the tidal range inside the bay. For backbarriers containing large intertidal areas, such as marsh and tidal creeks or tidal flats, calculation of the tidal prism is more difficult and it must be determined from tidal current and channel cross section measurements.

12.3 Inlet morphology

A tidal inlet is specifically the area between the two barriers or between the barrier and the adjacent bedrock or glacial headland. Commonly, the sides of the inlet are formed by the recurved ridges of spits, consisting of sand that was transported toward the backbarrier by refracted waves and flood-tidal currents. The deepest part of an inlet, which is termed the **inlet throat**, is normally located where spit accretion of one or both of the bordering barriers constricts the inlet channel to a minimum width (Fig. 12.6). This constriction is similar to placing your thumb over the nozzle of a hose to increase the velocity of the water flowing from the hose. Likewise, the minimum cross section of the inlet throat is the site where tidal currents reach their maximum velocity. Commonly, the strength of the currents at the throat causes sand to be removed from the channel floor, leaving behind a lag deposit consisting of gravel or shells or in some locations exposed bedrock.

12.3.1 Tidal deltas

Closely associated with tidal inlets are sand shoals and tidal channels located on the landward and seaward sides of the inlets. These sand deposits develop in response to tidal inlet and backbarrier processes. Waves breaking along adjacent beaches deliver sand to the inlet, dumping some of it into the main channel. Depending upon the tidal cycle, this sand is transported seaward by the ebb currents or landward by the flood currents. As the tidal waters flow beyond the constriction of the barriers, the currents expand laterally, losing their velocity and their ability to transport sand. The sand that is deposited landward of the inlet forms a flood-tidal delta and the sand deposited on the seaward side forms an ebb-tidal delta.

Flood-tidal deltas

The presence or absence, size, and development of flood-tidal deltas are related to a region's tidal range, wave energy, sediment supply, and backbarrier setting. Tidal inlets that are connected to one broad backbarrier channel and tidal marsh system (mixed energy coast) usually contain a single relatively large flood-tidal delta lobe (Fig. 12.7a). Contrastingly, inlets such as Drum Inlet along the Outer Banks of North Carolina (wave-dominated coast), which are backed by large shallow bays, may contain flood-tidal deltas with numerous lobes (Fig. 12.7b). Along some microtidal coast, such as Rhode Island, flood deltas form at the end of narrow inlet channels cut through the barrier. Temporal and spatial changes in the locus of deposition at these deltas produce a multilobate morphology resembling a lobate river delta (see Chapter 16). The small tidal range of this region prevents their reworking by ebb-tidal currents, as occurs on mesotidal coasts.

To some extent, delta size is related to the amount of open water area in the backbarrier and the size of the tidal inlet. Along the mixed energy coast of Maine, where tidal inlets are comparatively small (width <100 m), flood-tidal deltas are correspondingly small and stacked in alternating patterns along the main tidal creek. Tidal inlets along the barrier coast of central South Carolina have no flood-tidal deltas because the backbarrier has almost completely filled with fine-grained sediment and marsh deposits, resulting in tidal channels that are too narrow and deep for delta development. In some cases, deltas may have become colonized and altered by marsh growth, and are no longer recognizable as former flood-tidal deltas. This may be the case for some inlets in central South Carolina. At other sites, portions of flood-tidal deltas are dredged to provide navigable waterways and thus they become highly modified.

(a)

(b)

Fig. 12.7 The morphology of flood-tidal deltas is a function of inlet size, open-water area of the bay, tidal range, and other factors. (a) View of single flood-tidal delta in New Zealand. (b) Multiple delta lobes at Chatham Inlet, Massachusetts (photograph taken by Albert Hine in the early 1970s).

Flood-tidal deltas are best revealed in areas with moderate to large tidal ranges (1.5–3.0 m), because in these regions they are well exposed at low tide. As tidal range decreases, flood deltas become largely subtidal shoals. Mixed energy flood-tidal deltas have similar morphologies consisting of the following components (Fig. 12.8):

1 Flood ramp. This is a landward-shallowing channel that slopes upward toward the intertidal portion of the delta. The ramp is dominated by strong

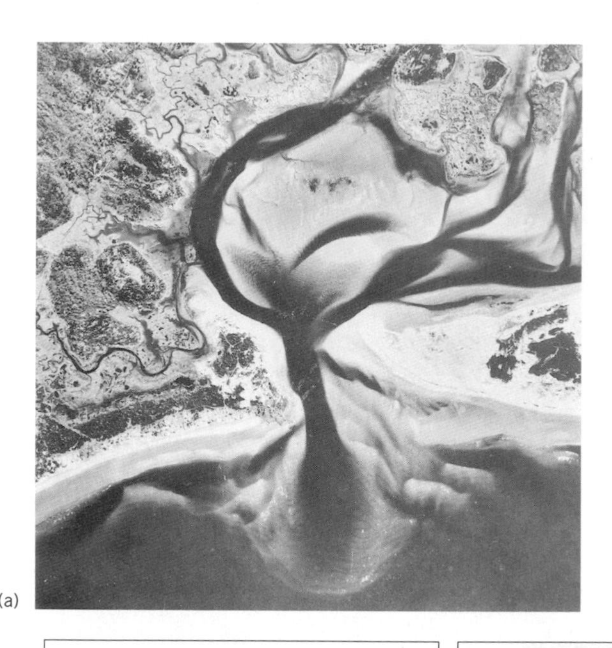

(a)

Fig. 12.8 Miles Hayes conceived mixed energy models of tidal deltas in the early 1970s working along the coast of New England. (a) Vertical aerial photograph of Essex River Inlet. (b) Model of flood-tidal delta (after M. O. Hayes, 1975, Morphology of sand accumulations in estuaries. In L. E. Cronin (ed.), *Estuarine Research, 2*. New York: Academic Press, pp. 3–22). (c) Model of ebb-tidal delta (after Hayes, 1975).

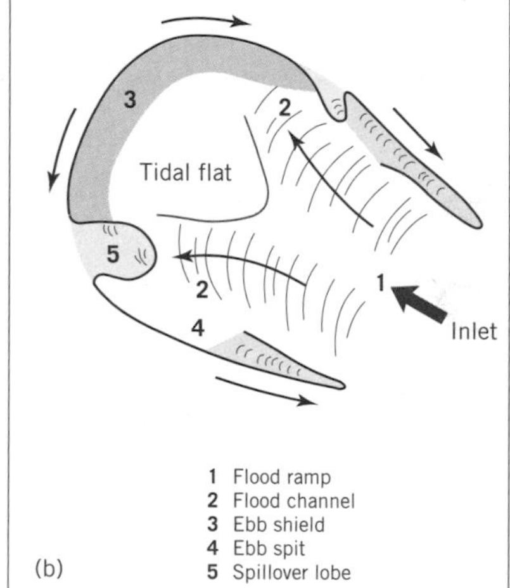

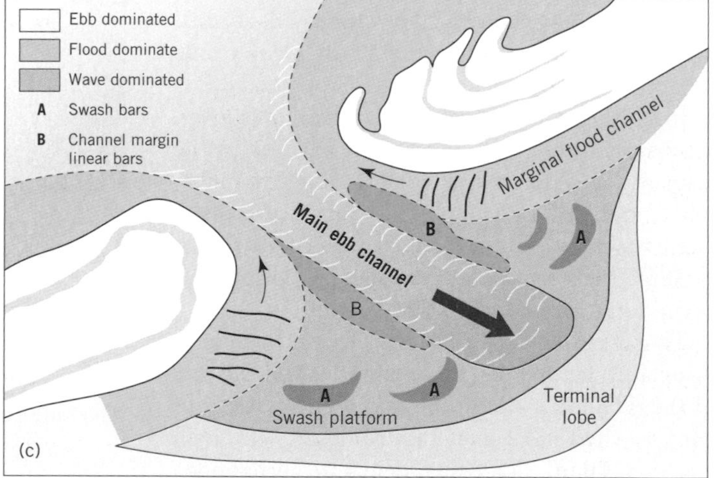

flood-tidal currents and landward sand transport in the form of landward-oriented **sandwaves**.[1]

2 Flood channels. The flood ramp splits into two shallow flood channels. Like the flood ramp, these channels are dominated by flood-tidal currents and flood-oriented sand waves. Sand is delivered through these channels onto the flood delta.

3 Ebb shield. This defines the highest and landwardmost part of the flood delta and may be partly covered by marsh vegetation. When the ebb currents

[1] Sediment moving along the channel bottom by tidal currents is commonly organized into repetitive, elongated packets of sand called bedforms. Like any wave form, each bedform has a crest and trough. A group of bedforms covering any one area tend to parallel one another and their crests are aligned perpendicular to the flow. Sandwaves are a type of bedform usually ranging in height (vertical distance from crest to trough) from 0.5 to 3 m, with a spacing (distance from crest to crest) from 10 to more than 100 m. Sandwaves commonly floor channels and extend across the width of the channel.

reach their strongest velocity in the backbarrier the tide has fallen such that the ebb shield is out of the water. Thus the ebb shields protects the rest of the delta from the effects of the ebb-tidal currents.

4 Ebb spits. These spits extend from the ebb shield toward the inlet. They form from sand that is eroded from the ebb shield and transported back toward the inlet by ebb-tidal currents.

5 Spillover lobes. These are lobes of sand that form where the ebb currents have breached through the ebb spits or ebb shield, depositing sand in the interior of the delta.

Through time, some flood-tidal deltas accrete vertically and/or grow in size. This is evidenced by an increase in areal extent of marsh grasses, which require a certain elevation above mean low water to exist. At migrating inlets new flood-tidal deltas are formed as the inlet moves along the coast and encounters new open water areas in the backbarrier. At most stable inlets, however, sand comprising the flood delta is simply recirculated. The transport of sand on flood deltas is controlled by the elevation of the tide and the strength and direction of the tidal currents. During the rising tide, flood currents reach their strongest velocities near high tide, when the entire flood-tidal delta is covered by water. Hence, there is a net transport of sand up the flood ramp, through the flood channels and onto the ebb shield. Some of the sand is moved across the ebb shield and into the surrounding tidal channel. During the falling tide, the strongest ebb currents occur near mid to low water. At this time, the ebb shield is out of the water and diverts the currents around the delta. The ebb currents erode sand from the landward face of the ebb shield and transport it along the ebb spits and eventually into the inlet channel, where once again it will be moved onto the flood ramp, thus completing the sand gyre.

In some locations, such as Shinnecock Inlet on Long Island, New York, and Ogunquit River Inlet, Maine, flood-tidal deltas have been mined for their sand, which is pumped onto eroding beaches. However, this practice may actually create a sediment sink in the backbarrier, which, in turn, may contribute to the erosion of beaches along the adjacent inlet shoreline.

Ebb-tidal deltas

These are accumulations of sand that have been deposited by the ebb-tidal currents and that has been subsequently modified by waves and tidal currents. Ebb deltas exhibit a variety of forms dependent on the relative magnitude of wave and tidal energy of the region as well as geological controls. Despite this variability, most ebb-tidal deltas contain the same general features, including (Fig. 12.8c):

1 Main ebb channel. This is a seaward-shallowing channel that is scoured in the ebb-tidal delta sands. It is dominated by ebb-tidal currents.

2 Terminal lobe. Sediment transported out through the main ebb channel is deposited in a lobe of sand forming the terminal lobe. The deposit slopes relatively steeply on its seaward side. The outline of the terminal lobe is well defined by breaking waves during storms or periods of large wave swell at low tide.

3 Swash platform. This is a broad shallow sand platform located on both sides of the main ebb channel, defining the general extent of the ebb delta.

4 Channel margin linear bars. These are bars that border the main ebb channel and sit atop the swash platform. These bars tend to confine the ebb flow and are exposed at low tide.

5 Swash bars. Waves breaking over the terminal lobe and across the swash platform form arcuate-shaped swash bars that migrate onshore. The bars are usually 50–150 m long, 50 m wide, and 1–2 m in height.

6 Marginal-flood channels. These are shallow channels (0.2–2.0 m deep at mean low water) located between the channel margin linear bars and the onshore beaches. The channels are dominated by flood-tidal currents.

As stated above, the deepest section of an inlet occurs at the inlet throat, where depths exceeding 8 m are common. Moving out through the inlet channel depths gradually shallow to the point up to one or two kilometers seaward of the inlet throat, where water depths may be less than 2 m. Waves breaking over the terminal lobe lead to numerous boating accidents each year, including the loss of lives. Boaters may be caught unaware of the breaking wave conditions because in the deeper, landward portions of the main ebb channel the waters may be

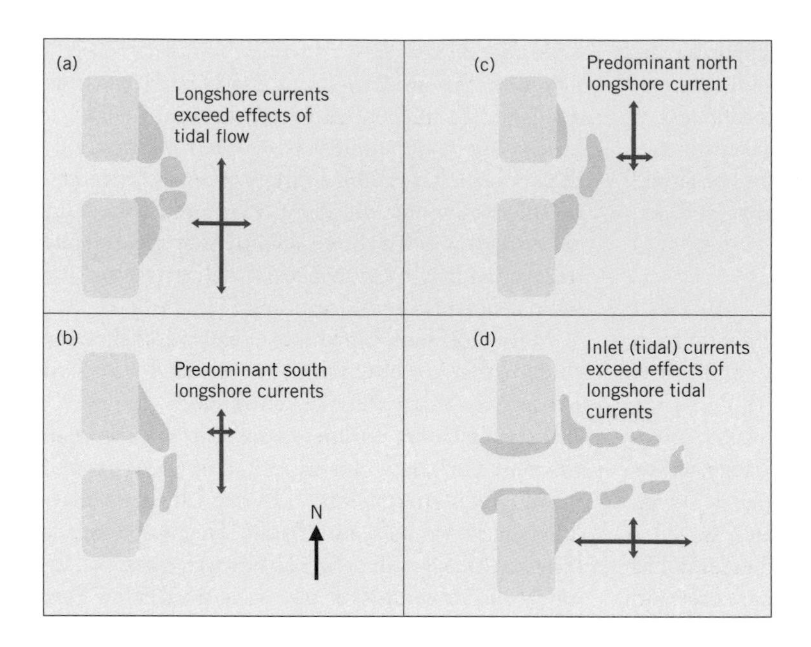

Fig. 12.9 The morphology of an ebb-tidal delta indicates the relative influence of wave versus tidal energy as well as the dominant direction of longshore sediment transport. (From G. Oertel, 1975, Ebb-tidal deltas of Georgia estuaries. In L. E. Cronin (ed.), *Estuarine Research, 2*. New York: Academic Press, pp. 267–76.)

relatively calm. Breaking waves along the periphery of the ebb delta are usually due to a combination of near low tide conditions that produce the shallow water depths, large waves, and ebb-tidal currents. The ebb currents cause a shortening of the distance between the incoming waves. This stacking phenomenon produces steep waves, leading to breaking waves.

12.3.2 Ebb-tidal delta morphology

The general shape of an ebb-tidal delta and the distribution of its sand bodies tells us about the relative magnitude of different sand transport processes operating at a tidal inlet (Fig. 12.9). Ebb-tidal deltas that are elongate, with a main ebb channel and channel margin linear bars that extend far offshore, are tide-dominated inlets. Wave-generated sand transport plays a secondary role in modifying delta shape at these inlets. Because most sand movement in the inlet is in the onshore–offshore direction, the ebb-tidal delta overlaps a relatively small length of inlet shoreline. As will be demonstrated, this has important implications for the extent to which the inlet shoreline undergoes erosional and depositional changes.

Wave-dominated inlets tend to be small relative to tide-dominated inlets. Their ebb-tidal deltas are pushed onshore, close to the inlet mouth, by the dominant wave processes. Commonly, the terminal lobe and/or swash bars form a small arc outlying the periphery of the delta. In many cases the ebb-tidal delta of these inlets is entirely subtidal. In other instances, sand bodies clog the entrance to the inlet, leading to the formation of several major and minor tidal channels.

At mixed energy tidal inlets the shape of the delta is the result of tidal and wave processes. These deltas have a well formed main ebb channel, which is a product of ebb-tidal currents. Their swash platform and sand bodies substantially overlap the inlet shoreline many times the width of the inlet throat due to wave processes and flood-tidal currents.

Ebb-tidal deltas may also be highly asymmetric such that the main ebb channel and its associated sand bodies are positioned primarily along one of the inlet shorelines. This configuration normally occurs when the major backbarrier channel approaches the inlet at an oblique angle or when a preferential accumulation of sand on the updrift side of the ebb delta causes a deflection of the main ebb channel along the downdrift barrier shoreline. Both conditions occur at Parker River Inlet along the North Shore of Massachusetts and thus its ebb-tidal delta significantly overlaps the downdrift shoreline of Castle Neck, whereas very little of the ebb delta overlaps the updrift shoreline of Plum Island.

12.4 Tidal inlet formation

The formation of a tidal inlet requires the presence of an embayment and the development of barriers. In coastal plain settings, the embayment or backbarrier was often created through the construction of the barriers themselves, like much of the east coast of the United States or the Friesian Island coast along the North Sea. In other instances, the embayment was formed due to rising sea level inundating an irregular shoreline during the late Holocene. The embayed or indented shoreline may have been a rocky coast, such as that of northern New England and California, or it may have been an irregular unconsolidated sediment coast, such as that of Cape Cod in Massachusetts or parts of the Oregon coast. The flooding of former river valleys has also produced embayments associated with tidal inlet development. The coastal processes responsible for the formation of tidal inlets are described below.

12.4.1 Breaching of a barrier

Rising sea level, exhausted sediment supplies, and human influences have led to erosion along much of the world's coastlines, including its barrier island chains and barrier spit systems. This condition has caused a thinning of many barriers such that they are vulnerable to **breaching** during storms. Breaching occurs when a barrier is cut, forming a channel (Fig. 12.10). It is by far the most common mechanism by which tidal inlets form today. The breaching process normally occurs during storms after waves have destroyed the foredune ridge and storm waves have overwashed the barrier, depositing sand aprons (washovers) along the backside of the barrier. Even though this process may produce a shallow overwash channel, seldom are barriers cut from their seaward side. In most instances, the breaching of a barrier is the result of the storm surge heightening waters in the backbarrier bay. When the level of the ocean tide falls, the elevated bay waters flow across the barrier toward the ocean, gradually incising the barrier and cutting a channel. If subsequent tidal exchange between the ocean and bay is able to maintain the channel, a tidal inlet is established.

The breaching process is enhanced when offshore winds accompany the falling tide and if an overwash channel is present to facilitate drainage across the barrier. Along the Gulf Coast of the United States hurricanes have been responsible for the development of numerous tidal inlets (e.g. Hurricane Pass, Florida). Many of the tidal inlets that are formed through breaching are ephemeral and may exist for less than a year, especially if stable inlets are located nearby. Barriers that are most susceptible to breaching are long and thin and wave-dominated. For example, although there are only four stable inlets along the Outer Banks of North Carolina today, historical records indicate that at least 26 former inlets have opened and closed at various locations in the past. The reason why inlets close is discussed below.

12.4.2 Spit building across a bay

The development of a tidal inlet by spit construction across an embayment usually occurs early in the evolution of a coast. The sediment to form these spits may have come from erosion of the nearby headlands, discharge from rivers, or the landward movement of sand from inner shelf deposits. As discussed in Chapter 8, most barriers along the coast of the United States and elsewhere in the world are less than 5000 years old, coinciding with a deceleration of rising sea level. It was then that spits began enclosing portions of the irregular rocky coast of New England, the US west coast, parts of Australia, and many other regions of the world.

As a spit builds across a bay, the opening to the bay gradually decreases in width and in cross-sectional area (Fig. 12.11). It may also deepen. Coincident with the decrease in size of the opening is a corresponding increase in tidal flow. The tidal

Time 1 ~20 years prior to breaching

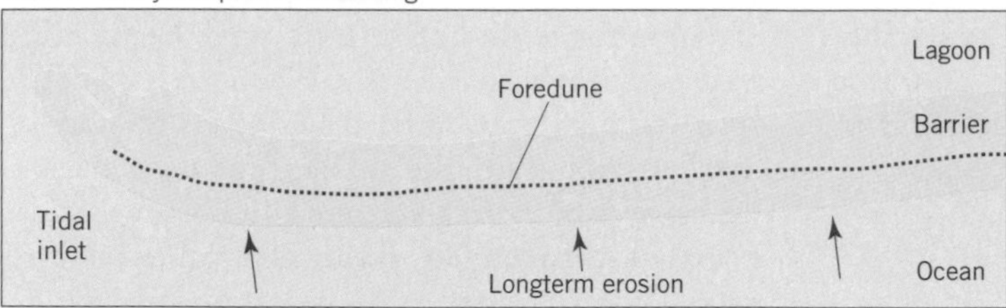

Time 2 ~2 years prior to breaching

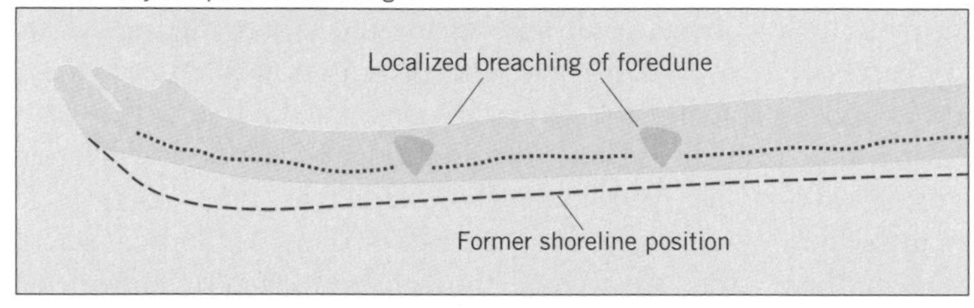

Time 3 Storm overwash

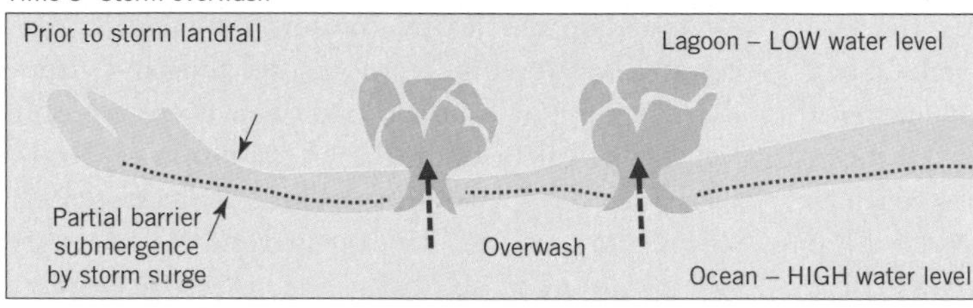

Time 4 Storm ebb-surge and breaching

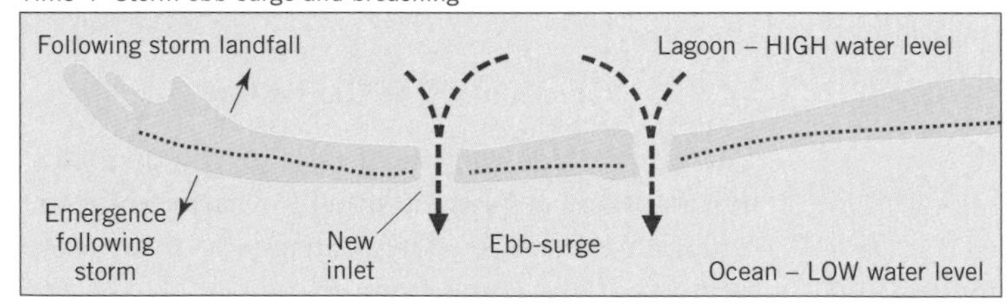

Fig. 12.10 Generally the formation of a new tidal inlet is associated with the breaching of a barrier during a storm. The stages in this process involve a thinning of the barrier through long-term erosion, destruction of the foredune ridge, storm overwash, and finally a deepening of a channel through the barrier.

prism of the bay remains constant, so as the opening gets smaller, the current velocities must increase. Again, this is similar to gradually placing your thumb over an increasingly larger portion of the nozzle of a hose. For the flow out of the hose to remain constant, the velocity has to increase. The tidal inlet is formed as the bay reaches a stable configuration.

The equilibrium size of a tidal inlet can also be explained in terms of sediment transport. Waves and flood-tidal currents are responsible for delivering sediment to the inlet and dumping a large portion of the sand into the inlet channel. The inlet responds to this deposition and decrease in cross-sectional area by increasing the tidal flow, thereby increasing the transport capacity of the tidal currents. Thus, the tidal inlet reaches an equilibrium state when the amount of sand dumped into the inlet equals the volume removed by the tidal currents.

In fact, there are some tidal inlets, such as Barnstable Harbor Inlet in Cape Cod Bay, Massachusetts, that are still developing because Nauset Spit continues to build across Barnstable Bay. As the width of the inlet decreases, the equilibrium throat cross

Box 12.1 Breaching of Nauset Spit and the formation of New Inlet, Cape Cod, Massachusetts

The Town of Chatham, which is located along the outer coast of Cape Cod, Massachusetts, is protected by a sandy barrier known as Nauset Spit. The ancestors of this town were not foolhardy when they chose to build their homes and establish their community along the glacial uplands across Pleasant Bay. Normally, northeast storms that wreak havoc along this coast have less effect on the mainland coast due to the shelter afforded by the barrier. However, all that changed following the January 2, 1987 northeast storm that breached Nauset Spit, establishing a new opening to Pleasant Bay, which was given the name New Inlet (Fig. B12.1). The once idyllic coastal community was now threatened by storm waves, shoreline erosion, shoaling of its navigation channels, closure of its harbors, and tidal inundation of its lowland areas.

The Nauset barrier is part of a spit system that has accreted southward, forming lagoons and bays along the irregular southeast mainland coast of Cape Cod. The chain is broken by several tidal inlets. The sand forming the barrier complex is sourced from eroding glacial bluffs along upper Cape Cod and is transported southward by the dominant northeast wave climate. Prior to the breaching event, Nauset Spit was 14 km long, extending southward from a glacial headland to where it overlapped the northern end of Monomoy Island. At this time Pleasant Bay was connected to the open ocean through a long circuitous route of shallow channels, shoals, and a wave-dominated tidal inlet. Scientists studying the long-term history of outer Cape Cod, including Charles McClennen at Colgate University in New York and Graham Giese at Woods Hole Oceanographic Institute on Cape Cod, discovered that the barrier spit system experiences a long-term cycle of growth and decay. They showed that with a frequency of about 100–150 years the barrier maintains a period of southerly accretion, followed by a destructional phase when the spit becomes segmented and portions of the barrier migrate onshore (Fig. B12.2).

Segmentation of Nauset Spit is related to a gradual reduction in tidal exchange between Pleasant Bay and the ocean, which is caused by a restriction in tidal flow through the existing inlets. This hydraulic inefficiency produces large differences in both tidal range and times of high and low tide on opposite sides of the barrier. These differences lead to certain times of the tidal cycle in which the water level in the ocean is more than a meter higher than in the bay. Under these conditions the barrier is susceptible to breaching, particularly during storms when the storm surge increases the height of the ocean tide level. It should be noted that the periodicity and location of breaching along Nauset Spit is also dependent on the morphology and overall width of the barrier. If the spit is wide and has a well developed frontal dune ridge and secondary dune system, breaching of the barrier is difficult, regardless of the potential hydraulic head. In contrast, destruction of the foredune ridge and thinning of the barrier facilitates barrier overwashing, channelization of the return flow, and inlet formation.

Fig. B12.1 Aerial photographs of Nauset Spit before and after the formation of New Inlet in January 1987.

Box 12.1 (*cont'd*)

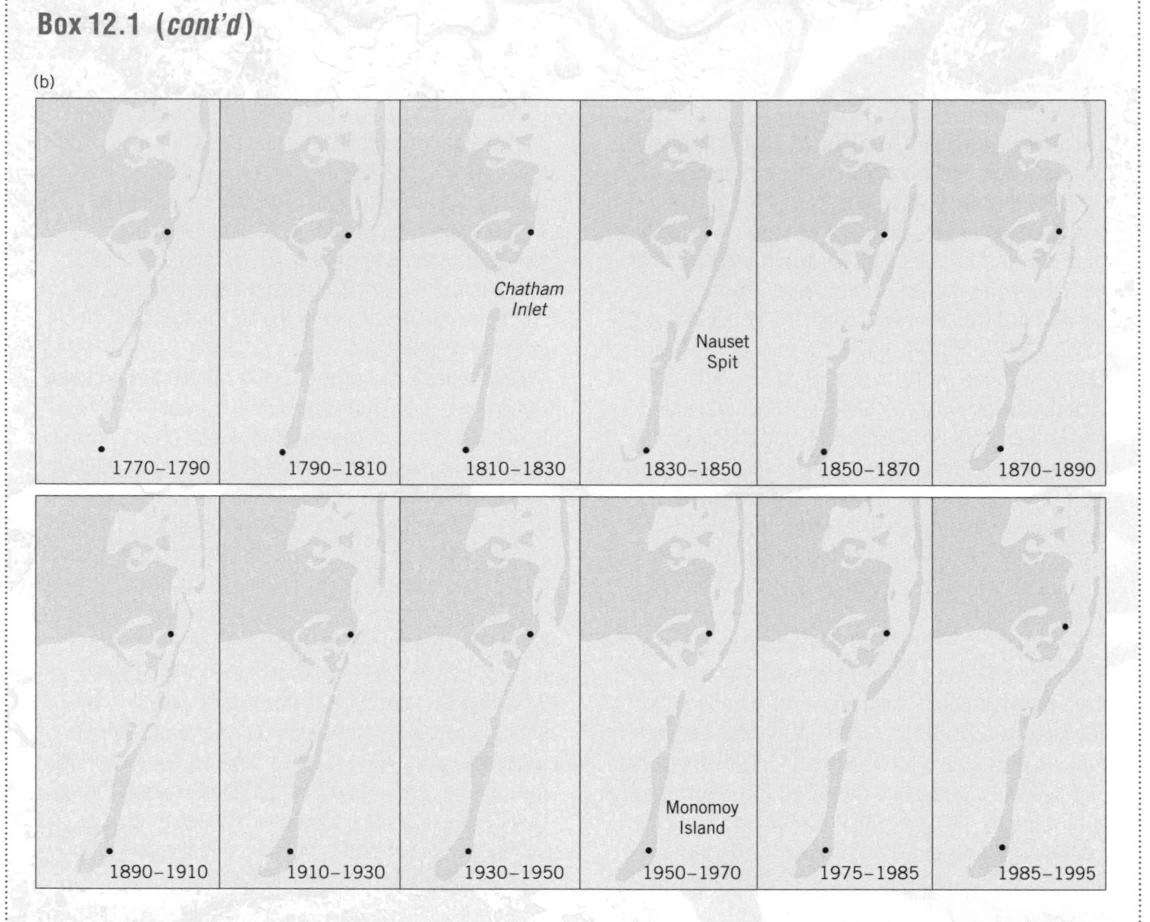

Fig. B12.2 Historical shoreline changes to Nauset Spit during the past 200 years. (From G. S. Giese, 1988, Cyclic behavior of the tidal inlet at Nauset Beach, Chatham, MA. In D. G. Aubrey & L. Weishar (eds), *Hydrodynamics and Sediment Dynamics of Tidal Inlets*. Berlin: Springer, pp. 269–83.)

On January 2, 1987 high water levels associated with perigean spring tides (see Chapter 11) and a storm surge produced by a strong northeaster (see Chapter 5) allowed storm waves to carve away the last vestiges of the frontal dune ridge along central Nauset Spit. As waves continued to overwash the barrier, eventually an overwash channel was created that allowed tidal exchange between the ocean and bay. The day after the storm the channel was several meters wide and about a meter deep. This marked the beginning of a new tidal inlet. As the channel captured an increasingly larger portion of the Pleasant Bay tidal prism, the inlet grew in size from 0.5 km after two months, to 1.0 km in six months time, and by early 1988 the inlet reached almost 2.0 km in width.

The opening of New Inlet drastically changed the hydraulic setting and sediment transport patterns in Pleasant Bay. In the process of enlarging the inlet channel, tidal currents washed much of the sand from the eroding barrier into the bay. Here, the sand was reworked into shoals, bedforms, and other deposits. Some of the eroded sand was also transported seaward, forming a large ebb-tidal delta. After the breaching event the tidal range in Pleasant Bay

Box 12.1 (*cont'd*)

increased by 0.3 m from 1.2 to 1.5 m. The increased tidal fluctuation generated stronger tidal currents in the backbarrier channels, which changed the sand dynamics in the bay. The influx of sand and its movement within the bay resulted in the migration of bedforms and shoals, which in turn led to the closure of certain channels and the opening of others. One dramatic example of changes that took place in Pleasant Bay was the movement of a large flood-tidal delta 0.7 km long and 0.5 m wide. Over an eight-year period from 1988 to 1996 the flood delta marched northward into the bay, moving about a half kilometer (Fig. B12.3). As a result of this migration, the access channel to the town's main harbor was temporarily closed and had to be dredged, the buoys in the navigation channel on the west side of delta had to be repositioned, and a major shellfish bed was destroyed as the deltaic sands drowned a large mussel community.

These were not the only changes to the bay. After inlet formation the inner harbor shoreline landward of the inlet was subjected to wave erosion, endangering millions of dollars worth of properties (Fig. B12.4). The owners responded by constructing expensive seawalls and revetments, but not before acres of valuable real estate were lost and numerous structures had to be relocated. The havoc wreaked by the formation of the new inlet will continue because New Inlet is not stable in its position but is migrating downdrift in response to the dominant southerly longshore transport system. Thus, in the foreseeable future many other erosional and depositional problems will occur as the inlet migrates southward.

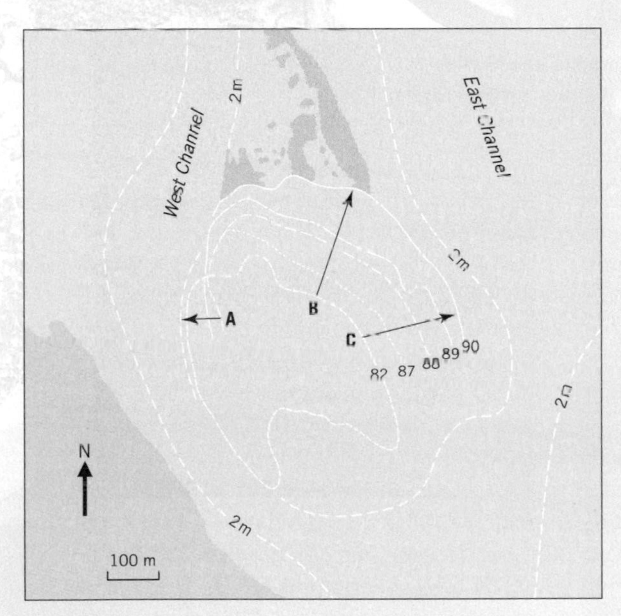

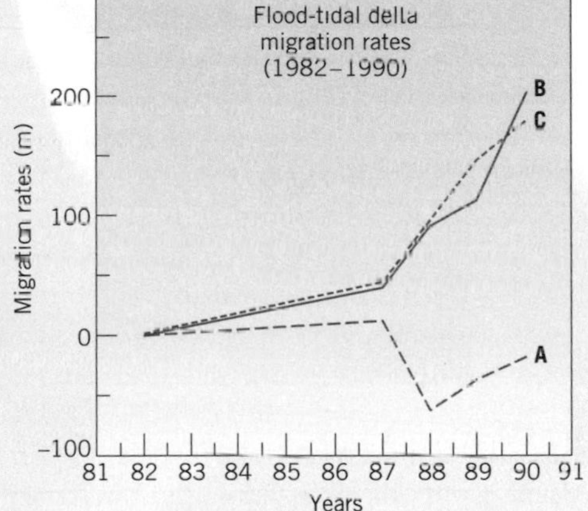

Fig. B12.3 Landward movement of a flood-tidal delta.

Box 12.1 (*cont'd*)

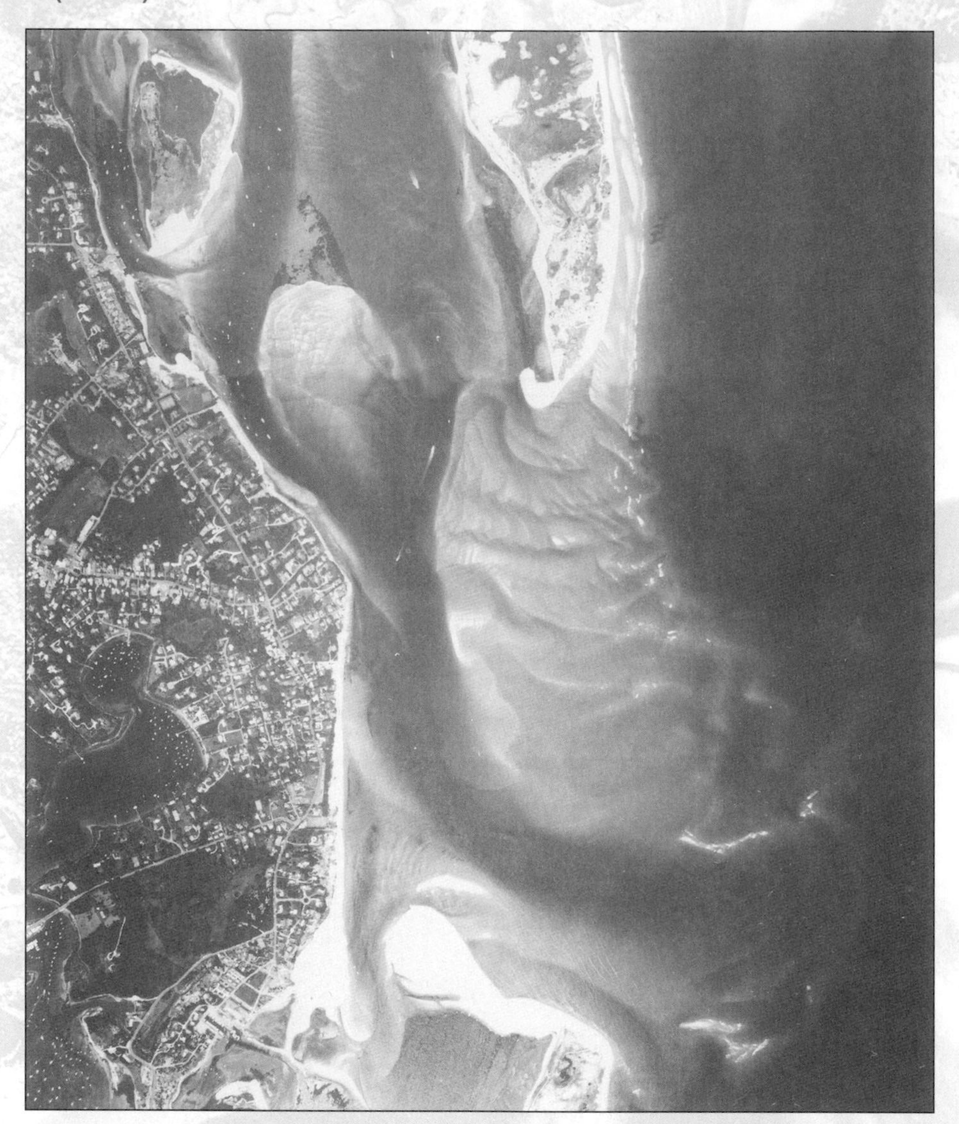

Fig. B12.3 (*cont'd*)

Fig. B12.4 Photograph of revetments constructed inside Pleasant Bay to combat erosion.

(a) Time 1 Spit growth across a bay

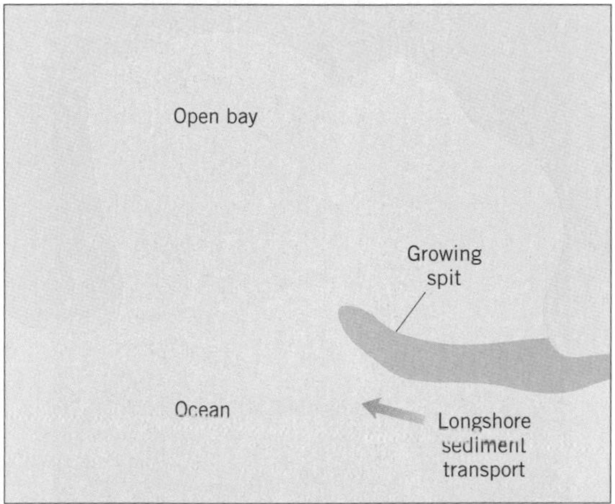

Time 2 Spit extension and inlet formation

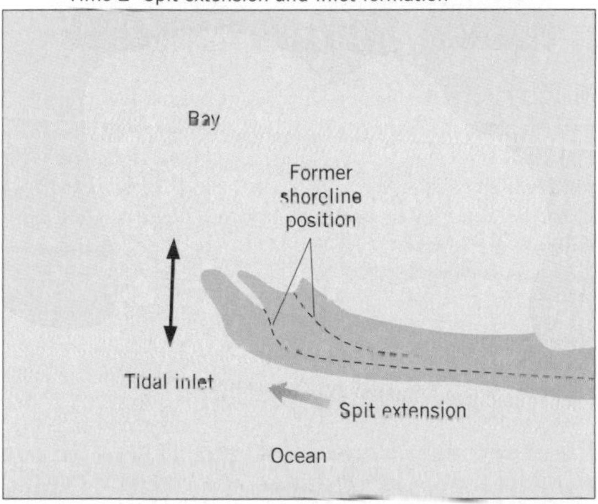

(b)

Fig. 12.11 Spit construction across an embayment can create a tidal inlet. (a) Model of inlet formation due to spit accretion. (b) Aerial photograph of spit building and inlet development in Slocum Embayment, Buzzards Bay, Massachusetts.

section is maintained by the inlet channel deepening. This spit and inlet system has been evolving over the past 3500 years.

12.4.3 Drowned river valleys

In many locations tidal inlets are located at the sites of drowned river valleys. A drowned river valley is a valley that was enlarged when sea level was lower and rivers extended their pathways across the continental shelf to shorelines that were many miles seaward of where they are today. Sea-level lowering was in response to the growth of continental glaciers during the Pleistocene Epoch. When the ice sheets retreated northward and water from the melting ice was returned back to ocean basins, rising sea level flooded the enlarged valleys, forming drowned river valleys. Due to the freshwater discharge and saltwater mixing at these locations, most drowned river valleys are estuaries.

Tidal inlets have formed at the entrance to drowned river valleys due to the growth of spits and the development of barrier islands, which have served to narrow the mouths of the estuaries (Fig. 12.12). They are delineated as tidal inlets when the dimensions of the inlet throat and overall sediment transport trends are a consequence of the saltwater tidal prism and the reversing tidal currents. Thus, the entrance to Chesapeake Bay is not a tidal inlet because its mouth has not been constricted through barrier construction, whereas the entrances to Mobile Bay in Alabama and Grays Harbor in Washington are tidal inlets due to barrier development.

It has been shown through stratigraphic studies, particularly along the east coast of the United States, that in addition to drowned river valleys, many tidal inlets are positioned in paleo-river valleys in which there is no river leading to this site today. These are old river courses that were active during the Pleistocene when sea level was lower and they were migrating across the exposed continental shelf. Tidal inlets become situated in these valleys because the sediment filling the valleys is easily removed by tidal currents. Once a tidal inlet migrates to one of these former valleys the inlet channel scours vertically, excavating the former riverine sediments and

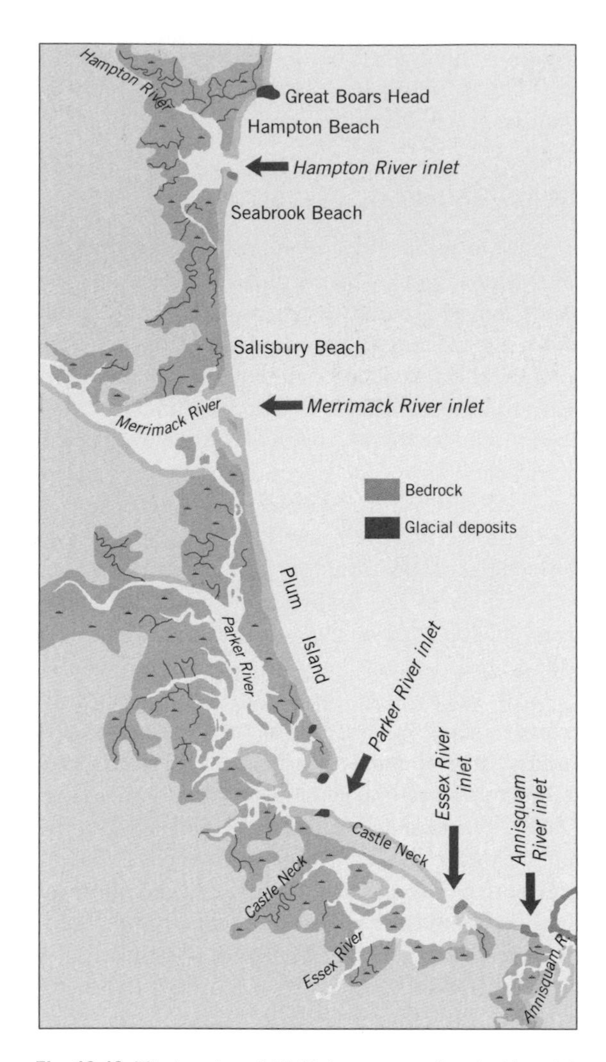

Fig. 12.12 The location of tidal inlets commonly coincides with former river valleys. This situation is exemplified by the inlets that occupy drowned river valleys along the Merrimack barrier system in northern New England.

becoming anchored. Commonly, the sediment layers on either side of the paleo-valley are more resistant to erosion than are the valley-fill sediments. Therefore, after a tidal inlet occupies a paleo-valley further migration of the inlet is impeded. Drowned river valleys and paleo-river channels comprise at least 25% of tidal inlet locations today, especially deep inlets (depth > 8 m).

Fig. 12.13 Several ephemeral inlets were opened along the northeast coast of Dauphine Island, Mississippi, in 1979 as a result of Hurricane Frederick. Hurricane passes, as they are called in Gulf coast region, usually close shortly after they are formed because they are unable to capture a significant portion of the bay tidal prism. (Photograph taken by Shea Penland, University of New Orleans.)

12.4.4 Ephemeral inlets

The most common type of ephemeral inlet is the one that is a product of hurricanes (Fig. 12.13). During Hurricane Alicia in 1983, 185 km h^{-1} (115 m.p.h.) winds and a 3 m storm surge were responsible for cutting 80 tidal inlets along the Texas coast. None of these **hurricane passes**, as they are called, lasted for more than a month. The empheral inlets were filled with sediment that was transported onshore and along the coast by wave action.

A special case of tidal inlet formation occurs at welded barriers along glaciated coasts, such as sections of Alaska, New England, and Canada. Welded barriers at these sites are usually short in length (<1 km), composed of sand and gravel, and are backed by small fresh- to brackish-water ponds and lakes. Freshwater inflow to the backbarrier is derived from small streams and precipitation. Under normal conditions the freshwater influx is insufficient to

Fig. 12.14 Ephemeral tidal inlets form along sand and gravel welded barriers on glaciated coasts. Inlets develop when the inflow of freshwater causes lake levels to overtop the barrier. As water drains across the barrier a channel is cut, forming an ephemeral tidal inlet. (a) Aerial photograph of welded barriers along the southern coast of New England. (b) Map of two ephemeral inlets illustrating how they close due spit accretion and the deposition of flood tidal deltas.

cause overtopping of the barriers because the sand and gravel comprising the barrier permit water to percolate through the barrier sediment and drain into the ocean. However, during intense rain storms and/or melting snow, stream discharge may increase substantially until water in the pond flows across the barrier, cutting a channel and forming a tidal inlet. These inlets are usually short-lived and last only a few months because spit accretion and the formation of flood-tidal deltas seal off the channel (Fig. 12.14).

12.5 Tidal inlet migration

Some tidal inlets have been stable since their formation, whereas others have migrated long distances along the shore. In New England and along other glaciated coasts, stable inlets are commonly anchored next to bedrock outcrops or resistant glacial deposits. Along the California coast most tidal inlets have formed by spit construction across an embayment, with the inlet becoming stabilized adjacent to a

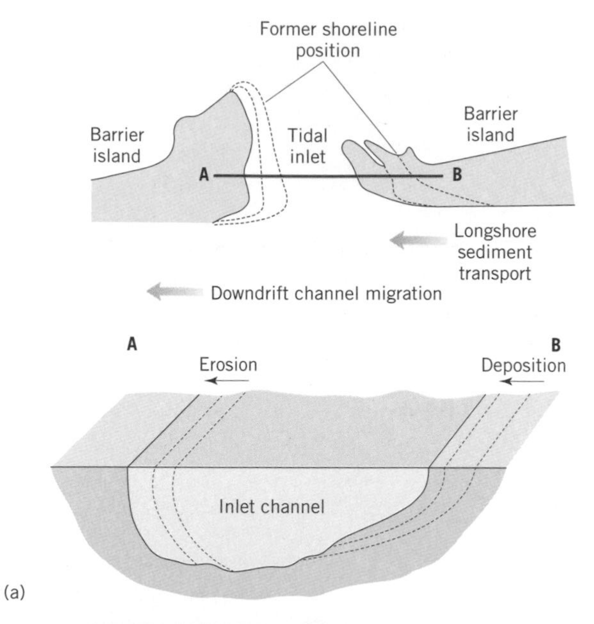

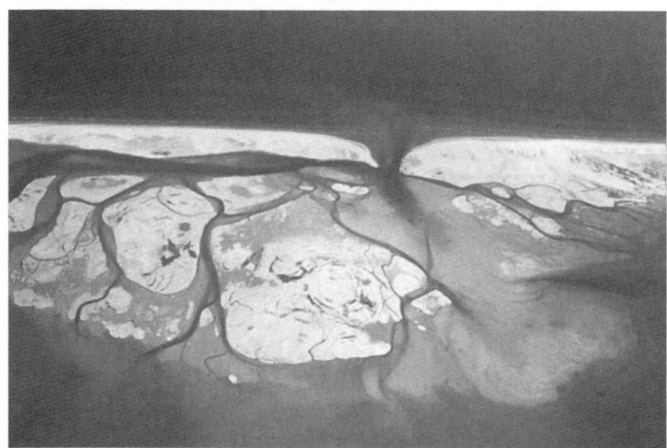

Fig. 12.15 Migrating tidal inlets generally occur along coasts having a dominant longshore sediment transport direction. (a) Model of a migrating inlet. (b) This inlet is migrating left to right. Flood-tidal deltas formed at former inlet positions are vegetated, whereas the flood-tidal delta immediately landward of the inlet is relatively immature and mostly intertidal to subtidal.

bedrock headland. As discussed above, in coastal plain settings stable inlets are commonly positioned in former river valleys. One factor that appears to separate migrating inlets from stable inlets is the depth to which the inlet throat has eroded. For example, along the South Carolina coast tidal inlets deeper than 8 m are stable, whereas inlets shallower than 3–4 m have histories of migration. Deeper inlets are often entrenched in consolidated sediments that resistant erosion. The channels of shallow migrating inlets are eroded into sand.

Tidal inlets migrate when the longshore transport of sand is added predominantly to one side of the inlet, causing a constriction of the flow area (Fig. 12.15). As the tidal currents scour the channel to remove this sand, the downdrift side of the inlet channel is eroded preferentially and the inlet migrates in that direction. Generally, the rate of inlet migration tends to be high along wave-dominated coasts, where the inlet channel is scoured into sand and there is abundant sediment supply.

Although the vast majority of tidal inlets migrate in the direction of dominant longshore transport, there are some inlets that migrate updrift. In these cases the drainage of backbarrier tidal creeks controls flow through the inlet. When a major backbarrier tidal channel approaches the inlet at an oblique angle, the ebb-tidal currents coming from this channel are directed toward the margin of the inlet throat. If this is the updrift side of the main channel, then the inlet will migrate in that direction. This is similar to a river where strong currents are focused along the outside of a meander bend, causing erosion and channel migration. Inlets that migrate updrift are usually small to moderately sized and occur along coasts with small to moderate net sand longshore transport rates.

12.6 Tidal inlet relationships

Tidal inlets throughout the world exhibit several consistent relationships that have allowed coastal engineers and marine geologists to formulate predictive models. These models are effective tools for undertaking tidal inlet projects and are used by engineers and coastal managers to plan jetty construction, channel dredging, and the use of ebb-tidal deltas for beach nourishment material. The models are based on field data collected at many different tidal inlet locations. Through statistical analysis (regression analysis), whereby inlet parameters are plotted against one another, two important correlations have been discovered: (i) inlet throat cross sectional area is closely related to tidal prism; and (ii) ebb-tidal delta volume is a function of tidal prism.

12.6.1 Inlet throat area–tidal prism relationship

It has long been recognized that the size of a tidal inlet is tied closely to the volume of water going through it. In 1931 Morrough O'Brien quantified this relationship for inlets on the US west coast (hence it is given the name the **O'Brien relationship**) by plotting the cross-sectional area of the inlet throat (measured at mean sea level) versus its tidal prism during spring tide conditions (Fig. 12.16). Because the data plot approximately along a line, he was able to derive a simple equation to represent this correlation. Later, in 1969, he showed that the relationship could be extended to inlets along the east and Gulf coasts and since that time other scientists have revealed that, with slight modification, the relationship exists for inlets all over the world.

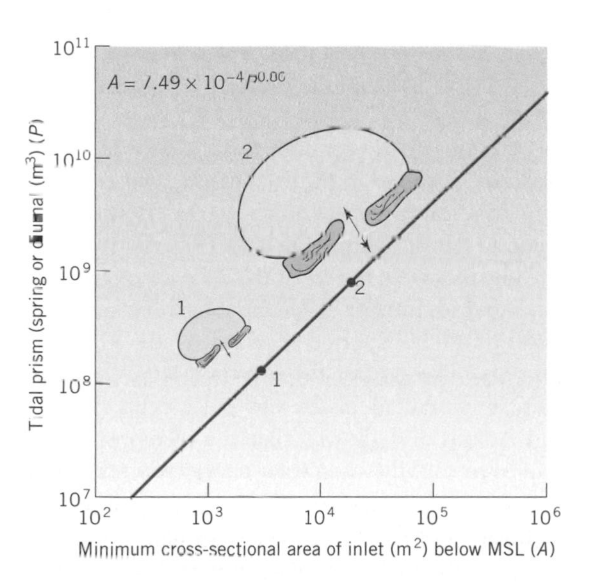

Fig. 12.16 O'Brien's relationship demonstrates that a strong correlation exists between an inlet's spring tidal prism (P) and its throat cross-sectional area (A). (M. P. O'Brien, 1931, Estuary tidal prisms related to entrance areas. *Civil Engineering*, **1**, 738; M. P. O'Brien, 1969, Equilibrium flow areas of inlets on sandy coasts. *Journal of Waterways, Harbors, and Coastal Engineering, ASCE*, **95**, 43–55.)

Although it seems very reasonable that the opening of an inlet should be controlled by its tidal prism, the reason why this correspondence exists globally is that the filling and emptying of the backbarrier are governed by the rise and fall of the ocean tides. Most barrier coasts experience semi-diurnal tides (two tidal cycles daily) and therefore the ocean tidal forcing of the filling and emptying of backbarrier areas worldwide has the same duration, approximately 6 hours and 13 minutes. This concept is illustrated well by comparing two inlets along the Gulf Coast: Midnight Pass (now closed) in Florida, with a tidal prism of 7.4 million cubic meters; and the entrance to Mobile Bay in Mississippi, which has a much larger tidal prism of 960 million cubic meters. If both inlets are to discharge their tidal prisms over the same time interval, it is easily understood that the opening to Mobile Bay (29,280 m^2) has to be much larger (100 times) than the inlet to Midnight Pass (300 m^2).

Although it has been stressed that inlet size is primarily a function of tidal prism, to a lesser degree inlet cross-sectional area is also affected by the delivery of sand to the inlet channel. For example, tidal inlets with jetties, which are stone or concrete structures built perpendicular to the entrance of an inlet, prevent the wave-generated transport of sand into the inlet. At these sites tidal currents can more effectively scour sand from the inlet channel and therefore they maintain a larger throat cross section than would be predicted by the O'Brien relationship for inlets with no jetties. Similarly, for a given tidal prism, Gulf coast inlets have larger throat cross sections than Pacific coast inlets. This is explained by the fact that wave energy is greater along the west coast and therefore the delivery of sand to these inlets is higher than at Gulf coast inlets.

Variability

It is important to understand that the dimensions of the inlet channel are not static: the inlet channel enlarges and contracts slightly over relatively short time periods (<1 year) in response to changes in tidal prism, variations in wave energy, effects of storms, and other factors. For instance, the inlet tidal prism can vary by more than 30% from neap to spring tides due to increasing tidal ranges. Consequently, the size of the inlet varies as a function of tidal phases. Along the southern Atlantic coast of the United States water temperatures may fluctuate seasonally by 16°C (30°F). This causes the surface coastal waters to expand, raising mean sea level by 30 cm or more. In the summer and fall, when mean sea level reaches its highest seasonal elevation, spring tides may flood backbarrier surfaces that normally are above tidal inundation. This produces larger tidal prisms, stronger tidal currents, increased channel scour, and larger inlet cross-sectional areas. At some Virginia inlets this condition increases the inlet throat by 5–15%. Longer-term (>1 year) changes in the cross section of inlets are related to inlet migration, sedimentation in the backbarrier, morphological changes of the ebb-tidal delta, and human influences.

Application

The O'Brien relationship is a very useful concept when designing modification projects for inlets. For example, when an inlet is to be jettied and dredged to provide a navigable waterway for large ships entering and leaving a port, the dimensions of the channel have to be planned (Fig. 12.17). If the channel is dredged to dimensions larger than what is in balance with the existing tidal prism, the channel will fill with sediment until the cross section decreases to the equilibrium area. Remember that tidal prism is primarily a function of the open-water area and the tidal range in the backbarrier, and under most conditions will not change if the size of the inlet is enlarged. Therefore, if the improved inlet has an equilibrium cross-sectional area of 12,000 m^2 and navigational constraints require a 12 m deep channel, then the jetties should be positioned approximately 1000 m apart.

1 Natural channel configuration

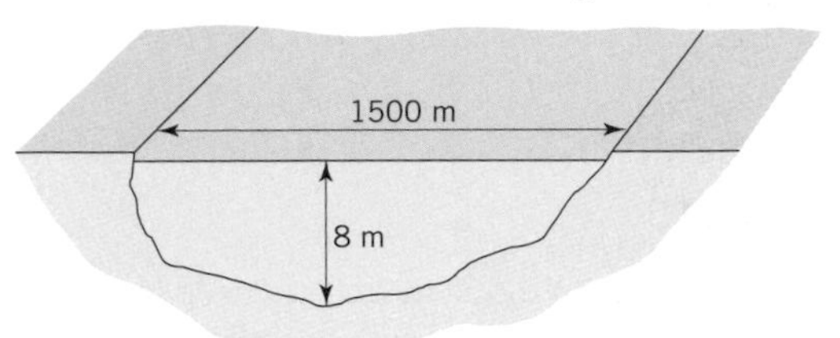

A = 1500 m × 8 m = 12,000 m^2

2 Stabilized channel configuration

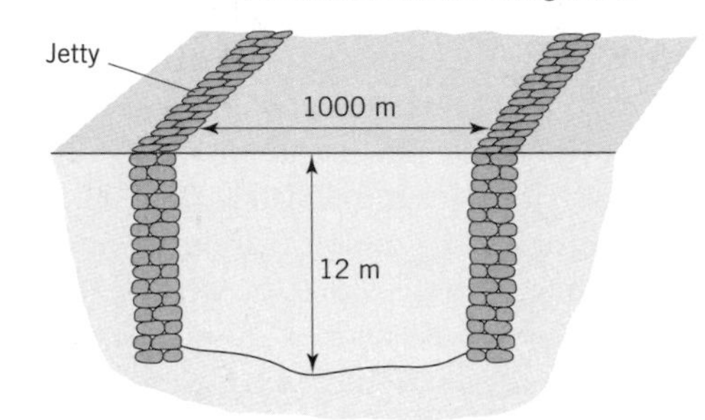

A = 1000 m × 12 m = 12,000 m^2

Fig. 12.17 Application of O'Brien's relationship. Disregarding the effects of friction and possible changes in tidal prism, if the jetties are positioned closer together, then tidal currents will scour the channel deeper.

12.6.2 Ebb-tidal delta volume–tidal prism relationship

In the mid-1970s Todd Walton and his graduate assistant William Adams did further statistical analysis (regression analysis) of various inlet parameters and discovered that, like inlet cross-sectional area, the volume of sand contained in the ebb-tidal delta was closely related to the tidal prism. This relationship has come to be known as the **Walton and Adams relationship** (Fig. 12.18). As we have already discussed, the ebb-tidal delta constitutes the sand that is diverted from the longshore transport system and transported seaward by the ebb-tidal currents. The greater the ebb discharge, the more sand is contained in the ebb-tidal delta. Walton and Adams also showed that the relationship was improved slightly when wave energy was taken into account. This was accomplished by separating the data set into three inlet classes based on their wave energy: (i) high wave-energy coasts, such as inlets along Oregon and Washington; (ii) moderate wave-energy coasts, including New Jersey, the Outer Banks of North Carolina, and Delaware, and (iii) low wave-energy coasts such as the Gulf coast. Waves are responsible for transporting sand back onshore, thereby reducing the volume of the ebb-tidal delta.

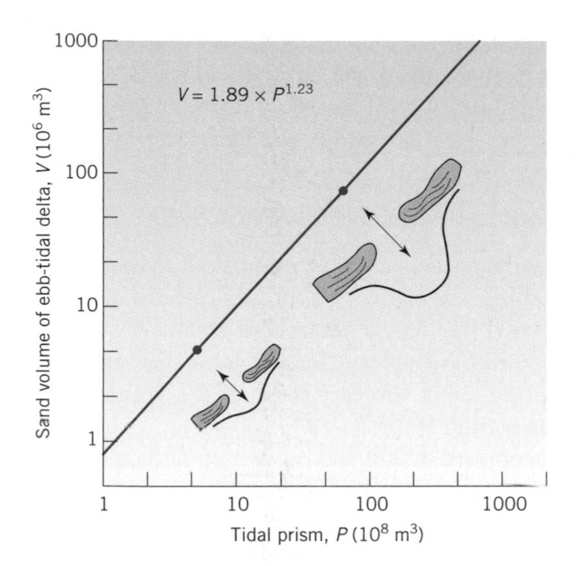

Fig. 12.18 The Walton and Adams relationship indicates that a strong correspondence exists between an inlet's tidal prism (P) and the volume of its ebb-tidal delta (V). (From T. L. Walton & W. D. Adams, 1976, Capacity of inlet outer bars to store sand. In *Proceedings of the 15th Coastal Engineering Conference*. Honolulu, HI: ASCE, pp. 1919–37.)

Therefore, for a given tidal prism, ebb-tidal deltas along the west coast contain less sand than do equal sized inlets along the Gulf or east coasts.

Variability
The Walton and Adams relationship works well for inlets all over the world. Field studies have shown, however, that the volume of sand comprising ebb-tidal deltas changes through time due to the effects of storms, changes in tidal prism, or processes of inlet sediment bypassing. When sand is moved past a tidal inlet, it is commonly achieved by large bar complexes migrating from the ebb delta and attaching to the landward inlet shoreline. These large bars may contain more than 300,000 m^3 of sand and represent more than 10% of the sediment volume of the ebb-tidal delta.

Application
Due to pervasive shoreline erosion, many barrier systems in the United States and elsewhere are being nourished with sand obtained from offshore sites, backbarrier and inlet dredging, and land sources. As these borrow sites become depleted, ebb-tidal deltas are also being mined for their sand. The Walton and Adams relationship helps engineers to compute ebb-tidal delta volumes and the effects to adjacent beaches. The relationship is also used to determine how nearby beaches will respond when a tidal inlet is formed due to storm breaching or if an artificial cut is made through a barrier. Immediately following inlet formation, the ebb-tidal delta grows until it reaches an equilibrium volume as predicted by the Walton and Adams relationship. The sediment that builds the ebb delta is sand that is removed from the longshore transport system, thereby causing erosion. The relationship helps coastal engineers to calculate rates of change.

12.7 Sand transport patterns

The movement of sand at a tidal inlet is complex due to reversing tidal currents, effects of storms, and interaction with the longshore transport system. The inlet contains short-term and long-term reservoirs of sand, varying from the relatively small sandwaves flooring the inlet channel that migrate meters each tidal cycle to the large flood-tidal delta shoals where some sand is recirculated but the entire deposit may remain stable for hundreds or even thousands of years. Sand dispersal at tidal inlets is complicated because in addition to the onshore–offshore movement of sand produced by tidal and wave-generated currents, there is constant delivery of sand to the inlet and transport of sand away from the inlet produced by the longshore transport system. In the discussion below the patterns of sand movement at inlets are described, including how sand is moved past a tidal inlet.

12.7.1 General sand dispersal trends

The ebb-tidal delta has segregated areas of landward versus seaward sediment transport that are controlled primarily by the way water enters and discharges from the inlet, as well as the effects of wave-generated currents (Fig. 12.19). During the ebbing tidal cycle the tidal flow leaving the back-

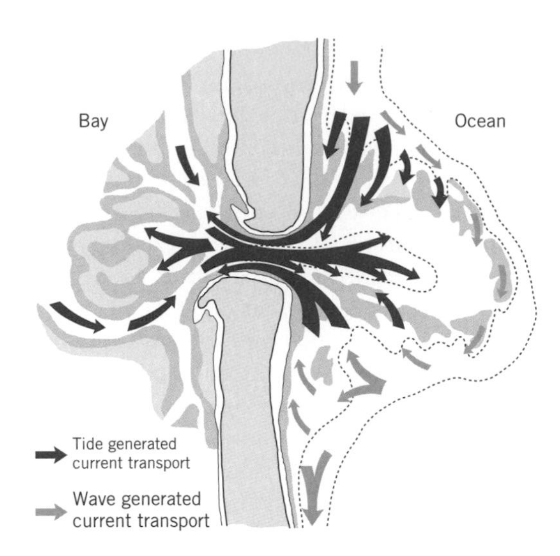

Fig. 12.19 Sand dispersal patterns at a tidal inlet reflect the dominance of wave and tidal processes.

barrier is constricted at the inlet throat, causing the currents to accelerate in a seaward direction. Once out of the confines of the inlet, the ebb flow expands laterally and the velocity slows. Sediment in the main ebb channel is transported in a net seaward direction and eventually deposited on the terminal lobe due to this decrease in current velocity. One response to this seaward movement of sand is the formation of ebb-oriented sandwaves with heights of 1–2 m.

In the beginning of the flood cycle, the ocean tide rises, while water in the main ebb channel continues to flow seaward as a result of momentum. Due to this phenomenon, water initially enters the inlet through the marginal flood channels, which are the pathways of least resistance. The flood channels are dominated by landward sediment transport and are floored by flood-oriented bedforms. On both sides of the main ebb channel, the swash platform is most affected by landward flow produced by the flood-tidal currents and breaking waves. As waves shoal and break, they generate landward flow, which augments the flood-tidal currents but retards the ebb-tidal currents. The interaction of these forces acts to transport sediment in a net landward direction across the swash platform. In summary, at many inlets there is a general trend of seaward sand transport in the main ebb channel, which is countered by landward sand transport in the marginal flood channels and across the swash platform.

12.7.2 Inlet sediment bypassing

Along most open coasts, particularly in coastal plain settings, angular wave approach causes a net movement of sediment along the shore. As we have learned in Chapter 7, the net volume of sand transported along the east coast of the United States varies from 100,000 to 200,000 m^2 yr^{-1}. Thus, there are upward of 200,000 m^2 yr^{-1} of sand delivered to tidal inlets along this coast on a yearly basis. If, over the long term, the sand reservoirs of these inlets remain approximately constant, then there must be mechanisms whereby sand moves past tidal inlets and is transferred to the downdrift shoreline. This process is called **inlet sediment bypassing**. There are multiple ways in which inlets bypass sand, including: (i) stable inlet processes; (ii) ebb-tidal delta breaching; and (iii) inlet migration and spit breaching. One of the end products in all the different mechanisms is the landward migration and attachment of large bar complexes to the inlet shoreline.

Stable inlet processes

This mechanism of sediment bypassing occurs at inlets that do not migrate and whose main ebb channels remain approximately in the same position (Fig. 12.20). Sand enters the inlet by: (i) wave action along the beach; (ii) flood-tidal and wave-generated currents through the marginal flood channel; and (iii) waves breaking across the channel margin linear bars. Most of the sand that is dumped into the main channel is transported seaward by the dominant ebb-tidal currents and deposited on the terminal lobe.

At lower tidal elevations waves breaking on the terminal lobe transport sand along the periphery of the delta toward the landward beaches in much the same way that sand is moved in the surf and breaker zones along beaches. At higher tidal elevations waves breaking over the terminal lobe create swash bars on

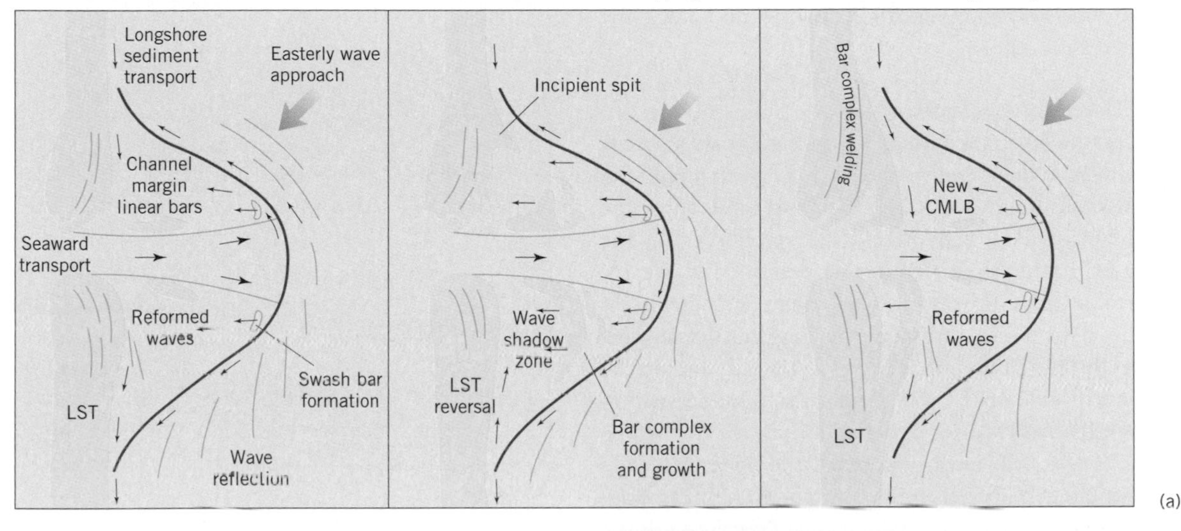

Fig. 12.20 Inlet sediment bypassing at stable inlets. (a) Model of stable inlet processes (from D. M. FitzGerald, 1988, Shoreline erosional–depositional processes associated with tidal inlets. In D. G. Aubrey & L. Weishar (eds), *Hydrodynamics and Sediment Dynamics of Tidal Inlets*. Berlin: Springer, pp. 186–225). (b) North Inlet along the northern South Carolina coast is an example where sediment bypassing occurs through stable inlet processes. Note the large bars migrating onshore to the downdrift inlet shoreline.

both sides of the main ebb channel. The swash bars (50–150 m long, 50 m wide) migrate onshore due to the dominance of landward flow across the swash platform. Eventually, they attach to channel margin linear bars forming large **bar complexes**. Bar complexes tend to parallel the beach and may be more than a kilometer in length. They are fronted by a steep face (25–33°) called a slipface, which may be up to 3 m in height. At midtide pleasure boaters often anchor behind the bars due to the quiet water and so that swimmers may dive off the bar slipface.

The stacking and coalescing of swash bars to form a bar complex is the result of the bars slowing their onshore migration as they move up the nearshore ramp. As the bars gain a greater intertidal exposure, the wave bores that cause their migration onshore act over an increasingly shorter period of the tidal cycle. Thus, their rate of movement onshore decreases. The growth of the bar complex is similar to cars on a highway all stacking up when they approach a toll booth.

Eventually the entire bar complex migrates onshore and welds to the upper beach. When a bar complex attaches to the downdrift inlet shoreline, some of this newly accreted sand is then gradually transported by wave action to the downdrift beaches, thus completing the inlet sediment bypassing process. It should be noted that some sand bypasses the inlet independent of the bar complex. In addition, some of the sand comprising the bar re-enters the inlet via the marginal flood channel and along the inlet shoreline.

Ebb-tidal delta breaching

This means of sediment bypassing occurs at inlets with a stable throat position, but whose main ebb channels migrate through their ebb-tidal deltas like the wag of a dog's tail (Fig. 12.21). Sand enters the inlet in the same manner as described above for "stable inlet processes." However, at these inlets the delivery of sediment by longshore transport produces a preferential accumulation of sand on the updrift side of the ebb-tidal delta. The deposition of this sand causes a deflection of the main ebb channel until it nearly parallels the downdrift inlet shoreline. This circuitous configuration of the main channel results in inefficient tidal flow through the inlet, ultimately leading to a breaching of a new channel through the ebb-tidal delta. The process normally occurs during spring tides or periods of storm surge when the tidal prism is very large. In this state the ebb discharge piles up water at the entrance to the inlet, where the channel bends toward the downdrift inlet shoreline. This causes some of the tidal water to exit through the marginal flood channel or flow across low regions on the channel margin linear bar. Gradually over several weeks or convulsively during a single large storm, this process cuts a new channel through the ebb delta, thereby providing a more direct pathway for tidal exchange through the inlet. As more and more of the tidal prism is diverted through the new main ebb channel, tidal discharge through the former channel decreases, causing it to fill with sand.

The sand that was once on the updrift side of the ebb-tidal delta and that is now on the downdrift side of the new main channel is moved onshore by wave-generated and flood-tidal currents. Initially, some of this sand aids in filling the former channel, while the rest forms a large bar complex that eventually migrates onshore and attaches to the downdrift inlet shoreline. The ebb-tidal breaching process results in a large packet of sand bypassing the inlet. Similar to the stable inlets discussed above,

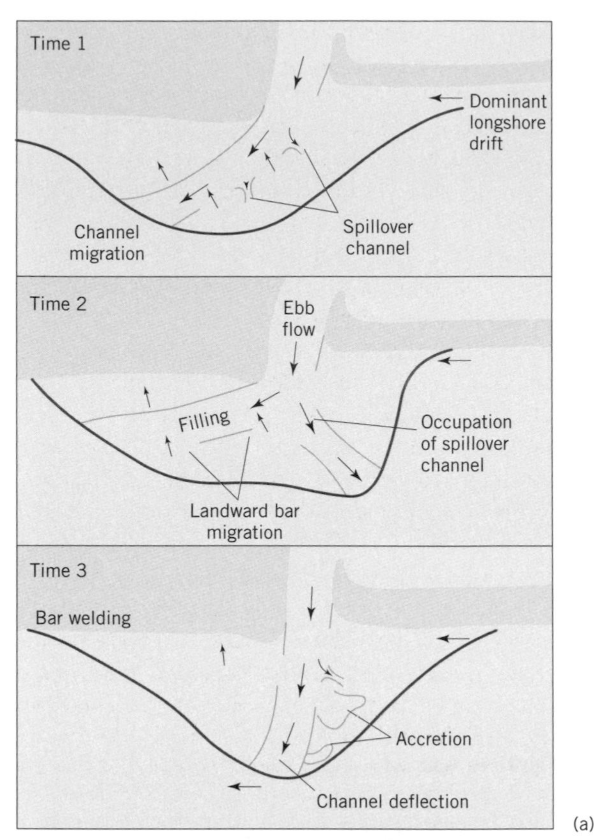

Fig. 12.21 Sediment bypassing at inlets whose main ebb channel migrates downdrift in response to wave energy and sand influx via the longshore transport system. (a) Model of inlet sediment bypassing by ebb-tidal delta breaching processes (from FitzGerald, 1988, details in Fig. 12.20). (b) View of Murrells Inlet, South Carolina, showing two large bar complexes welding to the beach after a recent channel was breached through the updrift portion of the ebb-tidal delta.

some sand bypasses these inlets in a less dramatic fashion, grain by grain, on a continual basis.

It is noteworthy that at some tidal inlets the entire main ebb channel is involved in the ebb-tidal delta breaching process, whereas at others just the outer portion of main ebb channel is deflected. In both cases, the end product of the breaching process is a channel realignment that more efficiently conveys water into and out of the inlet, with sand being bypassed in the form of a bar.

One of the largest scale ebb-tidal delta breaching processes takes place at Willapa Bay inlet on the Oregon coast. This inlet is 11 km wide and more than 12 m deep. Its outer channel is deflected south by a mostly submerged spit that builds 6 km southward from Cape Shoalwater. Every 8–27 years (16-year average) a new channel is breached back to the north, straightening the main entrance channel. The submerged shoal that is bypassed moves onshore, merging with inner bars. An additional interesting aspect of this cycle is that the breaching process correlates well with El Niño events, which cause water levels on the west coast to be elevated by 20–30 cm. Higher water levels cause areas within Willapa Bay that are normally above mean high water to be inundated, thereby increasing the tidal prism. In turn, larger tidal prisms lead to stronger tidal flow and a greater potential to cut a new channel through the ebb delta.

Inlet migration and spit breaching

A final method of inlet sediment bypassing occurs at migrating inlets. In this situation an abundant sand supply and a dominant longshore transport direction cause spit building at the end of the barrier (Fig. 12.22). To accommodate spit construction, the inlet migrates by eroding the downdrift barrier shoreline. Along many coasts as the inlet is displaced further along the downdrift shoreline, the inlet channel to the backbarrier lengthens, retarding the exchange of water between the ocean and backbarrier. This condition leads to large water-level differences between the ocean and bay, making the barrier highly susceptible to breaching, particularly during storms. Ultimately, when the barrier spit is breached and a new inlet is formed in a hydraulically more favorable position, the tidal prism is diverted to the new inlet and the old inlet closes. When this happens, the sand comprising the ebb-tidal delta of the former inlet is transported onshore by wave action, commonly taking the form of a landward migrating bar complex. It should be noted that when the inlet shifts to a new position along the

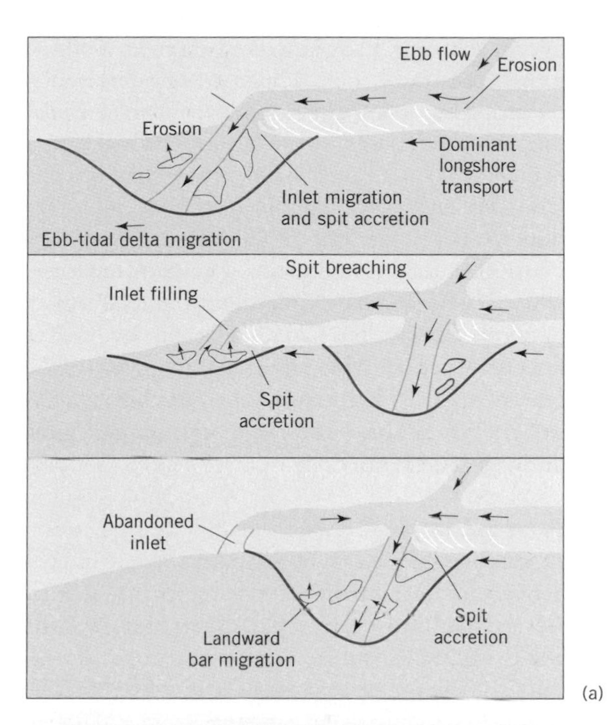

Fig. 12.22 Breaching of a spit allows a large quantity of sand to bypass the tidal inlet. (a) Model of inlet migration and spit breaching processes (from FitzGerald, 1988, details in Fig. 12.20). (b) Small inlet near the mouth of the Santee River, South Carolina, illustrating at least two episodes of spit breaching.

updrift shoreline a large quantity of sand has effectively bypassed the inlet. The frequency of this inlet sediment bypassing process is dependent on inlet size, rate of migration, storm history, and backbarrier dynamics. Nauset Spit along the outer coast of Cape Cod, Massachusetts, exhibits a cycle of spit accretion and inlet migration of 10–15 km, followed by multiple breachings occurring approximately every 100 years. Kiawah River inlet along the central coast of South Carolina has had a similar history, with at least three periods of southwesterly migration of the inlet of up to 15 km followed by breachings of the spit updrift at about the same position each time during a 150-year period.

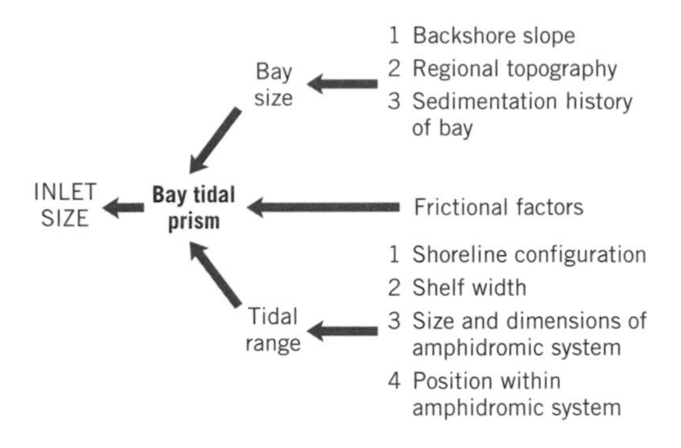

Fig. 12.23 Factors affecting the size and number of tidal inlets along a barrier shoreline. (From FitzGerald, 1988, details in Fig. 12.20.)

Bar complexes

Depending on the size of the inlet, the rate of sand delivery to the inlet, the effects of storms, and other factors, the entire process of bar formation, its landward migration, and its attachment to the downdrift shoreline may take from 6 to 10 years. The volume of sand bypassed can range from 100,000 to over 1,000,000 m^3. The bulge in the shoreline that is formed by the attachment of a bar complex is gradually eroded and smoothed as sand is dispersed to the downdrift shoreline and transported back toward the inlet.

In some instances, a landward-migrating bar complex forms a saltwater pond as the tips of the arcuate bar weld to the beach, stabilizing its onshore movement. Although the general shape of the bar and pond may be modified by overwash and dune building activity, the overall shoreline morphology is frequently preserved. Lenticular-shaped coastal ponds or marshy swales become diagnostic of bar migration processes and are common features at many inlets.

12.8 Tidal inlet effects on adjacent shorelines

Many people wish to live along waterways, particularly at tidal inlets, due to their scenic beauty, fishing opportunities, and boat access to backbarrier bays and the open ocean. For these reasons, property values are unusually high in the vicinity of tidal inlets and frequently there is considerable demand by the private sector to develop these areas. In conflict with these pressures is the instability of inlet shorelines. In addition to the direct consequences of spit accretion and inlet migration are the effects of volume changes in the size of ebb-tidal deltas, sand losses to the backbarrier, processes of inlet sediment bypassing, and wave sheltering of the ebb-tidal delta shoals. The manner in which these processes affect tidal inlet shorelines is presented below.

12.8.1 Number and size of tidal inlets

The degree to which barrier shorelines are influenced by tidal inlet processes is dependent on their size and number. As the O'Brien relationship demonstrates, the size or cross-sectional area of an inlet is governed by its tidal prism. This concept can be expanded to include an entire barrier chain in which the size and number of inlets along a chain are primarily dependent on the area of open water behind the barrier and the tidal range of the region. In turn, these parameters are a function of other geological and physical oceanographic factors (Fig. 12.23). As demonstrated in Chapter 8, wave-dominated coasts tend to have long barrier islands and few tidal inlets and mixed energy coasts have short stubby barriers and numerous tidal inlets. Correspondingly, along the wave-dominated, microtidal coasts of Texas and

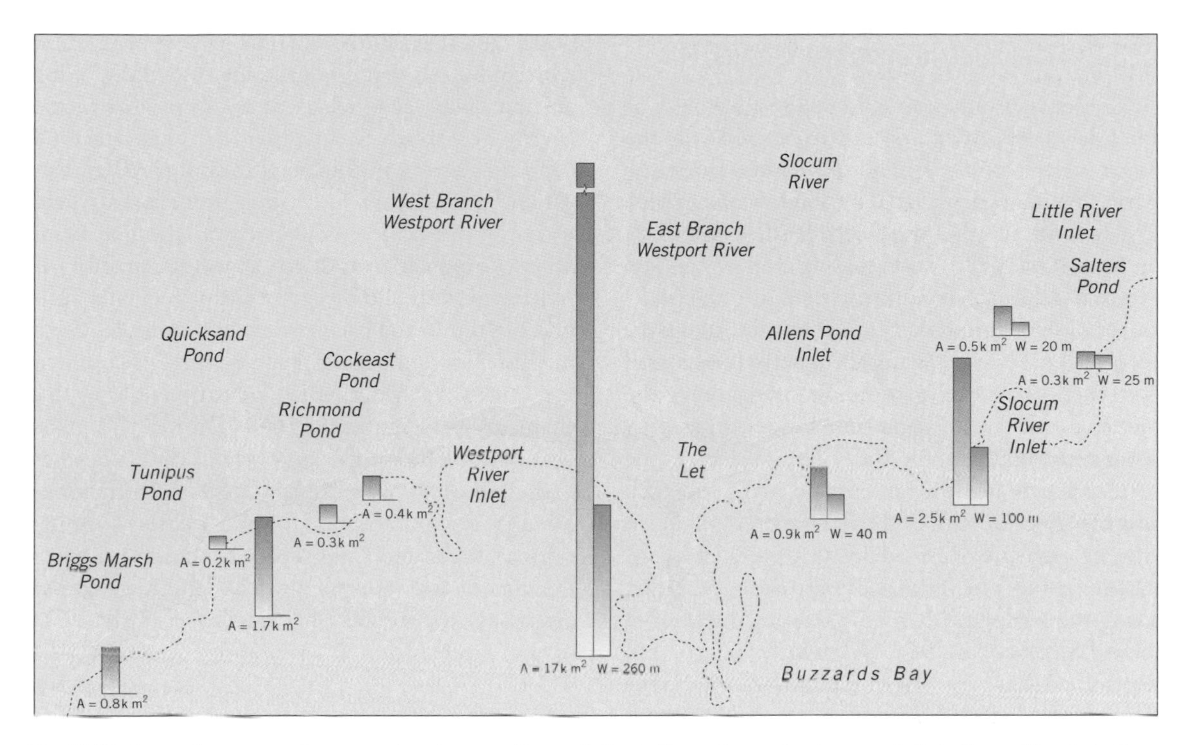

Fig. 12.24 Along the transgressive coast of northwestern Buzzards Bay in Massachusetts barriers have built in front of flooded valleys. Rising sea level and a scarcity of sediment have caused the barriers to migrate onshore, decreasing the size of the bays. Reduced bay areas and decreasing tidal prisms have led to closure of several inlets along this coast. Westport River Inlet is the largest inlet in this region and contains the largest bay area. (From D. M. FitzGerald, 1996, Geomorphic variability and morphologic and sedimentologic controls on tidal inlets. *Journal of Coastal Research*, special issue **23**, 47–71.)

eastern Florida tidal inlets occur every 40–50 km, whereas along the mixed energy, mesotidal coasts of Georgia, the East Friesian Islands of Germany, and the Copper River delta barriers of Alaska inlets are found every 10–20 km. Presumably, the mesotidal conditions produce larger tidal prisms than along microtidal coasts, which necessitate more holes in the barrier chain to let the water into and out of the backbarrier. Many coastlines follow this general trend but there are many exceptions due to the influence of sediment supply, large versus small bay areas, and other geological controls. For example, along the central Gulf coast of Florida the low wave energy of this region, limited sand resources, and large open water bays produce a coast containing numerous tidal inlets occurring about every 10–20 km.

Along the glaciated coast of southern Massachusetts in Buzzards Bay the influence of bedrock controls on the size and number of tidal inlets is well illustrated (Fig. 12.24). The peninsula and deep embayments of this region are a product of river erosion during the Tertiary (geological time period lasting from 66.4 to 1.6 million years BP) and repeated Pleistocene glaciations. As the valleys of this coast became flooded by rising sea level following deglaciation, erosion of the surrounding glacial deposits produced sediment for the construction of the barriers fronting the embayments. As the Holocene transgression proceeded along the shoreline, the barriers migrated landward at a faster rate than the bay shorelines were inundated. This resulted in increasingly smaller sized bays and the gradual closure of many tidal inlets due to decreasing tidal prisms.

12.8.2 Tidal inlets as sediment traps

Tidal inlets not only trap sand temporarily on their ebb-tidal deltas, they are also responsible for the longer-term loss of sediment moved into the back-barrier. At inlets dominated by flood-tidal currents, sand is continuously transported landward, enlarging flood-tidal deltas and building bars in the tidal creeks. Sand can also be transported into the back-barrier of ebb-dominated tidal inlets during severe storms (Fig. 12.25). During these periods increased wave energy produces greater sand transport to the inlet channel. At the same time the accompanying storm surge increases the water surface slope at the inlet, resulting in stronger than normal flood-tidal currents. The strength of the flood currents coupled with the high rate of sand delivery to the inlet results in landward sediment transport into the backbarrier. Along the Malpeque barrier system in the Gulf of Saint Lawrence in New Brunswick it has been determined that over 90% of the sand transfer to the backbarrier took place at tidal inlets and at former inlet locations along the barrier.

Sediment may also be lost at migrating inlets when sand is deposited as channel fill. If the channel scours below the base of the barrier sands, then the beach sand that fills this channel will not be replaced entirely by the deposits excavated on the eroding portion of the channel. Because up to 40% of the length of barriers is underlain by tidal inlet fill deposits ranging in thickness from 2 to 10 m, this volume represents a large, long-term loss of sand from the coastal sediment budget. Another major process producing sand loss at migrating inlets is associated with the construction of recurved spits that build into the backbarrier. For example, along the East Friesian Islands recurved spit development has caused the lengthening of barriers along this chain by 3–11 km since 1650. During this stage of barrier evolution the large size of the tidal inlets permitted ocean waves to transport large quantities of sand around the end of the barrier, forming recurves that extend far into the backbarrier. Due to the size of the recurves and the length of barrier extension, this process has been one of the chief natural mechanisms of bay infilling.

12.8.3 Changes in ebb-tidal delta volume

Ebb-tidal deltas represent huge reservoirs of sand that may be comparable in volume to that of the adjacent barrier islands along mixed energy coasts

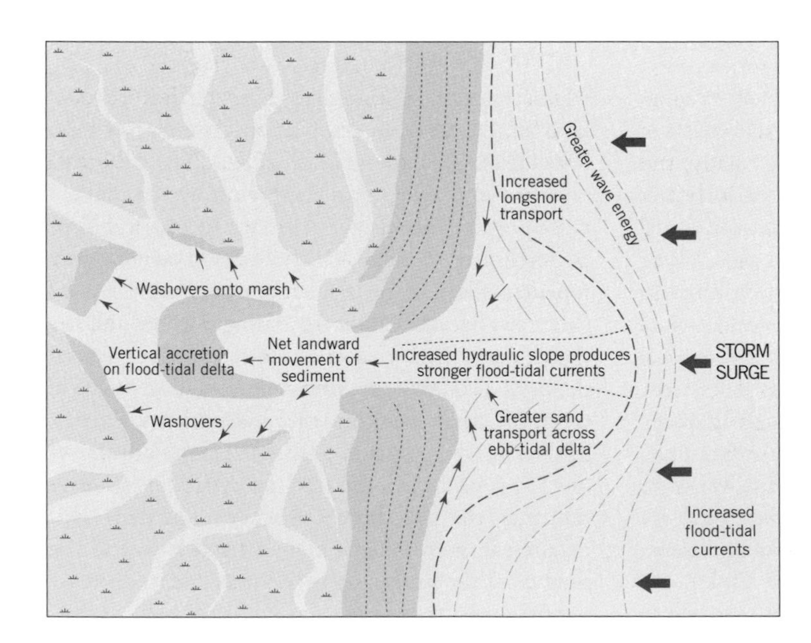

Fig. 12.25 Processes of sand addition to a flood-tidal delta during a storm. (From FitzGerald, 1988, details in Fig. 12.20.)

(e.g. northern East and West Friesian Islands, Massachusetts, southern New Jersey, Virginia, South Carolina, and Georgia). For instance, the ebb-tidal delta volume of Stono and North Edisto Inlets in South Carolina is 197×10^6 m^3 and the intervening Seabrook–Kiawah Island barrier complex contains 252×10^6 m^3 of sand. In this case, the deltas comprise 44% of the sand in the combined inlet–barrier system. The magnitude of sand contained in ebb-tidal deltas suggests that small changes in their volume dramatically affect the sand supply to the landward shorelines.

To illustrate this concept consider the consequences of changing hydraulic conditions at a tidal inlet along the central coast of Maine. It has been theorized that if a planned hydroelectric power plant is constructed in the Bay of Fundy, tidal ranges in the Gulf of Maine would increase by approximately 30 cm. At the Kennebec River inlet the larger tidal range would increase the tidal prism by a minimum of 5%. Using the Walton and Adams relationship, it is calculated that a potential 5% increase in tidal prism would ultimately add over 60×10^6 m^3 of sand to the Kennebec ebb-tidal delta. Although some of this sand would come from scour of the inlet channel, most of the sand would be eroded from the adjacent beaches, resulting in over 100 m of shoreline recession.

A similar transfer of sand takes place when a new tidal inlet is opened, such as the formation of Ocean City Inlet when Assateague Island, Maryland, was breached during the 1933 hurricane. Initially, the inlet was only 3 m deep and 60 m across, but it quickly widened to 335 m when it was stabilized with jetties in 1935. Since the inlet formed, more than a million cubic meters of sand have been deposited on the ebb-tidal delta (Fig. 12.26). Trapping of the southerly longshore movement of sand by the north jetty and growth of the ebb-tidal delta have led to serious erosion along the downdrift beaches. The northern end of Assateague Island has been retreating at an average rate of 11 m yr^{-1}. The rate of erosion lessened when the ebb tidal delta reached an equilibrium volume and the inlet began to bypass sand.

In contrast to the cases discussed above, the historical decrease in the inlet tidal prisms along the East Friesian Islands has had a beneficial effect on this barrier coast. From 1650 to 1960 the reclamation of tidal flats and marshlands bordering the German mainland, as well as natural processes, such as the building and landward extension of recurved spits, decreased the size of the backbarrier by 25% (Fig. 12.27). In turn, the reduction in bay area decreased the inlet tidal prisms, which led to smaller inlets, longer barrier islands, and smaller ebb-tidal deltas. Wave action transported ebb-tidal delta sands onshore as tidal discharge decreased. This process increased the supply of sand to the beaches and aided in lengthening of the barriers.

12.8.4 Wave sheltering

The shallow character of ebb-tidal deltas provides a natural breakwater for the landward shorelines. This is especially true during lower tidal elevations, when most of the wave energy is dissipated along the terminal lobe. During higher tidal stages intertidal and subtidal bars cause waves to break offshore, expending much of their energy before reaching the beaches onshore. The sheltering effect is most pronounced along mixed energy coasts, where tidal inlets have well developed ebb-tidal deltas.

The influence of ebb shoals is particularly well illustrated by the history of Morris Island, South Carolina, which forms the southern border of Charleston Harbor (Fig. 12.28). Before human modification, the entrance channel to the harbor paralleled Morris Island and was fronted by an extensive shoal system. The deflected southerly course of the main ebb channel was due to the preferential accumulation of sand on the updrift side (northeast side) of the harbor's ebb-tidal delta caused by the dominant southerly longshore transport of sediment. The shallow and constantly shifting position of the outer portion of the entrance channel made for treacherous navigation into the harbor, resulting in numerous ship wrecks along the outer shoals. In the late nineteenth century jetties were constructed at the harbor entrance to straighten, deepen, and stabilize the

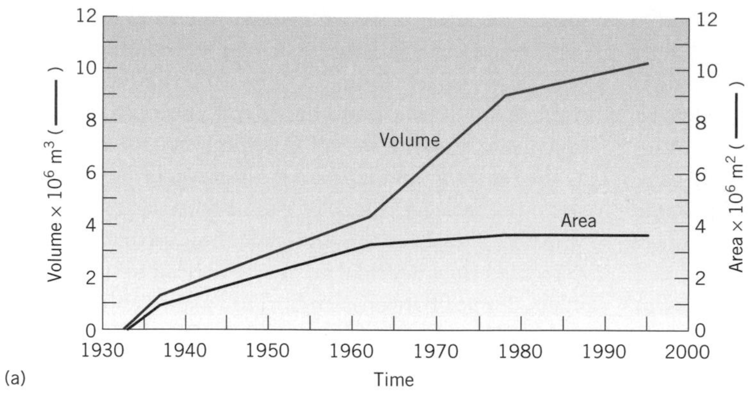

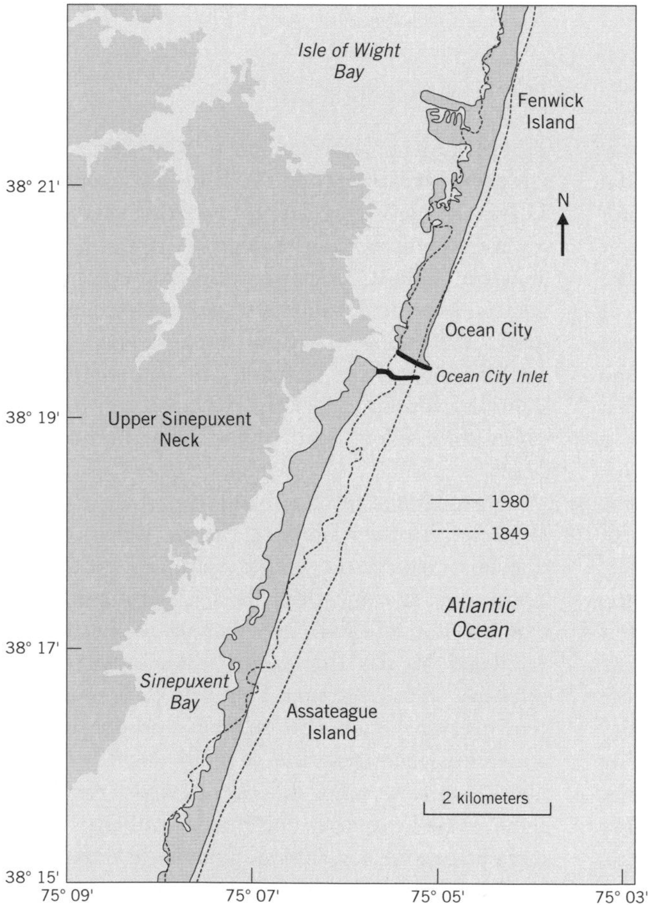

Fig. 12.26 Ocean City Inlet was opened along northern Assateague Island, Maryland during the 1933 hurricane. (a) The ensuing tidal exchange between the ocean and Isle of Wight Bay and Sinepuxent Bay produced a large tidal prism and led to sand trapping on the ebb-tidal delta (from S. P. Leatherman, 1984, Shoreline evolution of North Assateague Island, Maryland. *Journal of Shore and Beach*, 52(4), 3–10). (b) Growth of the ebb delta has starved the downdrift shoreline of sand, resulting in a landward migration of the northern end of Assateague Island by more than several hundred meters (from D. K. Stauble, 1997, *Ocean City Inlet, Maryland, and Vicinity Water Resources Study*. Baltimore District, US Army, Corps of Engineers).

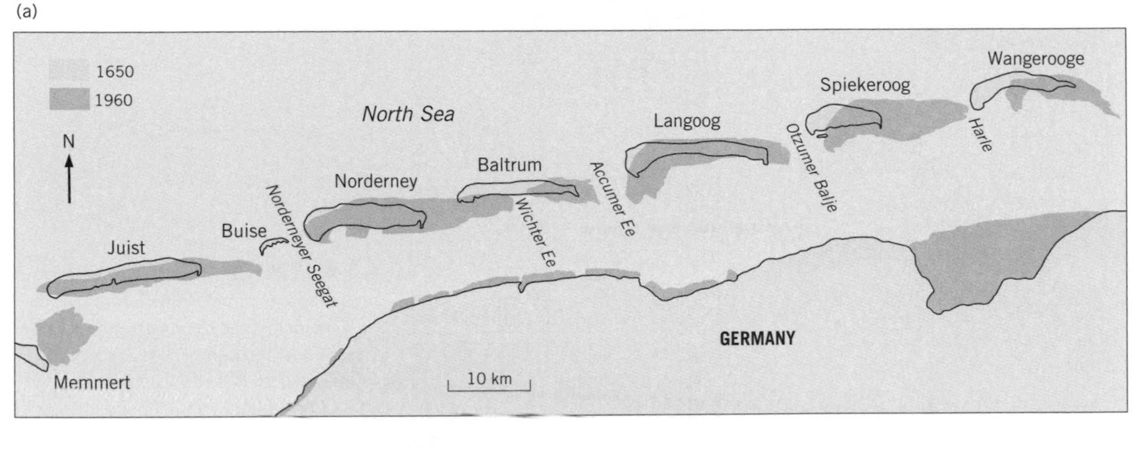

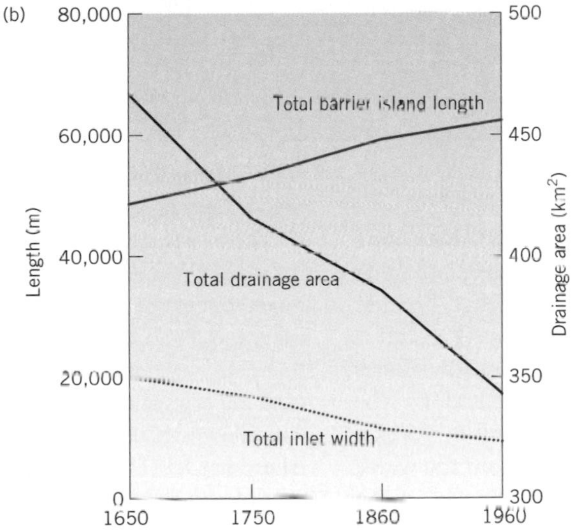

Fig. 12.27 Historical morphological changes to the East Friesian Islands along the German North Sea coast. (a) From 1650 to 1960 widespread land reclamation along the backside of the barriers and on the mainland increased the size of the barriers, while drastically decreasing inlet drainage areas. (b) Smaller drainage areas led to significantly smaller inlet tidal prisms and reduced size of the tidal inlets and their associated ebb-tidal deltas. (From FitzGerald, 1988, details in Fig. 12.20.)

main channel; the project was completed in 1896. During the period prior to jetty construction (1849–80) Morris Island had been eroding at an average rate of 3.5 m yr^{-1}. After the jetties were in place, the ebb-tidal delta shoals that bordered the old channel were cut off from their longshore sand supply. As the shoals eroded and gradually diminished in size, so did the protection they afforded Morris Island, especially during storms. From 1900 to 1973 Morris Island receded 500 m at its northeast end, increasing to 1100 m at its southeast end, a rate three times what it had been prior to jetty construction. One dramatic response to this erosion was the detachment of a lighthouse from the southeast end of the island. In 1900 the lighthouse was located 640 m onshore but by 1970 it was sitting in 3 m of water 360 m from the shoreline.

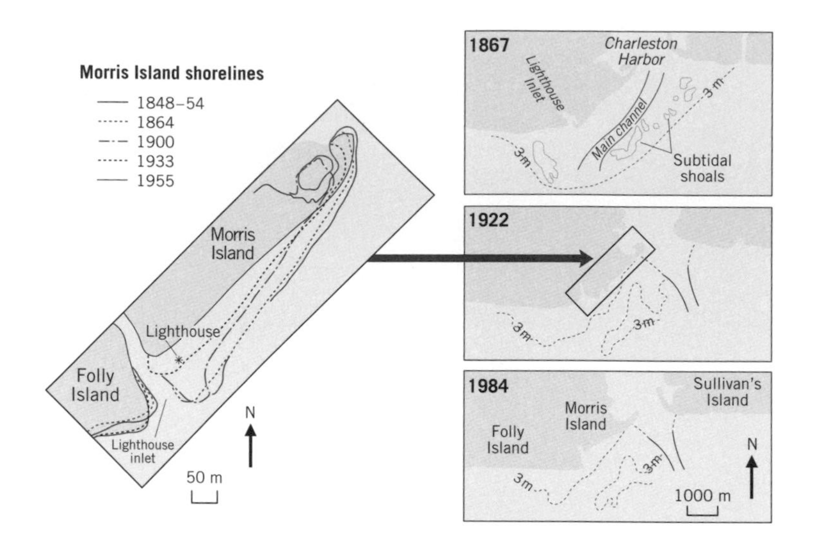

Fig. 12.28 Jetty construction at the entrance to Charleston Harbor, South Carolina, in the late nineteenth century significantly diminished the size and extent of sand shoals fronting the harbor, which in turn affected erosion and deposition along the landward shorelines. (From FitzGerald, 1988, details in Fig. 12.20.)

12.8.5 Effects of inlet sediment bypassing

Tidal inlets interrupt the wave-induced longshore transport of sediment along the coast, affecting both the supply of sand to the downdrift beaches and the position and mechanisms whereby sand is transferred to the downdrift shorelines. The effects of these processes are exhibited well along the Copper River delta barriers in the Gulf of Alaska. From east to west along the barrier chain the width of the tidal inlets increases, as does the size of the ebb-tidal deltas (Fig. 12.29). In this case the width of the inlet can be used as a proxy for the inlet's cross-sectional area. These trends reflect an increase in tidal prism along the chain, which is caused by an increase in bay area from east to west, while tidal range remains constant. Also quite noticeable along this coast is the greater downdrift offset of the inlet shoreline in a westerly direction. This morphology is coincident with an increase in the degree of overlap of the ebb-tidal delta along the downdrift inlet shoreline. The offset of the inlet shoreline and bulbous shape of the barriers are produced by sand being trapped at the eastern, updrift end of the barrier. The amount of shoreline progradation is a function of inlet size and the extent of its ebb-tidal delta. What we learn from the sedimentation processes along the Copper River delta barriers is that tidal inlets can impart a very important signature on the form of the barriers.

In an investigation of barrier island shorelines in mixed energy settings throughout the world, Miles Hayes at the University of South Carolina noted that many barriers exhibit a similar shape. Numerous tidal inlets studied by Professor Hayes and his students allowed him to formulate his **drumstick barrier island** model (Fig. 12.30). In this model the meaty portion of the drumstick barrier is attributed to waves bending around the ebb-tidal delta, producing a reversal in the longshore transport direction. This process reduces the rate at which sediment bypasses the inlet, resulting in a broad zone of sand accumulation along the updrift end of the barrier. The downdrift, or thin part of the drumstick, is formed through spit accretion. Later studies demonstrated that bar complexes migrating onshore from the ebb-tidal delta are an important factor dictating barrier island morphology and the overall erosional–depositional trends, particularly in mixed energy settings. The Copper River barriers conform well to the Hayes drumstick model.

Looking at the East Friesian Islands we see that, in addition to drumsticks, barriers can have many other shapes (Fig. 12.31). Inlet sediment bypassing

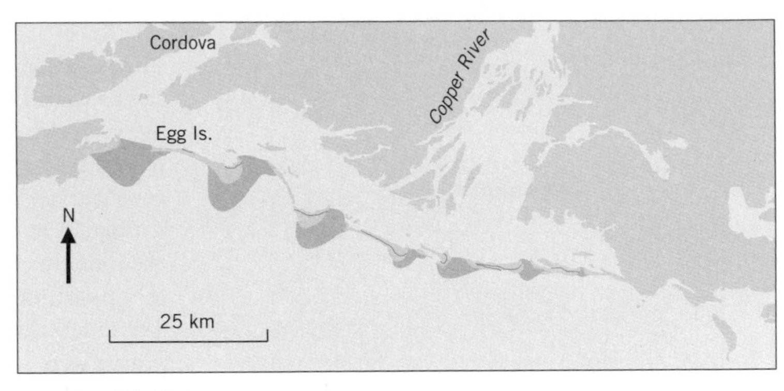

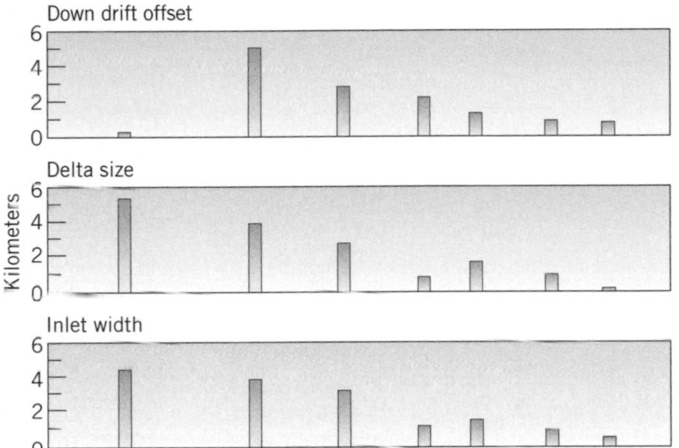

Fig. 12.29 The Copper River delta barrier chain in the Gulf of Alaska is a product of an abundant sand supply from the Copper River and high wave energy. (From Hayes, 1975, details in Fig. 12.8.)

along this barrier chain occurs, in part, through the landward migration of large swash bars (>1 km in length) that deliver up to 300,000 m^3 of sand when they weld to the beach. In fact, it is the position where the bar complexes attach to the shoreline that dictates the form of the barrier along this coast. If the ebb-tidal delta greatly overlaps the downdrift barrier, then the bar complexes may build up the barrier shoreline some distance from the tidal inlet. In these cases, **humpbacked barriers** are developed, such as Norderney or Spiekeroog. If the downdrift barrier is short and the ebb-tidal delta fronts a large portion of the downdrift barrier, such as at the island of Baltrum, the bar complexes weld to the eastern end of the barrier, forming **downdrift bulbous barriers**. Thus, studies of the Friesian Islands demonstrate that inlet processes exert a strong influence on the dispersal of sand along mixed energy barrier island shorelines and in doing so dictate barrier shape.

12.8.6 Human influences

Dramatic changes to inlet beaches can also result from human influences, including the obvious consequences of jetty construction that reconfigures an inlet shoreline. By preventing or greatly reducing an inlet's ability to bypass sand, the updrift beach progrades while the downdrift beach, whose sand supply has been diminished or completely cut off, erodes. There can also be more subtle human impacts that can equally affect inlet shorelines, especially those associated with changes in inlet tidal prism, sediment supply, and the longshore transport system. Nowhere are these types of impacts better demonstrated than along the central Gulf coast of Florida, where development has resulted in the construction of causeways, extensive backbarrier filling and dredging projects, and the building of numerous engineering structures along the coast.

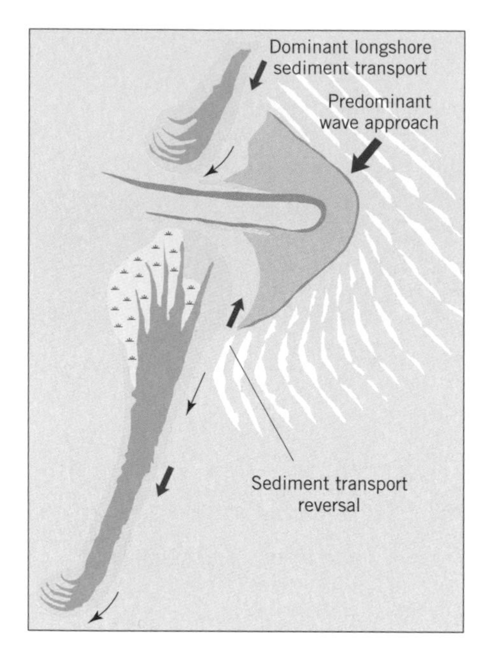

Fig. 12.30 Miles Hayes's drumstick barrier island model explains barrier shape as a function of wave refraction around the ebb-tidal delta trapping sand along the downdrift inlet shoreline. (From M. O. Hayes & T. Kana, 1976, *Terrigenous Clastic Depositional Environments*. Columbia: Department of Geology, University of South Carolina.)

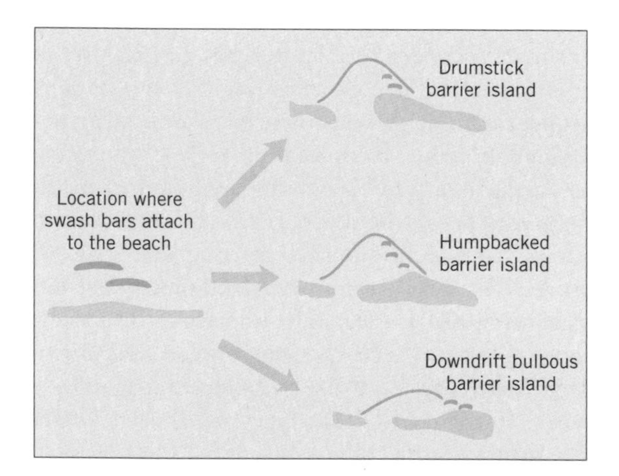

Fig. 12.31 The shape of mixed-energy barriers along the East Friesian Islands is primarily a function of wave inlet sediment bypassing processes. (From FitzGerald, 1988, details in Fig. 12.20.)

Of the 17 inlets that have closed along this coast since the late 1880s, more than half of the closures can be traced to human influences caused primarily by changes in inlet tidal prism. For example, access to several barriers has been achieved through the construction of causeways that extend from the mainland across the shallow bays. Along most of their lengths the causeways are dike-like structures that partition the bays, thereby changing bay areas and inlet tidal prisms. In some instances, tidal prisms were reduced to a critical value, causing inlet closure. At these sites the tidal currents were unable to remove the sand dumped into the inlet channel by wave action. Similarly, when the Intracoastal Waterway (protected inland canal built for barge and boat traffic) was constructed along the central Gulf coast of Florida in the early 1960s, the dredged waterway served to connect adjacent backbarrier bays, thereby changing the volume of water that was exchanged through the connecting inlets. The Intracoastal Waterway lessened the flow going through some inlets while at the same time increasing the tidal discharge of others. This resulted in the closure of some inlets and the enlargement of others. Improved access to the central Florida barriers led to their development, including the formation of marinas and finger canals along the backside of the barriers. These were formed by dredging small canals and then using the dredge spoil to build land peninsulas where there was once just water. As seen in a comparison of historical and present day maps of Boca Ciega Bay, this process can drastically reduced the extent of open water area in backbarrier, leading to smaller tidal prisms and smaller equilibrium-sized tidal inlets (Fig. 12.32). These examples demonstrate that altering the natural system can produce undesired consequences, which emphasizes the need to assess the potential affects of developmental projects before they are undertaken.

12.9 Summary

As we have seen in this chapter, tidal inlets occur along barrier coasts in coastal plain settings and in other regions where there has been a sufficient

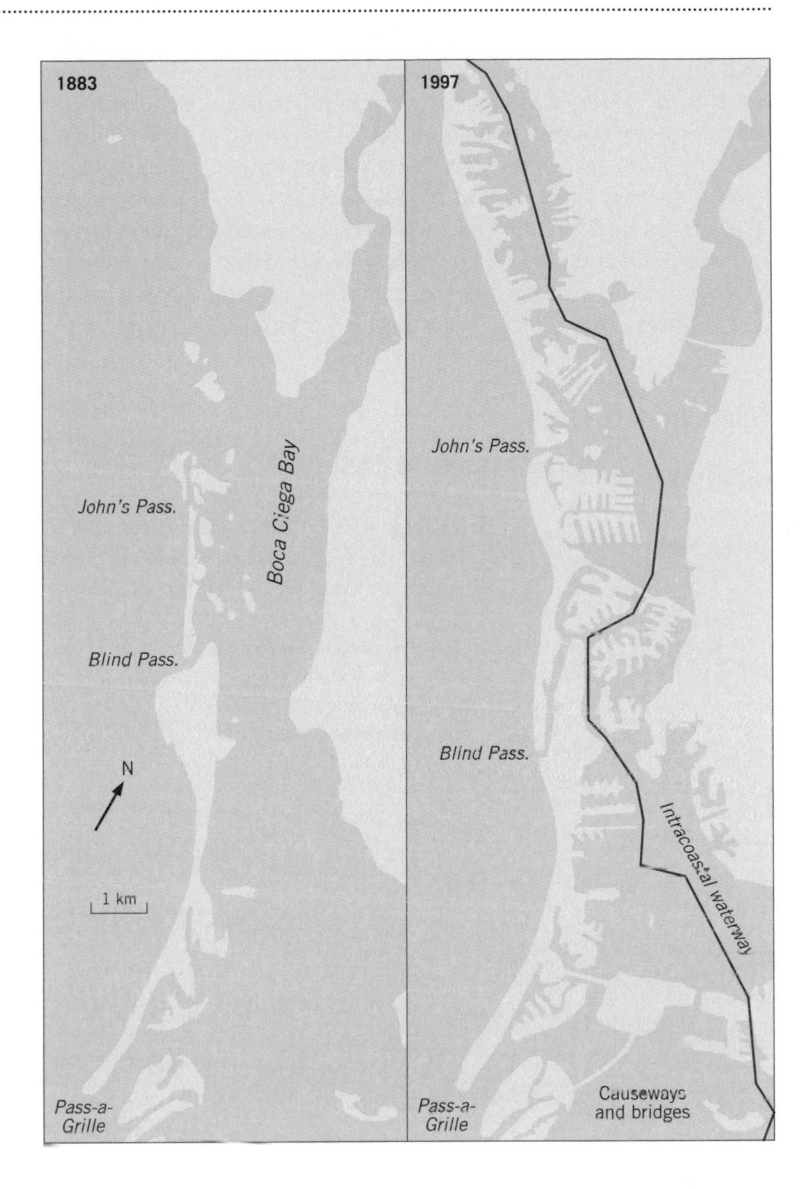

Fig. 12.32 Anthropogenic changes along the west-central coast of Florida as indicated by maps of the region in 1883 and 1997. The construction of finger canals and solid causeways have decreased open water areas and reduced water circulation in the bays. This condition has led to smaller bay tidal prisms and reduced sized associated tidal inlets. (From P. L. Barnard & R. A. Davis, 1999, Anthropogenic versus natural influences on inlet evolution: west-central Florida. In N. C. Kraus & W. G. McDougal (eds), *Coastal Sediments '99*. Reston, VA: ASCE, p. 1495.)

supply of sand for barrier spit construction across embayments. Inlet formation today occurs primarily when narrow, low barriers are breached during severe storms. Tidal inlets are narrowest and deepest at their throat section, where tidal currents and potential sediment transport reach their maximums. Many inlets have stabilized next to bedrock outcrops, in former river channels, or in resistant sedimentary strata. Migrating inlets are usually shallow and positioned in easily eroded sands. Flood-tidal deltas are horseshoe-shaped shoals situated on the landward side of an inlet and formed from sand entering the inlet channel and being transported into the backbarrier by flood-tidal currents. Sand transported seaward by ebb-tidal currents forms arcuate-shaped ebb-tidal deltas. The morphology of ebb deltas reflects the wave versus tidal energy that shapes them.

A direct correspondence exists between an inlet's tidal prism and: (i) its throat cross-sectional area;

and (ii) the volume of sand contained in its ebb-tidal delta. These concepts are very useful when planning jetty construction, channel dredging, and sourcing ebb-tidal deltas for beach nourishment programs. There are various mechanisms whereby sand bypasses unmodified tidal inlets. Wave-dominated inlets bypass sand along the periphery of the delta by wave action. At mixed energy inlets one of the end products of sediment bypassing is the formation of large bar complexes that migrate onshore and attach to the landward shoreline. Processes of inlet sediment bypassing, volumetric changes of the ebb-tidal delta, sand losses to the backbarrier, the sheltering effect of the ebb-tidal delta, and other inlet processes strongly influence the distribution of sand along inlet beaches. Along mixed energy coasts tidal inlets may dictate the shape of barrier islands.

Suggested reading

Aubrey, D. G. & Giese, G. S. (eds) (1993) *Formation and Evolution of Multiple Tidal Inlets.* Washington, DC: American Geophysical Union.

Aubrey, D. G. & Weishar, L. (eds) (1988) *Hydrodynemics and Sediment Dynamics of Tidal Inlets: Lecture Notes on Coastal and Estuarine Studies.* New York, Springer-Verlag.

Boothroyd, J. C. (1987) Tidal inlets and tidal deltas. In R. A. Davis (ed.), *Coastal Sedimentary Environments.* New York, Springer-Verlag.

Bruun, P. & Gerritsen, F. (1960) *Stability of Tidal Inlets.* Amsterdam: North-Holland Publishing.

Bruun, P. (1966) *Tidal Inlets and Littoral Drift.* Amsterdam: North-Holland Publishing.

Cronin, L. E. (ed.) (1978) *Estuarine Research, Volume 2.* New York: Academic Press.

Hayes, M. O. & Kana, T. (1976) *Terrigenous Clastic Depositional Environments.* Columbia: Department of Geology, University of South Carolina.

13 Intertidal flats

13.1 Introduction

Unvegetated intertidal environments that accumulate sediment and are occupied by specially adapted organisms are located around the margins of most coastal embayments and some open coasts. These environments are called intertidal flats or, simply, tidal flats. The width and extent of tidal flats are directly related to tidal range and to the morphology of the bay or other environment in which they are located. This term is typically reserved for those intertidal environments that are not exposed to significant wave energy. For example, many beaches have extensive intertidal components that are not vegetated but they are not considered to be tidal flats.

In some embayments, especially those that experience macrotidal conditions and are tide-dominated, much of the bay may be intertidal except for tidal channels that dissect the flats. The Wadden Sea along the Dutch and German coast of the North Sea is such a place, as are the Bay of Fundy in Canada and the Bay of Saint Malo on the north coast of France, the two places with the highest tidal ranges in the world. Each of these is discussed in a later section of this chapter.

Tidal flat surfaces and their associated tidal channels may be composed of mud, sand, or, more typically, a combination of both. Some channels may have high concentrations of shell debris on the floor. Most tidal flats and tidal channels have various types of bedforms, regular undulations, on their sediment surface. The nature and rigor of the tidal currents and waves tend to be the controlling factors in both the rate and nature of sediment accumulation and the types of bedforms that develop on the sediment surface.

In this chapter we discuss the morphology, sediments, and processes that characterize the tidal flat complex. This environment is by far most common along mesotidal and macrotidal coasts where low gradients characterize the shore zone.

13.2 Morphology of tidal flats

The standard appearance of a tidal flat is a gently sloping and fairly broad surface of unconsolidated sediments that is alternately inundated and exposed as the tide floods and ebbs. Typically the width is directly related to the tidal range, but the underlying geology and regional geomorphology can cause variations. Along broad coastal plains or other flat-lying areas the tidal flats tend to be very gently sloping (Fig. 13.1), but along some coasts the tidal flat slope tends to be relatively steep (Fig. 13.2).

Another factor in the size of tidal flats is the extent to which the estuary has been filled with sediment. Those that have not experienced significant infilling of sediment tend to have more narrow tidal flats and those that have much sediment have extensive tidal flats.

The most pronounced interruption of the nearly flat and featureless intertidal flat is the presence of tidal channels (Fig. 13.3), which dissect most tidal flats. These channels range from small and ephemeral ones that may be closely spaced to those that are large and deep. The latter commonly have water in them throughout the tidal cycle, even during spring tide conditions. These channels serve as

Fig. 13.1 Photograph of a wide, gently sloping tidal flat north of Adelaide, South Australia. The rough surface is produced by the mounds of burrowing organisms that are common on many tidal flats. A small tidal channel winds through the environment.

Fig. 13.2 Photograph of a narrow and steep tidal flat. Although the tidal range here in the Bay of Fundy, Canada, is very great, the shoreline area is quite steep, causing this narrow intertidal zone.

Fig. 13.3 Photograph of a typical tidal channel that cuts into the tidal flat. These channels serve as the main pathways for tidal flux into and out of intertidal areas. Some might be several meters deep and contain significant water even at low tide.

major conduits for sediment during both flood- and ebb-tidal cycles. Channel development on many tidal flat systems is quite similar in its pattern to a typical river system, with small tributary channels merging to serve a single major channel.

13.3 Sediments

Sand and mud with some scattered shells form the typical sediment package on tidal flats. There is generally a pattern of sediment distribution that is regular and predictable. This pattern is related to both the physical energy and the position within the intertidal flat. The highest energy occurs at the base of the intertidal zone, which is the lowest part of the tidal flat. Here there is typically a concentration of sand, with the specific grain size dependent on the spectrum of sizes available in the specific estuarine system. The grain size decreases landward and upward across the intertidal flat (Fig. 13.4), with the mud at the top or most landward portion of the intertidal environment. This is in contrast to most shoreline environments, where sediment grain size decreases away from the source, generally from the land area. Tidal flats receive their sediment from the ocean side and thus the grain size decreases toward land.

Generally, tidal flat sediments are quite well sorted at any specific location because they are subjected to similar conditions on a regular basis. The

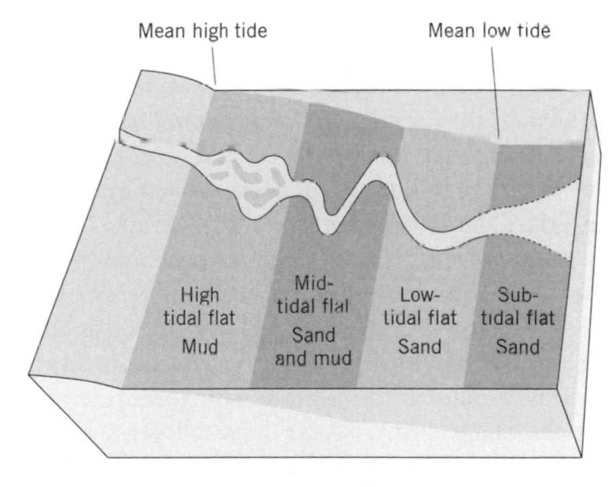

Fig. 13.4 Diagram showing the trend in grain size over a tidal flat. Although there is a continuum from coarse sediment near the low tide line to fine sediment at the high tide line, this diagram is subdivided into grain size bands. (After G. deV. Klein, 1972, Determination of paleotidal range in clastic sedimentary rocks. *XXIV International Geological Congress*, 6, 397–405.)

only common exceptions to this generalization are the presence of shells that may be scattered over various grain sizes because they are indigenous to the tidal flat environment, and mud that can settle from suspension during still water conditions. Some tidal flat environments may be adjacent to bedrock exposures that, when eroded, will provide large rock fragments to adjacent tidal flats.

13.4 Organisms

Although tidal flats are a rather harsh environment because of the regular and continual exposure and inundation, there is a community of abundant organisms that inhabits this rigorous environment. This discussion is restricted to the benthic portion of the community because that is the only portion that is truly restricted to the tidal flats. There are two different living habits here: the vagrant benthos that move about and the sessile benthos that are fixed in their position.

13.4.1 Vagrant organisms

There are various animals that move over the tidal flat surface, such as snails, worms, and amphipods. The small snails feed on detritus that accumulates on the sediment surface, and are important fecal pellet producers. They are especially abundant on the upper part of the intertidal zone (Fig. 13.5). Pellets can be an important constituent of sediments in intertidal and shallow marine environments.

Other vagrant types of snails are carnivorous and live on burrowing bivalves and oysters. These creatures have the ability to bore holes in the shells and then ingest the soft parts of the organism for food. So-called oyster drills may wipe out an entire oyster population.

13.4.2 Sessile organisms

Those creatures that cannot move about tend to be the most important and most abundant on the tidal flat overall. They include both epifaunal organisms that live on the sediment surface, such as oysters and cyanobacteria, and those that are burrowers, such as some worms and bivalves. Oysters and mussels (Fig. 13.6) are both intertidal and subtidal. They tend to occur in clusters of many individuals. As sessile filter feeders, they produce many pellets, which become part of the tidal flat sediment accumulation.

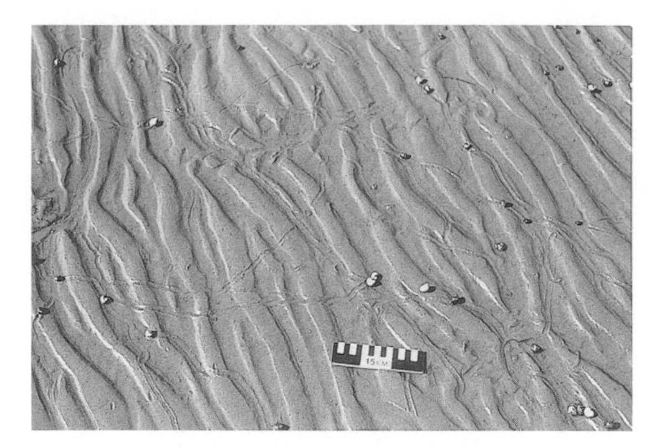

Fig. 13.5 Abundant detritus-feeding snails on the upper part of a tidal flat. These snails graze over the tidal flats and ingest fine sediment, then digest the organic material that it contains, and excrete the rest in the form of pellets.

Cyanobacteria or blue-green algae exist as microscopic filamentous organisms that produce a mat-like coverage of the upper parts of intertidal flats in lower latitudes. These mats (Fig. 13.7) are important sediment stabilizers, in that they protect otherwise non-cohesive sediments from deflation by winds or from wave erosion under non-storm conditions.

Filter-feeding infaunal bivalves and worms produce abundant fecal pellets on tidal flats. These pellets of mud commonly accumulate in troughs of ripples and produce flaser bedding when they become buried (Fig. 13.8). The worms are typically much more abundant than the bivalves and they cause the destruction of most laminations in the sediment due to their burrowing activities. The abundant worms may have tubes or they may not.

13.4.3 Limiting factors

Exposure to the atmosphere and subsequent desiccation is a problem, especially for soft-bodied organisms

(a)

(b)

Fig. 13.6 (a) Oysters, which are filter feeders but may live in the intertidal zones, and (b) crabs, which are detritus feeders on tidal flats. Both are major producers of pellets.

like worms. Most can tolerate exposure for only a short period of time. The higher the position in the tidal flats, the fewer sessile organisms will be present, because this is where exposure can last for at least several hours during each tidal cycle. Most of the shelled invertebrates, such as oysters, clams, and snails, are able to seal their soft parts from the atmosphere and can withstand fairly long periods of exposure during each tidal cycle.

Another limiting factor to benthic organisms is a large concentration of suspended sediment particles in the water column. Most of the infaunal and some epifaunal organisms that live on the tidal flat obtain

Fig. 13.7 Algal or cyanobacterial mats that are composed of a network of filamentous organisms. These microscopic organisms are woven in a thin mat that can stabilize the upper part of the intertidal zone.

their nourishment by filtering organic debris from the water column. These filter feeding organisms do not have the ability to select specific suspended particles for ingestion into their filtering system. As a consequence, when large concentrations of suspended sediment are present the organisms will ingest too much undigestible material, their siphons will become clogged, and they will die.

The other important limitation is a mobile sediment bed caused by waves and/or strong tidal currents. Many burrowing organisms, especially sedentary ones, need a reasonably stable sediment base in which to burrow and maintain an existence. It is obvious, therefore, that the problems confronting benthic organisms are numerous in estuaries. Waves may cause significant sediment mobility in the large estuaries and tidal currents can mobilize the bottom sediment in many locations. The floor of tidal channels is probably the most prohibitive environment for benthic organisms because sediment is moving nearly throughout the tidal cycle. In many areas, the low portion of the intertidal flats can also be subjected to vigorous substrate mobility during most of the tidal cycle.

13.4.4 Bioturbation

Many of the numerous benthic organisms that live in estuaries are infaunal: they burrow into the

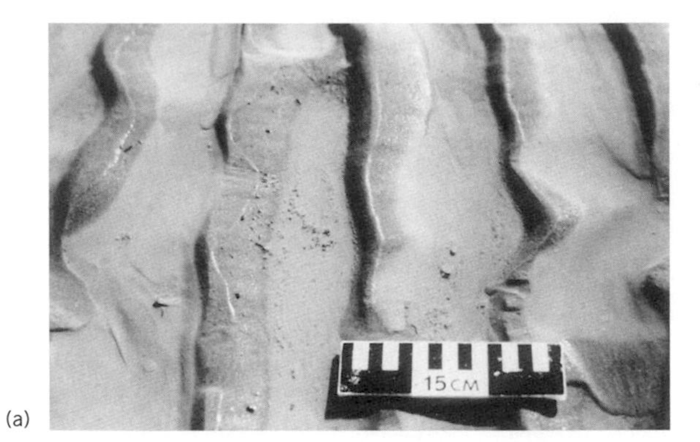

(a)

(b)

Fig. 13.8 (a) Muddy sediments accumulating in the troughs of ripples, and (b) the small lenses of mud within sand called flaser bedding that result from this association.

Fig. 13.9 Sediments completely homogenized by bioturbation. This homogenization might be produced by worms, snails, crabs, or any other infauna.

sediment both for protection and for feeding. It is these same animals that take in suspended particles and produce most of the pellets that accumulate in estuaries, but the activity of interest here is the actual burrowing process. As a bivalve or worm burrows into and through sediment it destroys the layering by essentially homogenizing the sediment (Fig. 13.9). This churning of sediment by burrowers, which may number thousands of individuals per square meter, also destroys the ability to recognize the sediments as having been deposited in a tidal environment. Extensive tidal flats of the Georgia and South Carolina coasts fall into this category. In fact, in the German Wadden Sea the combination of both waves and bioturbation work to destroy the laminations in tidal sediments.

It is only in places where there are few burrowing organisms that stratification is typically preserved. This lack of benthic infaunal organisms can result from a variety of conditions, such as: (i) exposure; (ii) too much suspended sediment; and (iii) substrate mobility.

13.5 Sedimentary structures

As might be expected, this special environment also contains some special types of sedimentary structures, and also some that are not so special. What is meant by the latter comment is that some

(a)

(b)

Fig. 13.10 (a) Ripples and (b) megaripples, which are typical of tidal flats. These bedforms tend to migrate and change both orientation and direction with flooding and ebbing currents.

Fig. 13.11 The lugworm, *Arenicola*, which is a common burrowing worm on sandy tidal flats. These organisms may extend more than 30 cm into the sediment and their population can be more than 100 per square meter.

of the sedimentary structures that are present, some even common, on tidal flats are not unique to that environment. Included among these are ripples, megaripples, and sandwaves (Fig. 13.10).

Of most importance are those features of tidal flats that are indicative of the environment because they become key factors for geologists who interpret sedimentary depositional environments from the ancient stratigraphic record. Included among these are both physical structures and biogenic structures. The many burrowing organisms leave characteristic markings on and in the tidal flat sediments, especially such examples as the lugworm, *Arenicola*, which is widespread on sandy tidal flats (Fig. 13.11). Mudcracks or desiccation features are important physical structures that occur on the upper, muddy portions of the intertidal zone. These develop as the result of significant exposure in high places on tidal flats, such as between neap and spring high tide (Fig. 13.12).

Fig. 13.12 Mudcracks formed from the desiccation of fine sediments in the upper elevations of tidal flats where mud is common and exposure is up to several hours each tidal cycle.

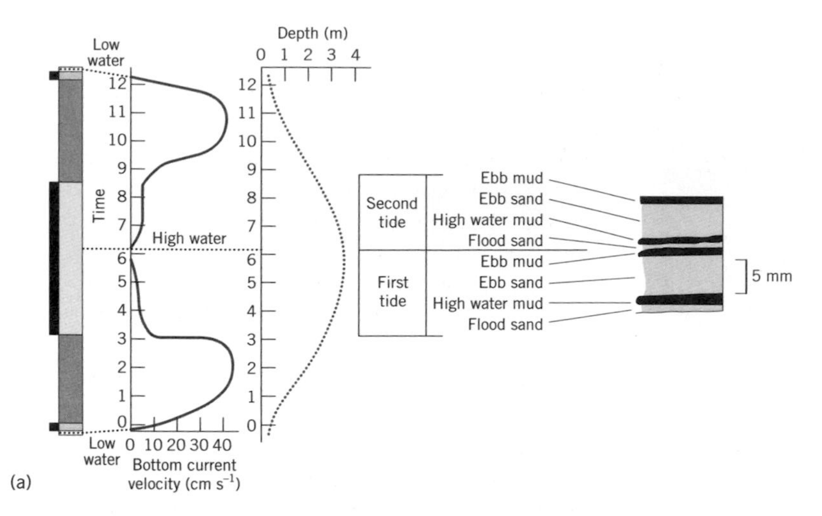

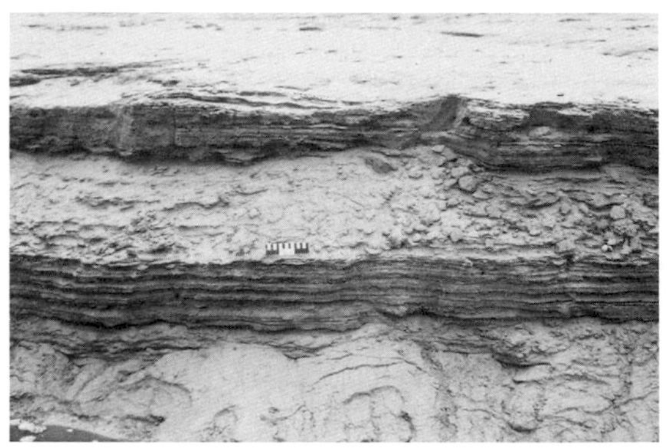

Fig. 13.13 (a) Diagram (from Klein 1972; details in Fig. 13.4) and (b) photograph showing the layering in tidal bedding up close and (c) along a tidal creek bank. Interpretation of these packages of fine and coarse sediment permits identification of spring cycles and neap tidal cycles.

Probably the most characteristic feature of tidal flats is a sedimentary structure called **tidal bedding** (Fig. 13.13). This is a special type of sediment accumulation that is developed as the result of tidal cycles. The alternating energy levels produced by the flooding and ebbing of the tides cause thin layers of alternating sand and mud. These **rhythmites**, as they are called, record the rise and fall of the tides, along with slack tide conditions. They also preserve the neap and spring cycles, with neap conditions producing thinner layers than spring conditions (Fig. 13.14).

13.6 Tidal flat processes

Tide-generated processes tend to dominate most sediments that accumulate on tidal flats, although waves can be important at certain times and in certain places. The rise and fall of the tides as the tidal wave is forced into estuaries and then back creates significant tidal currents, particularly in the tidal channels. They may range from only a few centimeters per second near slack tides on the flats to about a meter per second in the channels. These

Fig. 13.14 Tidal bedding from the Bay of Fundy showing neap and spring cycles. These tidal beds are generally fine sediment and they occur in the mouths of tidal rivers at the east end of the bay.

currents distribute sediment throughout the tidal flats. For purposes of explanation, the tidal flat is best viewed as a smooth and gently sloping surface. Certain key horizons are noted on this surface based on the position of the water level at given tidal stages. Spring high tide is the highest position of regular and predictable inundation of the sediment surface by water and spring low tide is the lowest (Fig. 13.15). Wind tides may cause water to be pushed up to the supratidal environment or down below low tide depending upon the direction and strength of the wind. Neap tidal range may be only about half of the spring range, causing considerable variation in the intertidal zone depending upon the lunar condition.

13.6.1 Tides

The scheme of sediment transport on tidal flats has been best described from studies of the tidal flats on the Wadden Sea. The model produced shows a combination of the settling lag effect and the scour lag effect on the paths of sediment particles as they are transported up on to the tidal flat surface (Fig. 13.16), and also shows how, over a long time, sediment builds up and out into the estuary. The distance–velocity curves are asymmetrical and show where a given particle is entrained, transported, and deposited.

During flooding tides, the sediment particle at location 1 on curve A–A′ is picked up at a tidal current velocity shown at point 2 and carried landward until that velocity is again reached at point 3 on the curve. The particle then begins to fall and reaches the tidal flat surface at point 5 when the current velocity is at position 4. The difference in the velocity of entrainment, which is greater than at settling, is quite important. This provides the settling lag effect.

Ebbing tidal currents follow curve B–B′, which represents a more landward water mass and achieves less of a maximum velocity at this position on the tidal flat. The same particle is picked up at location 5 when the velocity at point 6 on the curve is reached. It is carried until point 7, when it begins to fall and eventually settles to the bottom at point 9. This diagram shows that the net result is movement of a sediment particle from location 1 to location 9 during a single tidal cycle. This is obviously an

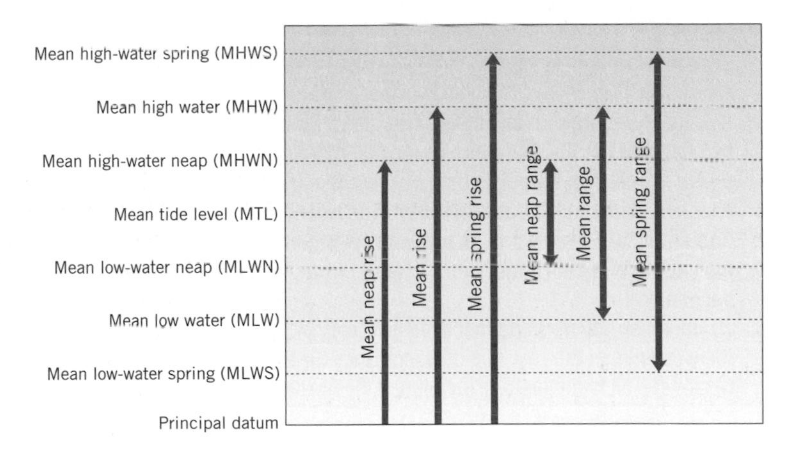

Fig. 13.15 Diagram showing the key horizons in the tidal range. The absolute values range widely depending upon the tidal range and the amount of variation from spring to neap tides.

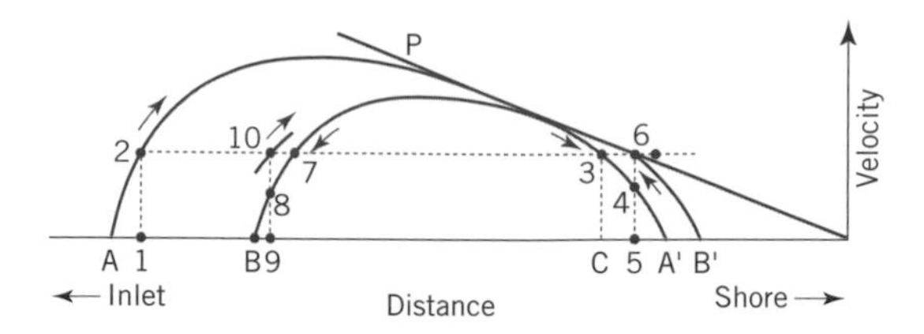

Fig. 13.16 Diagram showing the classic interpretation of settling lag and scour lag. Although it appears to be complicated, the basic model shows that there is a net transport of sediment across the tidal flat from near the low tide position toward the high tide position. (From H. Postma, 1961, Transport and accumulation of suspended matter in the Dutch Wadden Sea. *Netherlands Journal of Sea Research*, **1**, 170.)

oversimplified model because there are many perturbations on the tidal flat that interfere with the processes. It does, however, serve to illustrate the basic mechanism by which sediment particles are transported up onto the tidal flats.

The size of the sediment particles is an important variable in the above scheme. Bigger particles will be transported a shorter distance over a given tidal cycle than smaller particles. The greater the time an area is covered and the deeper the water, the more tidal energy is expended on a given location on the tidal flat. As a consequence, there is a regular decrease in sediment particle size up the tidal flat toward the spring high tide level.

Conditions in some areas produce discontinuous mud or sand layers due to some combination of sediment availability and tidal current strength. These can be formed through tidal processes and represent what is essentially discontinuous tidal bedding, or they may represent alternations in current energies that are not produced by flooding and ebbing tides. These are the conditions that produce flaser bedding (Fig. 13.8b), which commonly forms in the troughs of bedforms. Discontinuous sandy lenses within a mud sequence are referred to as lenticular bedding, a feature generally associated with limited sand availability, but one that may also reflect variations in tidal current velocities.

Another type of tidally produced stratification is tidal bundles, which are a type of stratification typically associated with tidal channels or relatively strong tidal currents and large bedforms. The alternation of flooding and ebbing tides is generally accompanied by significant differences in current velocity. This commonly produces pulses in the migration of large bedforms, which are characterized by medium to large-scale cross-stratification. The dominant current moves the bedform and the recessive current commonly produces a mud drape over the bedform, producing a muddy seam between each sand cross-stratum. In many tide-dominated areas the sequence contains readily distinguishable sets of cross-strata that change in thickness and sand–mud ratio in packages of 14 bundles. These tidal

Box 13.1 Tidal flats of Jade Bay, northern Germany

Jade Bay is a tidal estuary along the German Friesian coast of the North Sea. It is located near the tide-dominated "elbow" of this coast. This fan-shaped bay has been diked along its entire coast. Much of its areal extent is intertidal mud flats that are dissected by tidal channels. The mud is transported into the bay by flooding tidal currents and its cohesive character makes it difficult to remove once it settles from suspension.

The small city of Wilhelmshaven is located on the west side of the bay. It is of significance for two reasons: (i) it is the historic location of the German navy going back to the time of Kaiser Wilhelm; and (ii) it is the home of the Senckenberg Institute for Marine Geology, one of the premier institutions of its kind in Europe.

Scientists of the Senckenberg Institute have been conducting detailed research on tidal flat environments since the beginning of the twentieth century. They pioneered the study of the relationships between tidal flat sediments and the organisms that live there. Much of their work has been very important to geologists who study ancient rocks that originated as tidal flat sediments. Several new techniques for investigating modern tidal flat environments were also part of the contributions of these prominent scientists.

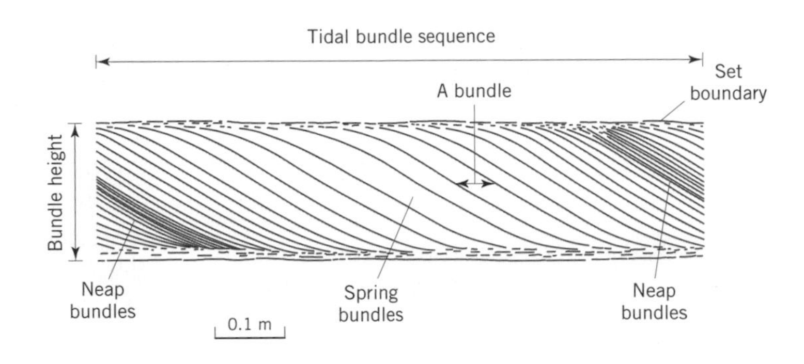

Fig. 13.17 Diagram of tidal bundles showing spring and neap cycles. These form in large bedforms and show thick accumulations during spring tide and thin ones during neap tide. (From C. S. Yang & S. D. Nio, 1985, The estimation of palaeohydrodynamic processes from subtidal deposits using time series analysis methods. *Sedimentology*, **32**, 42.)

bundle packages (Fig. 13.17) represent a spring and neap tidal cycle, and when preserved, they are found on the floors and margins of tidal channels.

13.6.2 Waves

We typically associate tidal processes with tidal flats but there are some locations where waves play an important role in the dynamics of tidal flats; enough to destroy all of the signatures of tidal processes. In order for this to happen the energy imparted by waves onto the tidal flat must exceed that of tides. The most common conditions under which this can occur are in places where extensive shallow water covers the tidal flat for long portions of each tidal cycle. Waves formed in this shallow environment will move large quantities of sediment through the back and forth motion they produce, thereby destroying any tidal signature or preventing it from happening in the first place.

The Wadden Sea area on the German coast of the North Sea is a good example of such wave-influenced tidal flats. Here, broad, sandy intertidal flats cover most of the area between the barrier islands and the mainland, several kilometers wide. The muddy tidal flats are mostly near the mainland. The sandy flats are flooded for 4–6 hours of each tidal cycle and the fetch of several kilometers permits the commonly strong winds to generate modest sized waves. This combination of strong wave action coupled with modest tidal currents prevents tidal bedding from forming over much of the tidal flat environment. In contrast, the tidal channels that disect the tidal flats display tidal bundling, showing that tidal currents are dominant in this environment.

13.7 Tidal channels

Much like the flat upland plain environments of the Midwest or the high plains, tidal flats develop a drainage network that contains a range of small to relatively large channels. Unlike the upland environments, however, these tidal channels carry water in two directions: the flood and ebb as the tide rises and falls. These channels are narrow and shallow in the upper reaches of the tidal flats, where muddy sediments are dominant or at least common, and extend to the lower elevations of the intertidal zone, where they might be large and have floors in the subtidal zone.

Slight and subtle undulations in the tidal flat surface cause water to be concentrated in the low areas after emergence of the tidal flat surface begins during the ebb phase of the tidal cycle. These small channels increase in size as each tidal cycle passes, until an equilibrium condition is established between the tidal flat/channels surface and the tidal flow. Where these channels are cutting through muddy and cohesive sediments, they tend to have relatively steep channel walls (Fig. 13.18), but where they cut through sand sediments the channels are broad, with more gently sloping walls.

The floors of these tidal channels are the sites of important sediment transport. This sub-environment is characterized by sand-sized sediment that lacks cohesion. This sediment is, therefore, susceptible to movement by both flood and ebb currents during each tidal cycle. As the tide floods over a tidal flat complex, the forced wave of the rising tide produces currents that may be strong enough to move the

Fig. 13.18 Steep tidal channel walls in the Wash, southeastern England. These deep channels with steep banks are caused by the transport of very large volumes of water during the tidal cycle.

Fig. 13.19 Megaripples at low tide from the Bay of Fundy, Canada. These bedforms have a length or spacing of about 1.5–1.7 m. Their crests have been smoothed off by the ebbing tidal currents.

sand on the channel floor. Once the water level has risen above the level of the channel margin, conditions are essentially like those of a flooding river. Water and suspended sediment spill over the channel walls onto the tidal flats, forming natural levees and causing tidal currents to slow.

During ebbing conditions of the tide there is a slow current as the water flows over the tidal flats in response to gravity. As soon as the tidal level is so low that part of the tidal flat surface becomes exposed, there is some channeling of the rest of the ebbing waters. These waters are fed into the tidal channels in large volumes, causing the ebbing currents to be rapid. It is typical for flow in tidal channels, like most main channels in tidal inlets, to be ebb-dominated.

The tidal currents that persist in these channels during the early parts of the flood cycle and the later parts of the ebb cycle transport considerable sediment, and in doing so, they develop a spectrum of bedforms along the channel floor. The size of these bedforms is partly related to the grain size of the sediments but is mostly due to the flow strength of the currents. Most channels with sand floors display what are called megaripples or small **subaqueous dunes**. These are asymmetrical bedforms that have a wavelength of about 1–5 m and a wave height of 20–50 cm (Fig. 13.19). The asymmetry of these bedforms is the result of the direction of current flow, such that we can tell the direction of the current that formed a particular group of bedforms by looking at them. Surveys of tidal channels show the nature of these bedforms: their wavelength and wave height as well as their orientation (Fig. 13.20).

Most channels display bedforms that show an orientation indicating that they were formed by ebbing tidal currents, further indication of the ebb-domination of the channels. In some instances these bedforms are modified by incoming flood currents but retain their ebb orientation. There are, however, some situations in which the bedforms reverse their orientation during each ebb and flood of the tides. Such a condition would be caused by nearly equal flood and ebb tidal currents.

13.8 Some examples

There are several places in the world where entire estuaries or even large portions of estuaries are intertidal. As might be expected, these tend to be macrotidal or at least high in the mesotidal range. In this section we take a look at some well known examples in order to demonstrate the profound influence tides have on estuaries and on tidal flats.

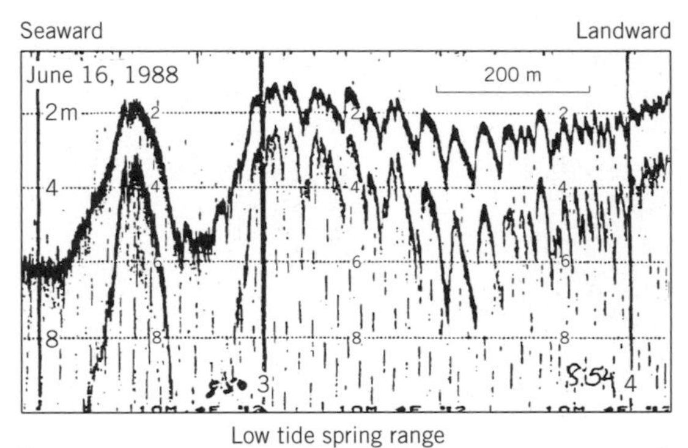

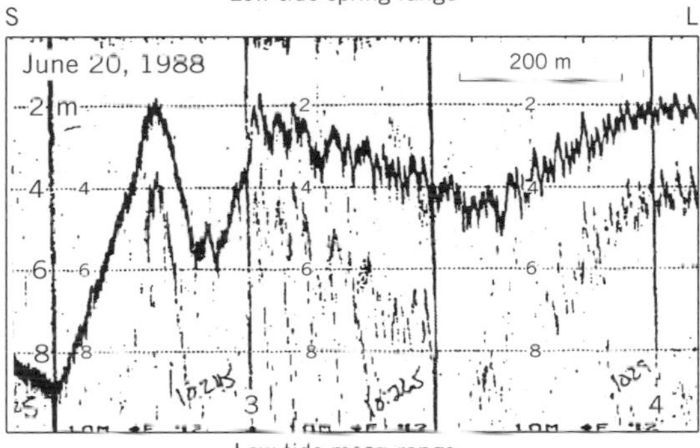

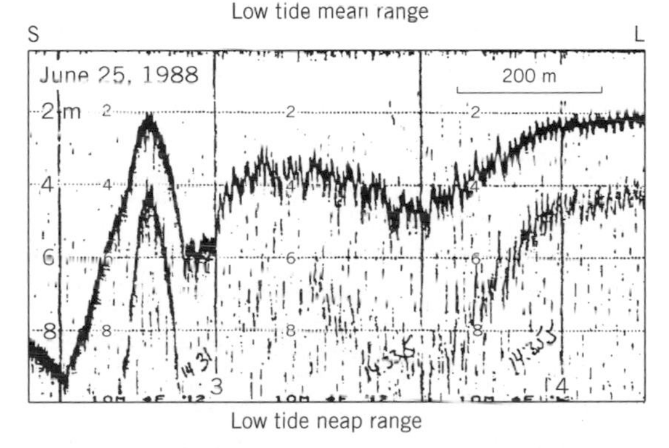

Fig. 13.20 Fathometer record of megaripples in channel floor of the German Wadden Sea. These diagrams demonstrate that although there are differences in the size and shape of these depending on lunar tidal cycle, they are present during all conditions. (From R. A. Davis and B. W. Flemming, 1991, Time-series study of mesoscale tidal bedforms, Martens Plate, Wadden Sea, Germany. In D. G. Smith, G. E. Reinson, B. A. Zaitlin & R. A. Rahmani (eds), *Clastic Tidal Sedimentology*. Canadian Society of Petroleum Geologists, Memoir 16, p. 278.)

13.8.1 German Wadden Sea and Jade Bay

The north coast of Germany includes some of the most studied tidal flat complexes in the world. This coast is a continuation of the north coast of the Netherlands and is comprised of short barrier islands with very large tidal inlets. The barrier system is separated from the mainland by the Wadden Sea, at the southern end of the North Sea (Fig. 13.21) near the apex of the German Bight, where the coast ranges from mixed energy to tide-dominated.

The Wadden Sea is essentially all intertidal flat except for subtidal channels that feed the tidal inlets. This extensive intertidal complex is subjected to spring tidal ranges of about 3 m. Drainage divides are located behind the middle of each of the barrier islands (Fig. 13.22) and represent locations of muddy sediments, as compared to the intermediate areas where sand dominates. There is also a fining of grain size toward the mainland, where a fringe of marsh borders the extensive tidal flats. Wadden Sea tidal flats actually tend to be dominated by wave action instead of tidal currents as might be expected. This is due to the fairly long fetch during high tide because the barrier islands are several kilometers from the mainland. Jade Bay is just around the corner from the Wadden Sea and is an enclosed tidal estuary that is largely intertidal (Fig. 13.23). Here mud-dominated tidal flats extend for several kilometers due to the combination of the low relief and the spring tidal range of nearly 4 m. The widespread tidal rhythmites are thinly laminated accumulations of fine sediments that are rather soft (Fig. 13.24). This tidal estuary was the location of some of the first detailed investigations into tidal flat sedimentology, particularly the interactions of organisms with sediments. The many burrowing species provided a wide range of these relationships.

13.8.2 The Wash

Although much of the coast of the British Isles has a fairly high tidal range, the embayment known as The Wash on the northern Norfolk coast (Fig. 13.25) has spring ranges of about 7 m. This primarily tidal estuary displays the typical transition from sand in

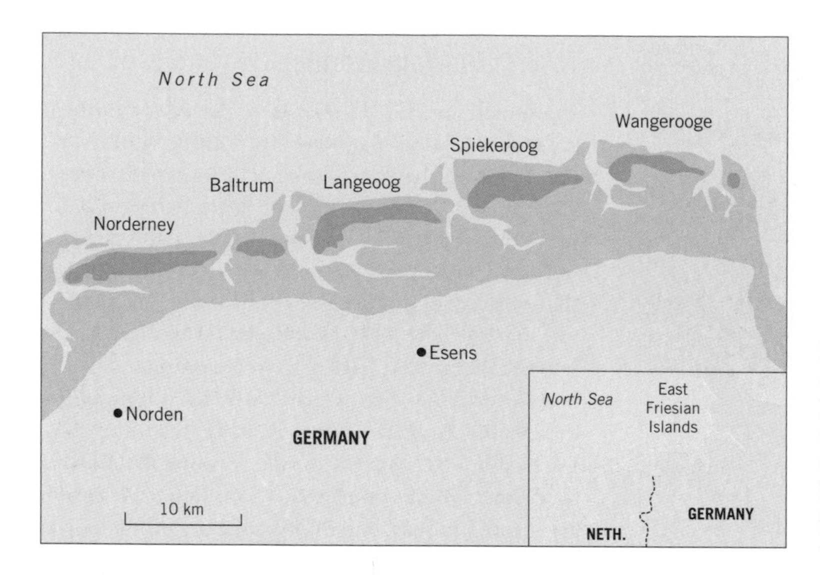

Fig. 13.21 Map showing the location of the Wadden Sea and adjacent East Friesian Islands on the German coast of the North Sea. The Wadden Sea is the mostly intertidal area that separates the barrier islands from the mainland. It extends across the Dutch, German, and Danish North Sea coasts.

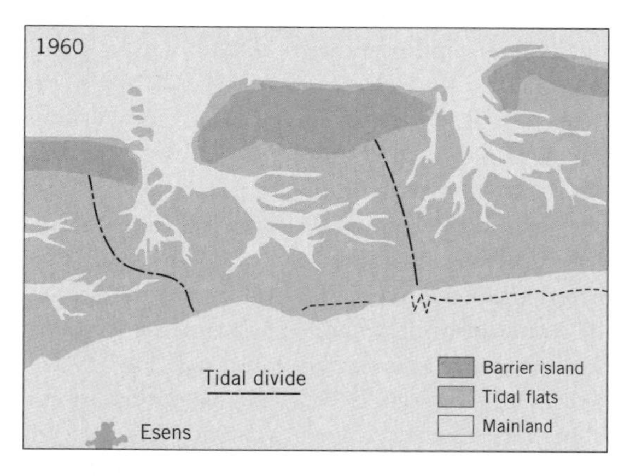

Fig. 13.22 Drainage divides in the German Wadden Sea. These lines are the approximate location of where the flooding and ebbing tidal currents move to and from. In reality, there is some water carried across these boundaries.

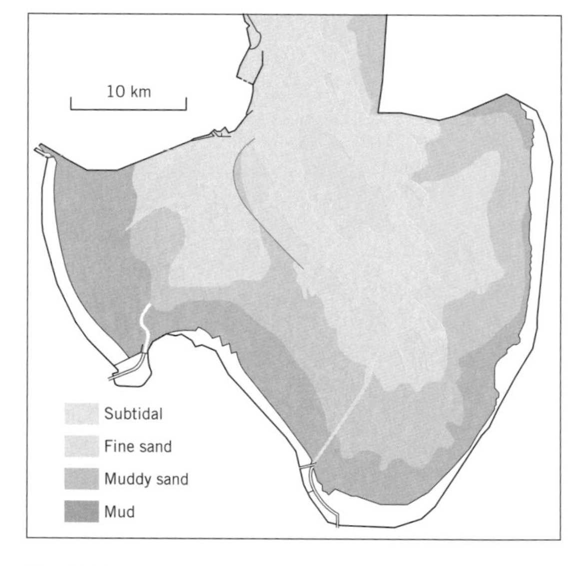

Fig. 13.23 Jade Bay, an intertidal estuary on the German coast. With a tidal range of near 4 m, much of this tidal bay is exposed during low tide. It is dominated by muddy tidal flats, with some several kilometers wide.

the outer parts to mud in the landward fringes. Marshes dominated by *Salicornia* extend across much of the upper and muddy levels of the Wash. These are dissected by tidal channels (Fig. 13.26) that carry mud clasts, which are eroded from the adjacent marsh areas.

As one proceeds in a seaward direction, there is an increase in the size of these tidal channels as they cut through the cohesive muds, and the bottom sediments become sandy. Eventually the entire intertidal system becomes sand-dominated and channels lose their definition. Here the sand is reworked and mobilized by the strong tidal currents associated with the macrotidal conditions of this tidal estuary.

Fig. 13.24 Muddy tidal flats with channels in Jade Bay. The soft and deep mud makes these almost impossible to traverse and study.

13.8.3 Bay of Saint Malo

The Bay of Saint Malo is a large embayment along the northern Brittany coast of France. This tidal estuary is one of the two areas in the world where spring tidal range is near 15 m. The inner portion of this huge estuary, called the Bay of Mont Saint Michel after the well known monastery and abbey that carries that name, is a major tourist attraction in northern France. Here, just down the coast from the area of the famous Normandy Invasion in the Second World War, the tides are so large that detailed study of the lower part of the tidal flat system has not yet been completed. Only the upper, muddy areas are accessible to investigation because

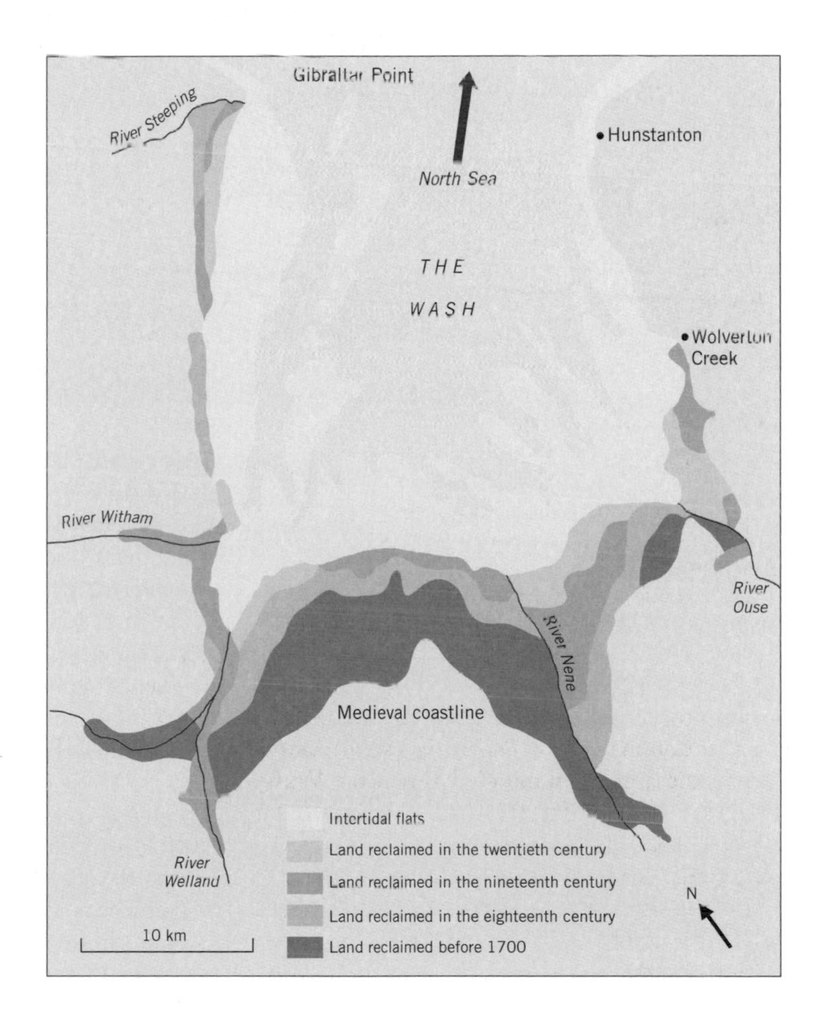

Fig. 13.25 Map of the Wash on the Norfolk and Lincolnshire coast of England. The tidal embayment has a maximum range of about 7 m on this North Sea coast.

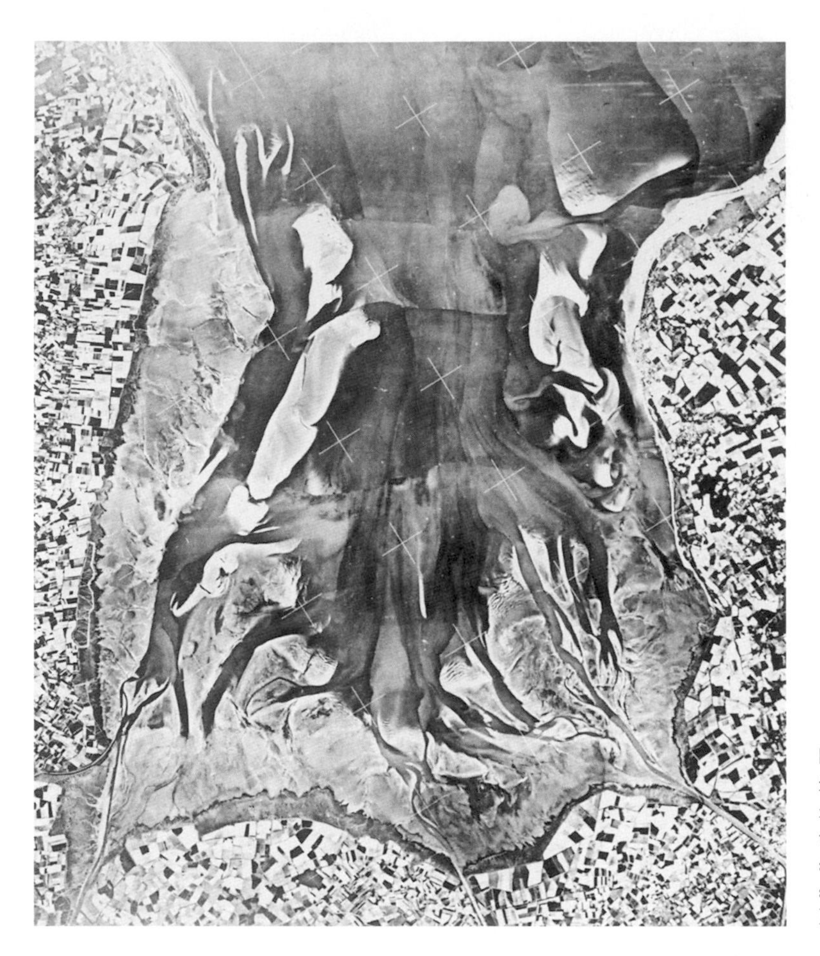

Fig. 13.26 Aerial photograph of tidal flats and channels on the Wash. As in most similar tide-dominated coastal bays, the channels are well defined and deep, and the sediment patterns mimic those shown in Fig. 13.4. (Courtesy of G. Evans.)

Fig. 13.27 Mont Saint Michel in the Bay of Saint Malo on the north coast of France, with extensive tidal flats. The island in the center is the one shown on many travel posters. It contains a small village that is built around a huge abbey.

the intertidal flats extend for tens of kilometers (Fig. 13.27).

Like most tidal estuaries there is a marsh fringe, and here it is dominated by succulent plants of the *Salicornia* type. The sediments under these plants and on the unvegetated tidal flats are typically muddy and show cyclic patterns of accumulation (Fig. 13.28). These rhythmites display neap and spring cycles, depending on the combination of the thickness of each lamination and the grain size of the sediments.

13.8.4 Bay of Fundy

The Bay of Fundy is a elongate coastal bay whose presence is the result of faulting in the lithosphere producing a structural basin called a graben. This bay is split into two tide-dominated, macrotidal

Fig. 13.28 Fine rhythmites, also called tongues, from the Mont Saint Michel area of the French coast. These are a special type of tidal bedding that has very thin but well defined layers.

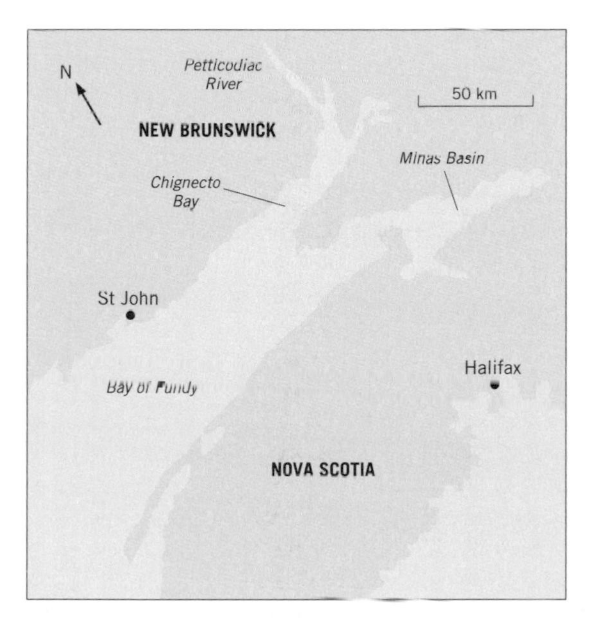

Fig. 13.29 Map of the Bay of Fundy, Canada, the location of the largest tidal ranges on Earth.

basins at its landward end (Fig. 13.29), the Minas Basin and Chignecto Bay. Here huge tidal ranges prevail, with spring ranges above 15 m – the highest in the world. These basins are largely intertidal, with shallow tidal channels. There is some fine sediment discharge from the rivers on the landward ends of

Fig. 13.30 Various sizes of bedforms in the Minas Basin, Bay of Fundy. This entire basin is exposed at low tide, showing this variety of sizes and orientations of bedforms.

these two basins but most of the intertidal sediment is reworked from the basins themselves or from erosion of the shorelines.

As in the other examples, sediments show the typical trend of fine sediment at the distal end of the basins with an increase in grain size toward the lower or open portion of the basins. Some steep slopes along the basin have marsh fringes that are only tens of meters wide or less. Most of the tide-dominated basins is covered with various scales of bedforms that are developed in sand. These range in scale from small ripples, through megaripples, to sandwaves of different wavelengths (Fig. 13.30).

13.9 Summary

The tidal flat environment appears at first to be one of little variation; just a flat surface over which the tide rises and falls with predictable regularity. In fact, however, there is great variation, but it is gradual and subtle in many places. Unlike many coastal environments that derive their sediment from land, tidal flats rely on waves and currents to provide the sediments that slowly accumulate on the surface. The cyclic nature of the processes and the relatively low energy conditions produce thin layers arranged in a predictable and recognizeable fashion. Abundant burrowing organisms can destroy this layered record in many areas.

Suggested reading

Alexander, C. R., Davis, R. A. & Hentry, V. J. (eds) (1998) *Tidalites: Processes and Products*. Tulsa, OK: SEPM Special Publication No. 61.

Amos, C. L. (1995) Siliciclastic tidal flats. In G. M. E. Perillo (ed.), *Geomorphology and Sedimentology of Estuaries*. Developments in Sedimentology No. 53. Amsterdam: Elsevier.

DeBoer, P. L., van Gelder A. & Nio, S. D. (eds) (1988) *Tide-influenced Sedimentary Environments and Facies*. Dordrecht, D. Reidel Publishing Company.

Eisma, D. (1998) *Intertidal Deposits; River Mouths, Tidal Flats and Coastal Lagoons*. Boca Raton, Florida: CRC Press (especially Chapters 6 to 9).

Reineck, H. E. & Singh, I. B. (1980) *Depositional Sedimentary Environments*, 2nd edn. Heidelberg: Springer-Verlag (lots of emphasis on tidal flats).

Smith, D. G., Reinson, G. E., Zaitlin, B. A. & Rahmani, R. A. (eds) (1991) *Clastic Tidal Sedimentology*. Calgary: Canadian Society of Petroleum Geologists, Memoir 16.

14 Coastal wetlands

14.1 Introduction

It is typical for a portion of the inner, protected margin of an estuary or a low-energy open coast to be covered by a vegetated intertidal environment. If covered with grasses or grasslike vegetation this environment is called a marsh. If covered with woody shrubs and trees typically called mangroves, this environment is a swamp or, more properly, a mangal. These environments may be normal marine in salinity or they may range through brackish toward freshwater. This discussion does not include the freshwater marshes along the rivers that may grade into the estuary. The proportion of the estuary that supports the salt marsh environment ranges widely: from essentially all of the estuary except for tidal channels, to only a border a few meters wide. The proportion of the estuary that is covered by vegetation tends to be an indication of the maturity of the estuary or the degree to which it has been filled in with sediment. For example, some of the estuaries on the Georgia coast have little open water except near the inlet between the barrier islands. A similar situation exists in coastal southwest Florida, where mangroves dominate. Only tidal creeks dissect the extensive vegetated environment in these sedimentologically mature estuaries (Fig. 14.1). By contrast, the German Wadden Sea is bordered by only a narrow marsh and the Bay of Fundy supports a narrow and discontinuous marsh environment where the gradients are steep (Fig. 14.2).

Fig. 14.1 Aerial overview of a marsh showing only tidal creeks interrupting the marsh vegetation. This condition of little open water landward of a barrier island is common where there is at least a modest tidal range.

Fig. 14.2 Narrow band of marsh vegetation along the margin of the Bay of Fundy in Canada. The steep sides of this bay result in a narrow marsh because of the limitations of marsh growth within the neap to spring high tide zone.

Both salt marshes and mangrove forests are special, vegetated intertidal environments and are discussed in detail. Some comparisons are made to demonstrate important differences between them.

14.2 Salt marshes

14.2.1 Characteristics of a coastal marsh

A marsh is really the portion of only the higher part of the intertidal environment that is covered by vascular plants. Above about neap high tide there is little energy to disturb the sediment substrate and the sediment that accumulates there tends to be relatively fine-grained with a fairly stable sediment surface. These factors provide the type of environment that supports vegetation: an undisturbed place of fine, organic-rich sediment. Various opportunistic and tolerant grasses thrive in this environment.

The marsh environment is commonly divided into the low marsh, which is approximately from neap high tide to mean high tide or slightly above, and the high marsh, which is from that level up to spring high tide.

14.2.2 Marsh plants

There are two genera that are particularly prone to establish dense stands on such substrates: *Juncus* (Fig. 14.3) and *Spartina* (Fig. 14.4). Although not the only marsh taxa, these are the most widely distributed in North America.

The specific type of vegetation that develops in marshes depends upon the elevation within the intertidal zone and the latitude; in other words, there is a climatic control. In the middle and southern coasts of North America, *Spartina alterniflora* is the typical low marsh grass (Fig. 14.5), not because

Fig. 14.3 Marsh showing *Juncus*, the primary high-marsh species in many North American marshes. These *Juncus* marshes may be several kilometers wide along the coasts of north Florida, Georgia, and South Carolina.

Fig. 14.4 Channel margin along a tidal creek showing high growth form of *Spartina alterniflora*. This species colonizes near the neap high tide level and commonly lines the tidal creeks in *Juncus* marshes.

Fig. 14.5 Outer margin of the low marsh with scattered oyster accumulations, a common situation in the southeastern United States. The oysters do well when they are exposed for only about half of each tidal cycle.

of the height of the plants but because of its elevation. It is typically found between neap and spring high tide. In most estuaries this zone has a narrow range in elevation of a few tens of centimeters, but this can be up to a meter or more in estuaries with very large tidal ranges. *Spartina alterniflora* is a coarse grass that grows in very dense populations. Individual plants are generally about knee-high but they display great variability, reaching up to more than 1 m in height depending upon the specific location within the marsh and the availability of nutrients. The highest plants tend to be on the highest elevations: the levees of the channel margins and near spring high tide. The *S. alterniflora* plants at the lowest part of the marsh may be quite small and discontinuous.

The high marsh in these areas is dominated by *Spartina patens*. This species is generally fine and small in contrast to *S. alterniflora*. It grows best on the upper flat surface of the marsh environment. *Juncus* is the high marsh grass in low to mid latitudes and is restricted to the elevation of about spring high tide. *Juncus roemerianus* is the species that is most common in southern North America and *J. gerardii* is most common north of Delaware and New Jersey. It is commonly called the needle

Fig. 14.6 *Salicornia*, a fleshy and short plant that is common in the high-marsh environment along many coasts. It is shown here along the southeast coast of Australia. Trace elements may change the color of this photosynthetic plant. In Australia it is commonly very red in color.

rush or black rush, it is as tall as a person, and it has a pointed end that has been known to penetrate shoes. This species attains its height throughout the extent of the spring tide position of the estuary margin. During the growing season the plant is a dark green color but it attains a silvery hue during the fall and winter.

Other high marsh plants include *Distichlis* and *Salicornia*. *Salicornia* (Fig. 14.6), also called a salt wort, is a fleshy plant that rises only 10–20 cm above the substrate. It is the only common marsh plant that does not look like a grass, although *Juncus* is not a grass either, but a rush. *Distichlis* looks very much like *S. patens*, and they may occupy the same part of the marsh.

Relief on the marsh is typically low, but there are numerous subtle variations in elevation, which cause distinct zonation of vegetation in the salt marsh environment because the plant species involved are quite susceptable to elevation differences. Quite subtle or local changes in relief or general morphology are reflected in the zonation of plant species and in their growth forms.

The low boundary of the marsh is the unvegetated tidal flat, or the margin of a tidal channel. The upper boundary can be a variety of environments, but is typically characterized by some type of upland vegetation.

14.2.3 Global distribution

The worldwide distribution of salt marshes can be organized into nine regions based on the vegetation communities (Fig. 14.7). In the northern high latitudes there is the Arctic region, which includes the northern portions of North America and Russia, along with Greenland, Iceland, and northernmost Scandinavia. Here marshes are fragmentary due to extreme weather conditions. Europe is divided into two regions: one along the north coast, including the Baltic and the coast of Ireland and Great Britain, and the other along the Mediterranean coast.

Another region extends along the Atlantic coast of North America from northeastern Canada through the United States, including the Gulf coast. It is this region that contains the most extensive marshes of North America and that is given the most consideration in this chapter. The south and east coast of South America is another region. The Pacific American region extends along the west coast of both North and South America. Australia and New Zealand comprise a region in the south Pacific. The eastern coast of Asia along with Japan completes the region in the Pacific basin.

The final region is really a special marsh environment that is restricted to high elevations in low-latitude areas that are otherwise dominated by mangroves. This situation exists in Florida and also in parts of Baja California.

14.2.4 Marsh characteristics

The physical environment of the marsh community is influenced by the degree to which it is protected from wave action, the tidal regime, the rate of sea-level rise, the topography of the coastal area, the sediment supply, and the nature of the substrate. The marsh environment is very similar to a river system (Fig. 14.8): (i) it is typically cut by meandering channels; (ii) the channels have point bars; (iii) there are natural levees along the channel banks; (iv) crevasse splays may form in breaches of the levees; and (v) there may be meander cutoffs and oxbow lakes. In addition, the marsh surface tends to be extremely flat and horizontal, just like a floodplain.

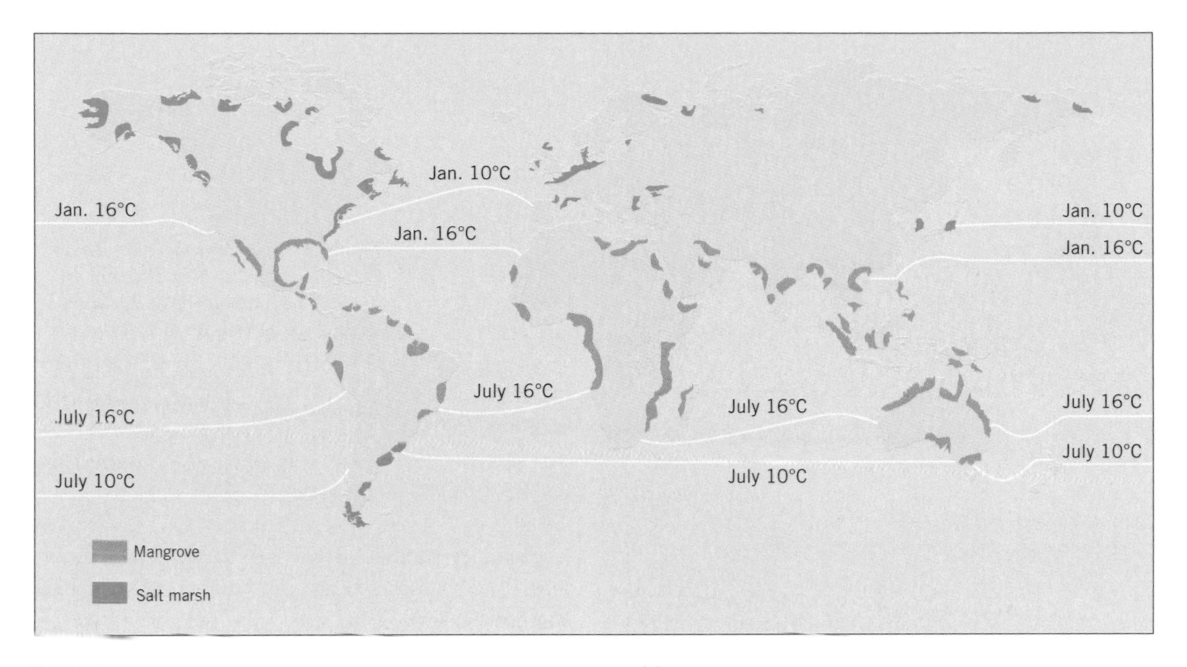

Fig. 14.7 World map showing the distribution of mangroves. Because of their inability to cope with freezing conditions, the mangroves are limited to low to moderate latitudes.

Because the zonation of vegetation is so closely tied to the elevation within the intertidal zone, it is practical to zone the marsh in a similar fashion. The most commonly used approach is simply to subdivide the marsh into the low marsh and the high marsh. The low marsh is that part of the marsh from the beginning of vegetation up to at least mean high tide. This is generally dominated by *Spartina alterniflora*. The high marsh extends from about the mean high tide up to the limit of tidal activity. This portion of the marsh is dominated by *Juncus roemerianis* and/or *Salicornia* depending on the overall setting.

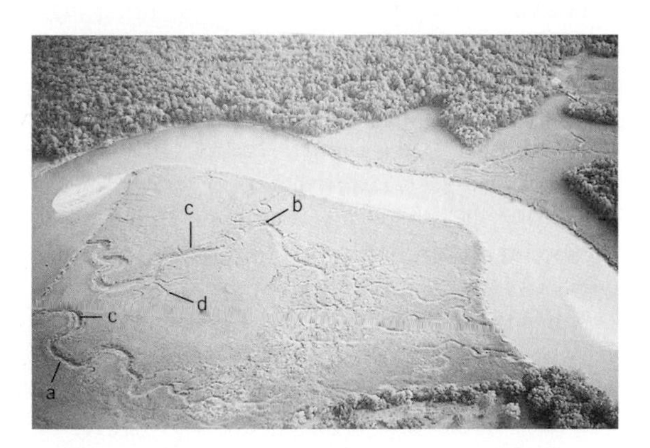

Fig. 14.8 Overview of a marsh showing the similarities with a river system. Included are (a) meandering channel, (b) point bar, (c) natural levee, and (d) crevasse splay.

There are, however, differences in marsh zonation and profiles depending upon the geographic location. For example, in New England (Fig. 14.9) the lower marsh includes *Spartina alterniflora* with the upper marsh being composed of *Salicornia*, *Distichlis*, and a fringe of *Juncus*. An upland scrub forest typically borders the marsh itself. Further to the south in Georgia and Florida, the typical zonation is a relatively narrow lower marsh of *S. alterniflora* and an extensive high marsh dominated by *Juncus roemerianus* (Fig. 14.10).

14.2.5 Marsh classification

A convenient way to consider marsh development is through its maturity. This can most easily be done by considering the relative distribution of the low

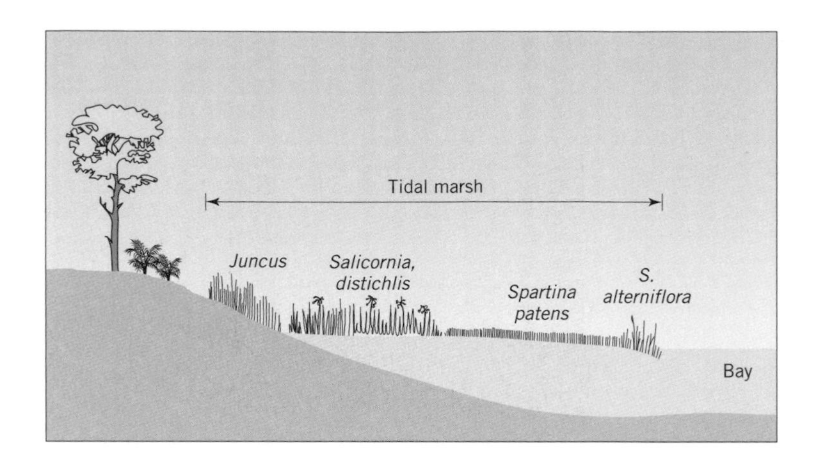

Fig. 14.9 Diagrammatic representation of the zonation on a New England marsh. The combination of climate and slope of the upper intertidal zone results in four primary species across the marsh environment. (From C. J. Dawes, 1998, *Marine Botany*. New York: John Wiley and Sons, p. 257, Fig. 9.11.)

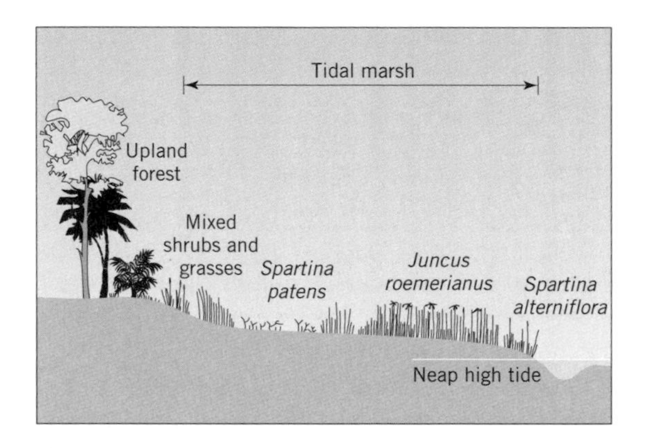

Fig. 14.10 Diagram showing typical zonation on a Georgia or Florida marsh. In a more mild climate there are different and fewer species across the salt marsh. (From Dawes, 1998, p. 258, Fig. 9.12; details in Fig. 14.9.)

and high marsh portions of the total vegetated environment (Fig. 14.11). Without using absolute ages, we can consider young, intermediate, and old marshes to reflect the progressive development of the marsh system, assuming that sea level has not changed substantially.

A young marsh is one that has mostly low marsh vegetation – that is, *Spartina alterniflora* – with perhaps only a fringe of high marsh around the outer edge (Fig. 14.11a). Tidal channels are abundant, providing good drainage and sediment supply. This stage of marsh development lasts until sufficient sediment has been delivered to the upper intertidal area to support a significant upper marsh community.

The intermediate stage of marsh development (Fig. 14.11b) has a near-equal distribution of high and low marsh. The tidal channels are fewer in number than in the young marsh. As the sediment continues to be delivered to the upper part of the intertidal zone, the marsh becomes more mature in its development. Much of the intertidal zone is covered by marsh vegetation, with only a few large tidal creeks interrupting an otherwise continuous marsh environment. Continued sediment accumulation will cause encroachment of land plants into the marsh as the estuary is reduced in overall size.

The end product of this scheme of succession of marsh development is complete infilling of the intertidal zone up to the level of near spring high tide. The marsh is essentially all high marsh, with only a fringe of low marsh, and tidal channels are widely spaced (Fig. 14.11c). Because marshes are sediment sinks, this is their eventual fate unless sea-level changes cause either enlargement of the estuary or abandonment at a high elevation. If this occurs, upland terrestrial vegetation will likely encroach into the highest part of the marsh.

14.2.6 Marsh sedimentation

We have noted that a marsh develops above the neap high tide level of the tidal flat, and that as the estuary fills with sediment, the marsh increases in its extent. In addition, there is an increase in the amount of high marsh as the marsh matures through

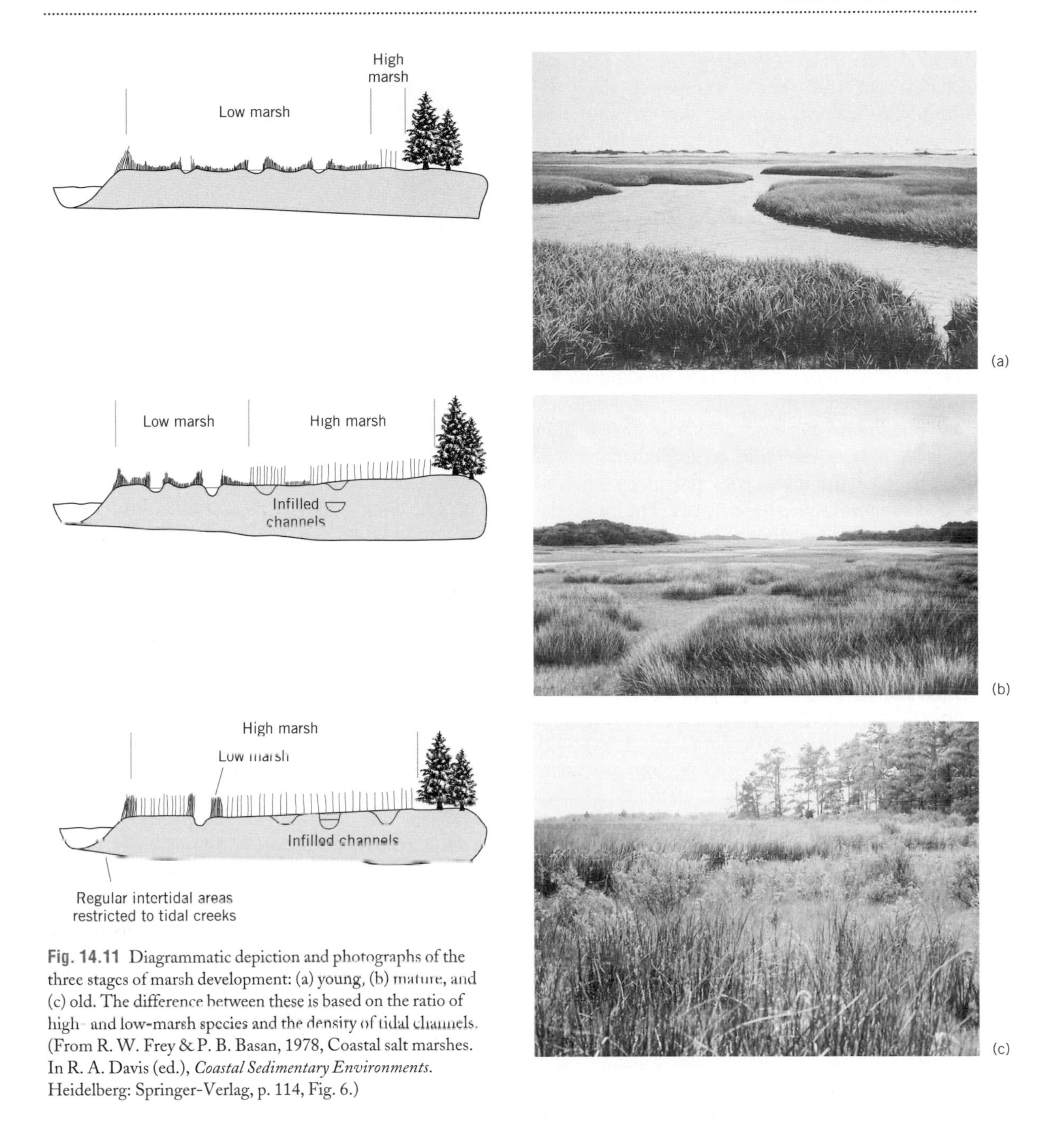

Fig. 14.11 Diagrammatic depiction and photographs of the three stages of marsh development: (a) young, (b) mature, and (c) old. The difference between these is based on the ratio of high- and low-marsh species and the density of tidal channels. (From R. W. Frey & P. B. Basan, 1978, Coastal salt marshes. In R. A. Davis (ed.), *Coastal Sedimentary Environments*. Heidelberg: Springer-Verlag, p. 114, Fig. 6.)

sediment accumulation with time. How does the marsh grow in this manner and what are the mechanisms for delivering sediment to the marsh or potential marsh environment?

There are various ways for sediment to reach the marsh environment but two are most prominent. One is the settling lag–scour lag mechanism for building up the tidal flat, discussed in Chapter 13. In this manner the aggradation and progradation of the tidal flat will result in the sediment surface increasing in elevation and thereby providing appropriate conditions for marsh vegetation to colonize the tidal flat. This building up of the tidal flat includes both sand that is transported along the substrate as bedload, and mud that is transported by suspension. Each tidal cycle, especially those between mean tide range and spring tide range, brings sediment up to the level where marsh vegetation can become established. This type of accumulation encourages marsh expansion in response to sediment accumulation.

Once marsh vegetation has been established, the primary mode of sediment delivery is via suspended sediment. This sediment is typically mud and is provided from normal high tide flooding of the marsh during near spring conditions and also during storm conditions. Most estuaries have some mud in suspension during each tidal cycle. Each of the high tide phases of the tidal cycles provides a small to modest supply of mud to the marsh. The longer the slack water period at high tide, the more sediment will settle out of suspension.

Storms provide the highest rate of sediment influx into the marsh environment. They do two primary things to help in this activity: (i) the waves and currents generated during storms cause large amounts of fine sediment to be carried in suspension; and (ii) many storms create storm surge or storm tides in the estuaries where the marshes occur. As a consequence, there is a great deal of sediment made available to the marsh environment. This sediment is delivered in two primary ways: (i) through the simple flooding of the marsh by sediment-laden water; and (ii) by breaching of the natural levees and deposition of a crevasse splay type sediment deposit. Both of these mechanisms provide considerable sediment to the marsh surface and both may produce enough sediment during a given storm to temporarily bury the marsh grass (Fig. 14.12). These storm layers may be several tens of centimeters thick. Because marsh grass is very resilient, it will not die when buried but will grow up through the storm layer in weeks to months. This type of high sedimentation rate on the marsh surface results in the eventual elevation of the marsh, rising up to above normal intertidal levels. The result is that the marsh environment disappears in favor of the upland environment.

Fig. 14.12 Overbank storm deposit (dashed line) covering marsh vegetation. Marshes must have sediment deposited on them to keep up with sea-level rise. Hurricanes and other severe storms flood the marsh with muddy water. The suspended mud is trapped by the marsh grass and some settles to the marsh surface.

Fig. 14.13 Marsh vegetation showing mud adhering to the blades of the grass and giving an indication of the high tide level. Rain will wash the mud off the grass and onto the underlying sediment surface.

Marsh vegetation tends to be quite dense and provides an excellent sediment trap in two ways. First, the grass slows the flow of tidal waters to permit settling out of fine suspended sediment particles to the floor of the marsh. Second, considerable amounts of fine sediment adhere to the marsh grasses (Fig. 14.13) as the sediment-laden water flows past. Both of these mechanisms provide for accumulation of generally muddy sediment on the marsh. Additional sediment accumulates on the marsh surface as the result of suspension feeders living within the marsh grass producing pellets that accumulate within the marsh and contribute to its aggradation. In high latitudes, ice can also be important in transporting sediment onto the marsh surface. This is very common in the New England area of the United States and along parts of the Wadden Sea on the North Sea coast of Europe.

14.2.7 Sediments

The general nature of salt marsh sediment is quite unlike that of other coastal environments, except for the upper part of the intertidal flats. It is commonly a subequal mixture of mud and plant debris, with small amounts of shell material, sand-size terrigenous particles, and large plant fragments. Overall, marshes typically contain the finest sediments of all coastal environments. This is not always true, especially for those marshes developed on washover deposits or flood-tidal deltas associated with barriers; most of these are dominated by sand-size sediments. It is also possible for the particular area to have little mineral-mud sized sediment throughout, thereby making it impossible for mud to be a major component of marshes. The Florida peninsula falls into this category, because marshes there form on sand-dominated substrates.

The coastal marsh accumulates a distinctive combination of sediment, structures, geometry, and biogenic features. Alhough there is some nearly universal similarity among marsh deposits, there may be striking contrasts. Most marshes accumulate much plant debris and typically develop peat. Numerous benthic invertebrates may live within the marsh. Foremost are infaunal organisms, such as various worms, burrowing crabs, and snails.

As a consequence of all of these burrowing organisms, together with the effects of the roots of the marsh vegetation, many marshes show considerable bioturbation in the substrate (Fig. 14.14). There are, however, many marshes that do accumulate well bedded marsh sediments (Fig. 14.15).

14.2.8 Sea level and marsh development

It should be apparent from the above discussions that the marsh environment is very delicately balanced

Fig. 14.14 Marsh sediment showing extensive bioturbation generally by crabs. The roots of marsh plants can also cause bioturbation.

Fig. 14.15 Well bedded marsh sediment suggesting that the sediment accumulates rapidly in the absence of burrowing organisms. This is generally limited to high-latitude areas.

with sea level. The entire marsh environment exists within much less than a meter of relief near high tide, except in places with extremely high tidal ranges. The high marsh environment is within only about 10–15 cm of relief. As sediment accumulates on the marsh, the elevation can reach above spring high tide. But this is without considering sea-level change; especially sea-level rise.

In Chapter 4 we discussed the current situation regarding sea-level change and noted, that globally, there is an annual rise of 1.5–2.0 mm. This is modest, but there are indications that the rate is increasing. If we consider the current rate, that means that a coastal salt marsh must accumulate 1.5–2.0 mm of

Fig. 14.16 Aerial view of typical marsh and interdistributary bay on the delta plain of the Mississippi Delta. These are probably the most extensive salt marshes in North America and they are in jeopardy because of the rapid rise in sea level in this area.

Fig. 14.17 Badly deteriorated marsh as the result of sea level rise along the Mississippi River Delta. (Courtesy of Lynn Leonard.)

sediment each year in order to maintain its current elevation relative to sea level. The desired situation is at least a balance between sea-level rise and sediment accumulation. In most coastal settings, this is not a significant problem: such a balance exists. However, if predictions of increased rates of sea-level rise come true, then we will have potential problems, with marshes being drowned by the rise in sea level. There is considerable concern about this scenario becoming a major problem for marsh stability. Because marshes are among the most productive environments of all, this situation could cause major problems for the coastal ecosystem.

Catastrophic conditions currently exist in the extensive marsh environment associated with the Mississippi River delta (Fig. 14.16) on the coast of Louisiana. We can see from the chapters on sea level (Chapter 4) and on deltas (Chapter 16) that this area is experiencing a relative sea-level rise of almost 1 cm each year. While sea-level rise is not a major problem along many coastal environments, it is a very big problem for a marsh. Remember, most of the marsh exists within a very small range in elevation. On the Mississippi delta, an area of less than a meter spring tidal range, the range is only about 10–15 cm.

As a consequence, a sea-level rise of nearly a centimeter may cause much of the marsh to be drowned (Fig. 14.17). If the rate of sediment influx amounts to a centimeter per year, then there is a balance between the rate of sea-level rise and the rate of marsh accretion. In the case of the Mississippi delta area, human interference with the discharge of the river, coupled with the withdrawal of fluids under the delta in the extraction of petroleum has contributed significantly to the high rate of relative sea-level rise. As a consequence, the delta is subsiding and the amount of sediment supply from flooding of the river has been greatly reduced. The bottom line is that much of the coast of Louisiana is drowning. The state is currently losing about 65 km^2 each year to drowning of coastal salt marshes.

14.2.9 Marsh summary

Although marshes are very diverse in their characteristics and their dominant vegetation, they have many common factors. There are some generalizations that can be made about marshes. Most of these are related to the position along the intertidal zone (Fig. 14.18). It cannot be stressed enough how important the elevation is within the marsh portion of the intertidal zone.

Marshes of all types are among the most important and most productive of all modern environments. They have high concentrations of photosynthetic organisms and they serve as a nursery ground for many animals. Because of their delicate position

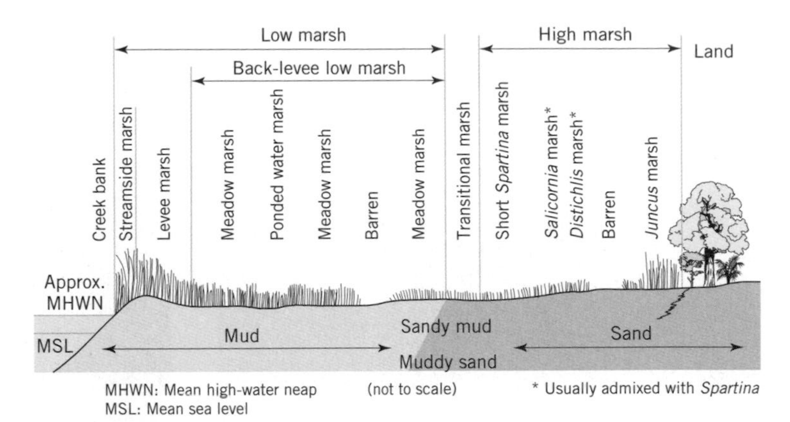

Fig. 14.18 Diagram showing the various specific environments across the coastal marsh. The primary zones are the high and the low marsh, characterized by different salt marsh plants.

within the intertidal zone, their existence is threatened both by human activity and by sea-level rise.

14.3 Mangrove coasts

Stands of mangroves, called mangals, are tidal forest ecosystems that exist in protected marine through brackish water to freshwater conditions, as long as there is some tidal influence. Although there are various environmental conditions that influence the nature and extent of mangrove development, the most critical is air temperature: mangroves cannot tolerate a hard freeze. This limits them to lower latitudes. Mangrove mangals are commonly considered as the low-latitude equivalent of coastal marshes. This comparison is not strictly correct in that there are two distinct differences between the two environments: (i) marshes are populated by grasses and mangrove mangals are dominated by trees and shrubs; and (ii) mangroves occupy different positions within the intertidal zone than do marshes. As mentioned in the previous section, there are two areas where mangroves and salt marsh vegetation occur together: parts of Florida and the Baja California coast. In Florida, this is primarily because of temperature. It is warm enough along the shoreline for mangroves to survive because of the water temperature, but even a few hundred meters inland the temperatures are too cold and salt marshes replace them as the wetland vegetation.

In this discussion we consider how mangroves are distributed, both globally and within specific coastal systems. The zonation of mangroves and their influence on coastal processes, especially sediment transport and stability, is also covered.

14.3.1 Mangrove distribution

Global distribution

More than 80 species of mangroves are recognized globally. The vast majority of these species are found in Southeast Asia and Oceania, in the Pacific and Indian oceans. This global distribution is controlled by winter temperature: hard freezes are not tolerated by these plants (Fig. 14.7). The Indo-Pacific Zone contains a tremendous variety of mangrove taxa, whereas the Atlantic Zone includes only ten species. As can be seen in the map of the mangrove regions, there is a distinct limitation to the low latitudes. In the United States, for example, only Florida, parts of the Gulf coast, and a little of southern California are home to mangroves.

Local distribution

Mangroves are restricted to protected waters where currents are sluggish and waves are small. These are typically associated with rather low-energy estuaries, lagoons, and backbarrier environments. The primary factor in this distribution is the nature of mangrove propagation. Their seeds drop from the trees and float with the currents until they come to rest at the shoreline. It is here that the propagules

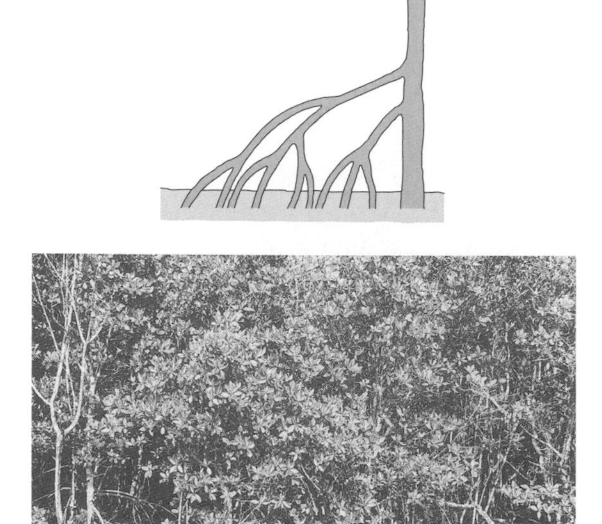

Fig. 14.19 (a) Sketch of roots of *Rhizophora mangle*, the red mangrove, and (b) photograph of the species. These prop roots provide outstanding stability for the trees.

Fig. 14.20 (a) Sketch of roots of *Avicennia germinans*, the black mangrove, and (b) photograph of the species. These root systems are pneumatophores and are thought to be primarily for respiration of the mangroves.

root and develop into seedlings. In order for this to take place the seeds must maintain a position for some time. Swift currents and wave action would prohibit this from happening.

Zonation

There is a zonation of the prominent mangrove species that is related to their position within the intertidal zone, in a fashion similar to that of the grasses within the marsh environment. The most seaward species is the red mangrove, *Rhizophora mangle* (Fig. 14.19), which commonly extends to below the low tide mark within the low part of the intertidal zone. Above this in elevation but intermixed to some extent is the black mangrove, *Avicennia germinans* (Fig. 14.20). This species is within the intertidal zone. The third typical mangrove of North American mangals is *Laguncularia racemosa*, the white mangrove, which inhabits the highest part of the intertidal zone and may extend up to the supratidal area. The zonation across the intertidal zone in Florida is basically in this same order, with the red mangrove being lowest and the white the highest, just above spring tide (Fig. 14.21). Unlike in marshes, it is common for mangrove species to be somewhat intermixed. There is no sharp boundary between species as there is in the salt marshes.

Mangrove communities have been classified into five types depending on the morphology of the coast, tidal influence, and river influence (Fig. 14.22). This classification is built around three extremes, with river-dominated, tide-dominated, and interior mangals as end members. The differences are as follows. The river-dominated types are those where there is considerable runoff of sediment, nutrients, and organic matter. The tide-dominated mangals are those where bidirectional flux of tidal flow is prominent. The interior mangals are those that are organic-rich and in which sediment sinks.

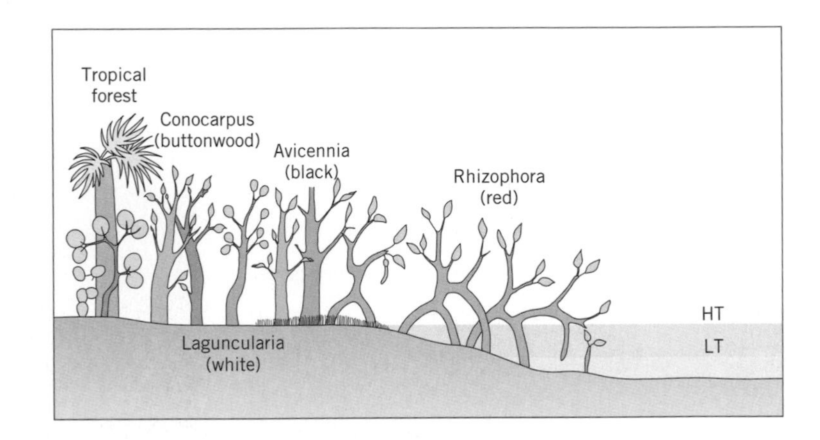

Fig. 14.21 Zonation of mangrove species across the intertidal zone as it typically occurs in south Florida. All four species of this area are shown. (From Dawes, 1998, p. 292, Figure 10.13; details in Fig. 14.10.)

Of the five types, the riverine, overwash, and fringe mangals are the most extensive. The fringing types occur along the protected open coasts of estuaries (Fig. 14.23) and other coastal bays. Overwash mangals develop on the landward side of barrier islands, where the low-lying overwash deposits accumulate. The riverine mangals line the lower parts of rivers, where they merge with the brackish waters of estuaries.

Basin mangals are generally limited in size and are located in depressions or small basins behind riverine or fringing mangals. They may be served by small tidal creeks. The hammock mangals are in inland tropical locations, where they are isolated by freshwater. The last type, scrub mangals, are located in areas where there is stress due to low water exchange, and therefore insufficient nutrient supply.

14.3.2 Mangroves and coastal processes

Mangroves have some influence on coastal processes because of their prominent size and dense network of root structures. Although most of their influence centers on physical processes, some biological processes may also be involved. Some of the mangrove species, especially those of the genus *Rhizophora*, have a significant influence on currents. The primary reason for this influence is the presence of the numerous, closely spaced, and resistant root structures that are possessed by nearly all mangrove species. In sites where open water currents may be as high as 100 cm s^{-1}, the currents within the dense mangrove root system may be as slow as 10% of those of the open water.

As the tide floods and ebbs, the prop roots and pneumatophores, along with substantial burrowing structures at the sediment surface, cause major increases in roughness and friction. This produces a significant decrease in the flow velocity of the tidal currents and thereby greatly affects sediment transport and accumulation. Additional roughness is caused by the algae, barnacles, oysters, and other organisms that may be growing on the root structures (Fig. 14.24). All of these factors have an effect on waves as well. They tend to attenuate wave energy and thereby minimize the role of waves in erosion of the mangrove substrate.

Mangroves also have a significant influence on the effect of storms along coastal environments. Because of their location in low-latitude regions, tropical storms and hurricanes are likely to impinge on mangrove coasts. These storms bring intense winds, large waves, and storm surge. Mangroves are able to withstand these forces very well. The relatively low trees with very dense root systems are adapted to resist such intense conditions. They also help to protect the sediment substrate in the mangal.

A good example of this situation is the passage of Hurricane Andrew across south Florida in August 1992 (Fig. 14.25). Many people are aware of the tremendous destruction that took place in the Miami area. Few people are aware of what happened on the other side of Florida, where mangrove mangals dominate the coastal zone. Here the mangroves

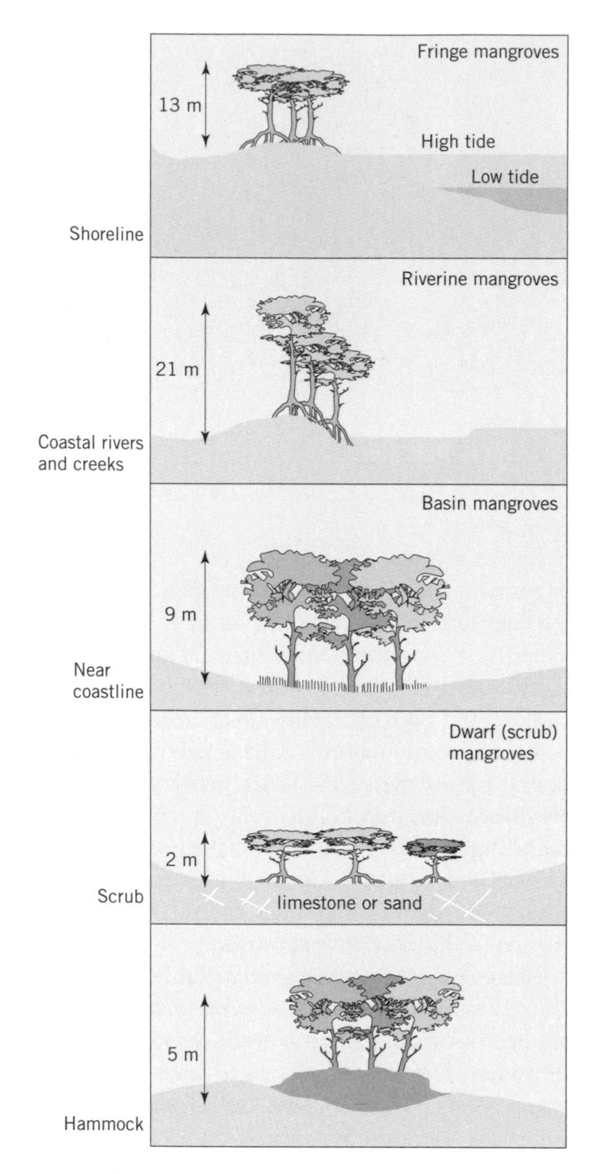

Fig. 14.22 Mangrove mangal classification that depends on the location and structure of the mangrove community.

Fig. 14.23 Photograph of typical fringing mangroves along the coast of an estuary. The mangroves will extend throughout coastal locations where there is a tidal influence.

Fig. 14.24 Oysters growing in association with prop roots of the red mangrove. The prop roots provide an excellent firm surface for these and other encrusting organisms, such as barnacles.

extend essentially to the open coast along this very low energy coast, where mean wave height is only about 15 cm. With large waves developing on a storm surge of 1.5–2.0 m and winds of about 150 km h^{-1} it would be expected that there would be major erosion on this undeveloped coast. Instead, the shoreline change was minimal: mangrove trees were broken off by the wind but the dense root systems prevented erosion. This is an excellent demonstration of how mangroves are adapted to withstand intense storms and prevent erosion of the coastline.

Another relationship between coastal processes and mangroves involves the dense network of exposed **prop roots** and **pneumatophores** that are associated with the red mangroves and the black mangroves respectively. These structures extend throughout the intertidal zone and below. Prop roots may be

Fig. 14.25 View of mangrove mangal in southwest Florida after the passage of Hurricane Andrew. Note that there has been little erosion, although the mangroves are defoliated and most have been broken off by the wind.

Fig. 14.26 Mud sticking to the prop roots of mangroves along a tidal creek on the Queensland coast of Australia. This type of sediment trapping can contribute to sediment accumulation on the mangal substrate.

meters high, generally related to tidal range. Pneumatophores are typically 20–30 cm above the sediment surface. The influence that they have on tidal currents has significant effects on sedimentation and erosion. The most obvious of these effects is the interference that the root structures present, causing a slowing of the currents. As a result, sediment that is entrained by these currents is allowed to come to rest due to the baffling effect of the roots. Additionally, the roots prevent sediment from being removed for the same reason. The net effect is that mangrove root systems are sediment traps.

Another aspect of the sediment trap effect of mangroves takes place primarily in the prop roots of the red mangroves. These structures are commonly at least a few centimeters in diameter, and they physically block suspended sediment, which adheres to the surface of these roots. This phenomenon can be important in muddy estuaries, especially those where tidal range is high (Fig. 14.26).

14.4 Summary

Dense vegetation on the intertidal zone represents one of the most important of all coastal environments. These diverse environments provide a tremendous level of productivity in the form of photosynthesis and as a food supply to many types of herbivores. These highly productive environments also provide a home and a place for reproduction for many organisms.

Another major impact of these environments is in the form of coastal protection. Both marshes and mangals are helpful in stabilizing sediment substrates and slowing erosion by waves and currents. This is especially the case for the mangrove mangals, which can withstand direct attack from hurricanes and experience limited erosion. These vegetated environments are also important sediment traps and substrate stabilizers.

Suggested reading

Chapman, V. J. (1976) *Coastal Vegetation*. Oxford: Pergamon Press.

Chapman, V. J. (ed.) (1977) *Wet Coastal Ecosystems*. Amsterdam: Elsevier.

Dawes, C. J. (1998) *Marine Botany*, 2nd edn. New York: John Wiley & Sons.

Frey, R. W. & Basan, P. B. (1985) Coastal salt marshes. In R. A. Davis (ed.), *Coastal Sedimentary Environments*. New York: Springer-Verlag.

Perillo, G. M. E. (ed.) (1995) *Geomorphology and Sedimentology of Estuaries*. Amsterdam: Elsevier (Chapters 10 and 11).

15 Estuaries

15.1 Introduction

Most coasts have embayments of various sizes, shapes, and origins. The differences displayed by the morphologies of these embayments is the result of a wide range of origins, many of which are diectly or indirectly related to plate tectonics. The embayments, or bays as they are most commonly called, that have been formed as a direct consequence of tectonic activity are generally located on leading edge margins like the west coast of the United States. Here faults and movement along these faults may produce bays that are typically long and narrow, such as Drakes Estero in California (Fig. 15.1), where the San Andreas Fault system provides the geologic setting for a coastal bay. Other bays may form as the result of rising sea level drowning drainage networks, producing branching systems such as Chesapeake Bay and Pamlico Sound on the central Atlantic coast of the USA. This common type of coastal bay is typical of trailing edge coasts with broad coastal plains and well developed river systems. Other varieties (Fig. 15.2) include: (i) fjords, which are elongate embayments excavated by glaciers; (ii) bays formed by barriers such as coral reefs and barrier islands; and (iii) embayments constructed by human activity, more commonly called harbors. This brief list provides some idea about the origins of coastal bays and the shapes that are related to those origins.

Fig. 15.1 Drakes Estero, California, an example of a fault-generated estuary in the San Andreas complex north of the San Francisco area. This rectangular estuary has formed over one of the faultlines.

The definition of an estuary given in Chapter 10 is that there must be fresh water input and significant tidal flux. This makes one wonder about the

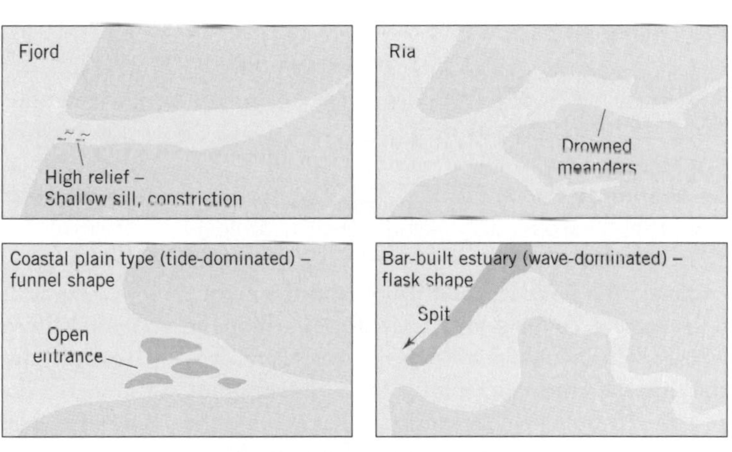

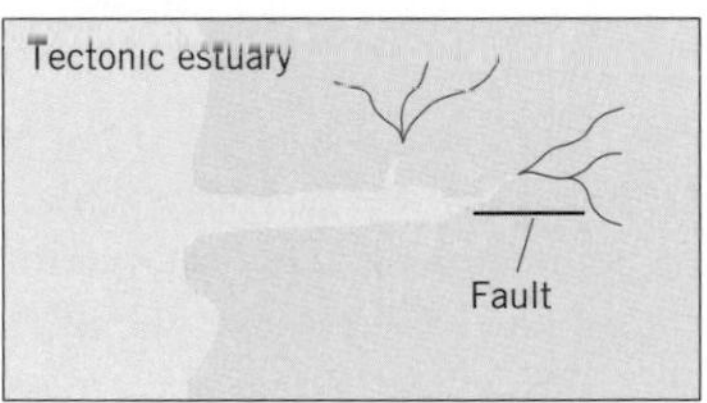

Fig. 15.2 Schematic diagrams of several types of estuaries showing their general shapes. Some are the result of rising sea level, some are glacial in origin, some are related to plate tectonics, and others are either wave- or tide-dominated.

difference between these and river deltas, which also have important freshwater input and which experience tidal influence. The simple difference is that an estuary is a coastal embayment but a delta protrudes into the ocean or other adjacent water body. There are several factors common to these two coastal environments: (i) both typically have tidal flats and marshes or mangrove swamps; (ii) both are influenced by rivers, waves, and tides; (iii) both are important sites of sediment accumulation; and (iv) both are geologically young features.

Estuaries are sediment sinks; that is, places where sediment tends to accumulate and stay for long periods of time. It is this characteristic that limits the geologic lifetime of an estuary. The runoff from rivers as well as the tidal flux transports sediment from both the landward and seaward directions into the estuary. The embayment provides a local basin for sediments to come to rest. Estuaries tend, therefore, to be filled in from the margins toward the middle in a manner that is commonly referred to as progradation. If sea level rises during this infilling process then the space available for sediment continues to increase. On the other hand, if sea level falls, the estuary is drained and the river(s) flows across its prior location leaving the "basin" essentially "high and dry."

15.2 Estuarine hydrology

In addition to the size, shape, and origin of these bays, their character also includes their hydrology. This is comprised of the characteristics of the water coming from both the land runoff and the marine environment, coupled with its circulation within the estuary. The hydrology controls the water chemistry, the biota, and the sediment that forms the substrate. It is the hydrologic characteristics that provide the best criteria for classifying coastal bays into broad categories.

The fact that both rivers and tides flow into the estuary means that both freshwater and seawater are being mixed in this environment. This interaction of different water types gives the estuary one of its most important characteristics along the coast: brackish salinities. Runoff from a river is generally continuous but the discharge may vary greatly depending upon the season, the overall climate, and other factors, similar to the situation for deltas. In contrast, the tidal influence from the open marine environment to the estuary is typically regular and predictable. The range in tidal fluctuation changes with spring and neap conditions but the periodicity remains fairly constant.

The salinity of seawater is about 35 p.p.t. or 3.5%, whereas that of freshwater is essentially zero. This contrast in salinity produces a significant difference in the density of the two water types. Remember how much easier it is to float in saltwater as compared to a freshwater pool. The difference in density is small in absolute terms (1.000 g cm^{-1} for freshwater and 1.026 g cm^{-1} for normal marine saltwater), but it is very important. In the absence of waves or strong currents, these different water types are layered, with the lighter freshwater "floating" on the heavier saltwater. This phenomenon demonstrates the potential complications that the merging of these water types present to the estuary in terms of circulation into and out of the embayment.

15.3 Classification of estuaries

A common classification of estuaries is based on the way that freshwater and saltwater interact. In the 1950s, Donald Pritchard, a scientist at the Chesapeake Bay Institute, recognized three types of circulation conditions in estuaries: (i) stratified; (ii) partially mixed; and (iii) mixed (Fig. 15.3). In a stratified estuary there is essentially complete separation between the fresh and saltwater masses due to a lack of mixing caused by waves or strong currents. Estuaries dominated by rivers may display stratified water masses, such as the Hudson River estuary in New York, where the saltwater wedge extends tens of kilometers up the river. In some estuaries some of the saltwater mixes with the freshwater to produce a transition zone of intermediate salinity between the fresh and saltwater. Places where tidal currents influence part, but not all, of the estuary, such as Chesapeake Bay, display this characteristic. Totally

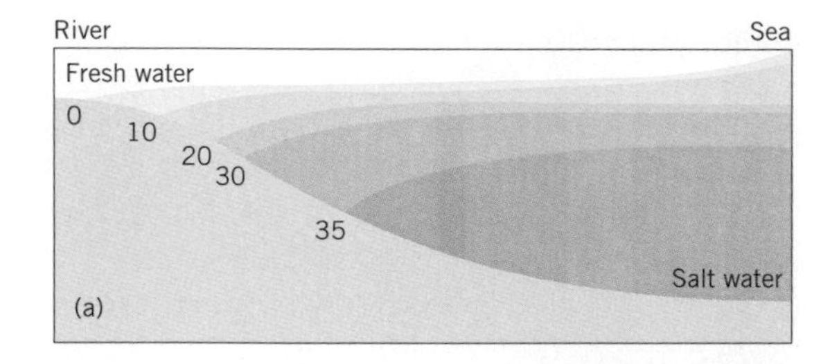

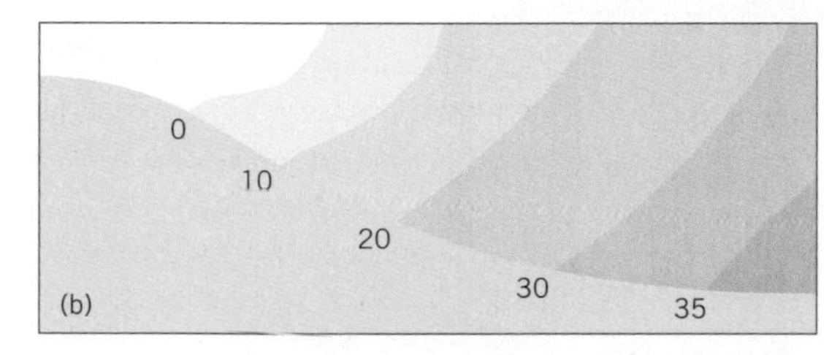

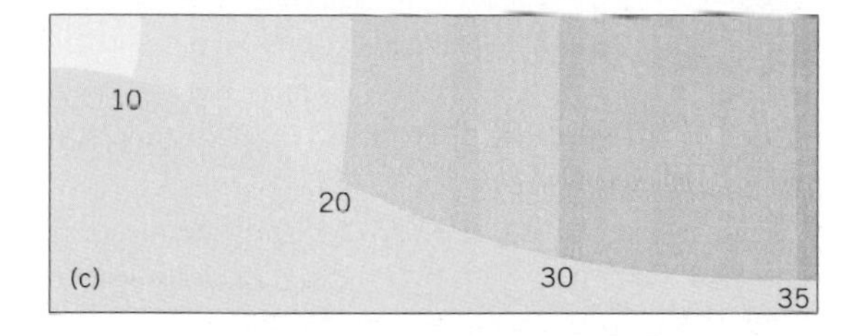

Fig. 15.3 Pritchard's estuary classification, which includes (a) stratified, (b) partially mixed, and (c) totally mixed categories. There is actually a continuum of salinity, waves, river influence, and tides that forms these different types. (Modified from D. W. Pritchard, 1955, Estuarine circulation patterns. *American Society of Civil Engineers, Proceedings*, **81**, 1 1.)

mixed estuaries produce a vertically homogenized water column with a gradient of increasing salinity toward the ocean. This could be the result of waves in a shallow estuary, such as Pamlico Sound, North Carolina, or Mobile Bay, Alabama, or it could be due to strong tidal currents, such as in the Bay of Fundy, Canada, or Delaware Bay on the Atlantic coast of the United States. Large and complicated estuaries such as Chesapeake Bay or San Francisco Bay can experience different conditions in different locations.

Many estuaries move from one hydrologic type to another depending upon seasonal variations in runoff, changes in wave climate, topographic variations of the estuary floor, or other phenomena that lead to variations in the amount of mixing. For example, a large but shallow estuary such as Mobile Bay or Pamlico Sound is susceptible to waves mixing the water column, thereby destroying any layering of water mass types. Waves tend to be absent or small during the summer, conditions that foster stratification. Near the other end of the spectrum, tidal currents in the Bay of Fundy are always strong enough to completely mix the water in this estuary.

15.4 Estuarine processes

Estuaries tend to be influenced primarily by river or tidal processes, with wave influence being dependent upon the size and depth of the estuarine basin. Freshwater and sediment are provided to the estuary by river discharge. The amount of both and the rate at which they are delivered is important to the character and longevity of the estuary. Some estuaries are supplied by a single river and therefore the sediment supply is essentially at one point. This situation tends to form a bayhead delta where much of the sediment delivered by the river accumulates (Fig. 15.4). Some of the large estuaries on the Texas coast have this characteristic: San Antonio Bay is fed by the Guadalupe River, Corpus Christi Bay by the Nueces River, and Galveston Bay by the Trinity

Fig. 15.4 Guadalupe River entering San Antonio Bay on the Gulf coast of Texas and forming a bayhead delta. As the river enters the estuary it drops its sediment load to form a delta much like deltas that form on the open coast.

River. Another good example is Mobile Bay in Alabama, where the Tensaw River forms such a delta. These estuaries tend to be river-dominated because of the strong influence of the stream processes and the absence of strong tidal currents and/or large waves. The tidal influence in these Gulf coast estuaries is diminished by the presence of the barrier islands across the mouth of the bays. It is also limited by the small tidal range around the Gulf: less than 1 m spring tide for all of them.

Some estuaries have multiple rivers emptying into them, with little or no development of a bayhead delta. Probably the best example is Chesapeake Bay (Fig. 15.5), which receives input from numerous large rivers but which has no significant bayhead deltas. Here the digitate nature of the many river valleys leading to the estuary traps most of the relatively coarse sediment before it reaches the open portion of the estuary. Many of the small west coast estuaries have similar conditions but they generally have only a single stream feeding them. Although this type of estuary can develop any of the three hydrologic styles mentioned above, nearly all are stratified or partially mixed. In addition to the presence of the bayhead deltas, terrigenous sediment accumulation in these river-dominated estuaries tends to be dominated by mud.

Fig. 15.5 Chesapeake Bay, an example of an estuary formed by the drowning of a complicated fluvial system producing a complex coastline.

At the other end of the spectrum are tide-dominated estuaries. This type is typically funnel-shaped and has no barrier or other constriction at its mouth. Such a configuration not only eliminates the dampening effect that barriers have on tidal flux but commonly amplifies the progressing tidal wave during flooding, producing high tidal ranges. Both conditions result in maximization of the influence of tidal flux, and they create fully mixed hydrologic conditions in the estuary. The combination of high tidal range and strong tidal currents generally results in a sand-dominated estuary floor because the mud tends to be carried out to sea in suspension or to be trapped at the low-energy, landward limits of the estuary. Good examples of tide-dominated estuaries are the Bay of Fundy and the Gironde Estuary on the western coast of France (Fig. 15.6).

The strong tidal currents in these tide-dominated estuaries move much sediment into the estuary, and they also move sediment that is already in the estuary back and forth during each flood and ebb of the tide. Sand can be moved by currents of as low as 20–30 cm s^{-1} depending on the size of the sand particles. These conditions are typically achieved or exceeded for several hours during each flood and ebb tidal cycle. As a result, the tidal currents produce numerous bedforms on the floor of the estuary; basically the same ones that we see exposed on tidal flats (Fig. 13.10). These bedforms are developed

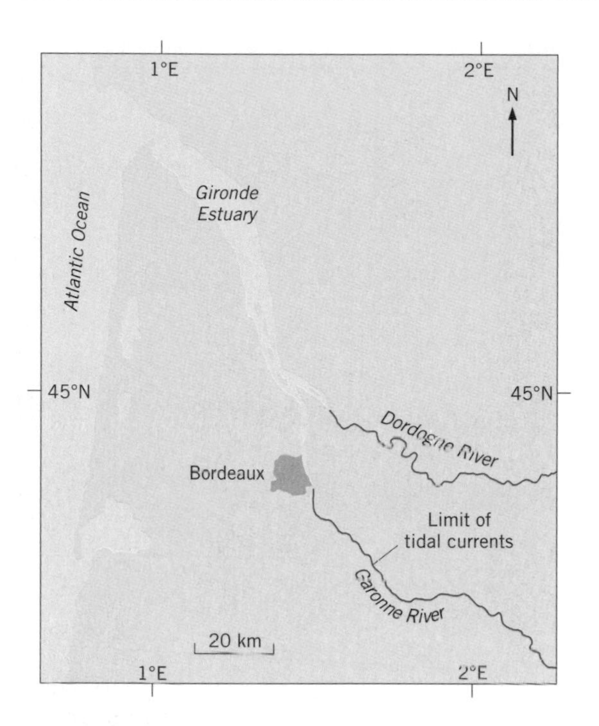

Fig. 15.6 The Gironde Estuary in France displays a funnel shape common among tide-dominated estuaries. Tidal influence extends beyond Bordeaux, more than 150 km from the ocean.

because of the shear between the bed (sediment) and the water column, thereby causing turbulence and producing these regular irregularities on the substrate. The bedforms range in size from ripples to sand waves. The size and shape of these features is controlled by both grain size and the velocity of the tidal currents.

15.5 Time–velocity relationships

The graphic record of the rise and fall of the tides shows the change in water level over each tidal cycle. This curve is essentially symmetrical. If we plot the velocity of tidal currents that are produced by this rise and fall of the tides we find that the curve has a very different shape. There is considerable asymmetry to the velocity data and the duration of flood and ebb may be different. This graph is called a time–velocity curve (Fig. 15.7) because it is a record of the

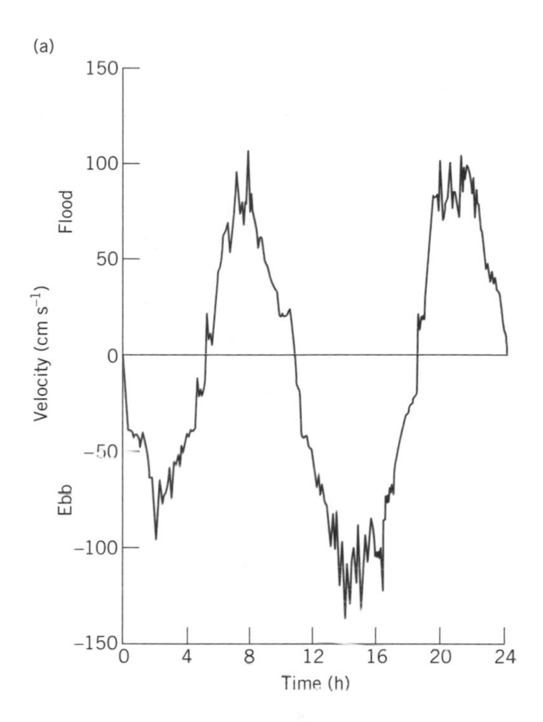

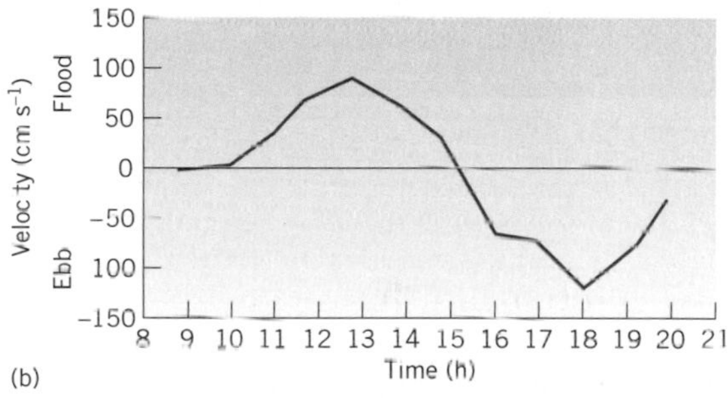

Fig. 15.7 Plots of two examples of time–velocity curves for different estuaries. These curves will have different shapes from one location to another, almost like fingerprints. (From H. Postma, 1961, Transport and accumulation of suspended matter in the Dutch Wadden Sea. *Netherlands Journal of Sea Research*, **1**, 170.)

velocity of the tidal current over time. The lack of symmetry is called time–velocity asymmetry.

Each location within an estuary displays its own characteristic time–velocity curve showing its own asymmetry. Changes in the asymmetry will occur within the lunar cycle from neap to spring conditions. If the estuary floor or tidal channels are changed, both of which strongly influence the flow

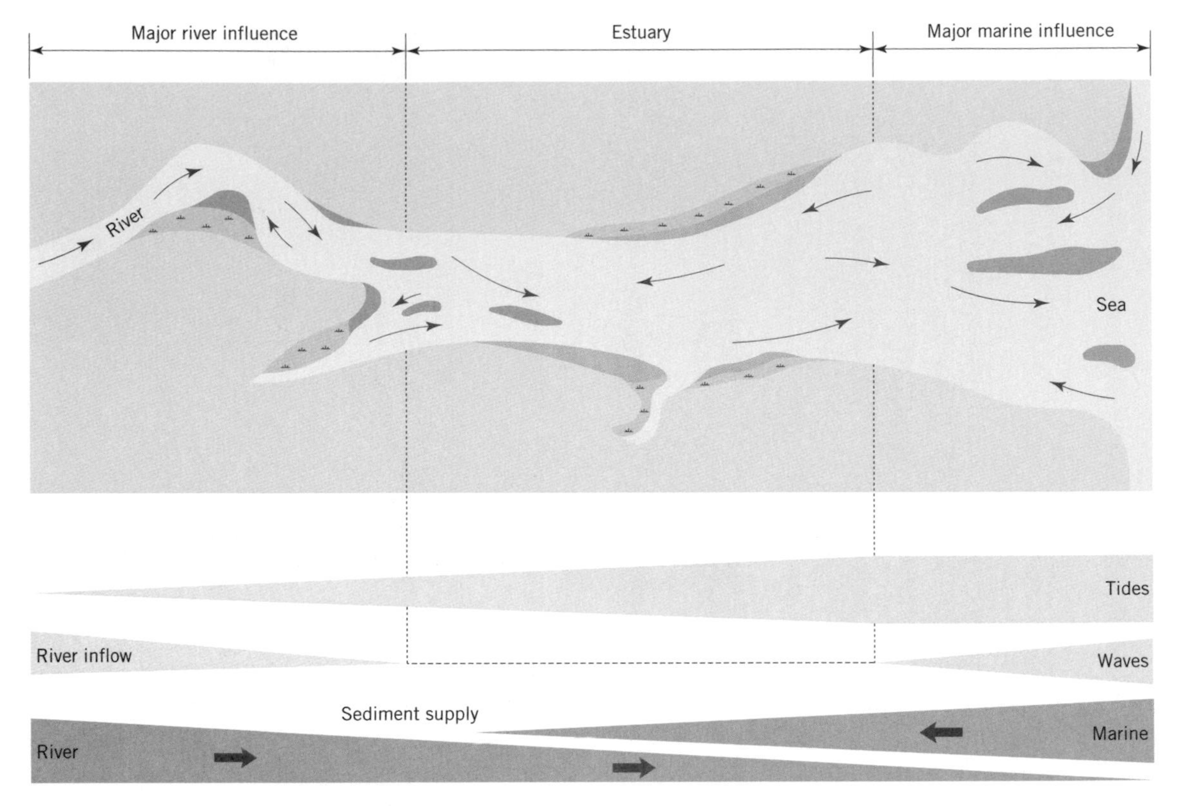

Fig. 15.8 General diagram of an estuary including the major sections and sediment varieties. This shows in simplistic fashion how the rivers and the ocean interact to form complex estuarine systems. (After M. M. Nichols & R. B. Biggs, 1985, Estuaries. In R. A. Davis (ed.), *Coastal Sedimentary Environments*. New York: Springer, p. 158).

of tidal current, then time–velocity curves may show steep or more gradual slopes or they may show distinct differences in the duration of the flood and ebb portion of the tidal cycle. This time difference exceeds an hour in many cases. These conditions may produce either flood-dominated locations or ebb-dominated locations and these different conditions may be adjacent to one another. It is common, for example, for a channel to be ebb-dominated but for the adjacent tidal flats to be flood-dominated.

15.6 Model estuary

A good way to conceptualize an estuary is through the use of a simple model. The estuary can be subdivided into three main parts: (i) the landward area of river influence; (ii) the middle, truly estuarine area; and (iii) the seaward area of marine influence (Fig. 15.8). The relative proportions of each vary with individual estuaries and with the influences of the major processes. Tides commonly diminish in their influence landward from near the mouth, but there are exceptions where the shape enhances tidal range, the Bay of Fundy being an example. Wave influence tends to be directly proportional to the size of the estuary or to the fetch of a particular portion of it. Riverine influence is likewise proportional to the amount and rate of river input relative to tidal flux.

There is nearly always significant overlap in the sediment supply from the river and from the marine sources, and the nature of the contribution is commonly different between these sources. River sediments are generally sand and mud, whereas marine

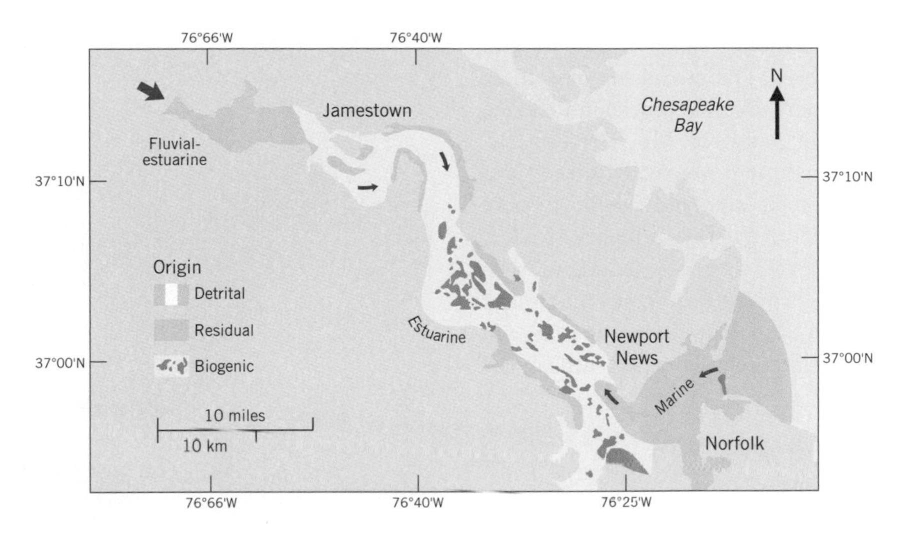

Fig. 15.9 Map showing major environments within the James River estuary, a branch of Chesapeake Bay. Compare this map with the diagram in Fig. 15.8. (From M. M. Nichols, 1972, Effect of increasing depth on salinity in the James River Estuary. In B. W. Nelson (ed.), *Environmental Framework of Estuaries*. Geological Society of America, Memoir 133, p. 209.)

sediments tend to be dominated by sand with some shell gravel; mud is rare (Fig. 15.8). The estuarine transport of sediments includes both bed load and suspended load. The latter is particularly important in low-energy estuaries, where it forms the bulk of the sediment that accumulates; less is carried out into the open marine environment with ebbing tides.

The zone of freshwater and saltwater mixing has a strong influence on suspended sediments because it is a place where water density changes significantly. Flocculation of the fine clay mineral particles (<2 μm in diameter) takes place here and floc size may reach up to 30 μm. This is also the zone of the turbidity maximum in both partially and fully mixed estuaries. Here suspended sediment concentrations are highest. This phenomenon is controlled by the mixing of freshwater with the leading edge of the saltwater. Some of the particles suspended in the overlying freshwater mass settle as currents diminish and are then entrained by the lower, more dense saltwater and carried landward to the turbidity maximum where suspended sediment is greatest. This process produces the high sediment accumulation rate associated with the turbidity maximum.

There is a third source of sediment in estuaries: the biogenic material that is produced or modified in the estuary itself. This sediment tends to be most abundant and accumulates most rapidly in the middle zone (Fig. 15.9). There are numerous organisms that thrive on the brackish salinities that characterize the central portion of most estuaries. The typical salinity range here is from about 5 to 20 p.p.t. Ostracods, foraminifera, various mollusks, and worms are the most common animals. In addition, various types of algae are present and some subtidal grasses are found on the fringe where water clarity is sufficient. Water clarity strongly inhibits photosynthesis in most estuaries due to the abundance of fine sediment and the waves and currents that can cause it to become suspended. Skeletal carbonate material from many of these organisms makes an important contribution to the sediment of the estuary. Oyster reefs are particularly abundant and widespread in many of the low-energy, muddy estuaries. In some areas mussels are also common.

The other aspect of the contribution made by bottom-dwelling organisms is the pelletization of suspended sediments by filter feeders and to a lesser extent by grazers. The biggest contributors to pellitized muds are oysters, mussels, and worms, all of which filter their nourishment from suspended particles provided by currents. Much of the suspended material is made up of fine mineral particles and other inorganic debris. The organisms pass these particles through their digestive tracts and excrete pelleted mud in large quantities. Much of the accumulated estuarine sediment is actually in the form of these pellets, which are sand-sized, cohesive aggregates of mud-sized particles. The filter feeders greatly increase the rate of benthic sediment accumulation by taking suspended sediment that might otherwise be carried to sea out of the water column and converting it to larger particles that settle to the floor of the estuary.

15.7 Estuary types

We can place most estuaries into one of two general morphodynamic types: wave-dominated and tide-dominated. If we consider the general estuarine model shown in Fig. 15.8 as a general morphologic model, it is possible to place it into each of these two types. The following discussion considers the differences and similarities, and gives various examples of each type.

15.7.1 Tide-dominated estuaries

The general configuration of the tide-dominated estuary is a funnel shape (Fig. 15.10), with tidal processes having a strong influence over most of the

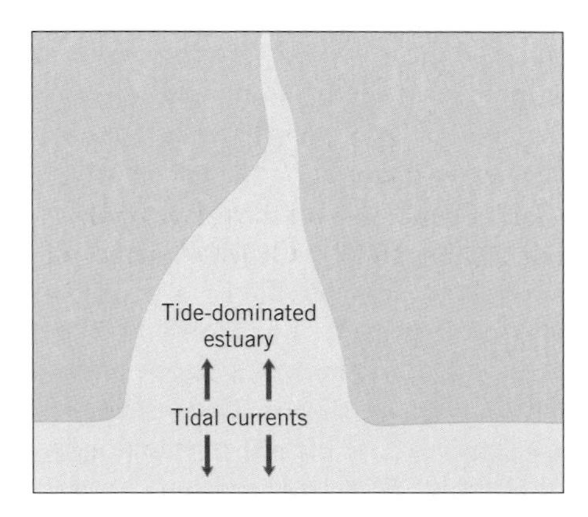

Fig. 15.10 Schematic diagram of a tide-dominated estuary. These funnel-shaped coastal bays are typically along coasts with little wave energy and high tidal ranges.

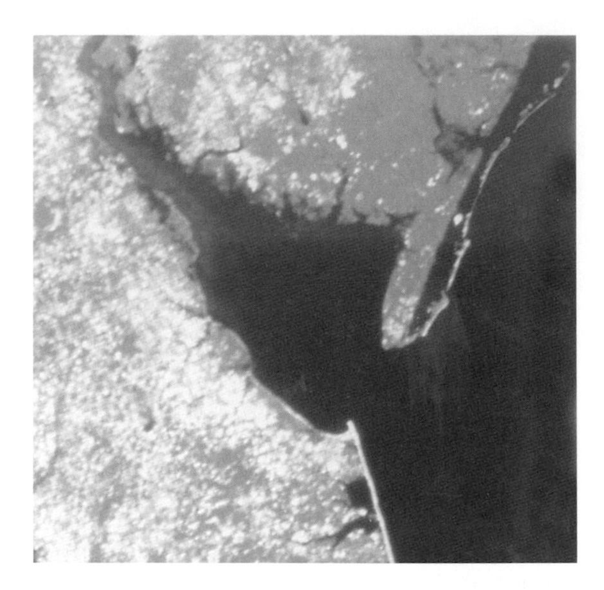

Fig. 15.11 Delaware Bay on the east coast of the United States, a good example of a tide-dominated estuary, although it does have a small barrier spit across part of its mouth.

estuary. This type of estuary develops along coasts that have some combination of high tidal range and large tidal prism with an absence of an energetic wave climate. As a consequence, we have an estuary that tends to be fully mixed throughout due to the influence of tidal currents. In addition, the sediments that accumulate in a tide-dominated estuary tend to be sand; most of the mud is swept away by the strong currents. This sandy estuary floor is characterized by linear sediment bodies that are aligned along the estuary by the flow of the tidal currents (Fig. 15.10). The mobility of the substrate caused by these currents tends to inhibit the colonization by benthic organisms.

The Gironde estuary along the west coast of France is a classic example of tide-dominated estuaries. Other examples include the Minas and Chignecto basins of the Bay of Fundy, the Wash on the Norfolk coast of England, and, to a lesser extent, Delaware Bay on the Atlantic coast of the United States (Fig. 15.11).

15.7.2 Wave-dominated estuaries

If we take the same funnel-shaped estuary as discussed above and greatly reduce the tidal flux,

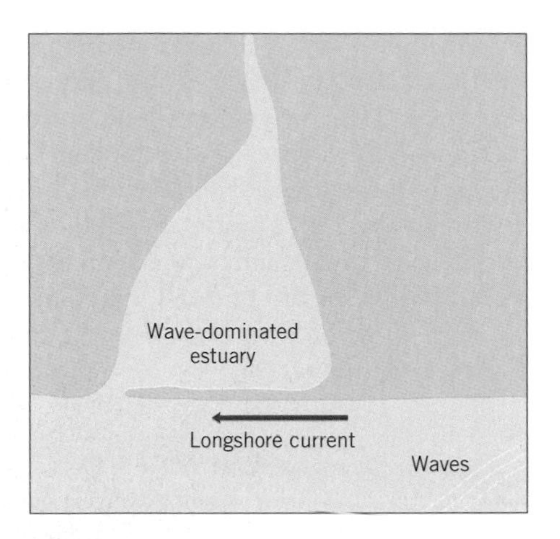

Fig. 15.12 Schematic diagram of a wave-dominated estuary, which is characterized by a barrier that limits tidal exchange between the open ocean and the estuary. These estuaries are characteristic of coasts with low tidal ranges and high wave energy.

while at the same time increasing the incident wave energy along the coast, the result is some type of barrier across the mouth of the estuary (Fig. 15.12). This is the general morphology of the wave-dominated estuary. In some the barrier is detached from the mainland, such as along the Texas coast, and in others it is attached to the mainland, generally as some type of headland. This is typical of the estuaries along the west coast of the United States, especially in the states of Oregon and Washington.

Such estuaries develop along coasts where waves and wave-generated currents dominate over tidal processes. The barriers between the estuary and the open marine environment inhibit tidal flux but still permit enough marine influx to produce brackish conditions when combined with freshwater runoff via rivers. Sediments in wave-dominated estuaries tend to be dominated by mud or muddy sand. The absence of strong currents permits the fine sediment to settle to the bottom and remain there. In addition, filter feeders are common on a relatively stable substrate, further contributing to the muddy sediments through the production of pellets.

These estuaries can have any of the three main hydrologic characteristics discussed above (Fig. 15.3), but are likely to be in the layered or partially mixed types more than in the fully mixed category. They will, however, become fully mixed in the event

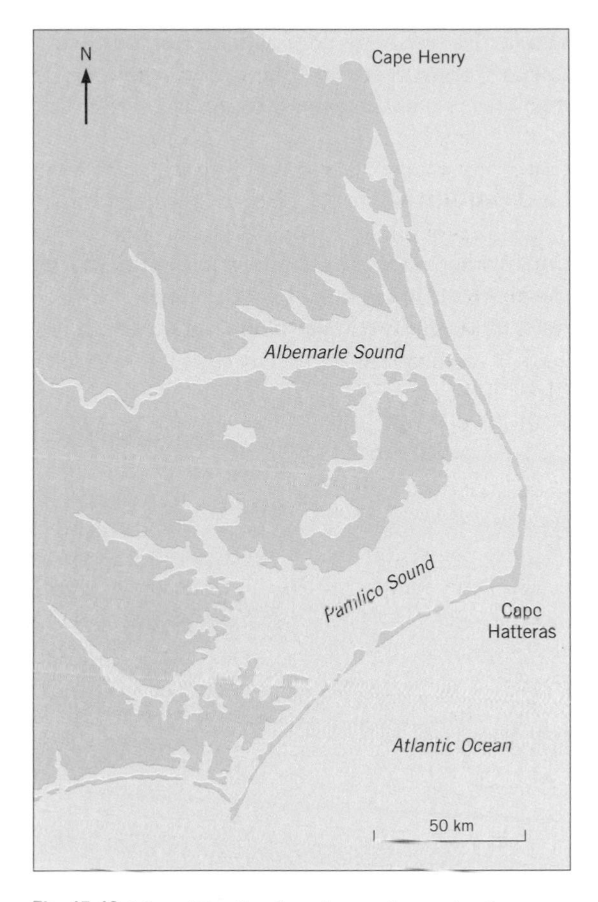

Fig. 15.13 Map of Pamlico Sound, a good example of a wave-dominated estuary on the east coast of the United States. After the rising sea level drowned river systems to form a very complex shoreline, waves reworked and redistributed sediment across the estuary to form the Outer Banks of North Carolina.

that waves are generated over extensive shallow estuarine water bodies. An example of this condition is Pamlico Sound, behind the Outer Banks of North Carolina (Fig. 15.13). Other examples of wave-dominated estuaries are found around the entire Gulf coast.

Some estuaries exhibit characteristics of both wave- and tide-dominated types. Typically these are estuaries that have a complex shoreline and an absence of a barrier across the mouth of the bay. Probably the best example of this type is Chesapeake Bay (Fig. 15.5) along the east coast of the United States. Most of this bay has low-energy conditions, with muddy sediments dominating. Near the mouth, however, there is strong tidal influence due to the large tidal prism that flows through it.

15.8 Summary

Estuaries are among the most complex of all coastal environments because of the wide range of hydrologic, biotic, and sedimentologic conditions. They can be formed by a variety of different factors, but most are the result of sea-level rise over the past few thousand years. The rivers emptying into estuaries contribute both freshwater and sediments. The tidal flux from the marine environment contributes sediments and saltwater. The estuary represents the mixing of these elements into a transition between marine and freshwater conditions. Organisms tend to be controlled by the nature of the water and by the substrate conditions.

Suggested reading

Dyer, K. R. (1973) *Estuaries: A Physical Introduction*. New York: John Wiley & Sons.

Dyer, K. R. (1979) *Estuarine Hydrography and Sedimentation*. Cambridge: Cambridge University Press.

Kennedy, V. S. (ed.) (1980) *Estuarine Perspectives*. New York: Academic Press.

Nelson, B. W. (ed.) (1972) *Environmental Framework of Coastal Plain Estuaries*. Boulder, CO: Geological Society of America, Memoir No. 133.

Perillo, G. M. E. (ed.) (1995) *Geomorphology and Sedimentology of Estuaries*. Developments in Sedimentology No. 53. Amsterdam: Elsevier.

16 River deltas: the source of most of our coastal sediments

16.1 Introduction

The rock cycle shows us that most of the sediment that is eroded from land is carried by streams and rivers to a water body that receives the discharge of these streams and rivers – both the water and the sediment that it carries. This is usually one of the oceans or another large water body associated with an ocean, such as the Gulf of Mexico or the Mediterranean Sea. In some situations large lakes may be the final destination of streams and rivers. The distance over which this sediment is carried by the river may be only a few kilometers, such as on much of the west coast of the United States, or it might extend thousands of kilometers, such as in the Amazon and Mississippi river systems. The reasons for these major differences in river length are related to the concept of global tectonics. The drainage system may be developed on a leading edge of a plate, such as the west coast of both North and South America, or on a trailing edge, such as the coastal plains of the United States and the stable crustal shield of Brazil.

Once the sediment is discharged at the mouth of the river along the coast, it might be carried out into deep water if it is fine-grained and suspended, such as silt and clay, it might be sand that is transported along the coast to be included in various coastal features, such as the nearshore bars, beaches, or dunes, or it might come to rest at or near the mouth of the river in the form of a large and complex sediment accumulation called a river or fluvial delta (Fig. 16.1). In this chapter we discuss these river deltas, how they develop, their characteristics, and what causes them to change. A delta is a coastal environment that is greatly influenced by human activities, which are also considered in this discussion.

Although we do not know how he recognized the shape of the Nile River delta, Herodotus is commonly given credit for coining the term delta in the fifth century BC in connection with this famous delta on the Mediterranean coast of Egypt. Given the date in history, it is a real puzzlement as to how he determined that the sediment accumulation at the mouth of the Nile was in the shape of the Greek capital letter delta. River deltas have historically been the site of human settlement because of their proximity to the sea and their abundant food supply in the form of waterfowl, finfish, and shellfish. In the past century they have become a major source of petroleum. The latter is a major economic aspect of many deltas and has led to extensive research on a wide range of environmental topics in addition to the geology of deltas.

Fig. 16.1 Satellite photo of the Nile delta, showing the generally triangular shape. This is the first river delta to carry the name. It is presently experiencing extensive, severe erosion. (Courtesy of EROS Data Center.)

Our scientific knowledge of deltas as a geologic sedimentary environment has been virtually all acquired during the past century. G. K. Gilbert (1885), a famous geologist with the US Geological Survey, investigated much of northern Utah, including what is known as Lake Bonneville, a large lake of the Pleistocene Epoch from which the Great Salt Lake was evolved. He recognized thick deltaic accumulations from rivers that emptied into this large lake.

Modern river deltas did not attract the attention of geologists until after the beginning of the twentieth century. Joseph Barrel of Yale University is generally given credit for writing the first research paper on the Mississippi delta in 1914. Little was done on the research of river deltas until the extensive work of H. N. Fisk on the Mississippi delta, which began in the 1940s. His research, combined with the interest in deltas generated by the production of oil and gas, initiated an explosion of activity beginning in the 1950s. This began on the Mississippi and then expanded to the Niger delta in Africa and the Orinoco delta in Venezuela; all are important oil producing deltas. Because of the extensive research and interest in the Mississippi delta, it became the primary model for interpretations of deltas throughout the world. In fact, however, the Mississippi delta is an extreme case, essentially one of a kind, and is a poor example with which to compare other river deltas.

16.2 How deltas develop

The presence of a river delta along any coast is an indication that the river is providing more sediment than can be removed and redistributed by coastal processes. This accumulation of sediments in riverine deltas may be quite temporary or it may be permanent. Some small deltas may be seasonal, appearing only in the spring when water and sediment discharge is at its highest. As the year proceeds, coastal processes remove that sediment and the delta is gone by the next spring, when it is formed again. There is considerable interaction of the riverine processes of sedimentation with the open coastal marine processes – especially waves, longshore currents, and tidal currents. The interaction of these processes, along with the sediment load of the river and the physical setting at and near the river mouth, determines the presence and the nature of the delta. The most important requirement for the formation of a delta is the discharge of sufficient sediment to produce a net accumulation above the amount removed and redistributed by waves and currents. The amount required is quite different from one coastal location to another. River mouths where the wave climate is characterized by large waves and/or where strong tidal currents persist require considerably more sediment to produce a delta than those locations where waves and tidal flux are small. An equally important factor is the geologic and bathymetric setting on the continental margin adjacent to the coast: the sediments discharged at the coast must have a place to accumulate and form a delta.

Plate tectonic history, regional geologic setting, and sea-level change are quite important in the development of large river deltas, as shown by their global distribution (Fig. 16.2). Trailing edge or passive margins foster the development of deltas but leading edges or active margins are difficult places for deltas to form. Extensive drainage basins typically form in areas where there is little relief, with no mountain ranges or other high-relief landforms blocking the path of rivers to the coast. Two of the best examples are the Mississippi River drainage system in the United States and the Amazon River in South America. The Mississippi system drains most of the country between the Rocky Mountains and the Appalachian Mountains. It includes the Missouri and Ohio river systems and other rivers that empty directly into the Mississippi. Most of the terrain in this system is the stable mid-continent area, known geologically as a craton, and the coastal plain that begins near St Louis. The Amazon River system drains most of the northern part of South America, with the continental divide in the Andes as the western boundary of the system, from where many small tributaries flow toward the east. Both of these river systems drain huge, relatively stable continental regions and deliver their sediment load onto a stable, trailing edge continental margin with a broad, gently sloping shelf – ideal geologic settings for the development of river deltas.

Marginal sea coasts are also good places for deltas to form. The marginal seas that receive the greatest volume of riverborne sediment are the Yellow Sea with the Huang River and the East China Sea with the Yangtze River, both of which form large muddy deltas. In this geologic setting the combination of huge volumes of sediment with the protection of a fetch-limited basin provide good conditions for delta development. Other examples of marginal sea coasts where deltas have developed include the north coast of Alaska (McKenzie), and the north coast of the Mediterranean Sea (Rhône and Ebro), although these rivers and their sediment load pale in comparison with those of China.

On the other hand, leading edge coastal settings do not permit the development of even modest-sized river deltas, e.g. the Columbia River mouth. The

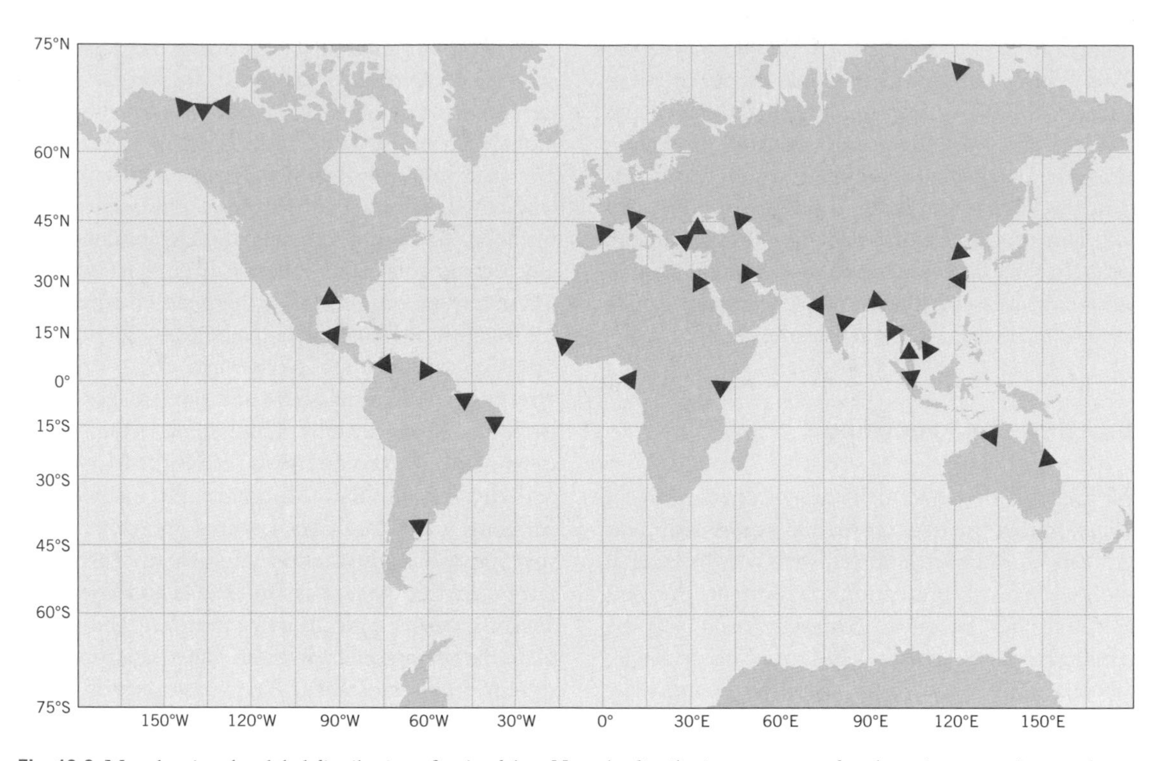

Fig. 16.2 Map showing the global distribution of major deltas. Note the distribution as compared to the various tectonic coastal types. Compare with Fig. 2.19. (From L. D. Wright, J. M. Coleman & M. W. Erickson (1974), Analysis of major river systems and their deltas: morphologic and process comparisons. Technical Report 156, Louisiana State University, Coastal Studies Institute, Baton Rouge, LA.)

first reason for this is the absence of large drainage systems in this type of geologic setting. These leading edge, active margins tend to be next to high relief, mountainous areas with drainage divides that are typically only tens of kilometers from the coast. While the gradients are steep and therefore there is significant erosion through downcutting of the flowing water, there is typically not much soil development that would produce sediment for transport by the rivers. A second major problem with delta development on leading edge coasts is the absence of a proper site for sediment accumulation. Typically the continental margin is narrow and steep; in some cases, it has multiple faults that create small basins. Further, this steep and narrow margin allows large oceanic waves to move very close to the coast without significant loss of energy because they do not feel bottom until almost at the shoreline. As a result, sediment can readily be removed from the mouth of a river, thus prohibiting delta formation. In summary, with rare exceptions, large deltas can only develop on trailing edge coasts because they provide abundant sediment, proper sites for accumulation, and appropriate physical conditions for their maintenance. The global distribution of the major deltas shows this relationship with plate tectonics quite well (Fig. 16.2).

There are some exceptions to this generalization about delta formation and leading edge coasts. On the west coast of North America we have two pretty good examples: the Fraser River delta near Vancouver, Canada, and the Copper River delta on the south coast of Alaska. The Fraser River has its tributaries near the continental divide in the Canadian Rockies in the province of Alberta. It flows for a few hundred kilometers and empties into

Table 16.1 Some large modern deltas.

River	Land mass	Receiving basin	Size (km^2)	Annual sediment discharge (tons × 10^6)
Amazon	South America	Atlantic	467,000	1200
Chao Phraya	Asia	Gulf of Siam	25,000	5
Danube	Europe	Black Sea	2700	67
Ebro	Europe	Mediterranean Sea	600	n.a.
Ganges–Brahmaputra	Asia	Bay of Bengal	106,000	1670
Huang	Asia	Yellow Sea	36,000	1080
Irrawaddy	Asia	Bay of Bengal	21,000	285
Mahakam	Borneo	Makassar Strait	5000	8
Mekong	Asia	South China Sea	94,000	160
Mississippi	North America	Gulf of Mexico	29,000	469
Niger	Africa	Gulf of Guinea	19,000	40
Nile	Africa	Mediterranean Sea	12,500	54
Orinoco	South America	Atlantic	21,000	210
Po	Europe	Adriatic Sea	13,400	61
Rio Grande	North America	Gulf of Mexico	8000	17
São Francisco	South America	Atlantic	700	n.a.
Senegal	Africa	Atlantic	4300	n.a.
Yangtze	Asia	East China Sea	66,700	478

n.a., not available.

somewhat protected waters on the coast of British Columbia, near Vancouver. Tides here are in the macrotidal range but the combination of sediment discharge and protection from the high wave energy of the Pacific Ocean has allowed a modest sized delta to develop despite the geologic setting.

The Copper River delta is located between Anchorage and Juneau on the south coast of Alaska. Special circumstances have permitted the development of a river delta along this coast even though it is one of very high wave energy and very severe winter storms. Most of the water and sediment discharge from the Copper River is derived from melting glaciers. This condition has produced a huge sediment discharge that has permitted the development of a significant delta. In addition to the unusual presence of a river delta on this coast, there are also several barrier islands. Both of these features are typical of trailing edge tectonic settings, but the huge volume of sediment has compensated for the high-energy conditions and other leading edge characteristics.

16.3 Deltas and sea level

Sea level is an important factor in the development and maintenance of river deltas. The deltas that we see around the world today are geologically quite young. Although these deltaic systems range from a few thousand to hundreds of thousands of years in age, the currently active delta lobe is typically at the young end of this spectrum. Deltas cannot exist without sediment supply from rivers. The sediment discharge of the river is partly a consequence of sea-level position and the rate of change of sea level.

At the time of widespread glaciers during the Pleistocene Epoch, sea level was much lower than it is at the present time. The large rivers flowed across what is now the continental shelf and discharged their sediment load near the edge of the continental shelf. The consequence of this was widespread and large density currents of suspended sediment called turbidity currents, along with other sediment gravity processes that transported most of the sediment

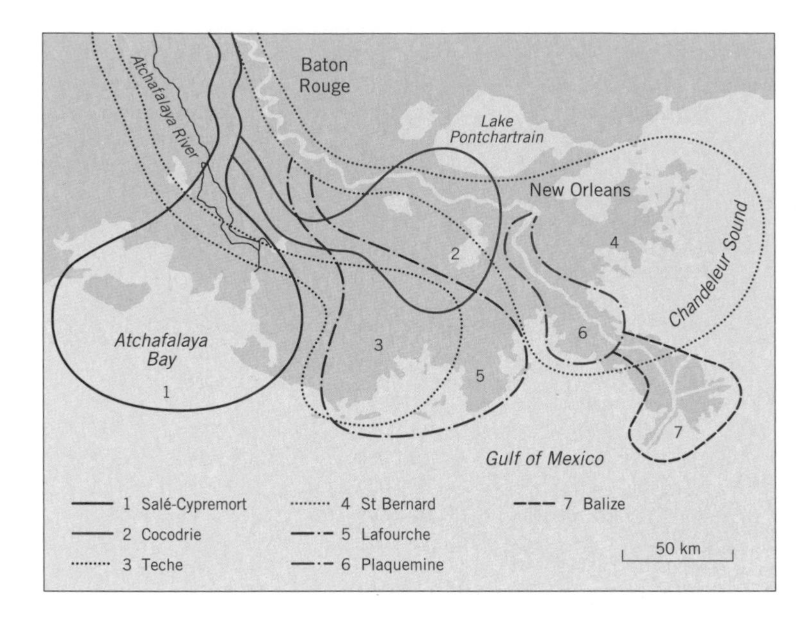

Fig. 16.3 Holocene deltaic lobes in the Mississippi delta showing the large number of shifting positions of sediment discharge into the Gulf of Mexico. All these lobes developed in only a few thousand years after sea level reached only a few meters below the present level. (From C. R. Kolb & J. R. van Lopik, 1966, Depositional environments of Mississippi River deltaic plain, southeastern Louisiana. In M. L. Shirley (ed.), *Deltas and Their Geologic Framework*. Houston, TX: Houston Geological Survey, p. 75.)

discharge of the rivers directly to the continental rise, where it accumulated in thick wedge-shaped deposits.

Under these sea-level conditions deltas were not being formed and those deltas that existed from previous sea-level highstands were being bypassed as rivers flowed across the continental shelf. The great ice sheets began to melt about 20,000 years ago, causing a rapid rise of sea level across what is now the continental shelf. The river mouth was essentially retreating across the shelf so rapidly that there was not enough time for deltas to accumulate. As the rate of sea-level rise slowed about 6000–7000 years ago the rate of migration of the shoreline also slowed greatly. This permitted deltas to actively accumulate large quantities of sediment because the time was available without the sediment being dispersed by waves or tidal currents.

This is not to imply that all deltas are geologically very young: some have existed for millions of years. They have not, however, been continuously receiving sediment from their associated rivers because they were abandoned by the shoreline as it moved in association with sea-level change. The Mississippi delta and the Niger delta in Africa are good examples among many old deltas. Both deltas are currently active and are underlain by deposits that are at least ten million years old.

The young portion of the Mississippi delta is 5000–6000 years old, coincident with the slowing of sea-level rise. This part of the Mississippi delta consists of 16 recognizable lobes. Each of these lobes represents sediment accumulation at the mouth of a different geographic location of the river. These different lobes are abandoned when the location of rivermouth deposition shifts due to channel switching, avulsion, or other natural causes. Although 16 lobes of sediment accumulation have been recognized, they can be combined into only a few (Fig. 16.3) based upon radiometric dating and location. The present lobe of the Mississippi delta began to form only about 600 years ago, not much before Columbus's first voyage to the New World. Most of the active portion has developed since the settlement of New Orleans by Europeans. The rate of sediment accumulation at the mouth of the Mississippi has been so great that nearly one-half of the State of Louisiana has been formed by the river since sea-level rise slowed about 6000 years ago.

We can see older deltaic deposits at the mouth of the Niger River too. As with the Mississippi delta, exploration for petroleum on the Niger has provided a wealth of information on the age and development of the delta. From these data it is possible to recognize sediment strata at least Miocene in age, i.e. up to 15 million years old. Differences of climate in western Africa over this extent of time have provided great quantities of sediment as the result of more humid conditions and associated rainfall.

16.4 Delta environments

Deltas are a transitional coastal environment located between terrestrial and marine conditions. Distinct landward or seaward boundaries do not exist on deltas; they grade continuously in both directions. This gradual transition is primarily due to the change from freshwater to seawater, and to the differences in sediment accumulation from the river through the open marine environment. The discharge of the river is carried from the main channel through a series of smaller channels that split off from the river to distribute the discharge of water and sediment. This splits up of the main channel into multiple distributaries that actually distribute the discharge of the river, both water and sediment, across the delta and into the marine basin. The result is a condition of overall progradation of sediment accumulation into the basin of deposition, which could be a lake, estuary, lagoon, or other standing body of water, as well as the ocean itself.

As a consequence of this type of setting, the delta includes subaerial, intertidal, and subaqueous sedimentary environments, as well as freshwater, brackish, and marine conditions. For the purposes of discussing sedimentary environments and their processes, we can best subdivide the delta into three major parts, each of which has its own specific environments. From landward to seaward these are the delta plain, the delta front, and the prodelta. This discussion emphasizes the first two; the prodelta is strictly subtidal and extends into fairly deep water of the outer continental shelf.

The delta plain is primarily influenced by the river and its processes, with tides and waves playing less of a role overall. There is, however, an increase in the influence of marine processes toward the seaward portion of the delta plain. The delta front is dominated by marine processes and tends to be subtidal, with a small intertidal portion in some deltas.

Fig. 16.4 Oblique photograph of a typical delta plain environment showing distributaries, natural levees, splays, and interdistributary marshes and bays. The interdistributary marshes are the dominant environment of the delta plain.

16.4.1 Delta plain

We can think of the delta plain (Fig. 16.4) as the coastal extension of a river system. The delta plain is dominated by channels and their deposits, and the associated overbank environments that receive sediment during flooding. This scenario is parallel to that of a typical meandering river system. In fact, all of the specific elements of a meandering river complex are typically present on many deltas. There are some deltas, however, that have only a portion of this spectrum of environments.

The distributary channels on a delta plain contain point bars formed as the channel migrates. In doing so, they form broad meander loops that may be cutoff leading to formation of oxbow lakes. As the channels migrate across the delta plain they produce scars of their former location that leave subtle but recognizable geomorphic and vegetation patterns, another parallel with the fluvial system. Adjacent to the channels are three major types of overbank or flooding deposits: natural levees, crevasse splays, and floodplains, in order away from the channel.

Natural levees (Fig. 16.5) are produced during flooding when the river overtops its banks and

Fig. 16.5 Natural levees along distributaries of a delta plain. These levees form as sediment-laden water overtops the banks during floods.

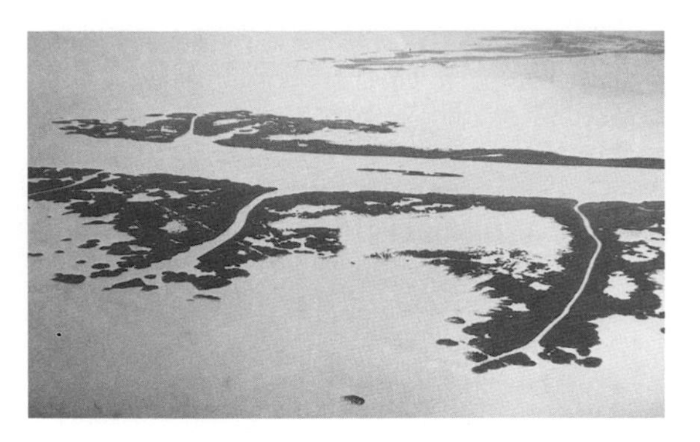

Fig. 16.6 Photograph of crevasse splay development on a delta plain showing the fan shape and the cut in the natural levee through which the sediment was transported.

immediately deposits much of its sediment load. The confinement of the channel, coupled with high discharge volume and rate, causes the river to carry considerable sediment. Sudden loss of this confining characteristic as overflow occurs results in a sudden loss of velocity and carrying capacity, causing much sediment to be deposited at the edge of the bank. This condition takes place each time a channel floods, and the levees build vertically. Although the natural levees may be only a meter or so high, they are important features of the distributary channel system.

Flooding may also cause a natural levee to breach due to a weakness in the accumulated sediment, perhaps a low area in the levee or even a human-induced cut in the levee. Any of these conditions can permit river flood waters to flow through the levee, depositing sediment on the flood plain in the form of a crevasse splay (crevasse refers to the cut in the natural levee). These are fan-shaped deposits (Fig. 16.6) that can cover up to many square kilometers with a sediment thickness that is typically less that that of the adjacent natural levee. These splays may be reactivated multiple times during successive flooding conditions and thereby they can grow significantly in elevation and extent (Fig. 16.6).

The most widespread but the thinnest of the overbank accumulations are the floodplain sediments. Even after losing sediment to natural levees and splay deposits, there is substantial fine sediment in suspension during flooding conditions. The spreading of the floodwaters beyond the channel causes important loss of velocity and thereby of capacity, resulting in the deposition of fine and extensive floodplain deposits. Commonly such floodplain sediments are draped over vegetation or other materials that occupy this environment. We have all seen many examples in the media of mud deposited by flooding, covering cars, carpets, and furniture in houses. The floodplain in the delta plain may take on a variety of characteristics. These include subtidal environments, such as interdistributary bays, intertidal marshes, swamps, and tidal flats, or, in the most landward areas, even subaerial environments of various types.

The upward and lateral growth of the delta plain portion of the delta is dependent upon flooding periods for sediment distribution to the overbank environments. The typical situation is that the channel and its associated levee extend seaward at the outer limit of the delta (Fig. 16.7). The levees may even be subaqueous at the most distal end of the channel. The initial subaerial portion of this distributary channel is the natural levee, followed by small splay deposits. Continued flooding will enlarge the splays until at least a portion is subaerial. Continued accumulation of these splays (Fig. 16.8), along with the slower but more extensive floodplain deposits, will eventually lead to the interdistributary

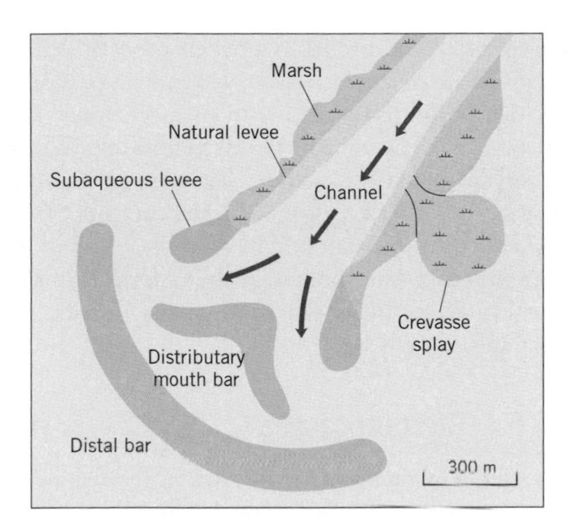

Fig. 16.7 Diagram of the terminus of a distributary showing the nature of subaqueous levees that are natural extensions of the subaerial levees.

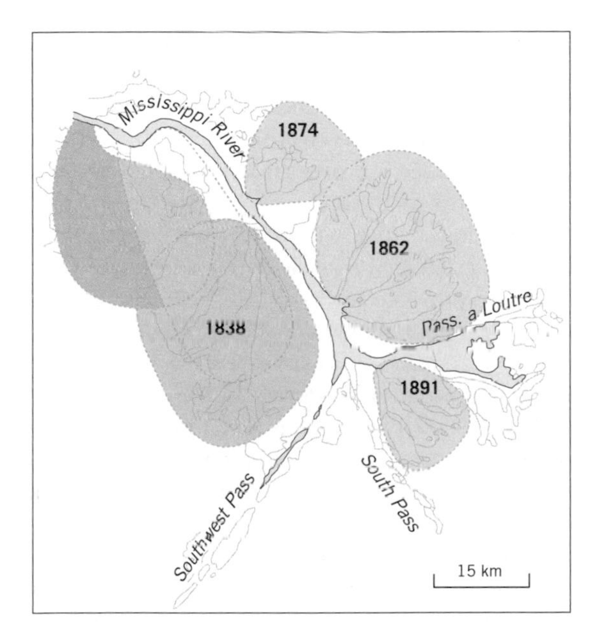

Fig. 16.8 Growth of the delta along a distributary channel through sequential building of the crevasse splay deposits. The crevasse splays are commonly reactivated during successive floods and eventually reach an elevation that supports salt marsh vegetation. (Modified from J. M. Coleman & S. W. Gagliano, 1964, Cyclic sedimentation in the Mississippi River deltaic plain. *Transactions of the Gulf Coast Association of Geological Societies*, **14**, 75.)

area being filled, producing a continuous delta plain system.

16.4.2 Delta front

The seaward edge of the delta plain merges with the generally continuous subtidal portion of the delta called the delta front. It is this part of the delta that is most affected by marine processes, especially the waves. Sediment empties out of the mouth of the distributary channels as both suspended load and bed load. The finer suspended sediments tend to be carried away from the mouth of the channel by currents, whereas much of the coarser bed load tends to accumulate near the channel mouth. The vast majority of the coarse sediment is sand, which comprises the delta front system.

The nature of the sand accumulations in the delta front depends upon the volume of sand transported to the distributary mouth and the relative roles of the interacting river currents with the waves and tidal currents. A common sand body is a distributary mouth bar (Fig. 16.9), which accumulates just seaward of the channel mouth and typically causes the channel to bifurcate. The isolated distributary bar with little or no sand on either side is not generally common because of the influence of waves, which spread the sand along the delta front on most

Fig. 16.9 Photograph of a distributary mouth bar during exceptionally low tide and offshore winds on the Mississippi delta. These present navigation hazards and are the result of waves reworking sand at the mouths of the major distributaries of the delta. (Compare with Fig. 16.7.)

Fig. 16.10 Photograph of a beach on the wave-dominated portion of a delta. These are similar to beaches in other areas, except that they tend to have very low elevations and are typically fine-grained sand with some mud.

deltas. As the waves approach the shallow part of the delta they refract and generate longshore currents, in the same fashion as they would along a beach. These currents carry the sand away from the mouth of the channel and distribute it along the outer delta plain, forming a nearly continuous delta front system. The degree to which this takes place is dependent upon the wave climate.

There is a wide range in the nature of the delta front sand bodies that comprise the outer part of the upper part of the delta. In some deltas where there are several distinct distributaries, as on the Mississippi delta, the delta front tends to be rather subtle, with distinct sand bars near the channels (Fig. 16.9). By contrast, on some river deltas there is considerable redistribution of the sand from the channel mouths across the outer delta plain margin. In these situations there may be beaches (Fig. 16.10) and dunes on this part of the delta, due to an abundance of sand and the appropriate wave climate to redistribute it. The Saô Francisco River in southern Brazil is a good example of this type.

16.5 Delta processes

The interaction of riverine processes with the wave- and tidal-generated marine processes is quite complicated and results in a wide variety of deltaic forms and features. River-generated processes include both confined flow in open channels and unconfined flow during flood conditions. Wave-generated processes include the waves themselves in a variety of scales, along with the currents developed by wave refraction. Tides produce important currents that not only distribute sediment but also influence the discharge from the distributary channels of the delta.

16.5.1 River processes

The fundamental role of the river in the delta system is providing the sediment. In this role the river is at the mercy of climatic conditions and seasonal changes in discharge. More recently, humans have played an important role in many river systems and have caused many problems on the delta. The variables that influence the nature, amount, and rate of sediment delivery include the geology, geomorphology, and climate of the drainage basin. Also involved are human influences, such as agriculture, navigational structures, and dams.

The combination of rock type and climate is a major control on the sediment provided to the river. Rainfall and its distribution over time are the fundamental factors of river discharge and therefore of sediment provided to the delta. Prolonged periods of drought place serious constraints on delta formation or maintenance. The typical climatic influence is the seasonal distribution of temperature and precipitation. There are at least two important aspects to these cycles, including the annual distribution of rainfall. In areas where monsoon conditions exist during the summer, such as in southeast Asia, there is tremendous discharge and typically devastating flooding during this two to three month period. The flooding in Bangladesh of the Ganges–Brahmaputra delta area is probably the most consistently dramatic case of flooding in the world. Most rivers, however, experience flooding during the rainy season (Fig. 16.11), and in some places this is especially problematic because the wet spring season coincides with the spring melting of snow. The Mississippi River provides an example. Most of the midwestern

Fig. 16.11 Photograph of a flood event on a delta showing widespread submergence of the delta plain. Only the higher parts of the natural levee area are visible above water level. This phenomenon occurs at least once each year and provides sediment to the marshes to help to compensate for the rapid rise in sea level in this area.

Fig. 16.12 Photograph of the high water mark on a building as the result of the 1993 Mississippi River floods. This level is more than 3 m above normal river level. (Courtesy of O. H. Pilkey.)

part of the United States has high rainfall in the spring during the same time that the snow in the Rocky Mountains and northern latitudes of the basin is melting. Flooding can be very severe for the people living along the river, as exemplified by the devastation of the 1993 floods (Fig. 16.12). Similar phenomena may take place in the delta area. On the other hand, these floods are highly beneficial to the delta, in that these conditions provide the highest rate of sediment delivery to the delta, a crucial source of mineral nutrients. Flooding also washes out soluble salts that have built up in delta sediments.

The annual distribution of sediment to a river delta varies greatly for each river because of the dependency on climatic conditions. Desert rivers tend to have little discharge of water and therefore transport little sediment. However, when there is rainfall, it typically is a large amount in a short time, a condition that delivers considerable sediment to the river; it is essentially a flood condition. Rivers like the Ganges–Brahmaputra discharge many times the normal rate during the monsoon season. Even the Amazon or Mississippi (Fig. 16.13) shows a marked difference in discharge during the wet season compared to the rest of the year.

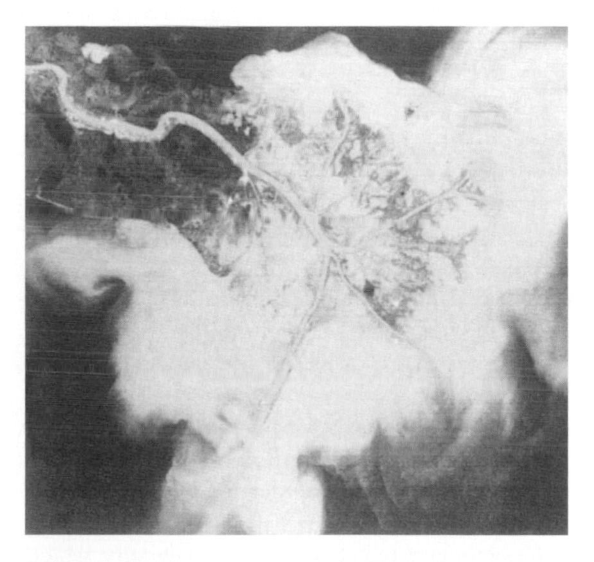

Fig. 16.13 Aerial photograph of the Mississippi River delta mouth area showing the abundant suspended sediment in the water. (NASA photograph.)

16.5.2 Human influence

As people began to populate the drainage basins of major rivers and the banks along the courses of these rivers, they profoundly influenced the delta in several ways. Most, but not all, of these influences have had detrimental effects on the deltas. In most countries, the earliest important human activities were agriculture and forestry. Both have tended to benefit the growth of the delta, although they have produced some important negative effects in the drainage basin. Cultivation and deforestation increase erosion of the soil and provide the river, and therefore the delta, with a high rate of sediment discharge. This has resulted in the accelerated growth of many deltas, with the prime examples being the Mississippi delta in the nineteenth century and the present Amazon River delta. As the rapid diminution of the rainforest in Brazil takes place, vast quantities of sediment are provided to the delta.

A more widespread human influence has the reverse effect, i.e. reducing the sediment supply and thereby causing the delta to shrink in size. There are three important ways in which this occurs: (i) diverting water from the river; (ii) navigation controls on the river; and (iii) damming the river. All reduce the discharge of the river and the latter two physically trap sediment and keep it from moving down the river.

There are several major cities that take large percentages of the discharge of rivers to use in the municipal water supply. The southern California area is dependent on water from various rivers in the southwest for its water, both for irrigation and for domestic use, including the Colorado, which flows through the Grand Canyon. Both activities greatly decrease the water discharge of the river affected and thereby diminish the sediment provided to maintain the river delta. They also decrease the frequency of flooding across the delta, causing delta sediments to become increasingly salty, which may affect the type of vegetation that grows there.

Locks for navigation on major rivers invariably have dams associated with them. On the navigational portion of the Mississippi River that begins near Minneapolis, Minnesota, there are many such structures. The small amount of water impounded is typically not a big problem, but the sediment that is trapped behind the dam is literally stolen from the system and, eventually, from the Mississippi Delta. More important, related impoundments are the huge dams built for reservoirs and/or hydroelectric power. They are extremely good sediment traps and some also serve as sources for water diversion. The bottom line is that the amount of water and sediment that the river has available and can transport is not being delivered to the delta.

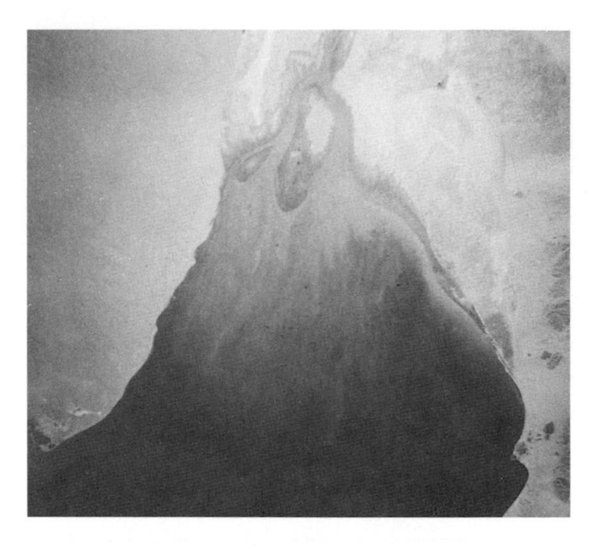

Fig. 16.14 Aerial photograph of the Colorado River delta at the north end of the Gulf of California. Very little sediment and water is now being discharged here due to the effects of human activities.

Two good examples of this problem are the Colorado River, which empties into the Gulf of California (Fig. 16.14), and the Nile River in Egypt. The headwaters of the Colorado are in the Rocky Mountains in the state from which it is named. Along the course of over a thousand kilometers there are numerous dams and reservoirs, as well as places of diversion. The result is that virtually no water and sediment are being provided to the Colorado River delta and it is rapidly being eroded by strong tidal currents.

The case of the Nile River is similar. The Aswan Dam, which was constructed to make the desert fertile, has been quite successful in trapping

Fig. 16.15 Photograph of erosion along the outer portion of the Nile delta as the result of the presence of the Aswan Dam. (Photo courtesy of D. J. Stanley.)

virtually all of the sediment carried by the Nile that was destined for the delta. As a consequence, the outer margin of the delta is being eroded rapidly (Fig. 16.15) by waves produced in the eastern Mediterranean.

We cannot continue to rob our rivers of their water and sediment load without experiencing the consequences that these circumstances produce for the deltas that they feed. At present the most viable alternative appears to be stopping development of any kind on deltas. The waves, tides, and their resulting currents interact with the riverine processes to prevent, mold, or destroy the deltas depending upon the specific local circumstances. At those river mouths where waves and tides carry all the sediment away there is no delta. At many places the delta is allowed to accumulate and prograde, but at some, such as the previously mentioned examples, the processes are now resulting in overall erosion. The relative role of the waves and the tides is also an important factor in the formation, maintenance, and overall morphology of deltas.

The primary direct marine processes are the waves, wave-generated currents, and tidal currents; the rise and fall of the tide has little direct effect on redistributing deltaic sediment. Waves impart energy along the delta and cause sediment to go into temporary suspension, whereas longshore and/or tidal currents transport it. During storms this distribution or removal of sediment reaches its maximum. The overall influence of the waves ranges from mostly longshore transport along the delta front, providing sediment to various parts of the delta, to actual offshore or alongshore removal of sediment from the delta proper. Waves and wave-generated processes work toward a smoothing of the outer delta shape.

Tidal currents usually have at least a shore-perpendicular component and move in and out of the delta complex as flooding and ebbing occurs. The stronger the current the more sediment is redistributed in this shore-perpendicular fashion. Some deltas have developed on coasts with high tidal ranges, such as the Colorado in the USA (3.5 m), the Ganges-Brahmaputra in Bangladesh (4.0 m) and the Ord River delta in Australia (7.0 m). These and similar deltas experience sediment being carried inland and deposited by flooding currents, as well as large volumes of sediment being carried offshore even beyond the delta, such as on the Amazon delta.

16.6 Delta classification

Both wave and tidal processes on the delta are essentially competing with the riverine processes to leave their imprint on the delta morphology. These competing processes and the resulting configuration of the delta provide a framework for classifying deltas. The three major processes that influence deltas provide convenient end members for a comprehensive organization of deltas by shape. This classification was first presented in published form by William Galloway of The University of Texas–Austin, and has become a standard.

The classification consists of a triangle-shaped diagram with riverine processes, waves, and tides at the three apices (Fig. 16.16). A delta that is clearly dominated by any one of the three processes is placed at the appropriate apex. The Mississippi delta is quite distinctly dominated by river processes in the form of sediment input due to the large volume of sediment discharge and little marine reworking. This gives it a so-called "bird's foot" configuration. By contrast, the Fly delta in Papua New Guinea and the Ord delta in Cambridge Gulf on northwestern Australia are fine examples of domination by tidal

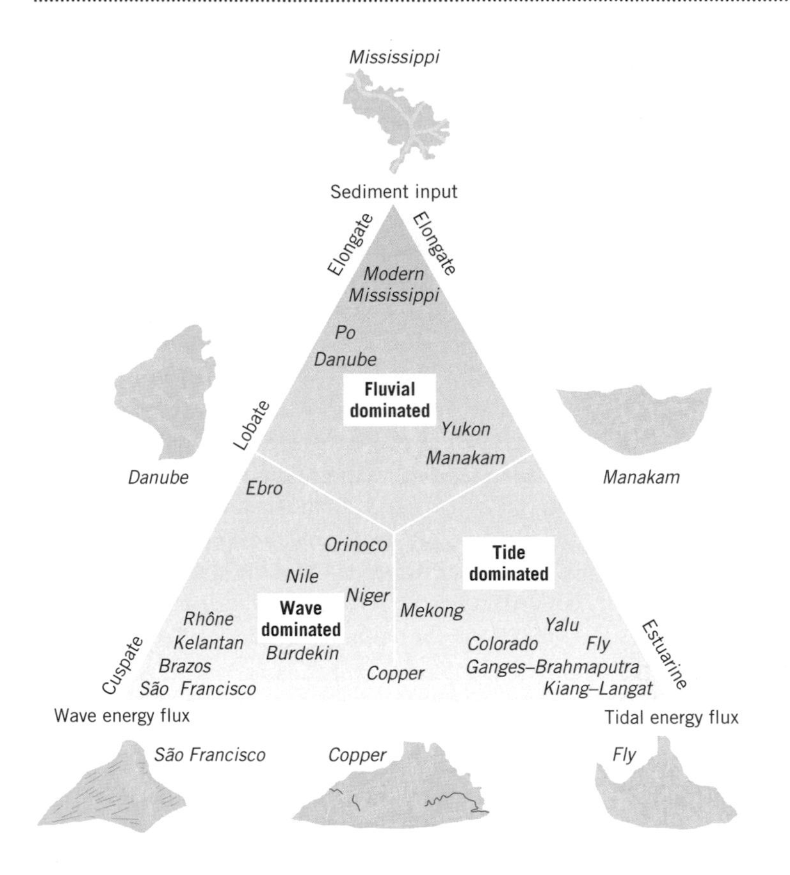

Fig. 16.16 Classification of river deltas based upon the scheme presented by William Galloway. The many examples included are based on qualitative assessments of the relative influence of river, wave and tide processes. (From W. E. Galloway, 1975, Process framework for describing the morphologic and stratigraphic evolution of deltaic depositional systems. In M. L. Broussard (ed.), *Deltas: Models for Exploration*. Houston, TX: Houston Geological Survey, p. 92.)

flux. At each of the sites sediment bodies of the delta are oriented essentially perpendicular to the trend of the coast. Lastly, the São Francisco delta in Brazil and the Senegal delta in Africa show distinct domination by wave processes. Both contain smooth outer margins caused by the distribution of sediment along the coast as the result of wave-generated longshore currents.

Most deltas fall somewhere nearer the middle of the classification than these examples because all deltas experience some influence from all three types of processes. The morphology of each tends to reflect these influences. In general, river influence produces a finger-like morphology with a well developed delta plain that has several distributaries. Tide-dominated deltas display a strong shore-perpendicular trend and have extensive tidal flats with little mud. Wave-dominated deltas typically have well developed beach and dune systems at their outer limits, with few distributaries. Neither the absolute values of the processes nor the size of the delta are important in determining the position of a given delta in the overall classification scheme. It is the relative influence of the interactive processes that gives the delta its character.

16.6.1 River-dominated deltas

Conditions that foster river-dominated deltas include high water and sediment discharge, with small waves and low tidal ranges in the receiving basin. A broad, gently sloping continental shelf provides the typical resting place for this large volume of sediments. These conditions are best fulfilled by trailing edge and marginal sea, tectonically stable coasts that are sheltered from large waves and have small tidal ranges. The Gulf of Mexico is a perfect setting, as it hosts the Mississippi delta (Fig. 16.17). Other similar settings and their example deltas are the Black Sea, with the Danube, the Adriatic Sea, with the

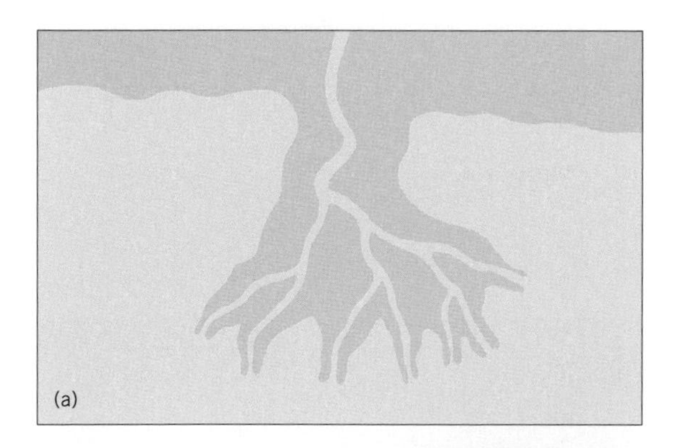

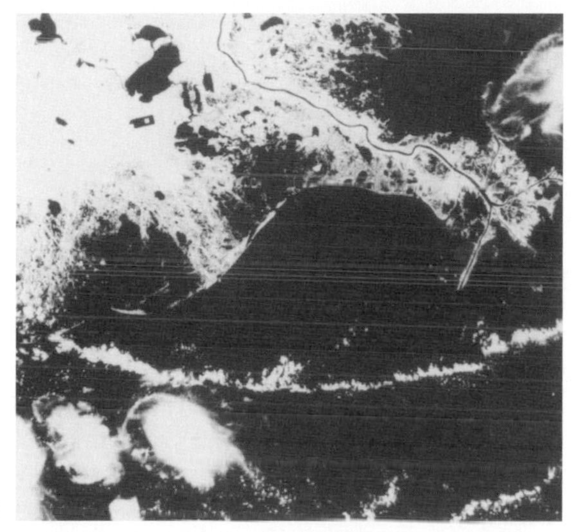

Fig. 16.17 (a) General diagram and (b) Mississippi delta, a good example of a river-dominated delta. (From L. D. Wright & J. M. Coleman, 1973, Variations in morphology of major river deltas as functions of ocean wave and river discharge regimes. *American Association of Petroleum Geologists*, **57**, 377.)

Po, and the Yellow Sea, with the Huang Ho. All have good sediment supplies and low tidal ranges, and all are sheltered from large waves. The Mississippi delta does experience exceptional wave energy when hurricanes pass through the area. Some of the most devastating of these storms have eroded large areas of the delta.

16.6.2 Tide-dominated deltas

Strong tidal currents in the absence of a substantial wave climate and strong river influence will produce a delta that is tide-dominated. Large tidal channels with intervening tidal sediment bodies dominate the delta and are generally more numerous than the river distributaries. Intertidal environments are widespread and are commonly partly covered with vegetation such as salt marsh or mangroves.

The Ganges–Brahmaputra (Fig. 16.18) is the largest of the tide-dominated deltas and is an area of great interest because of the common and devastating floods it experiences. This is a large river system that supplies huge quantities of sediment to a coast where the tidal range exceeds 3 m and waves are modest. One of the major factors in the development of this delta is the annual variation in discharge. During the monsoon season the amount of sediment delivered to the delta is orders of magnitude greater than during the rest of the year. Strong tidal currents redistribute the sediment in elongate bands that are separated by numerous large tidal channels.

16.6.3 Wave-dominated deltas

Some deltas don't really look like deltas because of the strong influence of waves. They may mimic the appearance of barrier island systems, with beaches, dunes, and wetlands landward of them. These wave-dominated coasts are deltaic in nature because the sediment is supplied directly by the river and then reworked by the waves and wave-generated currents. The distributary channels do not protrude into the basin, thus providing a smooth outer shoreline. Typically, wave-dominated deltas are small; in fact, they grade into conditions of no delta if wave processes are strong enough to carry away all of the sediment supplied by the river.

There are different styles taken by wave-dominated deltas depending upon the nature of the longshore current patterns. The São Francisco delta in Brazil and the Sabine delta at the Texas–Louisiana border (Fig. 16.19) are fairly symmetrical about a single large distributary with a smooth overall cuspate shape. This is due to the absence of a strong littoral drift in either direction caused by the wind patterns and related direction of wave approach. By contrast, the Senegal River in west Africa (Fig. 16.20) displays a very strong change in direction of its course

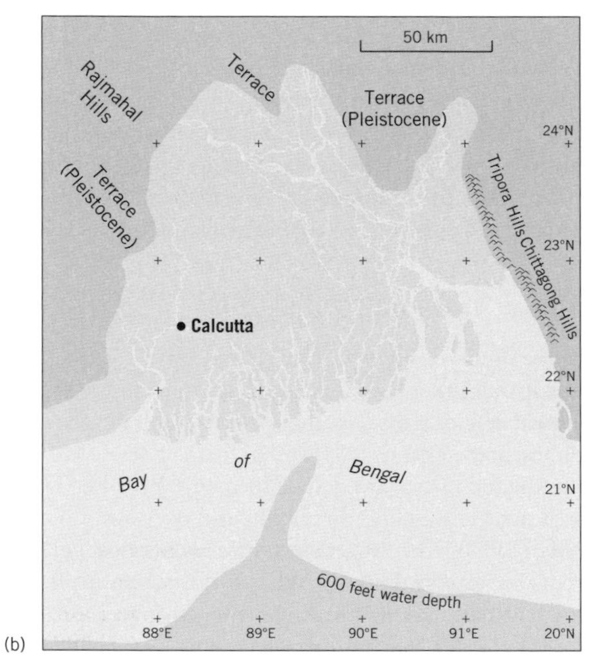

Fig. 16.18 (a) General diagram and (b) Ganges–Brahmaputra delta, a good example of a tidally dominated delta.

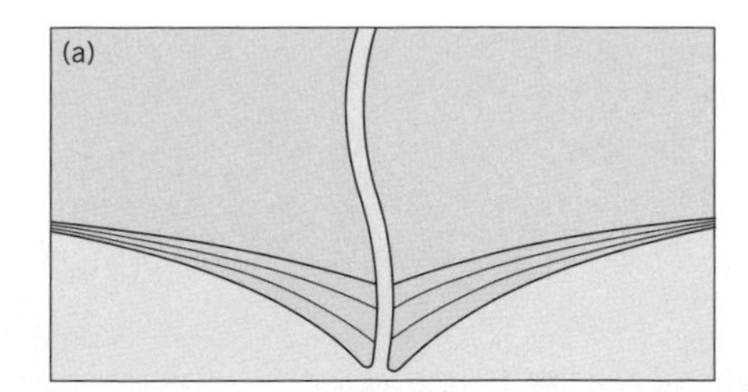

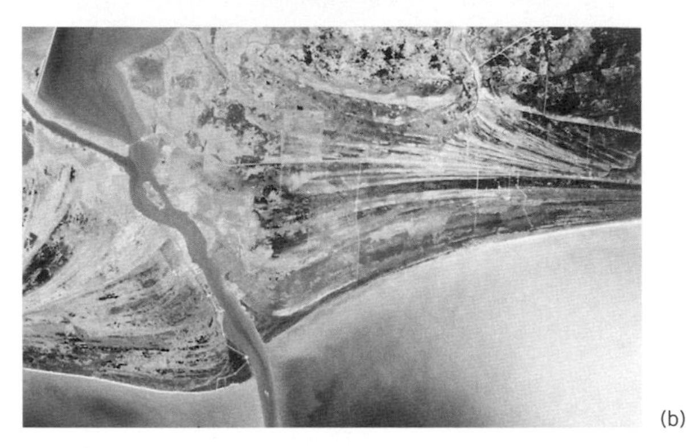

Fig. 16.19 (a) General diagram and (b) Sabine delta on the Texas–Louisiana border, a good example of a symmetrical, cuspate wave-dominated delta.

due to longshore currents and resulting littoral drift. The river course is shifted over 50 km by the longshore currents. The mouth of the river is marked by a distinct spit that mimics a coastal barrier, and extensive wetlands cover the delta plain.

16.6.4 Intermediate deltas

Intermediate types of deltas display features of both river influence and marine processes. The Mahakam delta on the coast of Borneo is small but has a shape that shows the important influence of the river and tidal currents. It has numerous distributaries and a well developed delta plain with distinct lobes that protrude into the receiving basin. Spring tides range up to 3 m and have currents of 100 cm s^{-1} that form distinct tidal channels between the distributary mouths.

The Nile delta is a good example of a delta that is intermediate between river- and wave-dominated deltas. Tides in the Mediterranean Sea are nominal and waves are modest. River input has historically been fairly high until the construction of the Aswan Dam in the 1960s. The delta plain is traversed by a modest number of well defined distributaries, each protruding into the sea. Between the distributary mouths the delta displays a relatively smooth outline, with beaches and other wave-dominated features.

Probably the best example of an intermediate delta is the Niger delta on the coast of Nigeria (Fig. 16.21). It falls in the middle of the classification, showing equal influence of the river, waves,

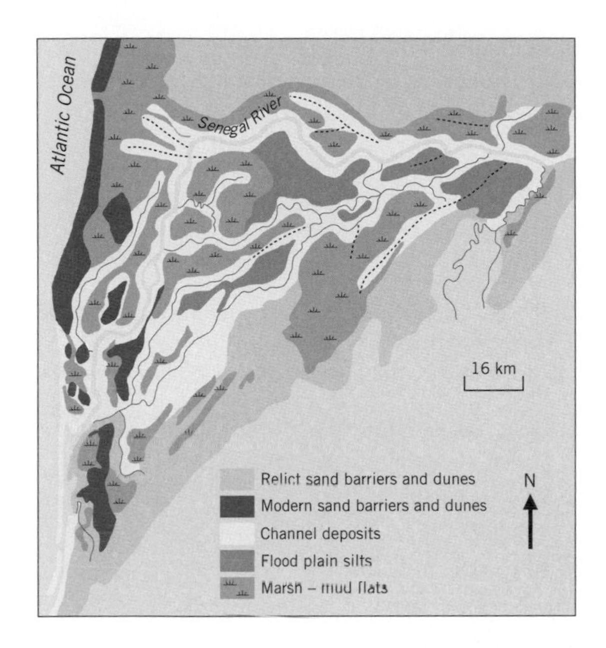

Fig. 16.20 Senegal delta in Africa, a classic case of the river mouth being shifted many kilometers due to a high rate of littoral drift. (From L. D. Wright & J. M. Coleman, 1973, Variations in morphology of major river deltas as a function of ocean wave and river discharge regimes. Amer. Assoc. Petroleum Geol. Bull., **57**, 377.)

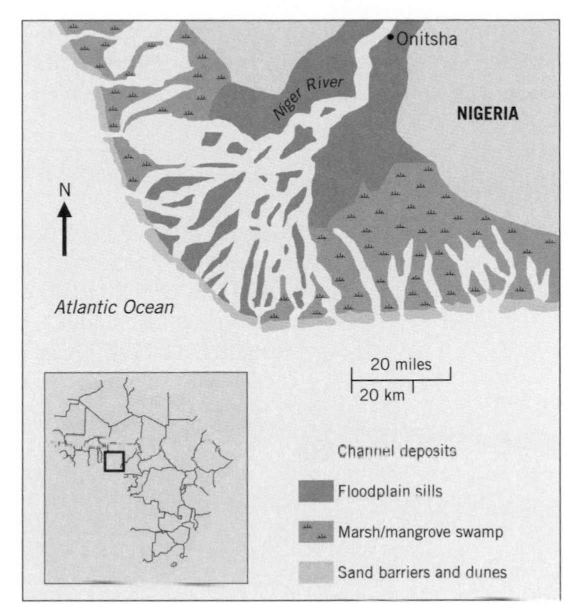

Fig. 16.21 Niger delta, a delta that is about equally influenced by river, wave, and tidal processes. (Modified from J. R. L. Allen, 1970, Sedimentation in the modern Niger delta: a summary and review. In J. P. Morgan & R. H. Shaver (eds), *Deltaic Sedimentation: Modern and Ancient*. Tulsa, OK: SEPM Special Publication 15, p. 140.)

and tides. It has a well developed delta plain with a complex network of distributaries and a spring tidal range of up to 2.8 m, and the delta is exposed to waves of the south Atlantic Ocean. The result is a delta that incorporates some features of each of the major processes that influence deltaic coasts.

16.7 Summary

In some ways river deltas may be considered as the most important of all coastal environments because they are the site of sediment introduction for most of the other parts of the coast. On the other hand, people rarely spend any time visiting a delta on vacation or going to and from places of work or play. Deltas tend to be remote, without traffic arteries, and are generally inhospitable due to insects. Because of their critical role in the overall scheme of the coastal zone it is important that we have an understanding of river deltas and their characteristics.

Deltas are among the most productive and valuable environments in the world. They contain highly productive ecological niches where a wide variety of both plants and animals thrive. Their marshes are among the most extensive and productive environments anywhere. Deltas represent very important nursery grounds for juvenile fish and marine invertebrates. Their marshes are also important filters that trap contaminants and pollutants during flooding of distributaries.

The size and shape of the delta is a consequence of the interplay between the river and the sediment it provides, and the wave and tidal processes of the marine coast. Deltas have developed quite rapidly in the context of geologic time and they can be destroyed just as rapidly. As we influence our environment more and more, we need to make a better job of considering the long-term consequences of our actions. The role of human intervention is critical to the maintenance of this coastal environment, as evidenced by what has happened on the Nile and Colorado deltas, as well as others.

Suggested reading

Broussard, M. L. (ed.) (1975) *Deltas: Models for Exploration*. Houston: Houston Geological Society.

Morgan, J. P. (ed.) (1970) *Deltaic Sedimentation*. Tulsa, OK: SEPM Special Publication No. 15.

Suter, J. R. (1994) Deltaic coasts. In R. W. G. Carter & C. D. Woodroffe (eds), *Coastal Evolution*. Cambridge: Cambridge University Press.

Wright, L. D. (1985) River deltas. In R. A. Davis (ed.), *Coastal Sedimentary Environments*. New York: Springer-Verlag.

17 Glaciated coasts

17.1 Introduction

Glaciated coasts exhibit diversity in both types of features and a changing landscape that is unparalleled in the world. The ability of continental glaciers (ice sheets) and valley glaciers to sculpture land surfaces, transport large quantities of rock and sediment, and eventually deposit these materials in a number of glacial features accounts for this variability. For example, the northern New England coast is mostly rocky, with small pocket beaches that range in composition from sand to gravel and even to boulder-sized sediment over distances of less than a kilometer. Interrupting this trend in northern Massachusetts and southern Maine are several extensive barrier systems (5–30 km long) that occur at the mouths of estuaries within arcuate embayments. The existence of these barriers along an otherwise sediment-starved coast is due to the large volumes of sand brought to the coast by rivers following deglaciation. Continuing northward, the central and northeast coast of Maine is rugged and highly irregular, with rocky peninsulas giving way to broader embayments and finally to bedrock cliffs. Here glaciers have stripped away most of the sediment overlying the bedrock. This morphology contrasts sharply with that of Cape Cod (Fig. 17.1), which has smooth coastlines consisting of mainland beaches, barrier spits, and barrier islands. Cape Cod is composed entirely of glacial sediment that was deposited during the deglaciation of this region approximately 17,000–18,000 years BP. Not only have the effects of glaciation produced very different sediment abundance along the New England coast, but glacial processes combined with the pre-existing bedrock geology of the region have also created a coast with numerous bays and sounds. In turn, this coastal morphology has resulted in highly varied physical settings, including wave-dominated, mixed energy, and tide-dominated coasts.

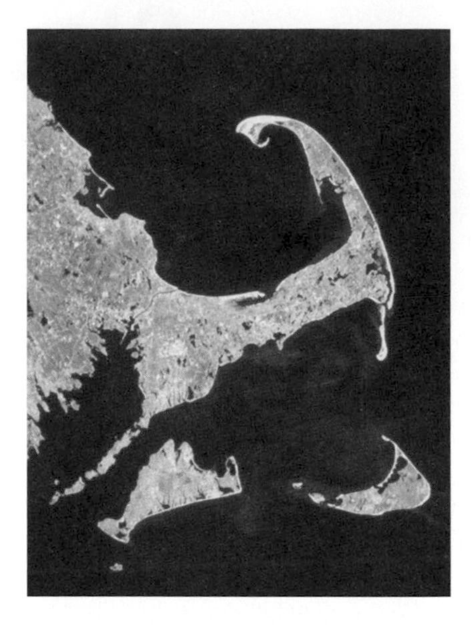

Fig. 17.1 Color infrared photograph of Cape Cod and the islands of Nantucket (right) and Martha's Vineyard (left). These regions were built from sediment carried south by the glaciers and deposited as end moraines, outwash plains, and other sedimentary features.

Fig. 17.2 Aerial view of the northern Alaskan Peninsula. Meltwater from nearby glaciers transports sediment to the coast via a braided stream. A large gravel and sand recurved spit has built from the abundant sediment supply.

The glaciated coast along the tectonically active Gulf of Alaska is even more diverse and more spectacular than that of New England (Fig. 17.2). In some locations, such as the Kenai Peninsula, valley glaciers are still found a short distance from the coast, having retreated into the bordering mountain valleys. The deep fjords of this region are a testament to the ability of glaciers to carve coastal landscapes. Along the Alaskan Peninsula, glacial meltwater streams feed sediment into broad embayments, forming extensive tidal flats, marshes, and barriers spits. During the summer, these streams burgeon with salmon seeking spawning grounds, thereby providing a tasty meal for the waiting brown bears. In yet another site along this active coast, several glaciers east of Prince William Sound have produced a wealth of sediment that has been carried to the coast via a myriad of meltwater streams. The deposition of these sediments and their reworking by waves and tides are responsible for forming the 80 km long Copper River delta barrier chain and an expansive backbarrier tidal flat system.

This chapter describes how glacial processes have produced diverse and dramatic landscapes along many high-latitude coasts. The manner in which glaciers excavate bedrock, transport large quantities of sediment and rock, and deposit these materials is discussed. The effects of sea-level changes associated with the enlargement and melting of continental ice sheets are also explained. Finally, the causes of repeated episodes of glaciation during the Ice Ages (past 2.2–2.4 million years) are explored.

17.2 The world's glaciers

Glaciers exist on almost every continent of the world (Fig. 17.3), including Africa, where retreating glaciers top portions of Mount Kilamanjaro and Mount Kenya. They are not present in Australia but are found in nearby New Zealand. Glaciers occur as narrow ribbons of flowing ice in high mountain regions or as thick ice sheets covering vast continental areas. Presently, glaciers cover about 10% of the continental landmass, but in the recent geologic past (several times during the past 2.2 million years) they extended over 30% of the land surface. This fact suggests that glacial processes have formed or strongly influenced a large portion of the world's coastlines, a concept that generally is not fully appreciated.

17.2.1 Glacier formation

A glacier is defined broadly as a large mass of ice that flows internally. The oldest glacier ice on Earth is a stagnant ice mass in Beacon Valley in the Dry Valleys region of Antarctica. This ice was part of an active glacier at least eight million years ago. The formation of glaciers is tied closely to certain climatic conditions, where cool temperatures and precipitation produce more snow accumulation during the colder months of the year than ablates during the warmer months (Fig. 17.4). Ablation is the term given to the collective processes of ice wastage, including melting, sublimation, and ice calving into water. Thus, glaciers tend to form in regions of high latitude or high elevation, and even in these regions slight changes in either temperature or precipitation can cause glaciers to advance or retreat.

The transformation of snow to ice and then to glacial ice is a progressive process in which air is gradually forced out, producing a dense mass of interlocking ice crystals (Fig. 17.5). Freshly fallen snow is commonly light and fluffy and may be 90% air, although those who live in northern regions and shovel snow during the winter might argue about this presumed weightlessness. As more and more snow accumulates, snow at the base is compacted by the overlying weight, changing the hexagonal snowflake crystals into smaller spherical structures called granular snow (the type of snow common during spring skiing). With greater weight added by more snowfall, along with some melting and freezing, the granular snow is transformed into a denser, recrystalized granular structure called firn. This is the type of ice comprising old snow banks along the sides of roadways during the close of winter. Under the pressure of more snowfalls and the development of thick firn layers, eventually all the air is expelled (except some minute air bubbles) and a mass of interlocking ice crystals is created. At a depth of about 50 m glacial ice is formed. This ice is highly compact and exhibits a vivid blue color, in sharp contrast with the surface ice, which is commonly whitish to gray in color due to the presence of air and sediment.

17.2.2 Glacier movement

Flow is a characteristic of all glaciers. Movement by glaciers is achieved by two major mechanisms: plastic flow and basal slip (Fig. 17.6). Most of our

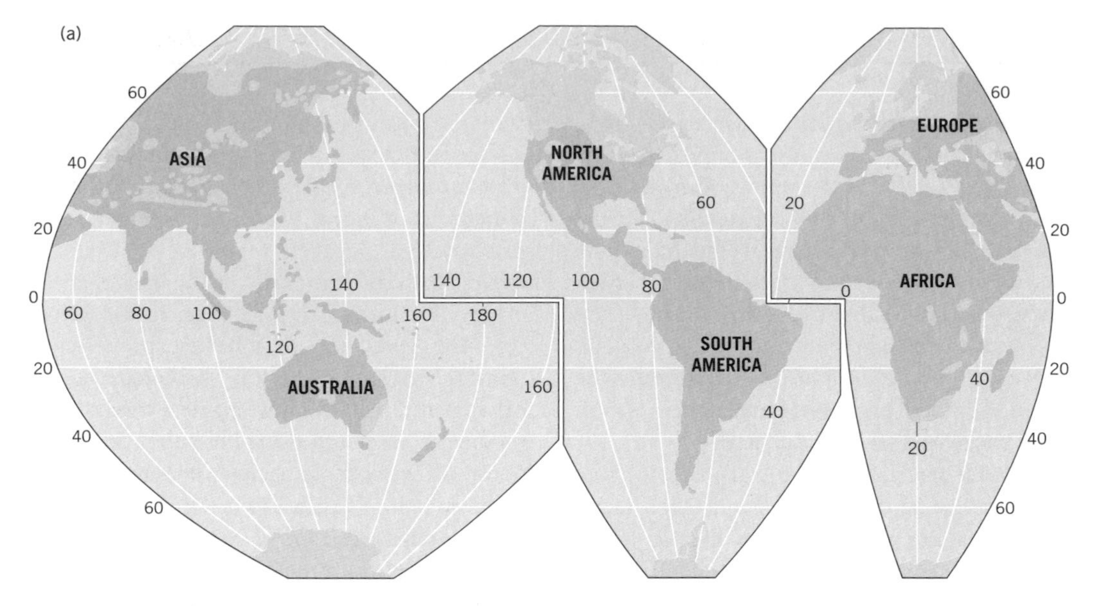

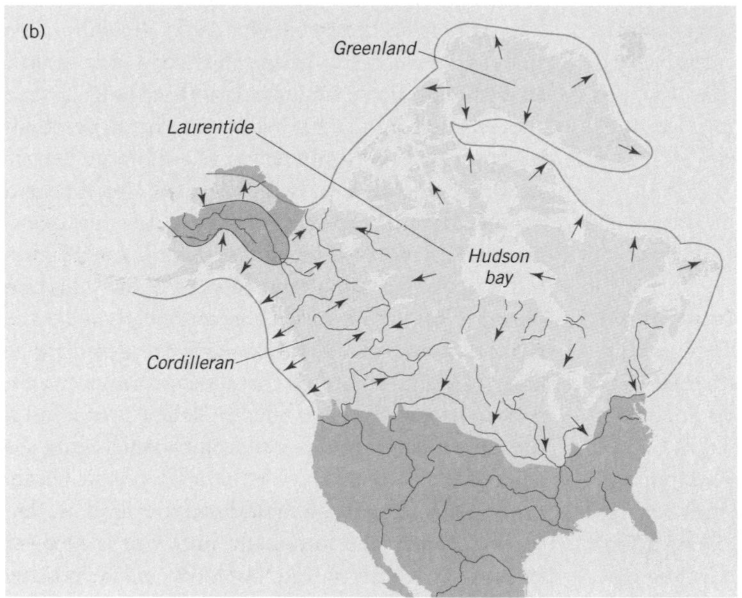

Fig. 17.3 Extent of glacial ice during maximum Pleistocene glaciation. (a) Ice covered portions of North America, Europe, and Asia and the high mountains in other regions 18,000–20,000 years ago. (b) In North America major ice sheets included the Laurentide, Greenland, and Cordilleran sheets.

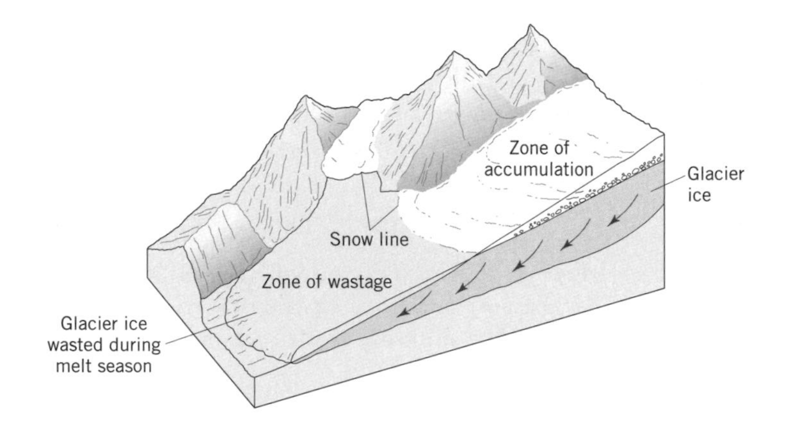

Fig. 17.4 Cross-sectional view of a glacier showing the annual accumulation zone and the ablation zone, also called the zone of wastage.

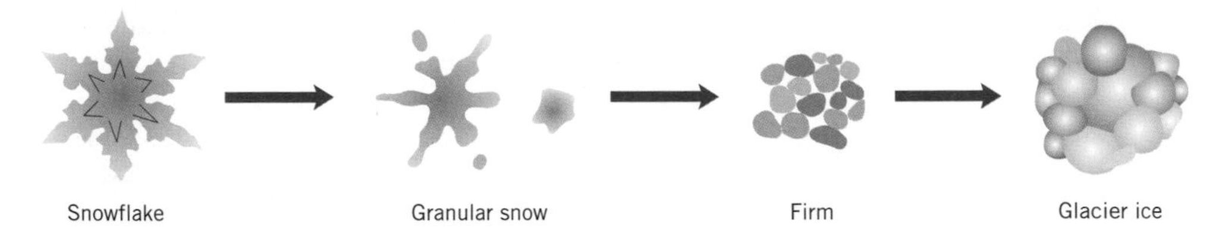

Fig. 17.5 Conversion from freshly fallen snow through several states to glacial ice.

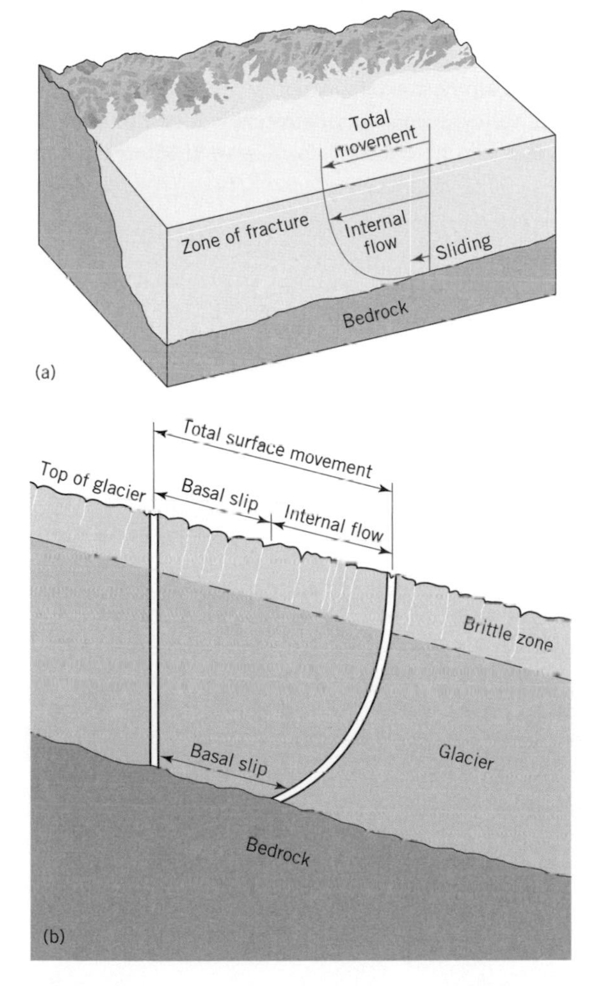

Fig. 17.6 Cross-section of a glacier, illustrating ice movement involving two methods. The first is by plastic flow below a depth of about 50 m. The second process is by sliding along the bottom, referred to as basal slip.

experience with ice leads us to believe that ice is brittle and will shatter when a force is applied to it, such as when we drop an ice cube on the floor and see it break into many pieces. However, at a thickness of 50 m or more ice behaves as a plastic material and can be deformed. Under these conditions the glacial ice below 50 m will flow downslope under the influence of gravity. Differential stresses in the overlying brittle ice may produce deep cracks in the ice surface called crevasses.

Basal slip is an equally important means of producing movement in glaciers. In this process water acts as a lubricant and reduces the friction between the base of the ice and the underlying bedrock or sediment surface. Basal slip allows the entire glacier to slide downslope along a layer, or in some cases a thin film, of water. Formation of meltwater at the base of the glacier is caused by several different mechanisms, including frictional heating, which is produced when the flowing glacier comes into contact with the bedrock or sediment surface. Heat rising from within the Earth's interior may also contribute to the warming and melting of ice at the base of the glacier. Additionally, meltwater is formed beneath the glacier due to the pressure exerted by the thickness and weight of the overlying ice.

17.2.3 Distribution and types of glaciers

Alpine glaciers

In mountainous regions where winters are long with abundant snowfall and summers are cool and short-lived, conditions are perfect for forming alpine glaciers, or valley glaciers, as they are also called (Fig. 17.7). Named from the Alps, where these glaciers are common, alpine glaciers occur throughout the world in all major mountain belts. They

Fig. 17.7 Aerial view of an alpine glacier terminating in a proglacial lake.

originate in the highest parts of mountains and flow down former river valleys under the force of gravity. Alpine glaciers in Alaska cover an area equivalent to half the size of New England (75,000 km^2). Most of these glaciers occur in southern Alaska, including many that have formed in mountain ranges along the Gulf of Alaska. Alpine glaciers are responsible for carving coastal landforms in some locations and for delivering large quantities of sediment to others.

Ice sheets

Unlike alpine glaciers, which are confined to mountain valleys, ice sheets stretch across millions of square kilometers and reach several kilometers in

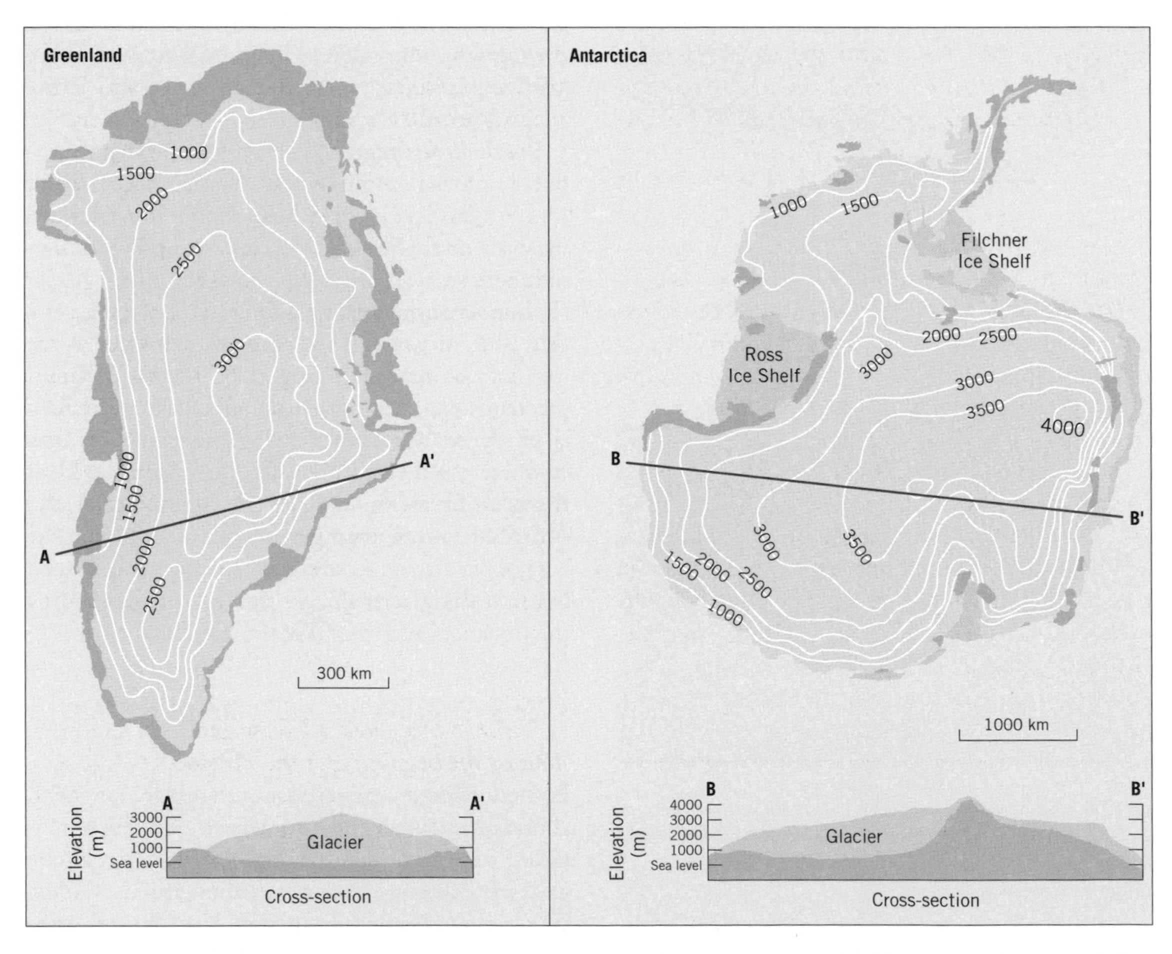

Fig. 17.8 The Greenland and Antarctica ice sheets have a combined area of almost 16 million square kilometers, which is slightly less than the area of South America. (From F. K. Lutgens & E. J. Tarbuck, 1999, *Earth: An Inroduction to Physical Geology*, 6th edn. Upper Saddle River, NJ: Prentice Hall, p. 296, Fig. 12.3. Reproduced with permission.)

thickness. Because of their great surface area, they are also called continental glaciers. Although ice sheets occupied vast regions of North America, Europe, and Asia as recently as 18,000 years ago, there are only two ice sheets left today, one in each of the two hemispheres. Ice covers over 80% of Greenland (1.7 million km^2), the world's largest island. The central portion of the ice sheet is 3000 m thick and thins toward the coast, producing a lens-shaped ice mass (Fig. 17.8). The thicker ice sheet in the interior of the island causes the ice to radiate outward and flow toward the coast. Here it is met by rugged mountain systems that fringe much of the Greenland coast, interrupting its passage to the sea. The coastal mountains act as dams, causing the ice to bulge and build pressure behind them. Ridge systems dissect the ice sheet while mountain passes allow individual lobes of ice to extend toward the sea. These glaciers are called **outlet glaciers** and are similar in appearance to alpine glaciers. Flow rates of the main ice sheet are on the order of 40–120 m yr^{-1}, in contrast to outlet glaciers, which may speed along by as much as a meter a day.

The Antarctic ice sheet is many times larger than the one in Greenland, with an area of just over 14 million km^2, which is about 1.5 times the size of the contiguous United States. The ice attains a maximum thickness of about 4200 m and overlies a mostly bedrock basement (Fig. 17.8). In several locations along the coast, the ice sheet extends across large embayments, forming **ice shelves**. These are regions where the ice thins and is no longer in contact with the land surface, but floats above the sea floor. The ice shelves, of which the Ross and Filchner are the largest, are fed by ice flowing from the Antarctic interior.

17.3 Pleistocene glaciation

17.3.1 Introduction

The cycles of glaciation, which began approximately 2.2–2.4 million years ago, marked one of the most dramatic periods of change in the Earth's recent history. During this time, fluctuations in the worldwide climate caused the periodic advance of huge ice sheets in high-latitude regions, followed by a general retreat of the glacial ice (Fig. 17.9). This period is commonly referred to as the Ice Ages. Through bedrock excavation and sediment deposition, glaciers significantly altered the landscape of large sections of North America, northern Europe, and Siberia, and lesser areas in the southern hemisphere. The vacillating extent of the ice sheet and shifting climatic conditions led to widespread changes in patterns of vegetation and the types of animals dwelling in northern regions. Glaciers created the Great Lakes, as well as Lake Winnipeg, Great Slave Lake, Great Bear Lake, and numerous other large and small lakes. The advance and retreat of the ice sheets is tied very closely to sea-level changes. The precipitation that falls on ice sheets causing their growth ultimately originates from water that is evaporated from the ocean surface. Thus, when an ice sheet enlarges, more and more water from the ocean is being stored in the form of glacial ice and sea level will correspondingly drop. During the most recent glacial maxima, sea level was lowered by at least 120 m.

Drastic changes such as ice sheet movement and sea-level fluctuations had a major impact on many of the world's coastal regions. In high latitudes, glacial action resulted in various types of erosional and depositional coastlines. In low latitudes, not affected by ice, the rise and fall of sea level produced numerous features that give clues to the timing and magnitude of the Ice Ages.

17.3.2 Defining the Pleistocene

The Ice Ages are intimately associated with the geologic time period known as the Pleistocene Epoch. Charles Lyell, a British geologist, originally defined the beginning of the Pleistocene. He based his designation on fossil-bearing sedimentary rocks in Italy, which have been dated at 1.65 million years old. During the past 700,000 years the glacial–interglacial cycles lasted approximately 100,000 years. Furthermore, analysis of fossils from deep sea sediment cores reveals that the Earth has experienced as many as 20 episodes of glaciation. Consequently, it is now accepted that although the lower Pleistocene

Box 17.1 The fate of the Antarctic ice sheet

With all the warnings of increased global warmth, it is sometimes difficult to remember that the Earth is still held firmly in the grips of a great ice age that began about 40 million years ago and intensified dramatically over the past three million years (3 Myr). Thirty million cubic kilometers of ice still reside in Antarctica (Fig. B17.1). Visitors there see the world much as it was 20,000 years ago, when similar sheets of ice, up to 4 km thick, covered much of the northern hemisphere. Over at least the past 2.5 Myr, in concert with changes in the geometry of the Earth's orbit, great ice sheets have expanded and contracted, reaching generally similar maximum dimensions every 100,000 years or so. During each of these glacial maximums, eustatic sea level falls because of water temporarily stored in the great ice sheets – only to be returned to the oceans, sometimes catastrophically, when the ice sheets again return to their present dimensions. If all the ice on Antarctica today were to melt, global sea level would rise by 60 m (about 195 feet). Will this ice melt with future greenhouse warming? Or is the remaining ice on Antarctica stable and well positioned to endure the worst case scenario of greenhouse-induced global warmth? Researchers are divided on this question. One reason for this division is that the ice on Antarctica is not everywhere equally stable. The marine-based West Antarctic ice sheet, which holds an equivalent of about 6 m of sea-level rise, is considered by many to be at risk to predicted global warming (about 3°C) with a doubling of atmospheric CO_2. However, most consider the much larger East Antarctic ice sheet, which is grounded above sea level, to be relatively impervious to postulated greenhouse warmth. Noting that researchers still don't have enough data to predict the behavior of the West Antarctic ice sheet, which leaves open the

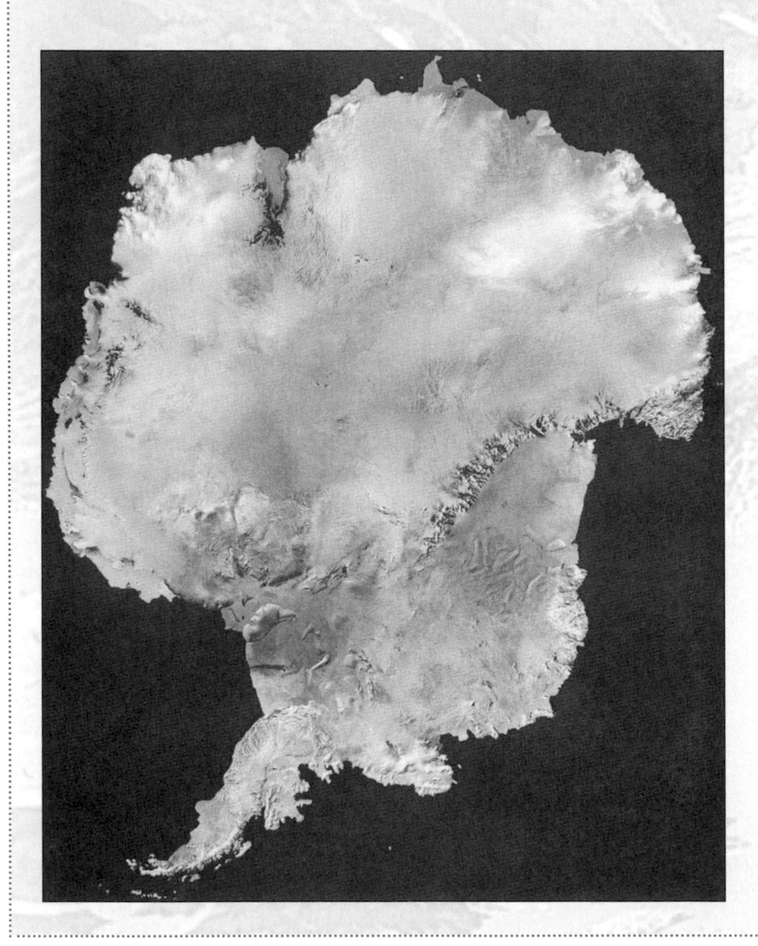

Fig. B17.1 Satellite mosaic image of Antarctica.

Box 17.1 (*cont'd*)

real possibility of up to 6 m of sea level rise with greenhouse warming, let us examine the possible behavior of the East Antarctic ice sheet, for which at present there are more data.

A key to unlocking the mystery of the future behavior of the East Antarctic ice sheet comes from examination of the past behavior of this ice sheet during periods of prior global warmth – particularly those that match or exceed predicted warming from greenhouse emissions. One such time interval is the early-to-mid-Pliocene, about 3–4 Myr BP. One research group, the "dynamicists," argues that most of the ice in East Antarctica melted during the Pliocene. They envision marine seaways crossing the South Pole and sea level 30–40 m higher than today. Indeed, there is some evidence from the Atlantic coastal plain and in Alaska for such high-level shorelines. The dynamicists base their arguments on the presence of marine diatoms in glacial deposits in a mountain range that crosses Antarctica (the Transantarctic Mountains). How were these diatoms deposited? One idea is that the diatoms must have lived in an ancient seaway in interior East Antarctica (at a time when Antarctica was largely deglaciated) and were subsequently picked up by glaciers that advanced across this seaway and carried high up in the mountains. The marine diatoms can be dated based on evolutionary lines, and it is now known that such diatoms lived during Pliocene time. Hence, the argument goes that East Antarctica was largely deglaciated during the Pliocene, succumbing to modest warmth (about 3°C above present values) that existed then and, therefore, likely to succumb again to future greenhouse warming. But the stabilists disagree. They argue that the marine diatoms in glacial sediments in the Transantarctic Mountains could have been deposited by the wind. Winds sweeping across the ocean can entrain diatoms, and once airborne such fine particles can be carried over great distances. In fact, diatoms are found blowing across the South Pole today. The stabilists argue that the East Antarctic ice sheet has been relatively impervious to periods of global warming over at least the past 10 Myr, and that it is unlikely to succumb to anticipated greenhouse warming. In fact, models suggest that the East Antarctic ice sheet might actually expand with modest warming. The reason is that such warming will increase snowfall over the continent, but will fall far short of bringing atmospheric temperatures above the melting point of ice, 0°C.

So what will happen to East Antarctica? The debate rages on. For now, the stabilists have the edge, but that may change as scientists gather more information from the far reaches of Antarctica. Wherever the truth lies, one thing is certain. Ice on Antarctica has the potential to completely change the world as we know it. Most of the world's great cities would be lost if its ice should melt (Fig. B17.2). It will melt sometime. The question is when.

David M. Marchant
Antarctic Scientist
Earth Science Department
Boston University

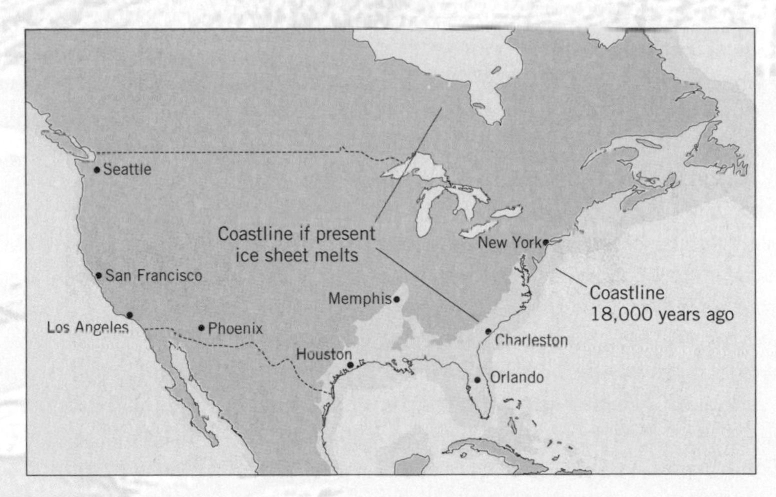

Fig. B17.2 Hypothetical shoreline of North America 18,000 years ago, during glacial maximum and after ice sheets melt. (From E . J. Tarbuck & F. K. Lutgens, 2002, *Earth*. New York: Prentice Hall.)

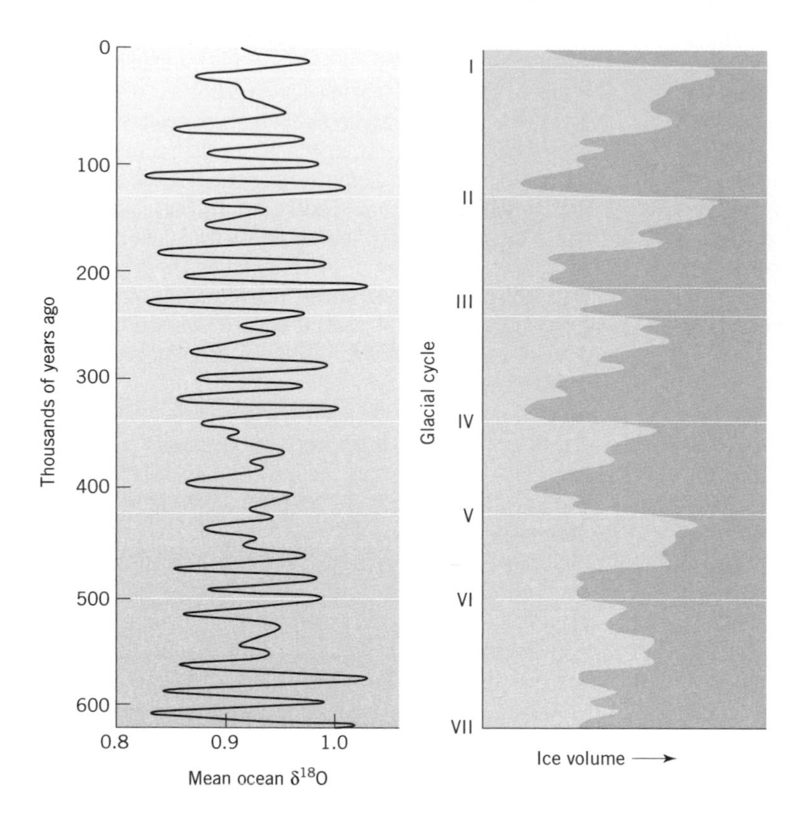

Fig. 17.9 Cycles of glaciation coincide with long-term changes in the intensity of summer sunshine in northern latitudes, which are driven by variations in Earth's orbital characteristics. (From W. S. Broeker & G. H. Denton, 1990, *Scientific American*, January, 48–56.)

boundary is dated at 1.65 million years, the Ice Ages began 2.2–2.4 million years ago.

17.3.3 Causes of the Ice Ages

In the mid-nineteenth century Louis Agassiz, a Swiss scientist, became a convert to the idea that large masses of glacial ice once covered much of Europe and extensive parts of North America. He eventually became the chief spokesman for the glaciation theory and through his studies and many lectures is credited with establishing glaciation as a major geological event. General acceptance of the glacial theory among scientists spawned numerous hypotheses to explain the cause of the Ice Ages. These ideas are still being debated today. Any satisfactory theory must account for the following:

1 Although glaciations have occurred in the geologic past, they have not been a common geological phenomenon.

2 During most of the Earth's history the climate was warmer than it is today. Beginning about 65 million years ago, global temperatures began cooling, eventually leading to the Ice Ages that commenced about 2.2–2.4 million years ago.

3 During the Ice Ages there was a repeated succession of ice sheet growth followed by ice retreat, coinciding with variations in global temperatures of about 5°C.

4 Periods of glaciation and interglacial climates occurred at approximately the same times in both the northern and southern hemispheres.

Effects of plate tectonics

When climatic conditions are favorable, ice sheets form in polar regions and advance to the mid-latitudes. Of course, this process can only occur if landmasses are present in the high and mid-latitudes. This requirement may explain why periods of glaciation have been rare events in geological history. Extensive glaciation occurred approximately 600 million and 250 million years ago. In the latter case, the landmasses had assembled over the South Pole in the super-continent of Pangaea (see

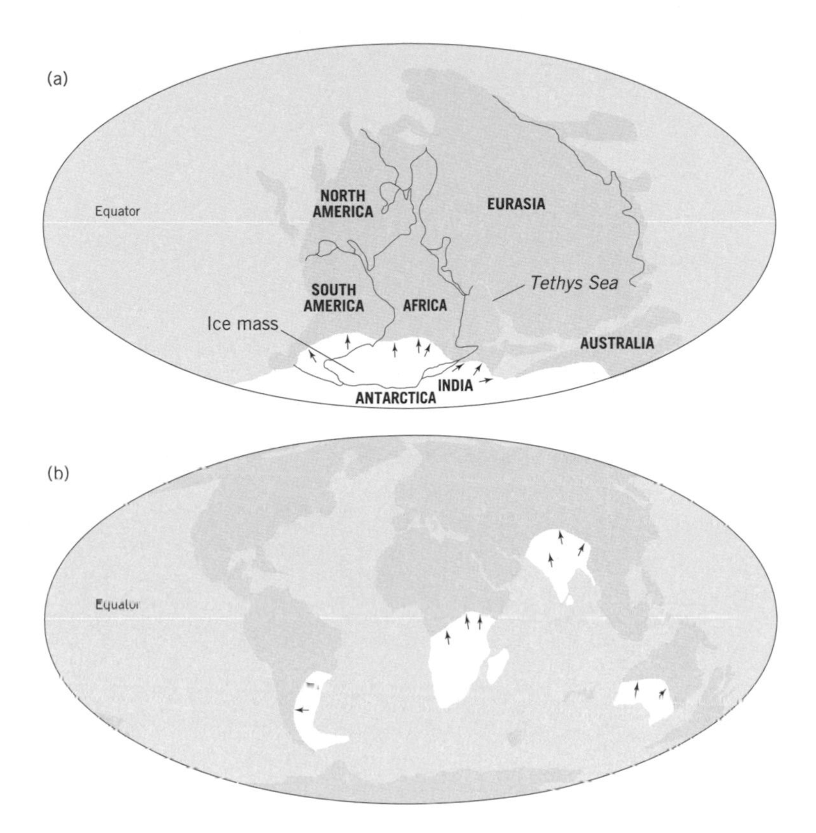

Fig. 17.10 Glaciations occur when climatic conditions are favorable and when continental landmasses are situated in polar regions. (a) View of Pangaea and the ice sheet that covered the Antarctic and surrounding region 300 million years ago. (b) Present day position of the continents and aerial extent of the former glaciated terrain.

Chapter 2). Glaciation ceased in these areas after the breakup of Pangaea as the continents moved to more equatorial regions. Thus, it is apparent that the rarity of glacial episodes throughout the Earth's history is a result of there having been few instances in which the continents have been in polar positions when climatic conditions were conducive for snow accumulation and ice sheet formation (Fig. 17.10).

Carbon dioxide abundance

As already mentioned, the Earth was considerably warmer in the geologic past and has cooled by as much as 10–15°C during the past 65 million years. Coincident with this cooling trend has been a dramatic decrease in the amount of carbon dioxide (CO_2) in the atmosphere, falling to a quarter of its level since the Cretaceous. Because CO_2 is an important greenhouse gas, a decrease in its abundance causes less trapping of solar radiation by the atmosphere, resulting in cooler climates. Because the amount of CO_2 in the oceans is many times greater than that found in the atmosphere (about 60 times), it would appear that the oceans must play a strong role in controlling atmospheric CO_2 and Earth temperature.

Milankovitch climatic cycles

Although plate tectonics explains well the long periods between glaciations (measured in hundreds of millions of years) that have characterized most of the Earth's history, the repeated glacial and interglacial climates that have occurred during the Ice Ages require a different mechanism. Plates move too slowly to account for the waxing and waning of ice sheets over periods of 100,000 years and less. In the early twentieth century Milutin Milankovitch, a Serbian mathematician, calculated seasonal changes in radiation received at various latitudes during the past 600,000 years. He linked cyclic variations in the

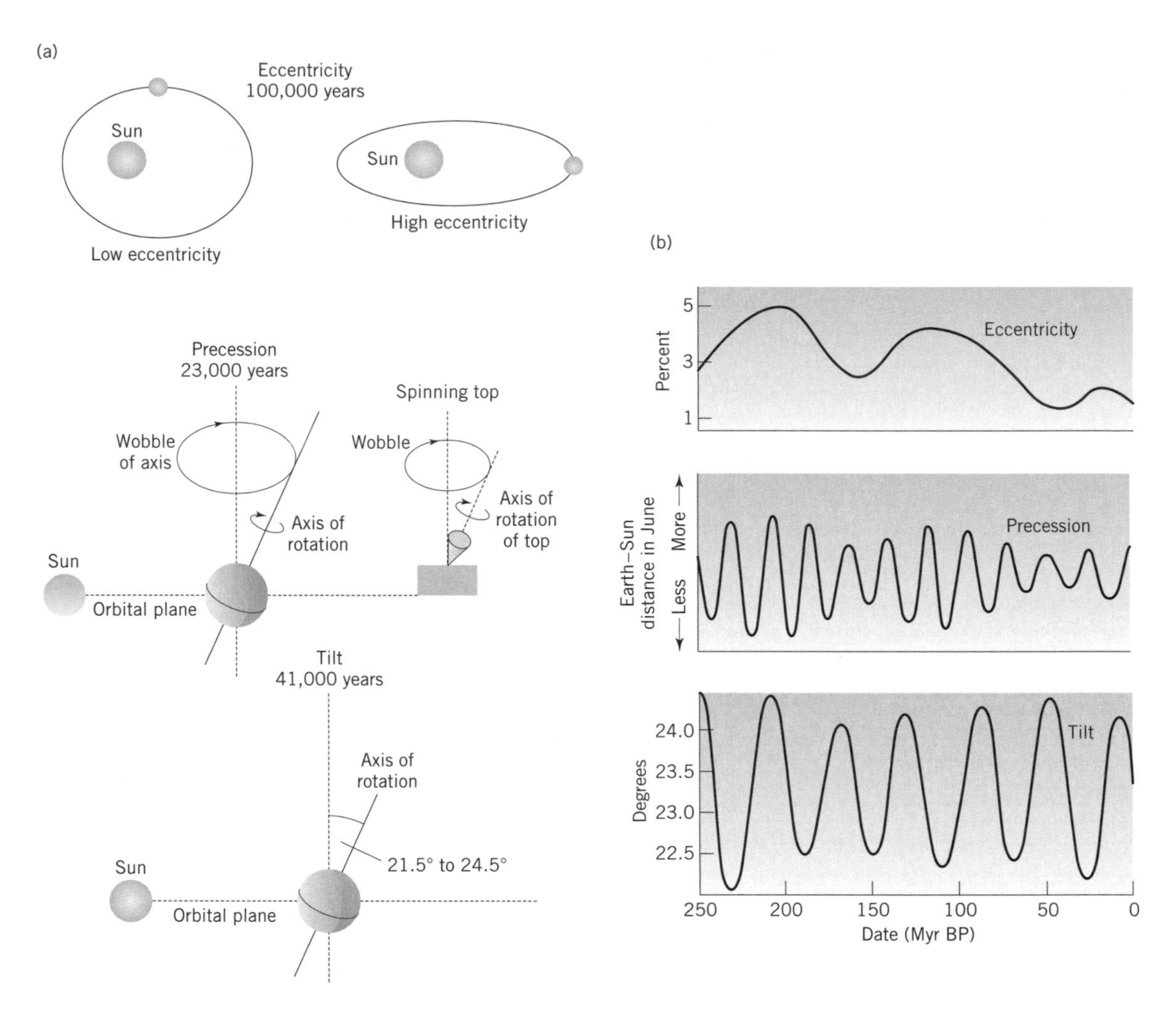

Fig. 17.11 Milankovitch cycles are a product of variations in the Earth's orbital behavior. (a) Orbital elements include eccentricity, precession, and tilt (from F. Press & R. Siever, 1998, *Understanding Earth*. New York: Freeman). (b) The cyclicity of orbital variations is responsible for the intensity of the seasons and is linked to climatic changes (from T. E. Graedel & P. F. Crutzen, 1993, *Atmospheric Change*. New York: Freeman).

Earth's orbital characteristics to changes in climatic conditions, principally the Earth's surface temperature. These variations are called Milankovitch cycles (Fig. 17.11) and are defined below:

1 Eccentricity. The Earth's present orbit about the Sun is elliptical, but at other times it has been almost circular. The span of time for the orbit to cycle from elliptical to circular and back to elliptical is approximately 96,000 years. The amount of eccentricity, which is a measure of how elliptical the orbit is, dictates changes in the distance between the Earth and Sun over the course of a year. Presently, the Earth is closest to the Sun when the northern hemisphere is experiencing winter and the southern hemisphere has its summer. The Earth is presently furthest away from the Sun during summer in the northern hemisphere. The opposite was true approximately 50,000 years ago.

2 Obliquity. The Earth's axis of rotation is inclined 23.5° with respect to a line drawn perpendicular to a plane containing the Earth's orbit. This is commonly referred to as the Earth's tilt. Every 41,000 years the Earth's tilt cycles between a minimum value of 21.8° and a maximum of 24.4°. The greater the tilt the more pronounced are the seasons.

3 Precession. Presently, the Earth's axis of rotation points toward the North Star. However, in 11,500 years the axis will point to the star of Vega, and 11,500 years after that it will once again be directed toward the North Star. This 23,000-year cycle describes a circular precession of the axis of rotation. It is likened to a spinning top that slows down and begins to wobble. The wobble is the precession of the spin axis. In the present configuration, summer in the northern hemisphere occurs when the Earth is inclined toward the Sun. It is tilted away from the Sun during winter. In 11,500 years the tilt will have precessed 180° such that in the northern hemisphere the present summer will become winter and vice versa.

Milankovitch showed that the interaction of these three cycles did not change the total amount of solar radiation reaching the Earth, but it did affect the contrast in the seasons. For example, if summers were cooler, then glaciers might be expected to enlarge due to less melting of ice, while at the same time slightly warmer winters might actually increase snowfall. This is an oversimplified scenario but it does demonstrate the type of climatic changes that are induced by the Milankovitch cycles. Present thought is that Milankovitch cycles are somehow propagated through changes in atmospheric and ocean circulation and these broad conveyer belts of heat and cold control global climatic fluctuations.

17.3.4 The late Pleistocene

Waxing and waning of ice sheets characterized the Pleistocene Epoch throughout the world. In North America the last of these major glaciations is referred to as the Wisconsin Ice Age, named after the state in which the deposits left behind by the ice sheet are easily studied. During the beginning of the Wisconsinan (70,000–90,000 years BP) the ice sheets began expanding, reaching their maximum southern extent about 20,000–18,000 years BP. The margin of the ice sheets is defined in many regions by particular types of glacial deposits, providing an ideal means of mapping the limit of the ice. Ice sheets covered all of Canada and the mountainous areas of Alaska. In western Canada coalescing glaciers flowed to the Pacific. An expansive ice sheet, called the Laurentide Ice Sheet, was centered over Hudson Bay in eastern Canada and flowed outward in all directions. In Hudson Bay the ice was almost 4000 m thick and in New England it reached more than 2000 m in thickness. The Laurentide Ice Sheet extended eastward to the Atlantic Ocean and as far south as southern Illinois and Indiana. Ice sheets also covered much of northern Europe and parts of Siberia. The only vestige of the Laurentide Ice Sheet today is the Barnes ice cap on Baffin Island in northeastern Canada. The end of the Pleistocene coincided with a period of abrupt global warming and rapid retreat of ice sheets in the northern and southern hemispheres. It should be noted that this and other sudden shifts in the Earth's climatic patterns are not easily explained by Milankovitch cycles and it is probable that other mechanisms are responsible, such as changes in ocean circulation. By definition, the Pleistocene ended 10,000 years ago, when the Holocene commenced. The beginning of the Holocene marks a period of rapid warming in North America and Europe, as indicated by pollen records.

17.4 Glacial effects on coastlines

The growth and decay of ice sheets dramatically affects the morphology of coastlines due to the ability of the glaciers to carve into bedrock, strip away loose materials overlying bedrock, deposit large quantities of sediment, and change the level of the world's oceans. Direct effects of glaciation include those due to glacial erosional and depositional processes and those resulting from elevation changes of the coast associated with ice loading and unloading the land. Indirect effects of glaciation are linked to sea-level fluctuations produced by volume changes of the ice sheets.

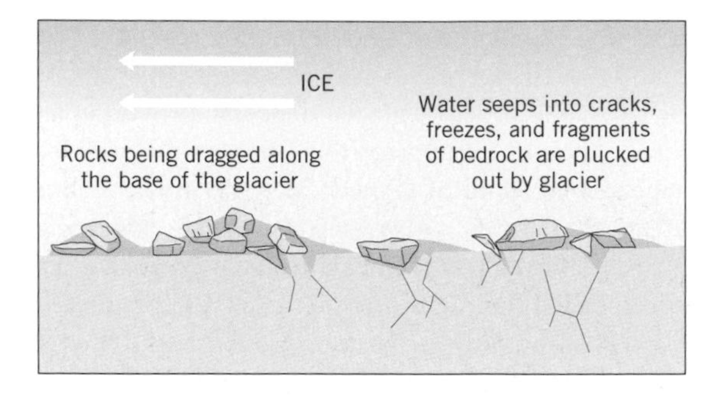

Fig. 17.12 Erosion by glaciers occurs through ice wedging and plucking and by abrasion.

Fig. 17.13 View of glacial striations on bedrock along the coast of Maine. Striations are oriented left to right.

17.4.1 General erosional processes

Erosional processes by glaciers have been responsible for the formation of the Matterhorn in the Swiss Alps, Half Dome in Yosemite National Park in California, and the deepwater port of New York City. Erosion by glaciers occurs by several processes, the most important of which are ice wedging, plucking, and abrasion (Fig. 17.12). Ice wedging occurs when meltwater flows into the cracks and crevices of the bedrock underlying the ice. As the meltwater refreezes, the expansion of water to ice causes appreciable pressure on the sides of the crack. As this process is repeated over and over again, pieces of the bedrock, large and small, are wedged free. The process is especially prevalent where the original bedrock surface is highly fractured and meltwater can readily penetrate numerous cracks. The excavation process (plucking) is completed when the pieces of rock are quarried from the bedrock and incorporated into the ice. This is achieved through material freezing to the base of the glacier. The plastic nature of glaciers at depth also allows the larger rocks and boulders to be enveloped by the ice. Once the material is incorporated within the ice, it flows with the glacier and is transported toward the ice margin. Along its journey, the rock fragments carried at the base of the ice abrade the underlying bedrock. Just as the sediment that is transported by the Colorado River accelerates the cutting of the Grand Canyon, the rock material carried by a glacier significantly increases the erosion process. The grooves and scratches cut into bedrock surfaces, called **striations**, are evidence of glacial abrasion (Fig. 17.13). Scientists use the orientations of these and other linear features to determine the flow direction of the ice.

17.4.2 Fjords

Formation

The erosional coastal landscapes exposed by the retreating ice at the end of the Pleistocene exhibited considerable variability. Their form was dependent on the original topography and bedrock structure of the region. Along mountainous coastlines in high-latitude regions, alpine glaciers deepened existing mountain valleys, producing U-shaped glacial troughs. In these settings the ends of the valleys were often eroded well below sea level. When the glaciers retreated from the valleys and sea level rose following deglaciation, the glacial troughs were flooded, creating fjords (Fig. 17.14). These features are found throughout the high-latitude coastlines of the world, including the Scandinavian countries, Iceland, Greenland, eastern and western Canada, Alaska, New Zealand, and Chile. Fjords are commonly spectacular features with steep, cliff-like sides and winding valleys that follow the original sinuous mountain valleys from which they were formed. They can be hundreds of meters deep and some reach more than 1000 meters in depth, which means that the creation of the fjords is attributed to more than just rising sea level resulting from melting

Fig. 17.14 Fjords are flooded coastal mountain valleys that are created when alpine glaciers excavate bedrock valleys below sea level.

ice sheets. In most instances, the depth of a fjord is chiefly a function of the amount of bedrock excavation that was accomplished by the alpine glaciers.

Kenai Peninsula, Alaska

The Kenai Peninsula is a 250 km long landmass that juts southwestward from southern Alaska. It is separated from the Alaskan Peninsula by the long, narrow, macrotidal embayment of Cook Inlet; Anchorage is located in the upper reaches of Cook Inlet. The mountainous landscape of this region is a product of tectonic processes associated with the subduction of the Pacific Plate beneath North America. During numerous episodes of alpine glaciation, ice flowed southeastward from the Kenai Mountains and excavated a series of deep fjords along the open coast, many of them more than 100 m deep (Fig. 17.15). Steep-walled bedrock slopes and cliffs that rise several hundred meters high characterize much of the seaward shoreline of the fjords. In contrast, the heads of many fjords contain outwash fans and fan deltas that are actively prograding into the deep water. Braided streams draining meltwater and sediment from retreating glaciers produce these sedimentary features and resulting low topography. During the 1964 Good Friday earthquake this region was downwarped a maximum of 2.3 m below sea level. One of the major effects of this event was the formation of drowned forests at the heads of some of the fjords and in other embayments along the Kenai coast.

Fig. 17.15 View of one on the many fjords found along the Kenai Peninsula of Alaska.

17.4.3 Rocky coasts

Formation

The effects of continental glaciers are strikingly different from those of alpine glaciers. Whereas alpine glacial processes tend to erode and deepen mountain valleys, thereby accentuating the rugged terrain, continental glaciers tend to reduce the relief along a coast, although they may produce a highly irregular coast. Continental glaciers are too thick to be confined to valleys, and spread over entire landscapes, including low mountain systems. Along many high-latitude, glaciated coasts in North America, Europe, and elsewhere the major effect of the Pleistocene ice sheets was to strip away the sediment cover and excavate several meters of the underlying bedrock. Although the glaciers left behind a thin layer of undifferentiated sediment, called till, these coastal regions are normally very rocky, with numerous embayments, bedrock promontories, and islands. Beaches and barriers are uncommon along these types of coastlines and, where they do occur, their sediment supplies are close by.

Northern New England

The coast of northern New England illustrates well the general effects of the Pleistocene ice sheets.

Fig. 17.16 Pocket beach located along a rocky sediment-starved coast. Sediment is derived from the reworking of local thin till deposits.

Fig. 17.17 Along glaciated coasts tidal inlets are commonly anchored next to bedrock exposures or till headlands (Westport River Inlet, Massachusetts).

From northern Massachusetts to northeastern Maine the coast exhibits a wide range of morphologies that are a function of isolated sediment sources, a highly variable bedrock fabric, and a land area that has been inundated by the sea during the past 11,000 years. The bedrock imprint on this region is particularly important, producing cliff coasts in extreme eastern Maine, broad deep embayments with numerous islands and peninsulas in central Maine, and finally a straighter coast along southern Maine, New Hampshire, and northern Massachusetts that contains bedrock promontories separated by gently curved embayments. The rocky nature of this coast is a direct consequence of glacial erosion, including the removal of sediment that had once covered the bedrock basement. Along much of this coast the only depositional landforms are small pocket beaches and barriers in protected embayments that have developed from the reworking of local glacial deposits (Fig. 17.16). The numerous bedrock islands that characterize this coast also led to the development of tombolos and cuspate spits. The composition of these depositional features is usually sand and gravel, reflecting the mixed sediment of the glacial sources.

Exceptions to the general trends cited above occur in regions where inland deposits of sand have been delivered to the coast in large quantities. In the lowlands of Maine, as in many other areas of New England, there are immense sand and gravel deposits that were produced as the ice sheet retreated northward from this region (discussed below). During and following deglaciation, these deposits were excavated by tributaries and trunk streams of major river systems and brought to the coast. This movement of sand down rivers, although less active, continues to the present time. The sediment deposited at the mouths of these rivers has been redistributed alongshore as well as onshore by wind, waves, and tides, forming sandy beaches and barriers chains, some of which reach 30 km in length. Most of these barrier systems are located within arcuate embayments and individual barriers are usually anchored to bedrock promontories or glacial headlands. Likewise, tidal inlets along these chains have stabilized next to bedrock outcrops or glacial deposits (Fig. 17.17). Depositional features in this region are usually isolated and directly linked to nearby sediment sources such as rivers.

17.4.4 General depositional processes

Glaciers not only carve mountain valleys and strip away the sediment cover from vast areas; they are also responsible for widespread sediment deposition. The rock and sediment that are removed and transported away by glaciers from one location are eventually deposited by the ice at another site. The large boulders that are moved by glaciers and laid down far from their origin are called erratics

Fig. 17.18 Large erratic along the Cape Cod shoreline.

(Fig. 17.18). They are commonly the size of a car, but the largest in North America, the Okotoks of southern Alberta, Canada, is larger than a two-story house. By matching the rock type of erratics to the bedrock from which they were derived, glacial geologists are able to determine the direction of ice flow and how far the erratics have traveled. For ice sheets, this distance may be as little as a few kilometers or in some cases as much as 1000 km. Erratics litter the landscape of New England, portions of the Midwest, and many areas within Canada and northern Europe. Louis Agassiz used these features as the primary evidence to advance his glaciation theory.

Glacial material comprises two major categories of sediment: (i) till, which is sediment deposited directly by the ice; and (ii) stratified sediment, which is layered sediment deposited by glacial meltwater (Fig. 17.19). Glaciers carry a variety of sediment sizes, from clay-sized material to large boulders. When a glacier retreats, the sediment it carries melts out from the ice and is deposited in an unsorted, chaotic mass called till. Anyone who has dug a hole in till to plant a bush or excavate a trench is familiar with its bouldery composition. In the farmlands that stretch across glaciated areas the boundaries of the fields are commonly outlined by

(a)

(b)

Fig. 17.19 Glacial sediment consists of: (a) till, undifferentiated sediment deposited by the glacier; and (b) stratified drift, layered sand, and gravel deposited by meltwater. (From Miles Hayes.)

stone walls. The rocks making up these walls, large and small, were placed there by farmers who wished to rid their land of the boulders that obstructed the tilling of their fields.

Unlike till, stratified drift is deposited by flowing water derived from the melting ice. Because the energy needed to transport different sized sediment is directly related to the velocity of the flow, water-laid sediments tend to be sorted and are deposited in layers. In stratified drift these layers commonly consist of sand and fine gravel. However, if the current is very strong, coarse gravel layers can form, or if the current flows into a standing body of water, layers of silt and clay may develop.

17.4.5 Depositional landforms

Along glaciated coasts, there are numerous examples

where glacial deposits formed the initial shoreline, and although coastal processes have subsequently modified these deposits, the original glacial features are still recognizable. These deposits may be large (>100 km) or small (<5 km) and may be composed of till or stratified drift, or, in the case of Cape Cod, Massachusetts, and Long Island, New York, they may consist of both types of deposits. The major types of glacial deposits along coasts include end moraines, outwash plains, and drumlins.

End moraines

As discussed above, ice sheets flow outward from interior regions of ice accumulation. Glaciers continue to advance as long as more ice is formed during the winter than melts during the summer. For ice sheets, the distance separating the area of net ice formation from the region of ice melting along the margin of the glacier may be over 1000 km. If the amount of ice that forms in the interior of an ice sheet equals the amount that is lost through melting and sublimation (the process whereby ice moves directly from the solid to the gaseous state), then the ice marginal position will remain stationary. Under these conditions the ice sheet, acting like a conveyor belt, continuously transports sediment to the terminus of the glacier, where it melts out from the ice and is deposited. As this process continues through time, the accumulating sediment forms a ridge of till called an **end moraine** (Fig. 17.20). The longer the ice terminus remains in the same position, the greater the amount of sediment that is delivered to the ice front and the larger the end moraine becomes. Because the conditions forming end moraines are so variable, their size and extent range widely, from prominent ridges 20–50 m in height and 100 km long to those that are only few meters high and extend discontinuously along the former ice margin. End moraines are useful to scientists studying past glaciations because they mark the furthest advance of an ice sheet, as well as its recessional positions, where the ice front remained stationary for a short period of time before continuing its retreat. In the Midwest a series of end moraines, hundreds of kilometers long, outline the southern borders of the Great Lakes.

Fig. 17.20 End moraines are formed when the margin of a glacier remains in the same position for a period of time. Under these conditions the flowing ice piles sediment at its terminus, forming a ridge of till called an end moraine. (From Miles Hayes.)

Outwash plains

As an end moraine is created at the terminus of a stationary ice sheet, the melting ice produces torrents of water choked with sediment. On warm summer days water flows everywhere from underneath, over the surface, and through tunnels within the glacier. Meltwater discharging from the ice front forms a broad network of shallow streams whose channels regularly divide and rejoin in a braid-like pattern. Within a short distance of the ice, much of the bed load transported by these braided streams is deposited due to the gentle slope and resulting decrease in current velocity. Meltwater streams sort the glacial sediment, leaving behind the largest sized material near or within the ice front. Sand and gravel are deposited in layers beyond the ice terminus and in time build an expansive **outwash plain** (Fig. 17.21). The finest sediment (clay and silt) is transported out of the system and often into the ocean or a lake. Thus, outwash plains are large sandy-gravel regions (some more than 100 km long with an area greater than 1000 km^2) with low topography, consisting of stratified drift. Examples include most of Cape Cod and Long Island, as well as the Skiederarsandur in southeast Iceland. These features are also formed at the end of some valley glaciers.

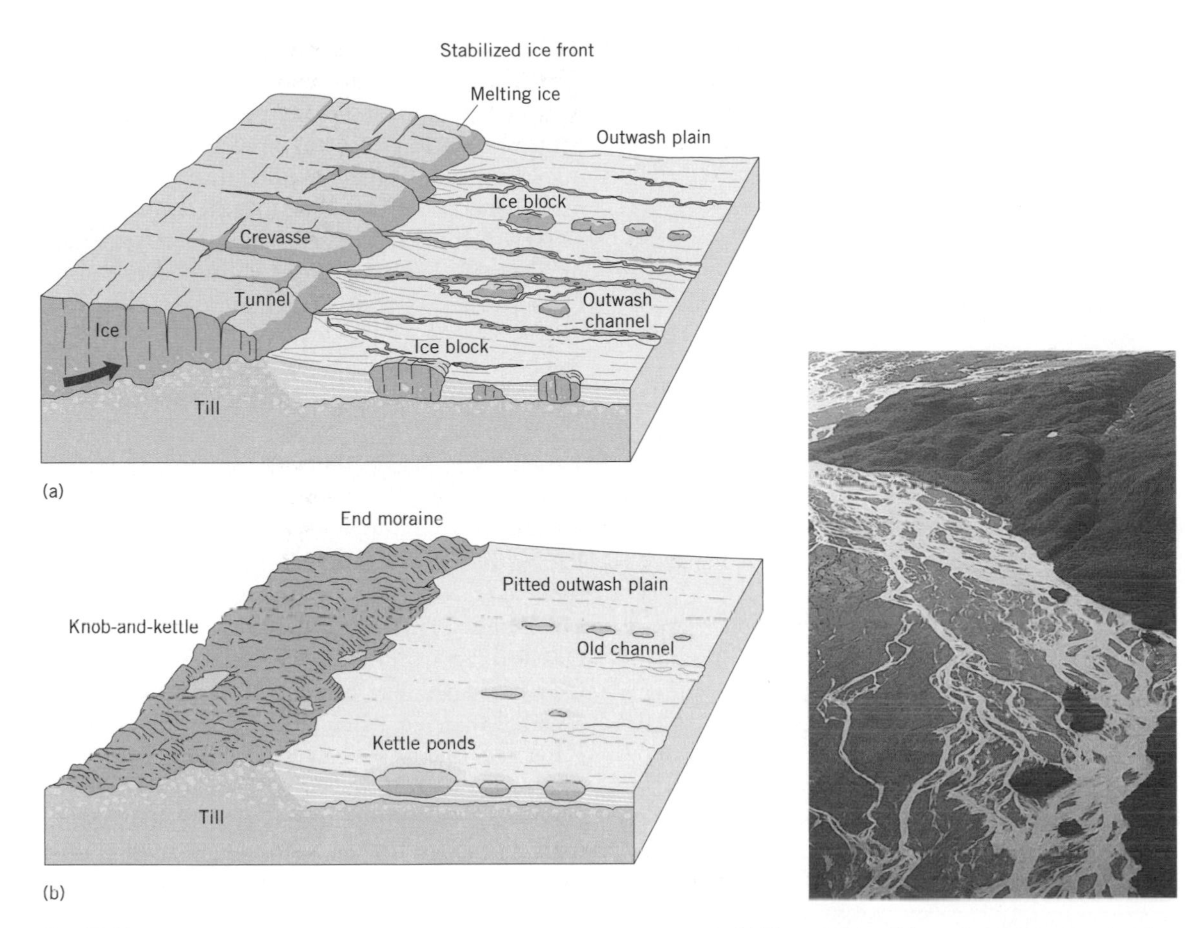

Fig. 17.21 An outwash plain consists of layers of sand and gravel (stratified drift). (a) They are formed through deposition by glacial meltwater streams (from A. N. Strahler, 1966, *Geologist's View of Cape Cod*. Orleans, MA: Parnassus Imprints). (b) The braided streams in this photograph are most active during the summer when nearby glaciers discharge abundant meltwater (from Dave Marchant).

Drumlins

One of the most distinctive glacial depositional features is a **drumlin**. Drumlins are teardrop-shaped accumulations of till ranging from 15 to 50 m in height and 0.5 to 2 km in length (Fig. 17.22). In some instances they may also be composed of highly contorted stratified drift, while others may be cored by bedrock. The blunt, steep side of a drumlin faces the direction from which the glacier advanced, and the down-glacier end is streamlined, with a gentler slope. This geometry is reminiscent of a spoon turned upside down. Because drumlins are elongated parallel to the flow direction, they are use by glaciologists to study patterns of ice sheet movement. Drumlins usually occur in significant numbers. In upstate New York over 10,000 drumlins exist in an area of approximately 12,000 km^2. A few hundred are located in the immediate vicinity of Boston, Massachusetts, including a drumlin where the famous Revolutionary War battle of Bunker Hill was fought in 1755. An interesting note of history is that while the Bunker Hill Monument commemorating the battle stands on the drumlin of Breeds Hill, the actual battle was fought on the adjacent drumlin of Bunker Hill. The

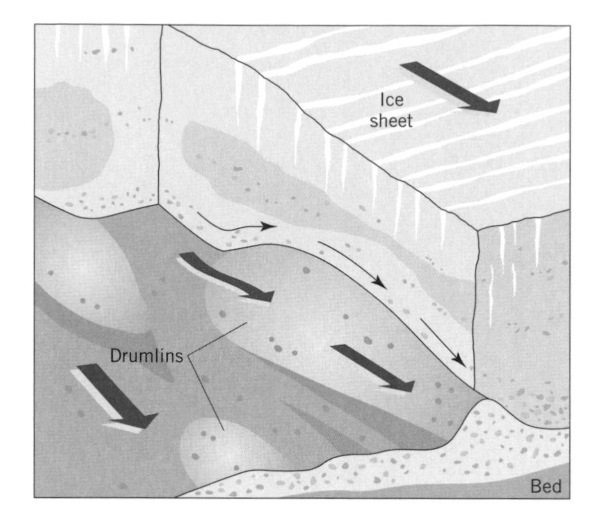

Fig. 17.22 Drumlins are composed of till and usually occur in groups. (a) They are formed beneath the ice and have the form of an upside down spoon (from S. Chernicoff, 1995, *Geology*. New York: Woth). (b) Erosion of a drumlin in Nova Scotia has left behind a boulder retreat lag, which can be used to trace the former extent of the landform.

fact that drumlins always occur in clusters with similar geometry and orientations leads scientists to believe that they represent an equilibrium bed configuration in which the flowing ice molds the underlying till. Other drumlins may be a product of glacial erosion. In addition to the drumlin fields of New York and New England others are found in England, Ireland, eastern Canada, Wisconsin, Michigan, and western Washington.

17.5 Examples of glaciated coastlines

17.5.1 Cape Cod

Formation

During the summer, one of the most popular vacation spots in New England is Cape Cod, including the two islands of Martha's Vineyard and Nantucket that lie directly off the Cape's southern shore. Cape Cod is shaped like an extended bent arm with a curled-fingered hand pointing northward. The Cape and islands were formed during the retreat of the Laurentide Ice Sheet, which reached its maximum extent approximately 20,000–18,000 years ago (Fig. 17.23). These landmasses consist almost entirely of unconsolidated glacial sediments (about 90% by area). Exceptions are the beaches, barriers, marshes, and tidal flats that outline the present cape, and even these features were formed from reworked glacial sediment. Although some large erratics exist on the cape, there are no bedrock exposures.

To discuss the glacial origin of Cape Cod and the islands we must go back to an interglacial period (between glaciations) when the climate was the same or slightly warmer than it is today, about 130,000–100,000 years ago. At this time there was no Cape Cod. Following this period, temperatures cooled and the Laurentide Ice Sheet grew in size, expanding outward from the Hudson Bay region. It reached its maximum southern extent in the Great Lakes region about 18,000 years ago and in eastern United States perhaps a little earlier (about 20,000 years ago). In the vicinity of southeastern New England, the terminus of the ice sheet consisted of several large

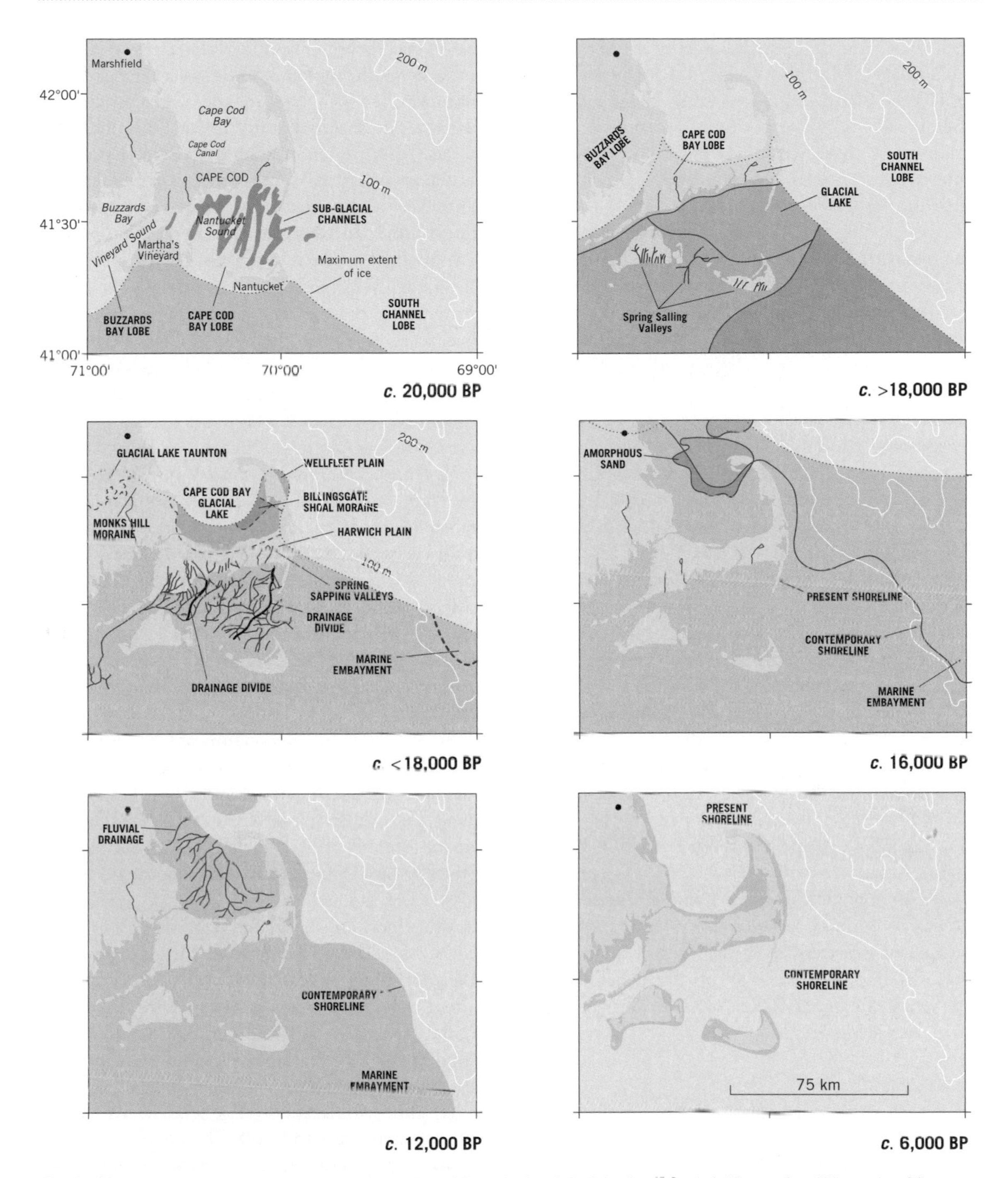

Fig. 17.23 Sequential diagrams depicting the formation of Cape Cod and the Islands of Martha's Vineyard and Nantucket. They consist of sediment that was carried south and deposited by the Laurentide Ice Sheet in the form of moraines, outwash plains, and other glacial deposits. (From E. Uchupi, G. S. Giese, D. G. Aubrey & D. J. Kim, 1996, The Late Quaternary construction of Cape Cod, Massachusetts: a reconsideration of the W. M. Davis model. Boulder, CO: Geological Society of America, Special Paper 309.)

lobes that coincide with the present-day locations of Nantucket, Martha's Vineyard, Block Island off the Rhode Island coast, and Long Island in New York. It was the overall position, configuration, and dynamics of these ice lobes that determined the shape and location of the cape and islands to the south and west.

Cape Cod and the islands developed during at least two periods when the margin of the Laurentide Ice Sheet stabilized and large quantities of sediment were deposited at these ice terminuses. At the ice sheet's southernmost position, an end moraine formed, corresponding to the northern third of Martha's Vineyard. This is a hilly, bouldery region and is relatively elevated compared with the rest of the island. A similar but less extensive moraine exists on Nantucket. During the same time that the moraines were formed, broad, sandy outwash plains were deposited by meltwater streams draining water and sediment from the ice terminus. These sand plains define the southern border of both islands. Following this period of ice-front stability, the ice sheet retreated 50 km northward and stabilized again. In this position, the Sandwich and Buzzards Bay moraines and a number of outwash plains were formed that together constitute the southern extent of Cape Cod. The forearm of the cape developed next after the ice sheet withdrew further northward. At this time, the lobate nature of the ice front left a large low area between the Cape Cod Bay lobe, filling much of Cape Cod Bay, and the South Channel lobe, which was located east of the present cape. This low area was filled with stratified drift (outwash) produced by meltwater braided streams that flowed westward from the South Channel Ice Sheet. With this final stage of glacial deposition completed, the general form of Cape Cod and the islands was achieved.

Modification

Cape Cod and the islands have undergone many important modifications during and after the ice withdrawal from New England. During the period of ice retreat, at least two large glacial lakes occupied what is now Nantucket Sound and Cape Cod Bay. The first of these lakes to develop, Lake Nantucket Sound, was dammed by the glacial deposits to the west and south, including Martha's Vineyard and Nantucket, and by ice to the north and east. Meltwater discharging from the ice terminus filled the lake. The unconsolidated, porous nature of the deposits that make up Martha's Vineyard and Nantucket led to water from the glacial lake being piped southward through these sediments, where it eventually flowed out along the surface. This process gradually formed channels that ate their way northward across the sand plains as sand and fine gravel were eroded at the heads of these channels and transported southward. These channels are called **groundwater-sapping channels** and can be seen on a small scale on a beach at low tide where groundwater leaks out along the beachface (Fig. 17.24). Along Nantucket and Martha's Vineyard groundwater-sapping processes have formed a series of semi-parallel channels that have since been inundated with seawater due to rising sea level. The exact same process was responsible for groundwater-sapping channel development along the southern shore of Cape Cod when a large glacial lake occupied Cape Cod Bay. Many of these elongated lagoons, as they have become, are important harbors along this part of the cape.

The other major modification of Cape Cod and the islands has been a general smoothing of the shorelines through erosional and depositional processes. The initial form of Cape Cod left by the retreating ice sheet was very different from how it appears today. Immediately following deglaciation sea level was as much as 120 m lower than the present position and thus the deposits comprising Cape Cod and the islands were simply regions of somewhat higher elevation. It wasn't until eustatic sea level rose to within 7–8 m of where it is today that the general configuration of Cape Cod became recognizable. The initial outline of the cape was very irregular, with numerous promontories, embayments, and islands (Fig. 17.25). However, wave action gradually eroded headlands and small islands, liberating sediment that was transported alongshore, eventually building spits across embayments. Examples of this process are seen in Cape Cod Bay, where Sandy Neck has accreted in front of

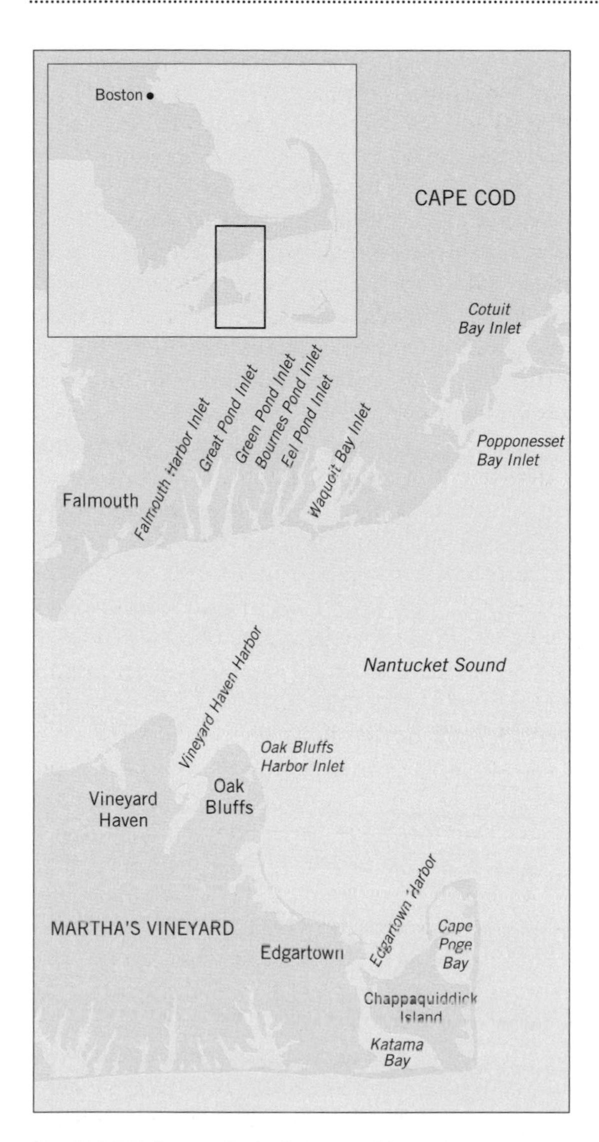

Fig. 17.24 Following the deglaciation of this region groundwater-sapping channels developed in the outwash plains of Cape Cod and Martha's Vineyard. When sea level rose to its present position, the channels were flooded, forming the highly indented lagoons along the southern shore of Cape Cod and the islands.

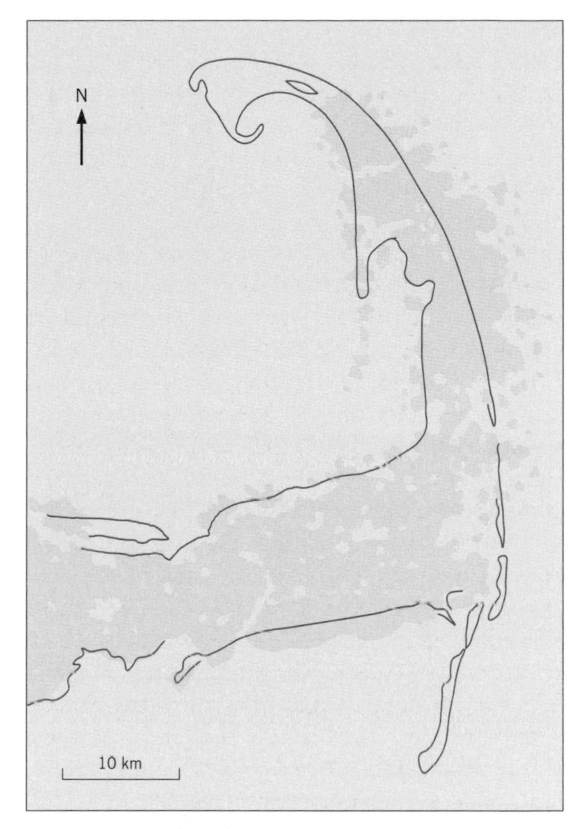

Fig. 17.25 The present smooth outline of Cape Cod is a product of wave erosion and the construction of spits in front of embayments. (From Strahler 1966, details in Fig. 17.21.)

Barnstable Bay, and along the outer cape, where Nauset Spit has formed in front of Pleasant Bay. The widespread glacial cliffs that occur along Cape Cod and the islands are evidence of the erosional process that has smoothed the coast. Some of the cliffs are more than 30 m high, such as those along the open-ocean coast of northeastern Cape Cod. Much of the sediment that has been eroded from these cliffs has been transported northward, forming the extensive spit of the Provincelands.

17.5.2 Drumlin coasts

In some glaciated terrains, notably the eastern shore of Nova Scotia, Clew Bay on the central west coast of Ireland, and Massachusetts Bay, drumlins and accretionary landforms that have developed from reworked glacial sediment dominate the coast. In Massachusetts Bay, for example, drumlins exert a strong imprint on the entire landscape. All the islands, save a few bedrock ledges, are drumlins and

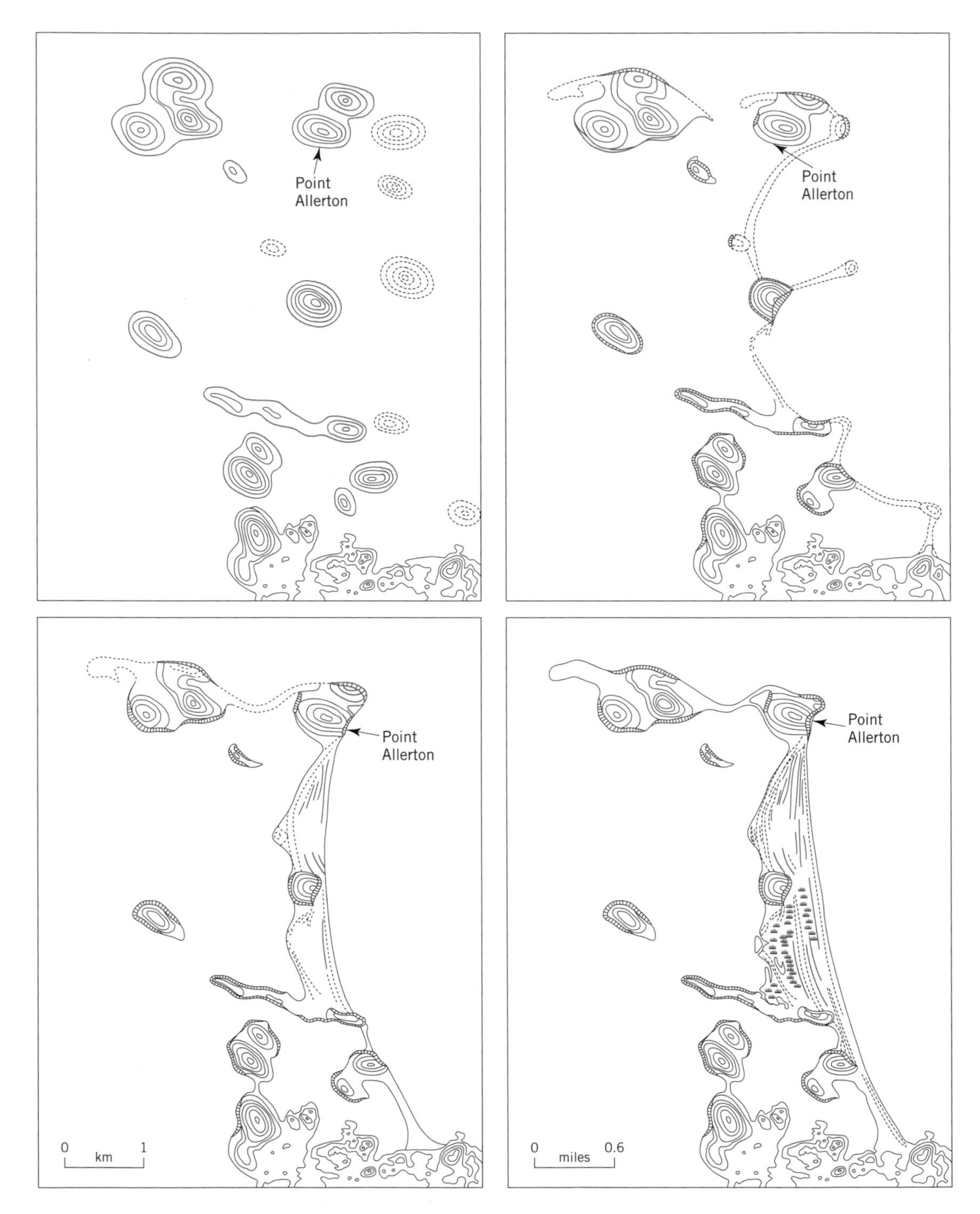

Fig. 17.26 Evolutionary model of Nantasket Beach, Massachusetts, as envisioned by D. W. Johnson, 1925, *The New England–Acadian Shoreline*. New York: John Wiley & Sons.

the northeast and southeast borders of Boston Harbor consist of drumlins and reworked drumlin deposits. D. W. Johnson, one of the first marine geologists to study glaciated coasts, presented an evolutionary model for this shoreline in 1910 (Fig. 17.26). Similar to the scheme presented for the eastern shore of Nova Scotia (see Chapter 7), the outer coastline of Boston Harbor represents the end product of sand and gravel eroded from offshore drumlins and moved onshore, where it formed beaches and barriers. Additional sediment was derived from the erosion of onshore drumlins and transported alongshore. When sediment is abundant, drumlins form pinning points for barrier and spit development. As sediment supplies begin to wane and the shoreline recedes, perhaps in response to sea-level rise, the drumlin anchor points erode and contribute new sediment to the system. With continued sea-level rise, eventually the entire barrier complex may narrow and become low enough for it to migrate onshore through rollover processes, until stabilizing next to landward drumlins. Alternatively, the sediment is moved onshore in the form of a subtidal sand sheet, feeding sediment to spit systems that build from nearby drumlins. Intertidal and subtidal boulder pavements identify offshore drumlins that have been exhausted of their sediment sources. These boulder accumulations are lag deposits and represent the sediment that was too coarse to be transported by storm waves. In some instances, the boulder pavements may be large enough to influence the distribution of wave energy and create shadow zones landward of the former drumlins. It is not uncommon to find gravel tombolos or subtidal bars developing in these types of regions.

17.5.3 Sand and gravel beaches

Beaches along glaciated coasts come in every size, shape, and composition. This diversity is unique to glaciated coasts and is attributed to the isolated nature and highly variable composition of the sedimentary deposits that occur along these coasts. The deposits left behind by glaciers may consist of stratified sands, stratified gravel and sand, boulder tills, sandy tills, and other types of sediment. It is customary to find glacial deposits with very different compositions occurring in close proximity. When waves and currents rework these deposits, they form very different kinds of beaches. In addition, because glaciated coasts are usually irregular, containing headlands and embayments, there is little exchange of sediment between embayments and therefore little mixing of sediment. Thus, it is not uncommon to find a pocket sandy beach or barrier within a kilometer of a boulder or cobble beach.

Fig. 17.27 Gravel beaches are common along glaciated coasts due to the prevalence of till deposits along these shores.

Sandy-gravel beaches are common along glaciated coasts, particularly where the beach material comes from erosion of nearby till deposits such as drumlins or moraines (Fig. 17.27). The exact make-up of these beaches is dependent on the composition of the glacial deposit. Gravel beaches often have multiple ridges or berms and their number tends to increase with increasing supply of gravel, tidal range, and exposure to storm waves. Each ridge is related to a particular magnitude storm and tide level. The highest ridge is formed by the highest elevation storm wave event and usually coincides with spring tide conditions. Each successively lower ridge corresponds to an increasingly lower magnitude storm and/or tide level.

Individual clasts making up gravel beaches have a variety of shapes and forms; however, they can be grouped into four general form classes. Using the Zingg diagram, which is based on ratios of the

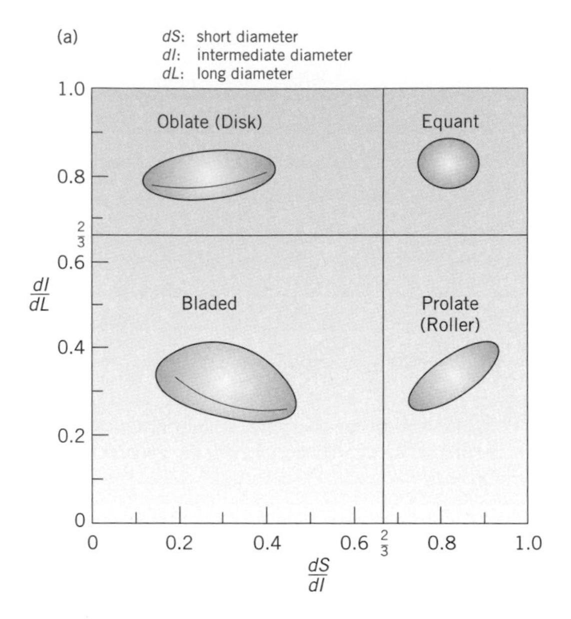

Fig. 17.28 Wave abrasion shapes pebbles and cobbles along gravel beaches. (a) Gravel clasts can be divided into four major types using the Zingg classification (Th. Zingg, 1935, Beitrage zur Schotteranalyse. *Schweizer Mineralogische Petrographologische Mitteilungen*, **15**, 39–140). These classes are based on the relative dimensions of their short, intermediate, and long axes. (b) Examples of a disk, sphere, roller, and blade.

short, intermediate, and long axes of a clast, the gravel forms can be separated into rollers (shaped like a rolling pin), disks (the best type of "skipping stone"), spheres, and finally blades, which have three different dimensions and look like "bricks" (Fig. 17.28). It is common to find disks high along the elevated portions of gravel beaches. Because of their form, they are relatively light for their size and are easily transported to the upper beach by storm waves.

17.5.4 Uplifted coasts

During the maximum extent of the Wisconsin glaciation (late Pleistocene), the Laurentide Ice Sheet varied from 1 to more than 3 km in thickness. Remember from the discussion on plate tectonics in Chapter 2 that lithospheric plates are floating on the semi-plastic region of the mantle called the asthenosphere. Thus, as the ice sheet grew in size and volume, the weight of the ice gradually depressed the lithosphere and the land surface sank. Depression of the lithosphere was accomplished by the semi-plastic asthenosphere flowing outward from underneath the portion of the continent that was covered by the ice sheet. Conversely, when the Laurentide Ice Sheet melted away and the weight of the ice was removed from the lithosphere, the land surface rebounded. This re-equilibration process that produces uplift is called glacial rebound.

The effects of glacial rebound are readily seen along the Hudson Bay and Baltic Sea shorelines, as both these areas were the centers of large ice sheets during the late Pleistocene. The Laurentide ice sheet was almost 3000 m thick in the Hudson Bay region, and since its disappearance the shoreline rimming the bay has rebounded almost 300 m. Scientists estimate that another 100 m of uplift is likely, with the highest rebound occurring where the ice was thickest (Fig. 17.29). The effect of this uplift has been the formation of over a hundred raised shorelines that are arranged in stair-step fashion, extending from the present bay shoreline to several kilometers inland. These features consist of sand and gravel beach ridges and intervening swales with a relief of 1–3 m. The beach ridges are a

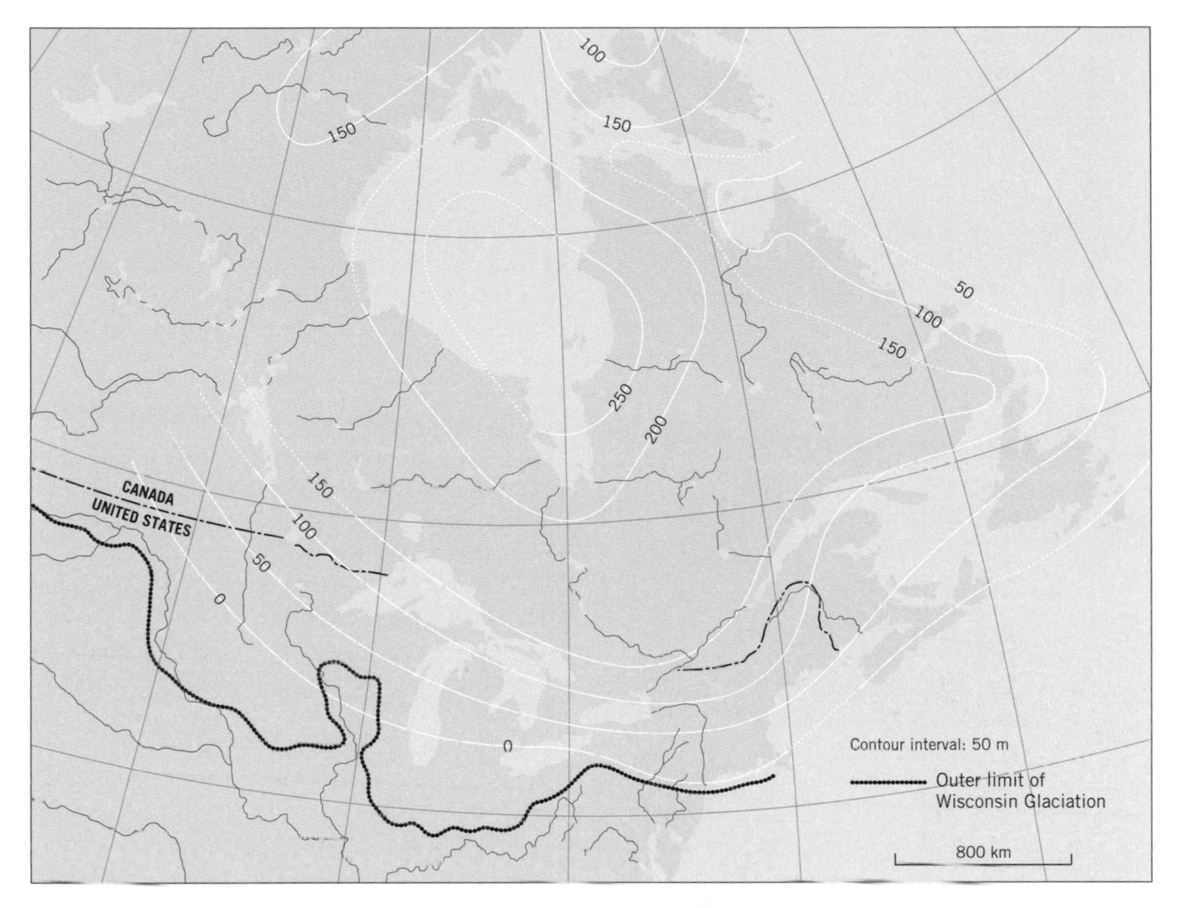

Fig. 17.29 Isostatic rebound in northeast North America resulting from retreat of the Laurentide ice sheet. (From P. B. King, 1965, Tectonics of Quaternary time in middle North America. In H. E. Wright & D. G. Frey (eds), *The Quaternary of the United States*. Princeton, NJ: Princeton University Press.)

product of storm waves reworking glacial and beach sediments, piling up gravel and sand along the rear of the beach. Glacial rebound eventually lifts the ridge, displacing it vertically and horizontally away from the shoreline. Thus, it is preserved from any further modification by storm waves. Similar processes of glacial rebound and the formation of raised shorelines have occurred along the Scandinavian Baltic coast. Present rates of uplift in this region are on the order of 6–9 mm yr^{-1}.

17.5.5 Drowned river valleys

In addition to the direct effects of glaciers, the waxing and waning of ice sheets caused the world's ocean levels to fall and rise numerous times during the Pleistocene Epoch. For example, during the last glaciation (Wisconsin glaciation) sea level dropped about 120 m and then rose back again to its present level. When sea level falls, rivers extend their courses across the exposed continental shelves to the new, seaward shoreline. During this process the original mouth of the river deepens and widens as the channel seeks to establish a new equilibrium with the lowered sea level. When the ice sheets melts and water is returned to the ocean basin, rising sea level floods these enlarged valleys, forming drowned river valleys (Fig. 17.30).

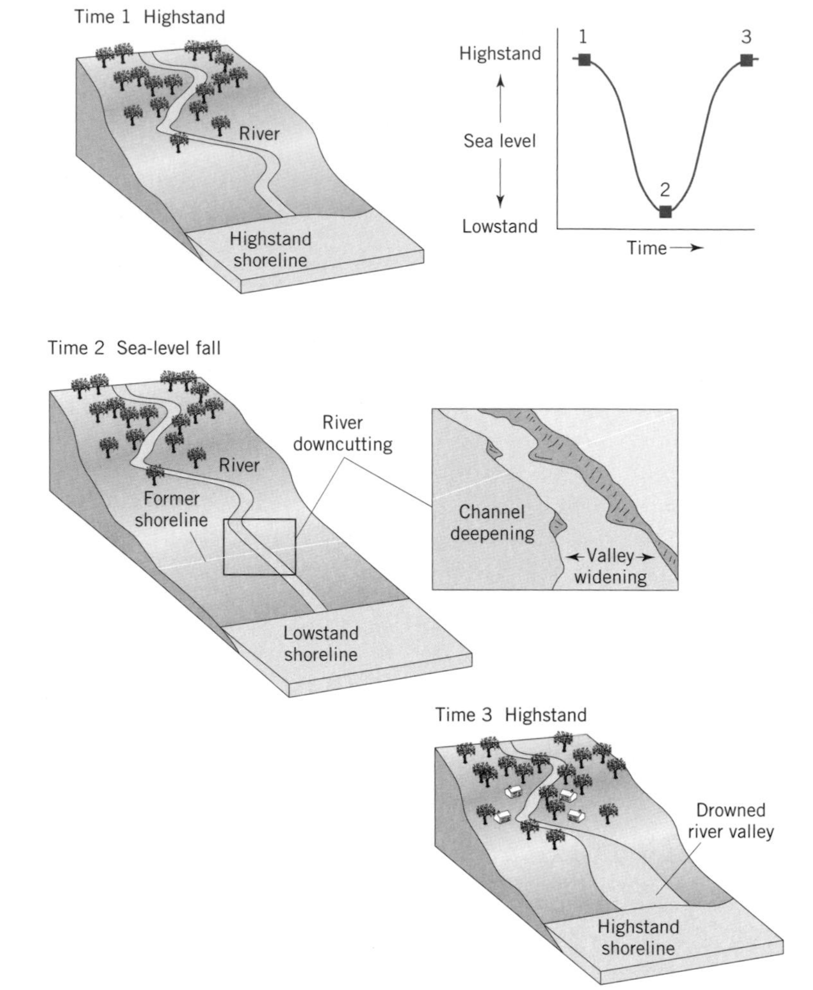

Fig. 17.30 Model of drowned river valley formation.

17.6 Summary

This chapter has emphasized the diversity of glaciated coasts and the wide range of glacial processes that produce these varied coastlines. The deep fjords of Alaska, Scandinavia, and many other high-latitude regions demonstrate the ability of alpine glaciers to sculpt and excavate coastal mountain valleys. The erosional effects of ice sheets are more subdued than alpine glaciers but far more reaching due to their size, which can extend across entire continents. The rocky coastlines of New England, northern Europe, and elsewhere attest to the ability of ice sheets to strip away sediment covers, exposing the underlying bedrock. While glaciers are impressive eroding agents, they also are important vehicles of sediment delivery to coasts. Long Island, Cape Cod, and several regions in Alaska are examples where glacial processes have essentially built the coastline. These deposits consist of end moraines that are composed of till, and outwash plains that are made up of stratified drift. Drumlins are a special type of glacial deposit, forming till headlands and intervening spits such as those along the eastern shore of Nova Scotia and in the Boston Harbor region. Beaches and barriers along glaciated coasts tend to be discontinuous, with a wide range of compositions, due to highly variable and often isolated glacial sediment supplies. Sand and gravel beaches are common.

The growth and decay of ice sheets have also produced changes in sea level and the position of the shoreline. During the most recent glaciation water removed from ocean basins and stored in huge ice sheets lowered sea level by 120 m. Through this process of sea-level lowering and rising back again, which occurred many times throughout the Pleistocene, drowned river valleys were formed along many coastal plain settings of the world. The enlargement and retreat of ice sheets were also responsible for loading and unloading the Earth's crust. In the Hudson Bay region of Canada and in the Scandinavian Baltic Sea, which were sites of thick ice accumulation during maximum glaciation, melting of the ice sheet and rebound of these coasts have produced an extensive set of raised beaches rimming the shoreline.

Formation of ice sheets occurs only when plate tectonics have arranged the continents in such way that landmasses are situated in polar regions. The onset of the Ice Ages began after the Earth underwent long-term cooling during the Cenozoic (past 65 million years), when average temperature decreased by about 5°C. The periodicity of glaciations during the past two million years appears to be linked to Milankovitch cycles, which are produced by variations in the Earth's orbital characteristics. These in turn affect the temperature contrast of the seasons. Present thought is that Milankovitch cycles are somehow linked to changes in atmospheric and oceanic circulation and these broad conveyer belts of heat and cold control global climatic fluctuations and the waxing and waning of ice sheets.

Suggested reading

Bennett, M. R. & Glasser, N. F. (1997) *Glacial Geology: Ice Sheets and Landforms*. New York: Wiley.

Coates, D. R. (ed.) (1974) *Glacial Geomorphology*. Binghamton: State University of New York.

Drewy, D. J. (1986) *Glacial Geologic Processes*. London: Edward Arnold.

FitzGerald, D. M. & Rosen, P. S. (eds) (1986) *Glaciated Coasts*. New York: Academic Press.

Hambrey, M. & Alean, J. (1992) *Glaciers*. Cambridge: Cambridge University Press.

Johnson, D. W. (1919) *Shore Processes and Shoreline Development*. New York: Hafner Publishing.

Patterson, W. S. B. (1981) *The Physics of Glaciers*. Oxford: Pergamon Press.

Ruddiman, W. F. (2000) *Earth's Climate Past and Future*. New York. W. H. Freeman

Sharp, R. P. (1989) *Living Ice: Understanding Glaciers and Glaciation*. Cambridge: Cambridge University Press.

Shepard, F. P. & Wanless, H. R. (1971) *Our Changing Coastlines*. New York: McGraw-Hill.

Sugden, D. E. & John, B. S. (1976) *Glaciers and Landscape*. New York: John Wiley & Sons.

18 Rocky coasts

18.1 Introduction

Rocky coasts offer some of the most striking ocean vistas in the world due to the exquisite beauty produced by waves crashing against their jagged shores (Fig. 18.1). These coasts are typically associated with cliffs and other erosional landforms, such as stacks, arches, and caves. Along many active continental margins, rocky coasts are bordered by majestic mountainous hinterlands. In contrast, low plains or hilly regions border the rocky shores of some mid-ocean islands. Despite the high wave energy and apparently harsh environment of rocky coasts, their intertidal zones commonly teem with life and are far more productive biologically than sandy shorelines.

Most people are unaware of the extent of rocky shorelines because many coastal cities, major coastal population centers, and even vacation sites are located in coastal plain settings, at the mouths of estuaries, and in other lowland areas. It is estimated that 75% of the world's shorelines are rocky, which includes beaches that are backed by bedrock cliffs or rocky uplands. Although rocky coasts are not confined to a single type of geological setting, they are more common along tectonically active coasts than they are on passive margins. For example, rocky coasts comprise much of the west coasts of North and South America, whereas beaches and barriers characterize most of the non-glaciated east coast of North America. Still, there are many sites throughout the world where rocky and rugged coasts are found along passive margins, such as South Africa, parts of Argentina and Brazil, eastern Canada, southern Australia, and sections of northwest Europe.

Fig. 18.1 Indian Ocean coast of South Africa, where wave heights are frequently in excess of 3 m.

Rocky coasts display a wide range of morphologies because they are composed of different types of rocks and have formed in a variety of geologic settings. For example, the Bahamian coast consists mostly of limestone that is derived from coral, shells, and coralline algae. In southern California and Victoria, Australia, cliff coasts have developed from the recession of relatively weak sedimentary rocks, including sandstones, siltstone, and mudstones. Marine erosion of volcanic rocks along the south coast of the island of Hawaii has produced dramatic cliffs. Differential weathering of folded metamorphic rocks intruded by granitic batholiths has formed the indented/island coast of central Maine. Tectonic processes control the rugged mountainous coasts of Alaska and Chile. Although erosional processes dominate most rocky coasts, depositional landforms, including pocket beaches situated between bedrock headlands and even barrier spits downdrift of river mouths and estuaries, are common in these regions.

This chapter explores the extent and variability of rocky coasts around the world. It will be shown that the morphology of these coasts is a function of the structure and type of bedrock, as well as the physical, chemical, and biological processes operating on the coast. The many different erosional features common to these coasts are discussed in terms of the processes by which they have formed.

18.2 Types and distribution

Three-quarters of the world's coasts are rocky and exhibit a lack of sediment accumulating along the coast. They also occur where the original sediment

Fig. 18.2 Subduction of the Pacific Plate beneath North America has produced the rugged mountainous coastline of Alaska, where earthquakes and volcanic eruptions are common geologic processes.

cover has been eroded away or where the sediment has been turned to rock through cementation. The most extensive rocky coasts are associated with mountainous regions and glaciated areas. Other shorter and/or discontinuous sections of rocky coasts occur in volcanic regions and in a variety of other geological settings.

18.2.1 Tectonic settings

The tectonic setting of the continental margin has an overriding influence on whether a coast is rocky or whether it contains beaches, barriers, tidal flats, river deltas, or other types of sediment accumulation forms. As we learned in Chapter 2, the convergence of oceanic and continental plates produces high-relief, continental borderlands such as the Andes Mountains along Peru and Chile or the North America Cordillera.[1] The rocky and rugged landscape of these mountains usually extends to the sea, giving the coast a rugged appearance as well (Fig. 18.2). In addition, the mountain chains of these continental margins act as dams and prevent rivers from delivering sediment from continental interiors to the coast. The rivers draining mountains along collision margins are typically short and usually transport coarse-grained sediment to the coast. However, due to their small drainage areas, these rivers discharge relatively small quantities of sand and mud, particularly in comparison to rivers along trailing-edge coasts. Waves tend to be large along collision coasts because the steep continental shelves of these margins dampen little of the deepwater wave energy. During storms, high-energy waves remove large quantities of sand from beaches and transport it alongshore and offshore. Along the California coast beach sands are permanently lost to the deep ocean basin due to the proximity of submarine canyons to the shoreline (<1 km) resulting from the narrow continental shelf (see Fig. 2.17). These canyons capture the longshore transport of sand at their heads and carry it to the deep ocean out of the coastal system. For all these reasons, collision coasts tend to be rocky, containing few depositional features. Because of their relative youth, neo-trailing edge coasts such as the Arabian coast along the Red Sea are also rugged and mostly rocky. There has been insufficient time to develop a mature continental margin with a wide shallow shelf to dissipate waves.

18.2.2 Glaciated regions

During the Pleistocene Epoch, continental ice sheets covered approximately 30% of the world's land surface. This means that the morphology of much of the world's coastline has been strongly imprinted by glacial processes. Although there are large sections of coastline that consist entirely of glacial sediment or reworked glacial deposits, such as Cape Cod, Massachusetts, Long Island, New York, extensive areas along the Great Lakes, and portions of the eastern shore of Nova Scotia (see Chapter 17), most glaciated coasts have widespread bedrock exposures. In mountainous regions, coastlines take the form of bedrock headlands, with intervening deep flooded valleys. The fjords of Scandinavia, Iceland, Chile, and western Canada are examples of where alpine glaciers carved and

[1] The North American Cordillera is a mountain system comprising the west coast of the United States, Canada, and Alaska.

Fig. 18.3 Glaciated coasts exhibit a wide diversity, but are dominated by a rocky landscape. Repeated Pleistocene glaciations have stripped much of the sediment from the Maine coast, leaving it rocky and irregular, with numerous bedrock islands but few beaches.

Fig. 18.4 Volcanic island chains, such as the Hawaiian Islands, form extensive rocky coasts in the Pacific Ocean and other parts of the world. Rapid physical and chemical breakdown of volcanic rock and nearby coral reefs produces sediment that forms pocket beaches.

deepened existing mountain valleys. At the heads of fjords in the Kenai Peninsula of Alaska and along much of Greenland, vestiges of the valley glaciers that formed these twisting water-filled valleys are still present. Repeated advancements and retreat of great ice sheets stripped away the sediment cover from the land, exposing the underlying bedrock along many other glaciated coasts. These coastlines are commonly embayed to deeply indented and contain numerous islands and bedrock ledges (Fig. 18.3). Beaches and barriers along these coasts are usually scarce and, where they do exist, they are adjacent to rivers, estuaries, or nearby isolated glacial deposits.

18.2.3 Other bedrock coasts

Volcanic coasts occur where hot spot activity in the mantle has produced island chains such as those found in the Pacific Ocean, including the Hawaiian and Marshall islands (Fig. 18.4). Outpourings of lava and welded tuff have also formed portions of rocky coasts along island arcs in the Caribbean, and the northern and western Pacific. In tropical regions, the seaward extension of these volcanic coasts forms the platforms upon which luxuriant coral reefs have developed. In some places, sinking volcanic islands have become coral atolls.

Another variety of rocky coast is formed from the shells of dead marine organisms. This type of coast is most common in low-latitude regions, including the Caribbean and Mediterranean seas, where calcium carbonate skeletal material is produced in high abundance in coastal waters. High rates of shell production may also occur in cooler, higher latitude regions, such as the south coast of Australia and South Africa, where inputs of other types of land-derived sediment are absent. Many of these rocky carbonate coasts were created during the Pleistocene when sea level fell and onshore winds blew carbonate sand onshore, building dunes and beaches. This sediment was turned to rock by a process called lithification. Individual grains of sand are cemented together when ocean spray, rainwater, and groundwater percolate through the sand, partially dissolving the calcium carbonate and precipitating it as cement. The high temperatures of these coasts produce rapid evaporation of wetted surfaces, which accelerates the lithification process. Dunes that are converted to rock are called eolianites. Wide

Fig. 18.5 Differential weathering of lithified carbonate dunes along sections of the Bermuda coast results in the formation of caves and embayments.

Fig. 18.6 View of beachrock that has formed along the shoreline of Kubbar, an island off the coast of Kuwait in the Arabian Gulf. In this region both terrigenous and biogenic sands are lithified by carbonate cement.

exposures of this rock with well developed dune layering can be seen throughout Bermuda (Fig. 18.5), in several Caribbean islands, along the Yucatan coast of Mexico, and in southeastern Australia.

Similar chemical processes in tropical settings produce rocky outcrops along the shore called beachrock (Fig. 18.6). In regions with high evaporation rates elevated concentrations of various salts occur in the seawater. Salt spray and tidal inundation constantly bathe the beach with this high salinity water. Partial evaporation of the water and interaction with groundwater cause the precipitation of calcium carbonate. This process cements sand grains, producing beachrock. The rapidity with which the process operates often creates beachrock containing modern day trash, including such articles as beer cans and soda bottles.

18.3 Erosional processes

Unlike on sandy shorelines, where the effects of a winter storm or the passage of a hurricane are readily visible, changes along a rocky coast occur very slowly and may be imperceptible over human lifespans. The present day appearance of rocky coasts is the product of physical, chemical, and biological processes that have been operating over thousands of years. The rate at which these processes act to weather the shoreline is a function of rock type and its structure, wave intensity, vegetation, climate, and numerous other factors. In most cases, physical processes operate over a larger scale and at a faster rate than do either chemical or biological agents.

18.3.1 Physical processes

Wave-induced erosion

The processes involved in eroding bedrock shorelines are numerous, but the most visible and important agent is wave action. Breaking and shoaling waves are responsible for several interactive processes, including wave hammer, air compression, quarrying, and abrasion. The first of these, wave or water hammer, occurs when a wave directly impacts a cliff face or sloping rock exposure, exerting a tremendous hydraulic force against the rock, particularly if the waves are large. As reported in numerous storm accounts, waves have been responsible for tearing the tops of seawalls and throwing huge 5 ton granite blocks tens of meters across roadways. However, in most cases it is not a single wave that finally causes a rock face to fail, but the accumulated effects of tens to hundreds or even thousands of years of wave action.

In some instances, breaking waves trap air between the water and the cliff face. As the wave collapses, extremely high pressures are instantaneously produced as the air is compressed on the rock surface

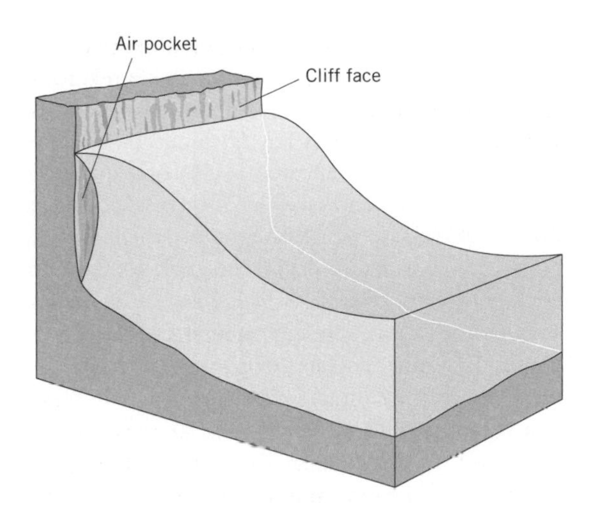

Fig. 18.7 Air is instantaneously compressed against a cliff when waves break. This compression of air can lead to the enlargement of cracks and failure of the rock face.

Fig. 18.8 Vertical sea cliffs along the southwestern coast of Victoria, Australia.

(Fig. 18.7). This process is particularly important when air pockets are compressed into the crevices of rocks, leading to the enlargement of the crack and ultimately to a shattering of the rock. Quarrying is the removal from a cliff face or bedrock exposure of pieces of rock, ranging from small grains to large blocks, that have been loosened or separated from the parent rock by biological, chemical, or other physical processes.

One of the most important mechanisms of physical erosion takes place when sand, granules, and larger-sized gravel are entrained by waves and washed, rolled, and scraped across the rock surface. Abrasion also includes the more violent process of large waves picking up pebbles, cobbles, and boulders and propelling them against the rock surface. The widespread cliffs along the British Isles and along the Alaska Peninsula are believed to have formed due to abrasion by gravel (Fig. 18.8). However, there are numerous other coastal sites around the world, including parts of New Zealand, much of Tasmania in Australia, and the Bay of Fundy in Eastern Canada, where cliff shorelines are fronted by sandy beaches. Thus, we can conclude that sea cliffs can be produced when the rock face is abraded by sand or any other available hard particles in conjunction with other wave processes.

The rate at which abrasion takes place is related to wave energy, the composition of the rock, and the type and abundance of abrading agents (gravel, sand, or other particles). The process of abrasion tends to produce a smoother rock surface than does quarrying. Exceptions occur when there are variable weaknesses in the bedrock caused by nonuniform rock type (mafic dike cutting through a metamorphic rock) or structure (presence of cracks or joints), resulting in differential abrasion rates. In these settings, abrasion locally produces grooves, ridges, and other surface irregularities (Fig. 18.9). Under special circumstances potholes measuring tens of centimeters deep and wide are formed. These rounded depressions are the result of pebbles and cobbles being swirled in the slight concavities of the rock by passing waves. With time, the gravel tools abrade and enlarge some of the depressions to become potholes.

Effect of freeze and thaw

There is another less visible physical process that contributes to erosion along rocky coasts in northern latitudes. This process involves the expansion and contraction of water as it changes from a liquid to a solid state. During moderate winter days melting snow may flow into the cracks, crevices, and joints of rocks, only to freeze at night as temperatures fall. Because water expands by 9% when it freezes, ice formation can exert great pressures in

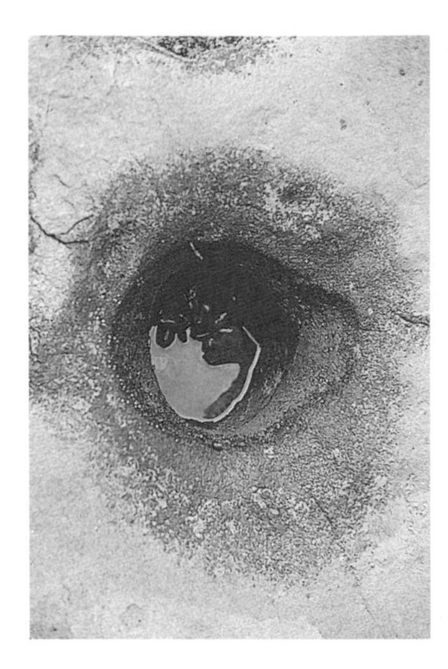

Fig. 18.9 Potholes are the result of wave action swirling gravel in a depression. The hole deepens and widens as the rock is preferentially abraded. This one is approximately 0.5 m in diameter.

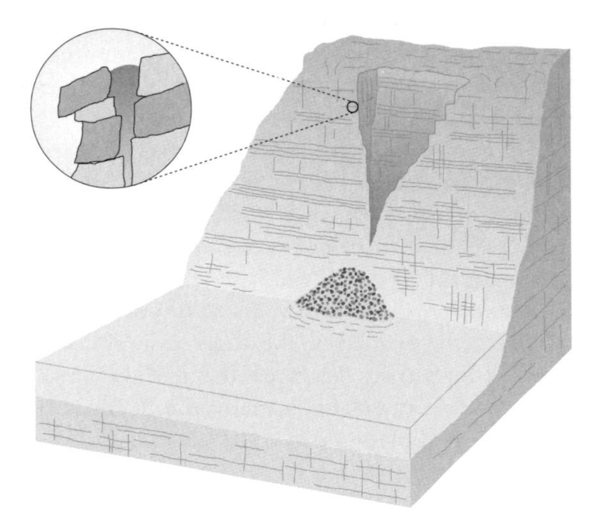

Fig. 18.10 Frost wedging occurs when water freezes in the cracks and joints of rocks. The expansion caused by ice formation dislodges rock from the cliff face.

confined spaces (Fig. 18.10). In high latitudes, freeze and thaw cycles can occur on a daily basis, especially in the intertidal zone, where temperatures are governed by the rise and fall of the tide. In addition to melting ice and snow, water can come from rain and groundwater. Although seawater may also freeze, the ice that forms from seawater is relatively soft and therefore is less capable of causing frost-wedging than freshwater ice. It should be emphasized that it is the repeated cycles of expanding ice applying force on the walls of the rock, and the subsequent release of this pressure when the ice melts, that eventually cause the rock to fail and break. The effectiveness of freeze and thaw on the erosion process is dependent on how well the rock is fractured (number and size of cracks) and its porosity. Rocks that are thinly layered, such as schists, or highly porous, such as sandstones, are much more susceptible to freeze and thaw weathering than are massive rocks such as granite or basalt. Another important weathering process contributing to rock failure is the growth of other substances in the cracks of rocks, including calcite, halite (rock salt), and clays. Calcite and halite expand and contract due to temperature changes, and some clay minerals do likewise when exposed to water. That latter process can be particularly effective in the upper intertidal zone, where these minerals are continuously being wetted and dried during the tidal cycle.

18.3.2 Biological processes

Bioerosion is a more subtle process than the mechanical wearing away of rocks, but in tropical regions where carbonate rocks commonly dominate, it is an important means of sculpturing the coastal landscape. There are numerous plants and animals that remove the substrate in search of food or shelter or both. The most effective bio-eroders of calcium carbonate rock are the microscopic blue-green algae that penetrate into the rock by as much as 1 mm. A microscope would reveal that as many as a million algae may colonize a square centimeter of rock surface. These organisms are able to bore into the rock by dissolving away the calcium carbonate. The algae break down the structure of the rock, causing individual grains to wash away by wave action. Other rock borers include certain sponges, worms, bivalves, and echinoderms.

The presence of algae as well as fungi and lichens in the intertidal and supratidal zones promotes

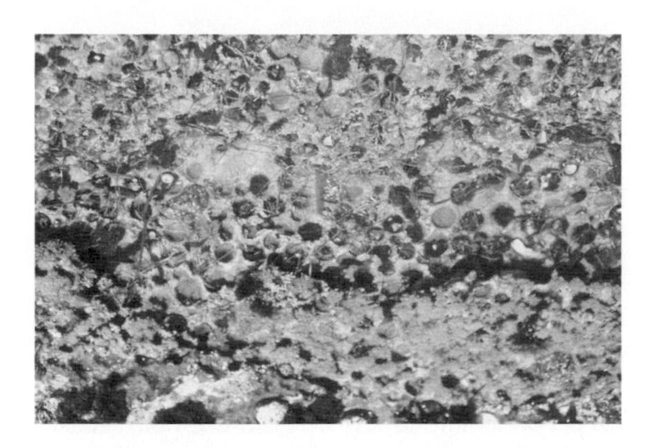

Fig. 18.11 Sandstone and siltstone outcrops along the shore of San Nicholas Island off the California coast are home to sea urchins, abalone and other intertidal fauna. These organisms bore into the rock, finding shelter from the high wave energy and protection from predators. The circular holes, formed by sea urchins, are approximately 10 cm in diameter.

additional erosion by larger grazing organisms, which seek out algae, bacteria, and other encrusting organisms as a source of food. Marine invertebrates, including snails, limpets, sea urchins, and chitons, abrade the rock surface as they feed on microflora. This process is not restricted to carbonate environments alone; the Channel Islands off the southern California coast are a good example. Along the shores of San Nicholas Island, located 160 km west of Los Angeles, mudstones and sandstones are home to a large community of sea urchins and abalone (Fig. 18.11) These organisms have fashioned homes by boring into the intertidal sedimentary rocks, accelerating the physical and chemical erosion of this rocky coast. The end result of the bioerosion is a honeycombed outcrop consisting of hollows and ridges.

18.3.3 Chemical processes

The chemical breakdown of rock in the coastal zone is a very slow process due to the slow rate at which rocks react chemically with water. The effects are difficult to notice over human life spans. However, it is a pervasive process that affects all rocky coasts from the subtidal to supratidal zones. Climate and rock type are the primary factors controlling the rate of chemical weathering. Because chemical reactions involve water and usually proceed more rapidly as temperature increases, hot and humid tropical climates experience much greater chemical weathering than arid temperate coasts or hot desert regions. Similarly, limestone is relatively soluble and those coasts may erode at a rate of 1 mm yr^{-1}, whereas coasts composed of quartzite are essentially stable because quartz is almost inert under surface conditions.

Chemical weathering of rocks encompasses many different chemical reactions, including **solution**, which is the dissolution of different minerals into water. **Hydrolysis** is another important reaction that converts feldspars to clays. Because feldspar is a major constituent of many igneous (granite), metamorphic (gneiss), and sedimentary (sandstone) rocks, hydrolysis is a significant chemical weathering process that accounts for much world's mud. Oxidation is a chemical reaction between iron and the oxygen in water whereby "rust" is formed. These reactions take from tens of thousands of years for limestone to be dissolved to millions of years for feldspar to be transformed to clay. Because of these slow rates, physical erosion processes usually obscure the effects of chemical weathering.

18.4 Factors affecting rates of erosion

Rocky coasts look pretty much the same visit after visit because the processes that change them work very slowly. Characteristics such as rock type and degree of fracturing, exposure to wave energy, tidal range, climate, relative sea-level changes, and other factors affect the rate of erosion and, ultimately, the morphology of the coast.

18.4.1 Rock type

The influence of rock type and structure is particularly well illustrated by comparing parts of the west and east coasts of the United States. In southern California waves have cut into relatively soft sandstones, mudstones, and other flat-lying sedimentary rocks, producing sea cliffs along much of the coast. In contrast, the metamorphic rocks and igneous

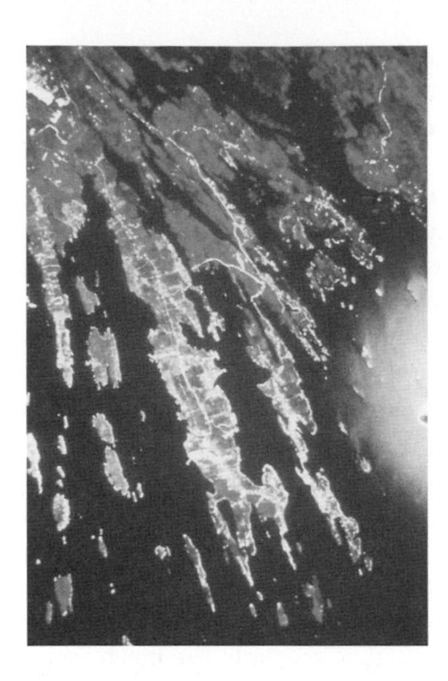

Fig. 18.12 False-color infrared photograph of the central peninsular coast of Maine. The highly indented nature of this coast is a product of differential erosion. The ridge and valley morphology has been accentuated by several episodes of Pleistocene glaciation.

intrusives (granites) of the central Maine coast are tough rocks that have resisted erosion. Here the coastline reflects the structure of the rocks and patterns of differential erosion. The folded metamorphic rocks are turned on end, producing an overall jagged, indented shoreline. Interrupting this coastal landscape are granite outcrops that, due to their size and resistance to erosion, have formed headlands and islands (Fig. 18.12).

18.4.2 Degree of fracturing

Fractures include the cracks, joints, and faults of a rock outcrop. Their number and trend are significant because erosion is usually concentrated at these sites. Fractures increase the area exposed to various weathering agents. Joints and faults can be indirectly responsible for the dissection of some rocky coasts and the formation of stacks, arches, and other features. Preferential weathering at joints and faults may also produce deep crevasses and the development of narrow embayments.

18.4.3 Wave energy

Wave energy is important because it controls the intensity of physical erosion and the removal of debris that is produced during the weathering process. Researchers have shown that waves exert their greatest force at a position just above mean high water. This is because water levels have a maximum duration at mean high tide and waves break for an extended period at this elevation. Second, waves are larger at high tide than at mid or low tide due to the deeper water and lower frictional resistance generated by the rocky intertidal zone. This condition is apparent when looking at seawalls and seeing that the greatest abrasion to the wall occurs near the high tide line. This level is also where chemical weathering processes are at a maximum. Although waves are even larger and reach higher elevations during storms, they have less effect due to their infrequent nature.

Waves and their attendant processes will continue to erode a rocky shore as long as the sediment that is produced during the weathering process is removed. For example, if a cliff erodes and produces enough sediment to form a wide beach, then the cliff effectively becomes protected from further wave action and erosion (Fig. 18.13). Also, if the rocky coast is barren of sediment, there is no material to provide for abrading the rock surface.

Fig. 18.13 Retreat of this cliff has produced a talus slope that protects the cliff face from storm wave erosion.

18.4.4 Tidal range

Rocky coasts are found in a wide range of tidal settings from the microtidal coasts in the Mediterranean (tidal range about 1 m) to some of the highest tidal range coasts of the world (tidal range greater than 10 m), including the Bay of Fundy and the eastern coast of the Bay of Saint Malo in France. Along most rocky shores the direct influence of tidal forces is minimal. However, tides do control the amount of time that waves erode different elevations along a rocky coast. Along steep or vertical cliffs wave action is focused in a relatively narrow zone, particularly along microtidal coasts. In contrast, along macrotidal settings the large excursion of the tide disperses wave energy over a relatively wide zone.

18.4.5 Climate

Climate influences the rate and type of weathering processes. Tropical areas tend to undergo greater chemical weathering, whereas temperate and polar regions experience more frost weathering. Climate also controls the patterns of storms and prevailing winds and therefore the deepwater wave energy along a coast. Generally, temperate regions have greater wave energy and physical processes operate more energetically than in tropical areas, where average wave energies are lower.

18.4.6 Relative sea level

This factor has an obvious effect on the development of marine terraces as they are a product of relative sea-level changes. They form when a coast rises tectonically and marine platforms are lifted from the water. They may also develop when eustatic sea level drops.

18.5 Morphology

Rocky coasts exhibit a wide variety of morphologies because there are so many factors that have influenced their development. One characteristic that they share is an overall scarcity of sediment. Thus, they are associated with tectonically active coasts and are also common in former glaciated settings where the sediment cover was removed by the ice. In addition, low rocky coasts occur in tropical areas where sediment has become lithified by chemical processes. There are too many types of rocky coasts to fully treat their range of morphologies. Instead, the major classes of rocky coasts are presented below, as well as some of the striking features of these coasts.

18.5.1 Sea cliffs

Cliffs have many different profiles and heights, reflecting their composition, structure, and weathering processes. There are the spectacular vertical cliffs at the entrance to Resurrection Bay leading to the deepwater port of Seward, Alaska (Fig. 18.14). These cliffs rise 200 m vertically from a deep glacially cut trough before ascending more gradually into mountains with elevations in excess of a kilometer. Other substantial vertical cliffs (height > 100 m) include the White Cliffs of Dover in England (Fig. 18.15), the Cliffs of Moher and the Giant's Causeway in Ireland, and the extensive cliffs along the Nullabor Plain of South Australia. The fact that these cliffs have similar morphologies despite

Fig. 18.14 The vertical sheeted dike complex in Resurrection Bay near Seward, Alaska, has produced a spectacular cliff face, which rises over 800 m above the water surface. These rocks were formed at a mid-ocean ridge system during production of oceanic crust. The ocean crust was subsequently raised to the surface by tectonism involved in the subduction of the Pacific Plate beneath North America. (Photograph from Tim Kusky, St Louis University.)

Fig. 18.15 The White Cliffs of Dover, England, are composed chiefly of coccolithophores, a planktonic alga with a calcium carbonate exoskeleton.

Fig. 18.16 Hanging valleys along the coast of Hawaii are the sites of waterfalls more than 200 m high. The cliff is probably the result of a huge landslide in which a massive block of volcanic rock slid into the abyss. The valleys were cut by groundwater sapping processes.

differing compositions and structure indicates that their development has been dominated by the same efficacy of erosional processes.

In some regions cliffs are not vertical, such as in sections of southern California, portions of south-western Wales, and much of the Alaskan Peninsula. Regional geology and intensity of various weathering processes control the steepness and profile of cliffs. When sea cliffs consist of relatively unfractured, massive rocks such as basalt, granite, or quartzite (metamorphosed sandstone), erosion proceeds in a uniform manner. However, when sea cliffs are formed through the erosion of horizontally layered sedimentary rocks, there are commonly large differences in how individual layers react to weathering processes. Under these conditions the cliff erodes unevenly, producing steps, notches, and other irregularities. Various patterns of joints, faults, and folds may also result in jagged slopes. Vertical sea cliffs and very steep slopes are created when the rocks are homogeneous and marine weathering processes clearly dominate subaerial erosion. In these situations marine agents cut the slope back at a faster rate than the upland can be eroded to form an incline. When wave abrasion undercuts the base of the sea cliff, causing eventual failure of the upper slope, a steep or vertical face is usually produced. Conversely, as erosion of the surface of the slope increases relative to marine erosion, sea cliffs become more gently inclined.

Surface processes coupled with marine erosion have created a spectacular coastal landscape along the southern shore of the island of Hawaii. Here surface and groundwater flows have cut large V-shaped channels into lava flows. Deep chasms have developed where large amounts of water have flowed to the coast. Along other sections of these sea cliffs, smaller drainage systems have formed dramatic hanging valleys, some with 300 m high waterfalls (Fig. 18.16).

The height of sea cliffs appears to be controlled by wave energy. On a worldwide basis there is a correspondence between cliff height and latitude. Most high-relief cliffs occur in mid-latitudes, where waves are relatively large. Modest and low cliffs are found in low- and high-latitude regions, where waves tend to be small. In polar regions, coasts are protected by sea ice during part of the year. Because wave energy is proportional to the square of the height, large waves exert a much greater force on the face of a cliff than do small waves. Abrasion rates are also greater in regimes of high wave energy due to the large amounts of sediment entrained by large waves. The greatest rates of abrasion coincide with regions of high waves and the presence of pebble, cobble, and small boulder abrading agents. In low latitudes, rock cliffs are limited in height because they have eroded into beach rock and lithified dunes. The complexities of coastal geology around the world, and the highly varied marine erosional processes that operate along these coasts, have produced many exceptions to the generalized trends presented here.

18.5.2 Horizontal erosional landforms

Platforms and benches

Wave erosion along rocky coasts not only creates vertical to steeply inclined landforms; horizontal features such as platforms (also called **benches** by early researchers) are also formed during the process (Fig. 18.17). Shore platforms are the flat bedrock ledges that border the base of sea cliffs. They vary in width from a few meters to more than a kilometer. Where platforms have formed on flat-lying sedimentary rocks their surfaces are relatively smooth. When they develop in regions of variable geology, such as highly jointed igneous rocks or dipping metamorphic or sedimentary rocks, they commonly exhibit irregular surfaces with relief up to a meter. Differences in resistance to weathering processes produce grooves, gullies, knobs, and other features.

Platforms may be almost horizontal to seaward dipping, with slopes ranging from a few degrees to more than 30°. In some areas, particularly where abrasion is the dominant erosion process, a correlation exists between tidal range and gradient of the platform, with increasing tidal range producing steeper slopes. Tidal range controls the distribution of wave energy over the platform and, to some extent, the wave erosional processes. Throughout the tidal cycle, waves shoal and break along the seaward edge of the platform. However, only at near high tide levels do waves directly affect the landward, highest portion of the platform. Thus, the greatest rates of vertical erosion occur at high tide, when wave processes are most effective.

Slightly elevated platforms are created in weak sections of rock cliff that erode at a faster rate than the rock below it. These benches are most common in areas of horizontally layered sedimentary rock, such as bedded limestones, sandstone, and mudstone, and occur at elevations above mean high water. They are usually fairly narrow and are connected to platforms by a step.

(a)

(b)

Fig. 18.17 (a) This shore platform extends for several kilometers along the shore in Alinchak Bay on the Alaskan Peninsula. (b) Close-up view of the retreating cliff, illustrating that the upper beach contains thin beach deposits. Note the brown bear for scale.

There are platforms in southern California and elsewhere in the world that extend to depths greater than 10 m. Because it has been shown that wave erosion here is essentially inactive below 10 m, the deeper portions of these platforms are viewed as relict and likely were drowned by rising sea level. Cliff retreat is slow and rates are measured in terms of centimeters per year along most rocky coasts, meaning that platforms were created very slowly as well. This suggests that platform development requires a stable or near stable sea-level position over an extended period of time. The widest shore platforms occur in regions of moderate to high wave energy where rock cliffs are highly susceptible to erosion, such as those that are composed of weakly cemented sedimentary rocks. Resistant rocks, such as quartzites, are commonly associated with narrow, steep platforms.

The formation of platforms involves wave processes and a type of chemical weathering called water-layer weathering. Wave-eroded platforms are closely related to sites of cliff retreat and encompass wave abrasion and quarrying. These processes are most energetic in areas of moderate to high wave energy where abrading tools are abundant. Wave energy must be sufficient to remove the debris that is produced during the erosion process, as otherwise the platform becomes insulated from further weathering processes.

Water-layer weathering includes all processes resulting from the wetting and drying of the bedrock, including salt crystalization. The process is most operative in areas of horizontally bedded, highly permeable rocks in tropical or subtropical regions with low tidal ranges, such as Hawaii, Australia, the Caribbean, or the Mediterranean coasts. High temperatures favor evaporation and chemical weathering. This type of platform is common along carbonate coasts and can be identified by small shallow-water pools that occur near mean high water. Granitic rocks along the coast of Maine may exhibit these same shallow pools of water at elevations from one to several meters above mean high water. These pools contain granules of quartz and feldspar that have weathered directly from the granite during the water-layer erosion process.

(a)

(b)

Fig. 18.18 (a) Marine terrace development along the southwestern coast of Victoria, Australia, and (b) in Bodega Bay along the central California coast (photograph taken by Rahm). Note that the marine stacks in Bodega Bay are morphologically similar to the rock outcrops atop the marine terrace. Indeed, the rock outcrops were marine stacks when the shore platform was developing.

Marine terraces

Marine terraces are platforms that have been displaced from their original position by tectonic forces, sea-level changes, or both. They are located above sea level where the land is rising and below sea level where it is sinking (Fig. 18.18). Along the coast of southern California there are at least six marine terraces in the Palos Verdes region and as many as ten on the islands of San Clemente and Santa Cruz. Extensive terrace development also exists along the tectonic coasts of New Zealand, New Guinea, the Alaska Peninsula, and the Pacific side of Central and South America. Steep slopes separate individual terraces. Elevation difference between successive terraces is a function of the uplift rate along the coast. Marine terraces are best developed in areas of moderate to high wave energy along rising coasts that are composed of easily eroded, horizontally stratified, sedimentary rocks. It should be noted that marine terraces also form along unconsolidated sediment coasts.

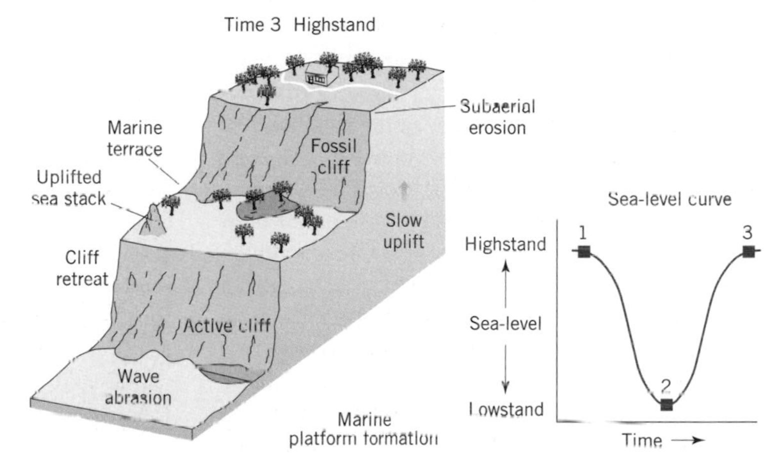

Fig. 18.19 Marine terraces form along coasts undergoing uplift. Time 1, terrace development begins during a sea-level highstand, with the formation of a marine platform produced by wave abrasion. Time 2, the marine platform is subsequently abandoned when sea level falls at the onset of a global glaciation event. Time 3, during the period of time in which sea level falls to a lowstand and then rises back to a highstand position, the marine platform is uplifted above the highstand position, becoming a marine terrace. The idealized sea-level curve shows the timing of marine terrace development.

The formation of marine terraces requires periods of sea-level stability in an overall regime of tectonic uplift (Fig. 18.19). Most scientists now agree that terraces represent platforms raised from the water during the Pleistocene Epoch. Remember from Chapter 17 that during the Pleistocene the advance and retreat of continental ice sheets produced coincident changes in sea level of as much as 120 m. It is believed that most platforms were formed while sea level stood at a highstand position (approximately present day sea level) that lasted for perhaps thousands of years. They were raised during the interval of time after the highstand when sea level fell to a lowstand position and then rose back to approximately the present sea level. This process was repeated during each period of glaciation, producing a series of increasingly higher marine terraces that record the times of Pleistocene highstands. Knowing the elevation difference between successive terraces and determining their age of formation by using various dating techniques provides a means of calculating the rate of tectonic uplift for that particular coast.

18.5.3 Sea stacks, arches, and erosional features

One of the striking features of rocky coasts is the erosional remnants that are left behind as cliffs retreat. Through differential weathering of bedrock, landforms such as sea stacks and arches are left sitting atop shore platforms. In other regions

Fig. 18.20 The Twelve Apostles are sea stacks that have formed from flat-lying limestone rock (Miocene age) in Port Campbell National Park along the coast of Victoria, Australia.

preferential erosion along cliffs produces notches and caves. The Twelve Apostles along southwestern Victoria in Australia are one of the most famous set of stacks in the world and are used widely to depict rocky coasts (Fig. 18.20). Some stacks, such as those along the Bay of Fundy shoreline of New Brunswick, Canada, are large and tall enough to have trees growing on top of them. Stacks exhibit many different forms but commonly are taller than they are wide, and are generally lower in elevation than the adjacent cliffs. Arches are equally impressive due to their unusual likeness to bridges that go nowhere.

The formation of these features is the result of differential marine erosion superimposed on specific types of bedrock geology. The development of a single stack is illustrated well where waves have eroded ash, welded tuff, and other rocks comprising the exterior of a volcano while its neck remains as a pinnacle. This is because the neck is composed of relatively strong basalt that resists erosion. Multiple stacks are generally related to horizontally layered sedimentary rocks, such as sandstones, limestones, and conglomerates that are highly jointed. These giant cracks are weaknesses in the rock that promote erosion due to increased surface area and greater exposure to marine weathering processes. As erosion proceeds, joints become indentations, then deep chasms, and with further wave attack finally embayments. Eventually, the headland is detached from the mainland and a stack is formed. Thus, stacks can be created if they are composed of rock that is more resistant than the cliff rock. Alternatively, stacks can form due to structural weaknesses of the cliff rock, which makes them more susceptible to differential erosion and retreat.

Arches are commonly found near sea stacks and are usually associated with horizontally layered sedimentary rocks. They develop after the cliff rock has been dissected and narrow headlands or elongated stacks have formed. The next stage in development occurs when wave action preferentially erodes the middle or lower section of the headland or stack. The "tunnel" that develops is a product of wave abrasion and other processes that concentrate their erosion at the high water line and attack both sides of the landform. As the tunnel widens, a sea arch is created. Eventually, continued erosion causes the bridge to collapse into the sea and two stacks are produced. London Bridge was the name given to well known sea arch located along the southwestern coast of Victoria in Australia (Fig. 18.21). In February 1989 the arch suddenly collapsed while two tourists were investigating its seaward end. Luckily for them they were not in the middle when the rocks gave way; they were later rescued by a helicopter and taken to shore. All in all they had a memorable day.

Along many rocky coasts the pounding of waves at the base of sea cliffs undercuts the rock, producing a notch (Fig. 18.22). In addition to abrasion and other wave processes, erosion is also attributed to freeze and thaw cycles in northern latitudes and salt crystal formation and other chemical and biological processes in tropical regions. Notches occur predominantly just below the mean high water line but may be found at lower elevations, particularly along coasts with large tidal ranges such as in the Bay of Fundy. Notches rarely extend more than several meters deep into the cliff face because as the overhang increases, there is less support of the overlying rock. Eventually, this leads to failure of the cliff and the production of debris on the shore platform. Stacks can also become notched, giving them a mushroom appearance.

Notches can be transformed into caves under

(a)

(b)

Fig. 18.21 (a) London Bridge was a famous sea arch along the southwestern coast of Victoria, Australia. (b) It collapsed suddenly in 1989, stranding two tourists, who were later rescued by helicopter.

Fig. 18.22 Notches usually develop near the high-tide line because wave energy is concentrated in this zone.

Fig. 18.23 Waterfalls are common features along tectonic coasts where the rate of uplift is greater than the rate at which the stream can downcut.

special geological conditions. This usually occurs where the bedrock is locally weak and susceptible to erosion. One of the authors will not forget walking along the shoreline of Puale Bay on the Alaska Peninsula and coming to a small headland that jutted into the bay, preventing further travel. We climbed several meters up the rock face, finding a 5 m wide cave that had been excavated into a conglomerate. The stronger, more resistant sandstones and mudstones formed the boundaries of the opening. The cave, which extended about 15 m into the cliff, broadened toward the rear and seemed to be lit by sunlight. Sure enough, there was a second passageway to the cave that opened to the other side of the headland. As we approached this opening a waterfall that rained down from the overhead cliff top obscured our view of the bay (Fig. 18.23). In fact, this was one of many spectacular landforms along this rocky coast, which features shore platforms and marine terraces, numerous waterfalls cascading to beaches below, and a wide variety of stacks, arches, and notches (Fig. 18.24). The stacks of this coast are favorite nesting sites of bald eagles due the protection afforded by the isolation of the stacks and the proximity to food sources.

Fig. 18.24 A sea arch created in layered rocks due to jointing patterns and a slight difference in rock type.

18.6 Summary

Rocky coasts exist in a wide variety of settings, but are most common along collision coasts due to the high relief and low sediment supplies. They also occur in glaciated regions where sediment has been stripped from the bedrock. In tropical regions rocky shores are found where sand has been lithified by calcium carbonate cement. Rocky coasts are resistant to erosion and retreat imperceptibly. They are slowly modified by a combination of physical, chemical, and biological processes. In areas of moderate to high wave energy where gravel and/or other sediments are present, wave abrasion is the dominant eroding agent. High-latitude coasts are affected by freeze and thaw, whereas in tropical areas chemical weathering and bioerosion tend to be important erosional processes.

Cliffs are common features of retreating coasts and are often fronted by shore platforms. These are best developed in regions where the bedrock consists of horizontally layered, easily eroded sedimentary rocks. Along tectonic coasts undergoing uplift, shore platforms have been raised, forming marine terraces. Downwarped coasts may have subtidal terraces. Erosional remnants, including stacks and arches, as well as notches and caves, are found along coasts that experience differential erosion of the retreating cliffs. This can be a product of differences in bedrock composition or structural weaknesses.

Suggested reading

Bird, E. C. F. (1993) *Submerging Coasts.* Chichester: John Wiley & Sons.

Bradley, W. C. & Griggs, G. B. (1976) Form, genesis, and deformation of some central California wave-cut platforms. *Bulletin of the Geological Society of America*, **87**, 433–49.

Carter, R. W. G. & Woodroffe, C. D. (eds) (1994) *Coastal Evolution.* New York: Cambridge University Press.

Davies, J. L. (1980) *Geographical Variation in Coastal Development.* New York: Longman.

Easterbrook, D. J. (1999) *Surface Processes and Landforms.* Upper Saddle River, NJ: Prentice Hall.

Griggs, G. B. & Savoy, L. E. (1985) *Living with the California Coast.* Durham, NC, Duke University Press.

Inman, D. L. & Nordstrom, C. E. (1971) On the tectonic and morphologic classification of coasts. *Journal of Geology*, **79**, 1–21.

Johnson, D. W. (1919) *Shore Processes and Shoreline Development.* New York: Hafner Publishing.

Shepard, F. P. & Wanless, H. R. (1971) *Our Changing Coastlines.* New York: McGraw-Hill.

Trenhaile, A. S. (1987) *The Geomorphology of Rock Coasts.* Oxford: Oxford University Press.

19 Reef coasts

19.1 Introduction

Reefs are many things to many people, depending upon the situation. Regardless of the specifics, reefs are a type of positive feature on the sea floor, generally in shallow water. A ship captain considers any abrupt shallow navigation hazard to be a reef, regardless of composition or origin. Many people, especially fishermen, apply the term reef to rock outcrops that rise above the bottom and are home to a wealth of fish and other marine life. In some areas, artificial reefs are constructed of concrete rubble and other waste, such as blocks, pipes, and other concrete construction elements. They provide an excellent substrate for benthic organisms and attract many fish. Coastal engineers construct underwater breakwaters that are commonly called reefs.

Most reefs are organic in nature and are composed, at least in part, of living communities. The preferred definition of an organic reef is a "wave-resistant, organic framework." Notice that there is no type of organism specified. Whereas most people typically associate the term reef with corals, many different organisms actually form reefal structures. The common edible oyster, *Crassostrea virginica*, some types of tube-secreting worms, and red coralline algae are all reef formers. A special type of gastropod, called vermitids, also may form a reef.

19.2 General reef characteristics

As mentioned above, reefs rise above the adjacent substrate, which provides at least the potential for being subjected to wave action. This represents the wave-resistant aspect of the reef. It is formed by the colonial or articulated nature of the skeletal part of the reef. Corals, algae, worms, and so on all grow in such a way that the individual organisms combine in a structure to form a framework much like the frame of a house prior to adding the siding. This framework is the wave-resistant part of the reef and only a part of the total community. Many organisms encrust on the framework and attach to it. There is also a large bottom community that is not part of the framework. The non-framework organisms may be attached or mobile, and they may live under the surface (infauna) or on it (epifauna). Many species of fish and other swimming animals are also included in most reef communities.

The size and shape of a reef range widely depending upon the nature, level, and distribution of physical energy. The framework organisms dictate the environment in which the reef can develop. Because of their attached nature all framework organisms are filter feeders that rely on waves and currents to carry suspended organic matter to them for nourishment. Some are present in low-energy estuaries or along low-energy open coasts where currents are sluggish and waves are small. Others thrive in very high wave energy areas, such as the coral reefs associated with Pacific islands.

Oyster reefs provide a good example of how the reef shape is controlled by physical energy. The tendency is for the oyster framework to extend itself across the tidal current (Fig. 19.1). This provides for the most efficient arrangement of the filter-feeding organisms, so that the maximum number of individuals can take advantage of the tidal currents that bring their food. Where currents are sluggish, such as in low-energy estuaries, there is no apparent elongation of the framework. Similar shapes are displayed by coral reefs. Those along the exposed coast where waves attack tend to be elongate and parallel to the shoreline. Reefs that are protected from wave action, such as in lagoons, tend to be

Fig. 19.1 An oyster reef at low tide near Beaufort, South Carolina, showing a typical amount of relief and size.

Box 19.1 Australian Great Barrier Reef

Not only is the Great Barrier Reef of Australia the most famous reef complex in the world, it is also the largest. It is probably the largest of all time because none of the known reefs systems from the geologic record are as extensive. We must note, however, that the Great Barrier Reef is really a complex of thousands of small and unconnected reefs across the outer part of the continental shelf adjacent to Queensland, the northeastern state of the Australian continent. The reef complex extends from near the Tropic of Capricorn northward for a distance of about 2000 km beyond the tip of Cape York.

Unlike most barrier reefs, the coralgal reefs of the Great Barrier complex are situated a long distance from the mainland. Most are 50–100 km out on the continental shelf. This is due largely to the fact that there is significant terrigenous sediment that is carried from the streams of the mainland across the inner portion of the shelf. Turbidity caused by this influx of land-derived sediment would prohibit coral reef development. As a consequence, the reefs are well seaward of this influence.

Most of the Great Barrier Reef is pristine, in fact most of the 2000 reefs have never been visited by humans. This is a major benefit of their distant location from land. Some of the reefs have islands associated with them, and a few of these have seen significant development, especially those fairly close to the mainland. Tourist areas for sightseeing and fishing, and small research laboratories, comprise most of the facilities that have been constructed on these small islands.

The reef is not, however, without its problems. In the 1970s there was a major scare for the long-term health of the reef because of an infestation of the crown-of-thorns starfish, a large and voracious species that feasts on the tissues of the living coral polyps. These starfish were proliferating and destroying the corals, without any natural predators. A variety of ideas to curb their spread included bringing various organisms from other parts of the world to prey on them, or introducing viruses that might eliminate them. Fortunately, cool heads prevailed and after a few years they were reduced in their numbers and in their influence through natural conditions.

At the present time the Great Barrier Reef is in good shape due largely to the many restrictions on visitations, pollution control from the mainland, and good overall conservation practices.

subcircular in shape. Water depth may also influence reef shape, such as around islands that rise from deep water.

The areal extent of reefs depends in part on the type of framework organism and in part on the environmental parameters at the site. Oysters, worms, and vermitid gastropods do not produce large reefs; most are less than 0.005 km^2 in area. By contrast, individual coral and algal reefs may extend for many tens of kilometers. The longest reef complex, the Great Barrier Reef in Australia, extends for about 2000 km.

The height and thickness of reefs is generally directly related to the area: the smaller reefs tend also to be the lowest in relief. Typically they are only a meter or two above the substrate. Coral reefs may be hundreds of meters thick overall and rise way above the surrounding sea floor, but the living portion is limited to the upper few meters.

19.2.1 Zonation of reefs

Regardless of the size or of the framework organisms, there is a common general zonation in reefs, which may be shown both areally and vertically. This pattern is the result of the response of the framework community to the distribution and amount of physical energy and nutrients throughout the reef. Typically there is a high-energy side of the reef and a low-energy side. The high-energy side is commonly produced by wave action and is on the windward side of the reef. It is the most rapidly growing and massive part of the reef and tends to develop a smooth outer boundary. By contrast, the opposite side receives little wave energy and tends to be narrower and have an irregular outer margin. In most reefs there is an inner part that is poorly developed, or that might even lack framework and have a lagoon instead.

Vertically there may also be a zonation that is related to both depth and energy. For estuarine and shallow reefs composed of oysters, worms, etc., there is no vertical zonation because they are thin and rise only a meter or two above the surrounding area. Coral reefs do display a significant zonation of growth forms and taxa.

19.3 Algal stromatolite reefs

Large colonies of cyanobacteria, also called filamentous blue-green algae, have the ability to construct small, wave-resistant structures called stromatolites. These structures are produced by the cementation of particles of calcium carbonate, carried by currents, that are attached to a sticky coating on the surface of the algae. As the sediment particles form a layer covering the algal filaments, the algae extend up to the surface in their effort to continue to stay exposed to light for purposes of photosynthesis.

The combination of the current energy and the upward growth of the algal layers produces mound-shaped reefal structures that become hard quickly because of the rapid cementation of the carbonate sediment grains. Individual structures well over a meter high and in diameter have been found in the ancient stratigraphic record extending back in geologic time to the Precambrian. Modern examples are typically less than about 0.5 m high and about the same in diameter. Most modern examples are in intertidal settings but some have been found in the shallow waters of the Bahamas. The best place to see widespread stromatolite development in various shapes is at Shark Bay (Fig. 19.2) on the western coast of Australia about 1000 km north of Perth.

Fig. 19.2 Modern intertidal algal stromatolites in Shark Bay, Western Australia. These are mound-shaped versions of cyanobacteria colonies. (Photograph courtesy of M. A. H. Marsden.)

19.4 Oyster reefs

The common edible oyster requires a firm substrate on which to establish its spat (the larvae). This may be a rock outcrop or it might even be dead shells. The oysters grow in clusters of individuals that attach to each other and eventually produce a mound-shaped reef structure. In some locations the individual oysters attach directly to bedrock and form very wave-resistant reef communities.

Oysters prefer lower than normal marine salinity and consequently are most common in estuaries, where they may occupy intertidal margins or various sites within the shallow subtidal areas. Regardless of their position, the oysters must be situated so that tidal currents will carry suspended detritus to them for nourishment. In some estuaries, the reef might be elongate and oriented perpendicular to the tidal flux to maximize its food capture.

Some oyster reefs develop in open water where salinities are low. Such environments are found along the low-energy coast of Florida, where open Gulf of Mexico environments are essentially estuaries in the hydrodynamic sense. Numerous freshwater springs with huge discharges empty into the open Gulf. This is also an area of significant tidal flux. The result of these interacting processes is a series of very long, narrow oyster reefs. These reefs are essentially parallel with the coast and may extend for kilometers (Fig. 19.3). Some of the reefs have small tidal channels cutting through the reef, with shell debris on both ends giving the general appearance of a tidal inlet with its associated tidal deltas.

Another important aspect of oyster reefs is the sediment that is typically associated with them. A

Fig. 19.3 Aerial photograph taken near Crystal River, Florida, showing elongate oyster reefs aligned perpendicular to the flow of tidal currents that provide food for the oysters.

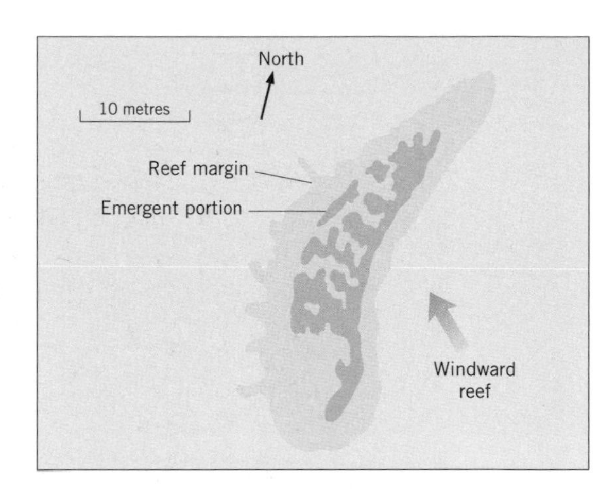

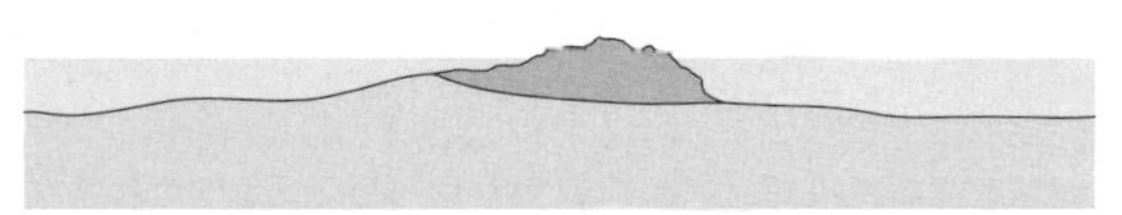

Fig. 19.4 Map of a serpulid worm reef from Alacran Bay in south Texas. These small reefoid structures develop in hupersaline bays.

look at one of these reefs shows that there are oyster shells and mud. The origin of the shells is obvious, but what about the mud? Remember that oysters are filter feeders. They ingest suspended debris without any ability to separate the digestible organic material from the fine detrital sediment particles. As a consequence, they pass considerable sediment through their digestive system and produce fecal pellets. These pellets are equivalent in size to fine sand and are soft. It is the mass of these pellets that we see as the mud component of oyster reefs.

19.5 Worm reefs

Although we typically think of worms as soft-bodied animals, there are many that produce tubes, some of which are hard. The hard tubes are composed of calcium carbonate that is secreted by the organism itself. The most common reef-forming worms of this type are the serpulid worms (family Serpulidae), which are a type of polychaete annelid. These tube worms construct massive colonies that rise as much as a meter above the adjacent substrate.

In contrast to oysters, serpulids prefer elevated salinities. They are typically found in lagoons and other areas of restricted tidal flux and freshwater input. Examples are in Baffin Bay, in the arid climate of south Texas landward of Laguna Madre, and along the Ten Thousand Island coast of southwest Florida. Baffin Bay salinities may reach the high forties in parts per thousand and southwest Florida is only a few parts per thousand above normal salinity.

These worms produce tubes that are about 0.5 cm in diameter and up to tens of centimeters long. The colonies of tubes are the framework for the small reefs that develop, usually over a few hundred square meters (Fig. 19.4). Serpulids also react to physical energy, especially currents. Those that live in areas of slow currents display a random organization of the worm tubes, much like a bowl of spaghetti. By contrast, the tubes that occupy places of strong currents show a distinct orientation, with all tubes lined up essentially parallel (Fig. 19.5). Worm reefs also tend to accumulate pelleted mud for the same reasons as in the oyster reefs.

19.6 Vermitid gastropod reefs

A certain type of gastropod looks much like a worm tube. These calcium carbonate shells are not coiled like a typical snail but are stretched out, almost straight. The shells are nearly a centimeter in

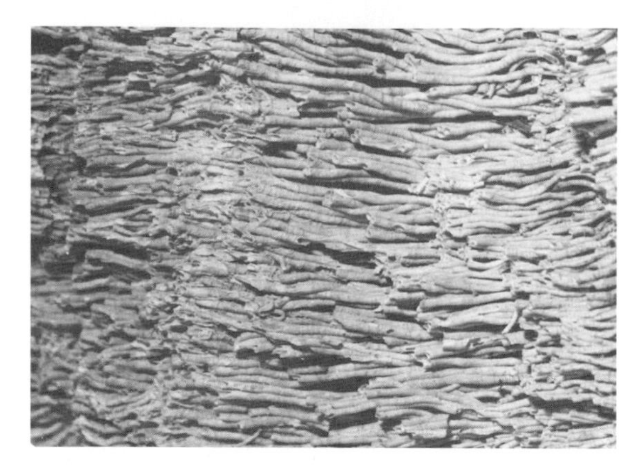

Fig. 19.5 Well aligned worm tubes that developed under strong currents. This kind of organization is not common. (Photograph courtesy of W. H. Behrens.)

Fig. 19.6 Vermitid gastropods that can form small reef structures. These snails lack the typical spiral coiling and may provide a small and local wave-resistant structure.

(a)

(b)

Fig. 19.7 (a) *Acropora cervicornis*, a fairly delicate branching complex that is a good example of a framework builder on coral–algal reefs. (b) *Acropora palmata*, a more robust framework coral. (Photographs by E. A. Shinn.)

diameter at the open end and are typically 10–15 cm long (Fig. 19.6). They grow in clusters that form small, wave-resistant framework structures that qualify as reefs. Such reefs are small and uncommon in their distribution. They are typically found in low-energy coastal areas such as the present coast of southwest Florida, where mangroves are extensive.

19.7 Coral–algal reefs

By far the most important and widespread type of reef has a framework built by hermatypic corals and red coralline algae. Hermatypic corals are those that contain zooxanthallae, a symbiotic brown algae, and are therefore restricted to the photosynthetic zone; that is, fairly shallow water. The coralline algae are limited to the same zone and are major contributors to the wave-resistant framework.

Hermatypic corals may be massive, branching, or encrusting. Some are rather fragile, such as the relatively delicate branching staghorn coral (*Acropora cervicornis*) (Fig. 19.7), whereas others may form large masses such as the typical brain corals, *Monastrea* and *Diploria* (Fig. 19.8). The algae can also occur as delicate branching varieties, such as *Goniolithon* (Fig. 19.9) and the encrusting red alga *Lithothamnion* (Fig. 19.10). This combination produces the rigid

Fig. 19.8 Massive or "brain" corals that are part of the reef framework. These coral structures are very resistant to wave action but may be "uprooted" during severe storms.

Fig. 19.9 The delicately branching red alga, *Goniolithon*, one of the framework contributors that tends to develop in the small open areas between framework corals.

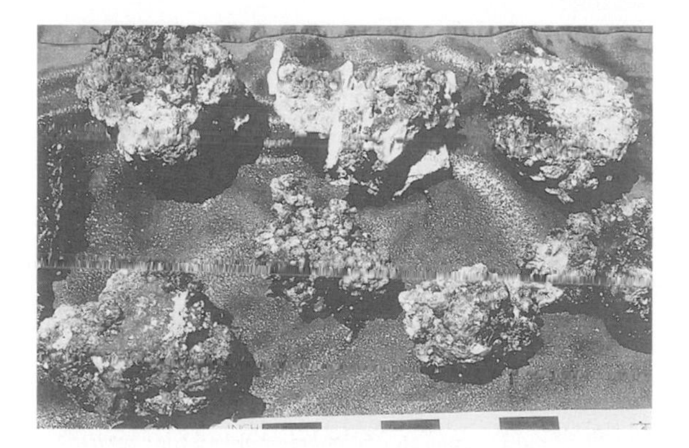

Fig. 19.10 The massive encrusting red alga, *Lithothamnion*, which provides much of the strength of the reef framework.

framework that supports the entire reef. This framework continues to persist and enlarge as new growth develops on the surface and the older part underneath dies off. This situation is found on all coral–algal reefs regardless of the general shape, location, and size of the reef.

The organisms that construct this reef framework are restricted in their distribution to shallow, clear water and warm temperatures. Reef-building corals are restricted to waters with temperatures of greater than 18°C, which limits their distribution to less than about 30° latitude in both hemispheres (Fig. 19.11). Water clarity is important because of the need for the zooxanthallae and other reef organisms to photosynthesize. Nourishment is provided to the filter-feeding framework of the reef via waves and currents. It should be noted, however, that although reefs themselves are among the most productive environments in the ocean, they cannot tolerate high levels of nutrients themselves. Recent research has shown that reefs will die with excessive levels of nutrients.

19.7.1 Reef classification

Although the basic composition and zonation of the coral–algal reef is similar regardless of size, location, age, and so on, there are various shapes that reefs achieve depending on environmental conditions. One of the first and best approaches to reef classification was that developed by Charles Darwin from data collected on his famous research cruise around the world on the *Beagle* during the mid-nineteenth century. He took profuse notes on a wide spectrum of reefs and then formulated a three-stage evolutionary approach to reef classification. These stages are: (i) fringing reef; (ii) barrier reef; and (iii) atolls. It should be noted that Darwin's observations were made in the south Pacific and were essentially confined to reefs associated with islands.

Fringing reefs are, as the name suggests, adjacent to the land (Fig. 19.12). The reef framework develops along the shore and may extend seaward for hundreds of meters. There is no lagoon or open water environment between the reef and the land. Such a reef is very good protection for the adjacent

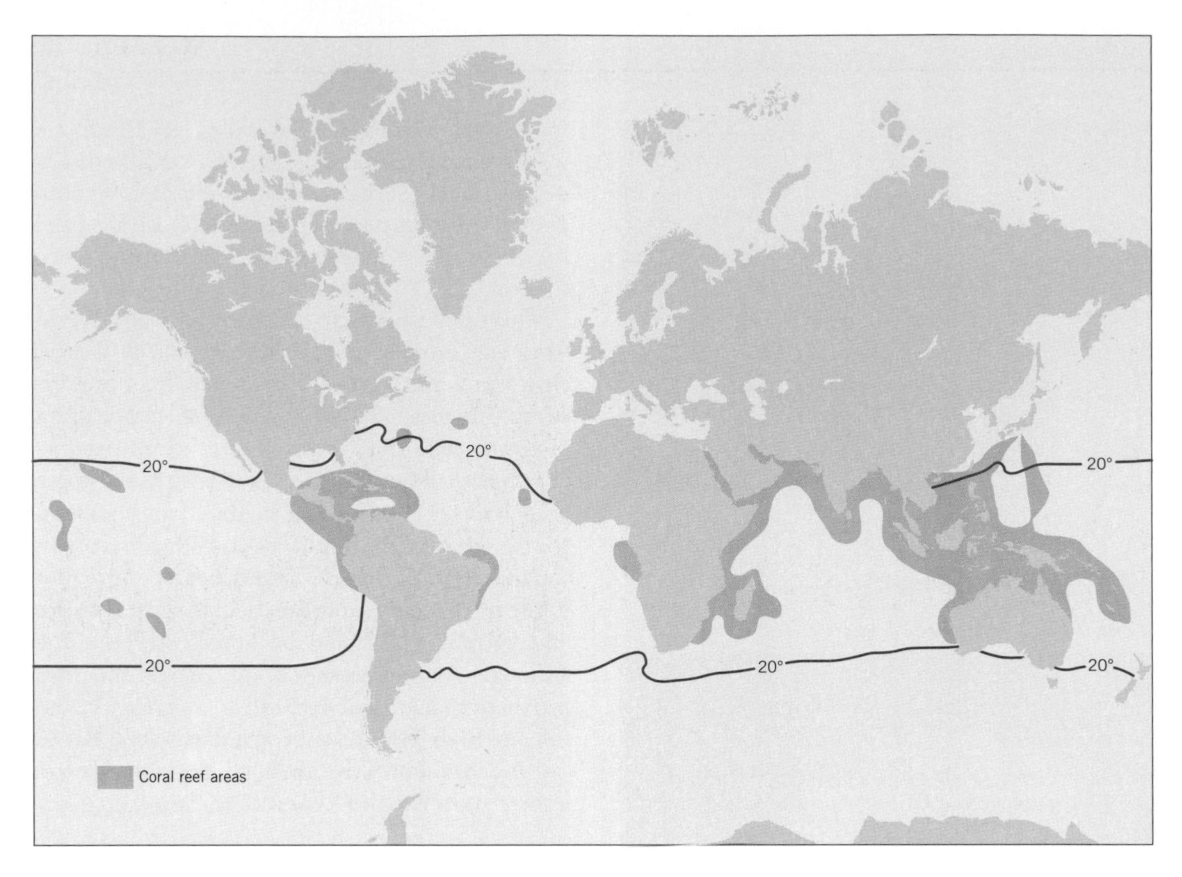

Fig. 19.11 World map showing the distribution of coral–algal reefs, which are controlled by water temperature; essentially the 20°C isotherm.

(a)

(b)

Fig. 19.12 (a) Diagram and (b) photograph of a fringing reef near Honolulu, Hawaii, that is essentially attached to the shore.

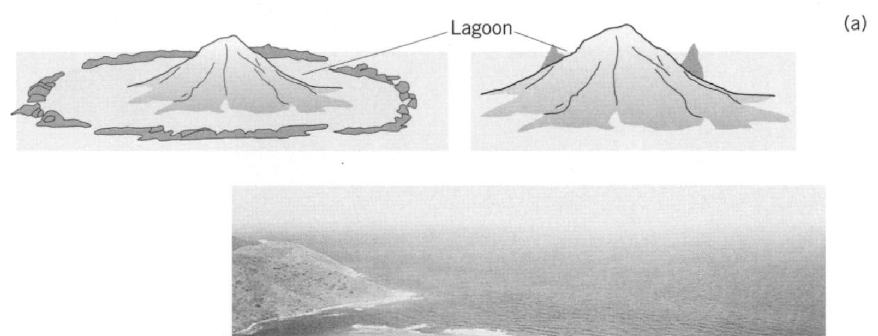

Fig. 19.13 (a) Diagram and (b) photograph of a barrier reef that has an open water lagoon between it and the adjacent shoreline on the island of St Croix in the Caribbean.

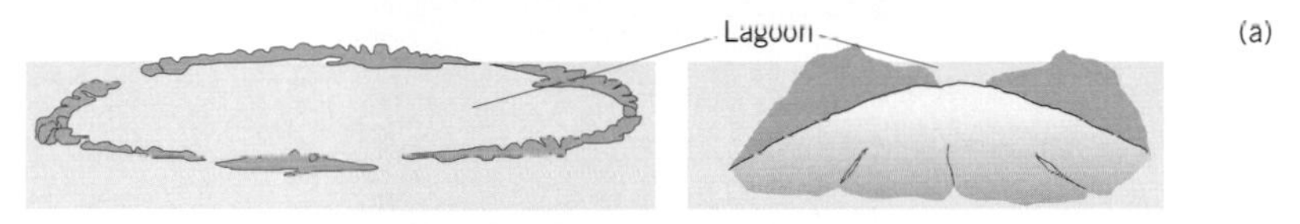

Fig. 19.14 (a) Diagram and (b) photograph of an atoll from the south Pacific. Notice that one side, the windward, is much wider than the other, the leeward.

land in that it presents a rigid and wave-resistant structure that will absorb wave energy and prevent or at least reduce erosion of the shoreline behind it. Reefs that are essentially parallel to the coast but separated from it by an open water lagoon are called barrier reefs (Fig. 19.13) because they act much like barrier islands (see Chapter 8). Most barrier reefs have open passes that act like tidal inlets and permit exchange of water between the lagoon and the open sea. Typically these reefs are at least a few kilometers in length but may extend to hundreds of kilometers. Surprisingly, the most famous reef complex in the world, the Great Barrier Reef, is not really a barrier reef at all. It is tens to hundreds of kilometers from land.

Atolls are associated with small islands or continental shelves as opposed to being adjacent to sizable land masses (Fig. 19.14). Darwin noted that there are many subcircular reefs in the Pacific that have no apparent land associated with them. He

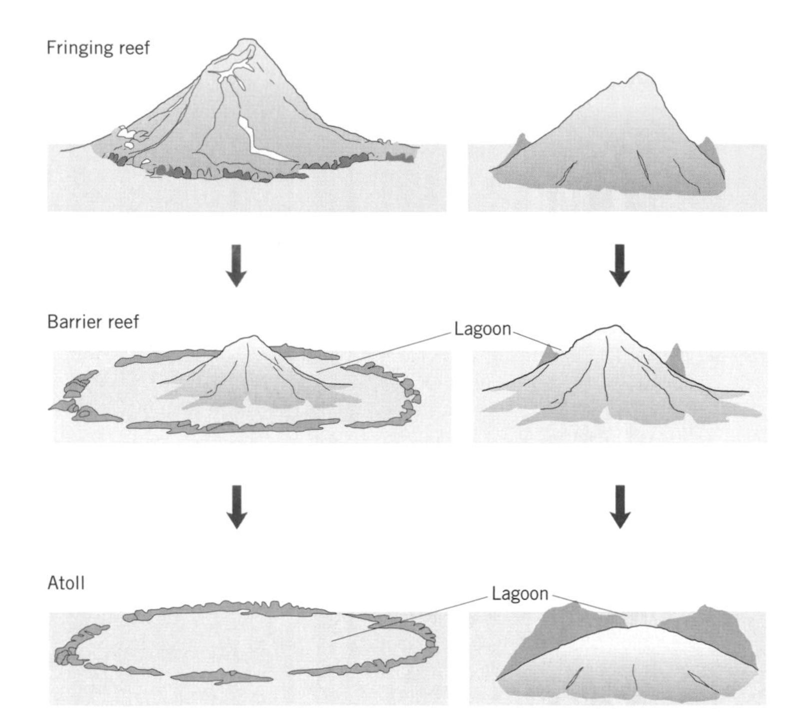

Fig. 19.15 The three-stage development of atolls as proposed by Charles Darwin. The development sequence is a continuum that takes place as volcanic islands subside over the oceanic crust.

reasoned that these reefs were the result of island subsidence and sea-level rise, with reef growth keeping pace with sea level. Darwin placed all three of the reef types in an evolutionary order, from fringing through barrier reef to atoll (Fig. 19.15). His ideas were challenged because people did not believe that oceanic islands would subside enough and sea level would rise enough to cause the island to be completely below present sea level as in the case of an atoll. Many years later atolls were drilled and cored to show that indeed Darwin was correct. The atolls were underlain by volcanic material that was part of the submerged island.

Although not part of any formal classification, there are other reef names that are widely used. These include pinnacle reefs, patch reefs, and table reefs. They are small reef structures that are associated with the other types mentioned above, generally in the lagoons of atolls or behind barrier reefs.

Another classification has been proposed much more recently, which incorporates Darwin's oceanic classification and combines it with a classification of shelf reefs. Oceanic reefs rise at least 100 m above the sea floor and shelf reefs develop on the continental shelf. Both major categories display a series of evolutionary stages, with atolls as the eventual morphology (Fig. 19.16). In oceanic reefs a land mass is surrounded by reef, whereas in the shelf reef types, wave refraction and bidirectional wave approach cause the reef to form atoll shapes, but they are noticeably asymmetrical.

19.7.2 Sea level and reef development

An important component of Darwin's evolutionary development of atolls is rising sea level. Oceanic islands of volcanic origin are composed of a large mass that causes significant subsidence. During the past 12,000 years eustatic sea level has risen about 130 m. This combination has necessitated that reefs must be able to grow upward to keep pace with such large changes in sea level and maintain optimum water depth for the reef to survive.

There are three styles of reef to sea level relationship, depending on the comparative rates of growth and sea-level change. One researcher, A. C. Neumann

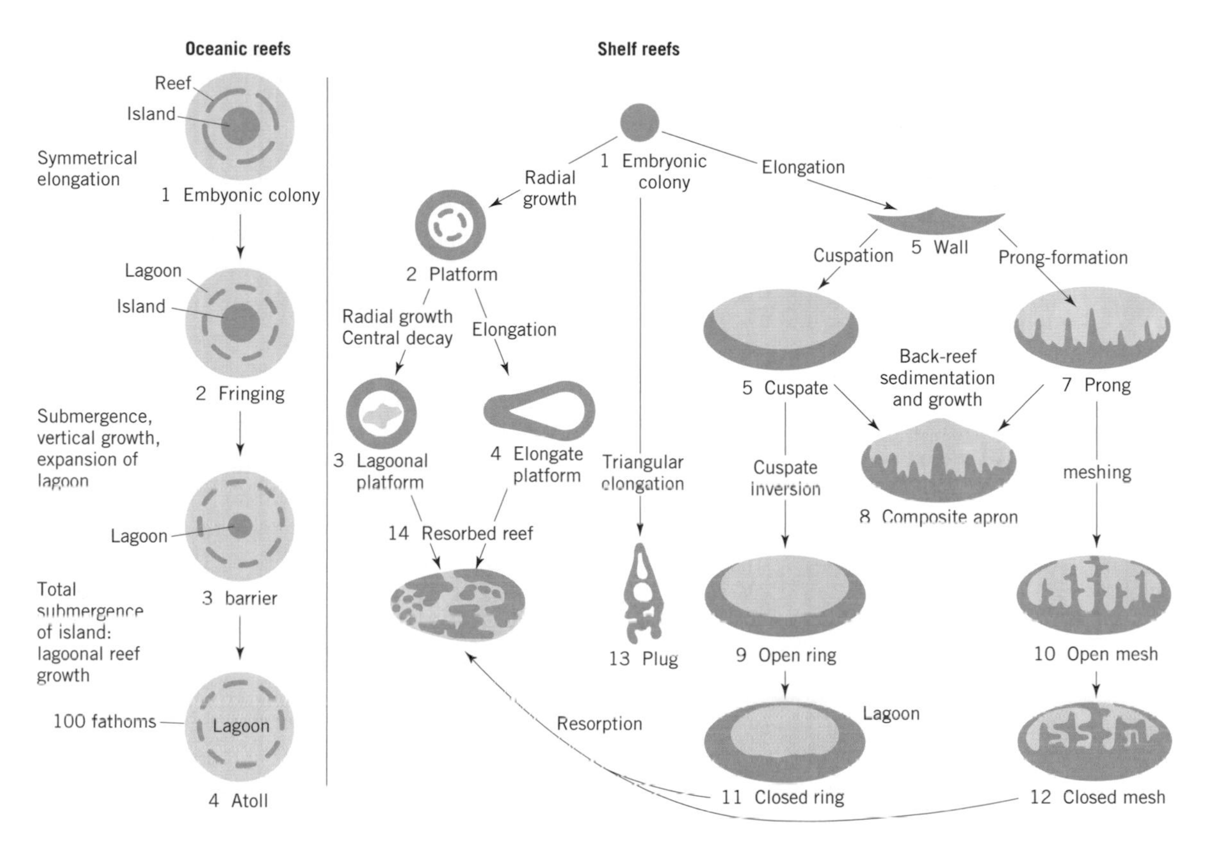

Fig. 19.16 Reef classification as proposed by Maxwell, based largely on Australian and Pacific reefs. This is a dual system of classification, which includes both oceanic reefs that rise from the ocean floor and the shelf reefs that develop on the shallow water of the continental shelf.

of the University of North Carolina, has referred to this combination as "catch up, keep up, and give up." Some reefs grow upward at rates greater than the rate of sea-level rise; they are catching up. Others keep pace with sea-level rise, so maintaining the equilibrium condition; this is the keep up situation. In some circumstances the reef growth cannot keep up with sea-level rise and the reef deteriorates and eventually dies; it gives up (Fig. 19.17).

During the past few thousand years there has been little change in sea level – only a few meters (see Chapter 4). In most places where reefs are present there has been a slow rise, but one with which reefs have generally been able to keep up. That is not to say that reefs are thriving globally, because that is not true. This topic is considered below.

19.7.3 Reef environments

The reef can be zoned into a series of rather discrete environments, each characterized by certain species and/or growth forms as a response to physical conditions (Fig. 19.18). Regardless of the size, shape, or basic type of reef, there tends to be a predictable organization to coral–algal reefs. The typical situation is one of three major zones: (i) reef slope; (ii) reef surface; and (iii) lagoon. In fringing reefs, the lagoon is absent.

Reef slope

This environment is partly a transition between the adjacent non-reef and the shallow reef environment. It is commonly quite steep, generally at least 30°,

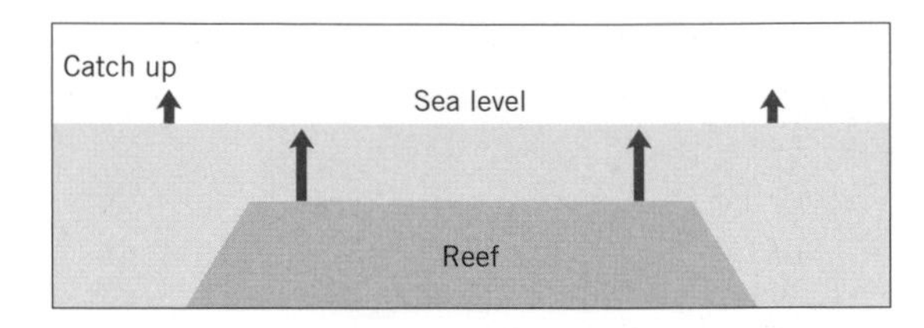

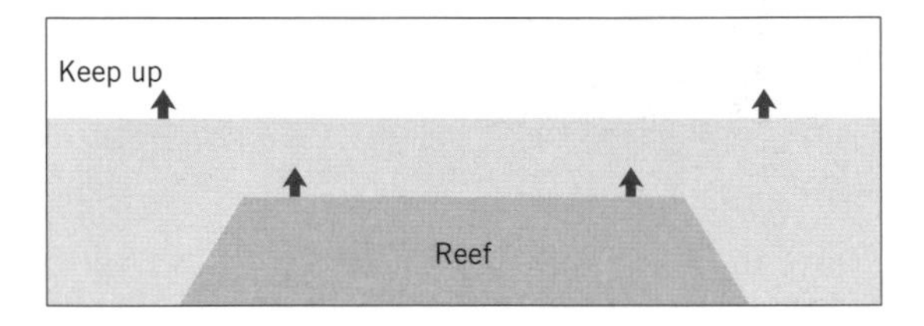

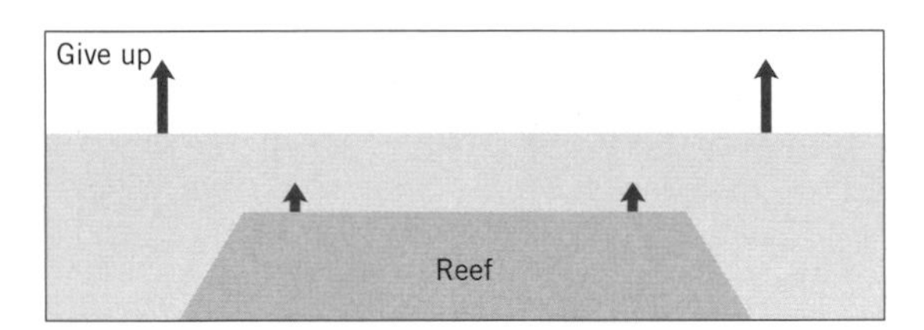

Fig. 19.17 The three stages in the sea level and reef relationship: catch up, keep up, and give up. All of this is aimed as describing the relationship between a reef and rising sea level. (After A. C. Neumann.)

and may be essentially vertical. Much of the reef slope consists of organic framework, although at least the lower part is not living. The lower part of the slope is generally more gently inclined than the upper part because it tends to be reef debris. This debris is the result of breakdown of the upper framework during storms and it forms a debris slope below the exposed upper framework. This material is angular and poorly sorted, with individual particles up to boulder size.

At depths greater than about 50 m there are few living reef corals and the slope is a mixture of debris and *in situ* dead framework. The living corals down here are small and delicate compared to those in shallow water. Commonly, reefs in high wave energy areas such as the Pacific develop a terrace at a depth 10–15 m, about where normal wave action extends. This surface is typically dominated by encrusting red algae such as *Lithothamnion*, and is generally swept free of sediment. Above this zone is the upper reef slope. It is characterized by typical reef corals, including branching and fan-shaped types such as the stinging coral (*Millipora*), staghorn coral (*Acropora cervicornis*), and moosehorn coral (*A. palmata*) (Fig. 19.19). *A. palmata* shows various growth forms depending upon wave energy. Where energy is low,

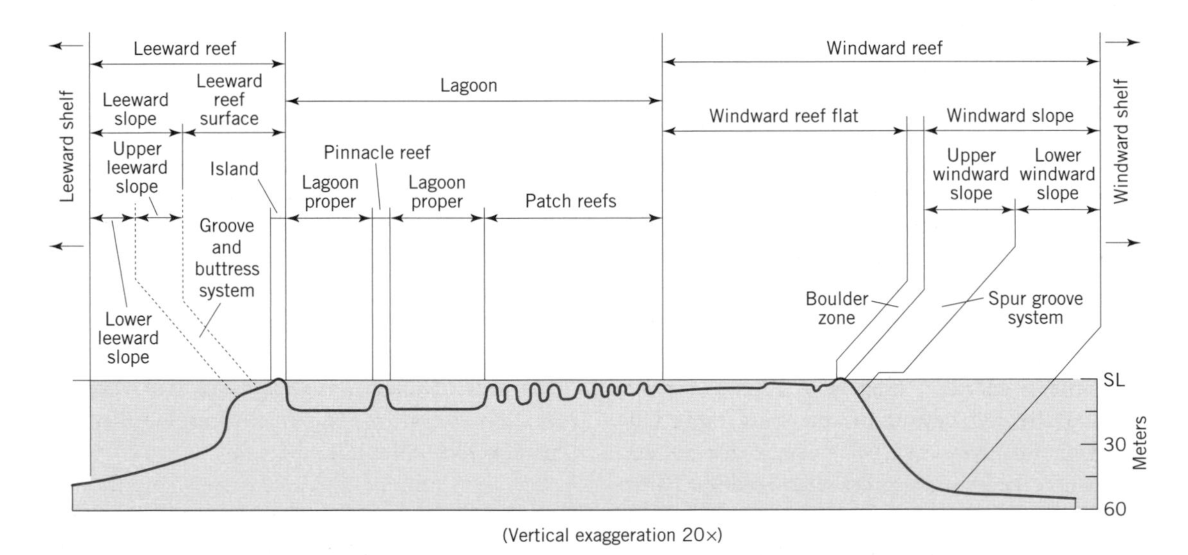

Fig. 19.18 Profile diagram across a shelf reef showing its various environmental zones. Note the similarities between the zonation across the windward and leeward portions, with the main difference being their width.

Fig. 19.19 Upper slope of the reef showing the common branching coral type *Acropora palmata* (moosehorn coral).

this coral develops broad, flat, and long branches that look a bit like a moose horn. Under high-energy conditions the same species has short and more massive branches. Light penetration is good and food is usually abundant, resulting in a diverse and dense population. Algae, bryozoans, encrusting sponges, and other sessile invertebrates are also abundant, in addition to corals.

The upper slope shows a large-scale morphology that is related to wave activity. The most prominent feature of this part of the reef is the spur and groove structures (Fig. 19.20) that are a response to wave action. The spurs, also called buttresses because of their appearance, are the positive reef ridges that extend essentially perpendicular to the reef trend. They are typically 2 or 3 m higher than the intervening grooves and are spaced a few meters apart. The spurs are dominated by coral–algal growth, whereas the grooves tend to be floored in reef debris. These spur and groove features are prevalent on the windward reef and much less developed on the leeward side.

Another aspect of the upper slope is the variety of growth form exhibited by the branching corals. This is especially shown by moosehorn coral and staghorn coral. In more protected areas or in areas near wave base, the growth forms show long branches (Fig. 19.21a), whereas in areas of high wave energy they are short or broad (Fig. 19.21b) to withstand the rigor of the environment.

(a)

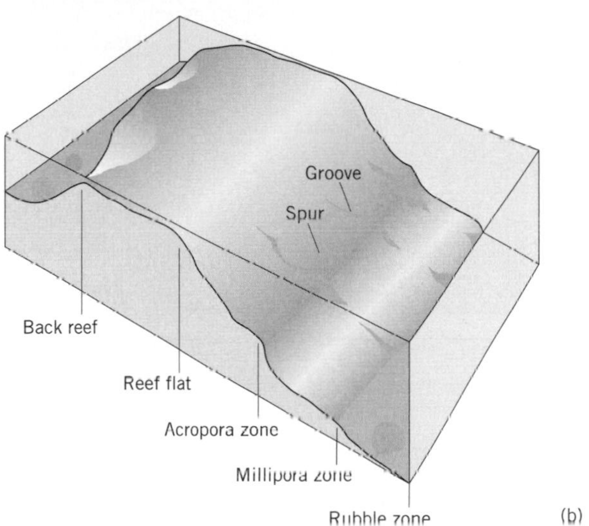

(b)

Fig. 19.20 Photograph and diagram of the spur and groove structures on the forereef slope. The spur is the living reef and the groove is the largely erosional trench between spurs. (Courtesy of E. A. Shinn.)

Reef surface

The upper surface of the reef tends to have as many as three fairly distinct zones: the *Lithothamnion* ridge, the boulder rampart, and any reef islands or cays that might be present. At the crest of the reef at many locations there is a relatively smooth and very wave-resistant zone formed by the encrusting coralline algae, *Lithothamnion*. This taxon covers the surface of the framework and provides the

(a)

(b)

Fig. 19.21 (a) Delicately branching corals in a low to modest wave energy environment, and (b) short and strong branching corals in a high wave energy environment.

Fig. 19.22 The *Lithothamnion* ridge on the windward side of the reef, displaying the flat and smooth pavement that forms in this high wave energy environment on the coast of a Fijian island.

Fig. 19.23 Storm-generated reef debris on the *Lithothamnion* ridge. Most of this is derived from the upper portion of the windward portion of the reef.

strongest part of the windward reef. In some areas where tidal range is at least mesotidal, this ridge may be intertidal. It can be almost as smooth as a paved road (Fig. 19.22) and provides excellent protection for the rest of the reef.

The windward reef receives the highest wave energy during storms and these large waves deliver enough power to the reef to break off pieces of the framework. Coralgal cobbles and boulders are thrown onto the *Lithothamnion* ridge, where they are concentrated into a linear accumulation (Fig. 19.23), the boulder rampart. This accumulation of very coarse sediment particles may display a certain level of sorting, and it fines away from the open water. The large clasts are quite angular because they are not agitated for any significant period of time but are broken and thrown up on the surface by storm conditions. Under normal wave conditions they are not moved.

Reef islands are common on many shelf atolls where debris is accumulated in small areas, generally just in from the boulder rampart on the windward reef. Sorted, biogenic sand comprises essentially the entire island. Depending on elevation, there may be small dunes, upland vegetation, and beach environments. It is very common for the islands to be somewhat stabilized by beachrock that forms in the intertidal beaches (Fig. 19.24). The calcium

Fig. 19.24 Beachrock forming on a dominantly carbonate beach in the low latitudes. Beachrock is very important in protecting small reef islands from erosion.

Fig. 19.26 Parrotfish feeding on coral. These fish are commonly about 30–40 cm long and have massive jaws for scraping and chewing on the coral, from which it obtains its nutrients.

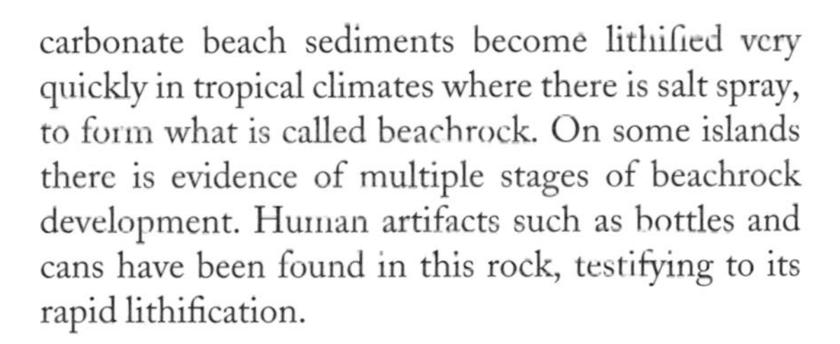

carbonate beach sediments become lithified very quickly in tropical climates where there is salt spray, to form what is called beachrock. On some islands there is evidence of multiple stages of beachrock development. Human artifacts such as bottles and cans have been found in this rock, testifying to its rapid lithification.

Reef lagoon

The protected area behind a barrier reef or in the central part of an atoll (Fig. 19.25) is the lowest energy part of the reef. Unlike the coastal lagoons that are a type of coastal bay, reef lagoons are protected low-energy water bodies, but they do not generally have elevated salinities. Tidal flux is common through breaks in the reef surface. This lagoon typically contains pinnacle reefs that rise from its floor to near sea level. The depth of the lagoon in some atolls may be so great that photosynthesis cannot take place on the floor, but the more typical depth is less than 10 m. The community present in a reefal lagoon includes many of the same species that are on the outer reef but in low-energy growth forms. There will also be some organisms in the lagoon that are restricted to the low energy conditions found there.

Fig. 19.25 Underwater photograph of a reef lagoon, showing the delicate, generally low energy forms of sea fans and sea whips that persist in that environment.

The absence of strong tidal currents and significant waves gives rise to very poorly sorted sediments that include mud and shells. The sediments here are the result of expiration of various algae and invertebrates. They also include a range of fecal pellets and disaggregated pellets that are produced by sea urchins, parrot fish (Fig. 19.26), worms, clams, and other animals that live on the lagoon floor. The urchins and especially the parrot fish actually eat the corals. A diver on a reef can hear the continual crunching of the parrot fish as they use their large, massive jaws to feast on the reef. Other reef organisms are filter feeders that rely on currents to bring their food and detritus feeders that ingest sediment on the floor of the lagoon. Many of the vagrant bottom dwellers that live in the lagoon cannot live on the outer reef because of its rigorous environment, but they are well adapted to the protection afforded by the lagoon. Delicate versions of various corals, both hard and soft, are a common constituent of reef lagoons. Sea whips, sea fans, and other soft alcynarian corals are examples.

Pinnacle reefs or patch reefs are small framework structures that rise from the lagoon floor. They are in many ways like miniature versions of the reef itself, except that they do not have a windward side or a coralline algal ridge.

19.8 Coral reef deterioration from human activities

Reefs are fragile environments that cannot tolerate a wide range in conditions. Turbidity, temperature, and water depth have already been mentioned as variables that must be tightly controlled in order to support healthy reefs. Other potential problems can be attributed to human activity. These include chemical and organic pollution, thermal pollution, turbidity, and physical destruction. The general trend is that reefs tend to be most protected along the coasts of developed countries and are in serious jeopardy in developing countries. One might ask about the value of coral reefs. The first things that come to mind are: (i) their protection of adjacent coastal areas much like that afforded by barrier islands; (ii) the abundant fishery that they support for both sport and commercial fishing; (iii) their tourist or recreational value; and (iv) their position as one of the most prolific and diverse biotic systems in the marine environment and as such deserving of protection.

Fringing and barrier reefs tend to be most vulnerable to human impact because of their proximity to land. Atolls typically have little or no population on or near them. There are, however, human-generated problems on atolls. One of the most well known is the nuclear bomb tests in the southwest Pacific by France and the United States. French tests have continued in the face of widespread protests from other countries and various environmental groups. Military activities in general have been a severe problem for atoll reefs because of the widespread operations on them. A recent accident in the Florida Keys caused severe damage to a reef as a research vessel ran aground.

Mining activities have also been conducted on reefs. The coral, sand, and gravel have been taken for various uses, especially for construction. The blasting that is frequently used also kills many of the fish. In some of the island reef areas dynamite is actually used as a means of fishing. As with the nuclear testing activities, French Polynesia has been the primary focus of these activities, although they have also taken place in the Caribbean. A somewhat related activity is the collection of corals for sale to tourists.

The biggest problems for reefs come from the water and its contents. Because reefs require warm, clear, pure seawater, any deviation from this high quality may cause problems. Suspended sediment can be a problem for all filter feeders on the reef and can reduce light penetration for photosynthesizers. This may be caused by dredging nearby, such as in borrow areas for beach nourishment. One of the most important considerations for dredging in low latitudes is the potential effect on reefs. Another source of turbid waters that might influence reef environments is runoff from the adjacent mainland. Rainfall in the tropics is generally high and summer or monsoonal rains can cause tremendous amounts of runoff that carries large quantities of sediment. Much of this sediment will be in suspension as it moves into the shallow marine environment and will cause serious problems for reefs. One place where such conditions occur is on the Gulf of Mexico coast of Belize in Central America. Here the reef location is controlled, at least partially, by the distribution of suspended sediment.

Chemical pollutants can be a major limiting factor in reef distribution or can kill existing reefs. Nutrients, typically thought of as aiding in reef development and growth, can be a limiting factor if present in large quantities. Recent work in the Caribbean has shown that overloading a reef with nutrients will actually kill the reef; a bit like overeating. There is currently concern about reefs in the Florida Keys because of the ever-increasing domes-

Box 19.2 Grounding of a research vessel on the Florida Reef tract at Looe Key

The largest living reef in the continental United States is the Florida Reef tract, which parallels the Florida Keys a few kilometers offshore. This reef systems extends for about 200 km from just south of Miami to the Key West area. The reef system is under severe stress from the combination of huge numbers of visitors throughout the year and pollution from the densely populated nearby Florida Keys.

On August 10, 1994, disaster struck the reef near the Looe Key National Marine Sanctuary, in the form of the University of Miami research vessel *Columbus Iselin.* Reports indicate that the wheelhouse was unoccupied at the time of the grounding. This 155 foot (47.4 m) ship ran over the outer spur and groove system of the western part of the sanctuary, severely damaging four of the spurs. The damage included fragmentation of the reef structure and loss of reef substrate. More than 400 m^2 of the reef surface was impacted and nearly 300 m^3 of volume was lost. The excavation on the reef extended more than a meter below the reef surface, leaving it looking like a road bed. The severity of this accident was increased the following month when Hurricane George passed over the same area, causing failure of some reef structure due to the weakness from the collision with the ship.

A large and comprehensive reconstruction was completed in 1999 in an attempt to return the reef to its pre-collision state by Coastal Planning and Engineering of Boca Raton, an engineering consulting firm. The habitat and reef topographic configuration have been fully restored using new techniques. The damaged spurs have been reconstructed to their original topography using large limestone boulders bound together by fibreglass reinforcing rods and a special type of poured concrete. Harold Hudson, a biologist with the Florida Keys National Marine Sanctuary (NOAA), has developed a special marine cement that will bind various benthic organisms to this artificial substrate. Large colonies of different taxa of corals, sea whips, sea fans, and sponges have been transplanted onto the reconstructed reef. Although these techniques have been successful in the past, this is the most extensive transplant project yet undertaken. It was completed in August 1999 and it will be a few years before the fate of the project can be accurately assessed.

tic sewage causing nutrients to be dumped into the marine environment. The same situation exists on some heavily developed areas of islands such as Tahiti. Other types of chemical pollution include various types of industrial waste that is accidentally or purposely released into the nearshore environment where reefs are present. Along the southern coast of St Croix in the US Virgin Islands there is permanent effluent from a rum distillery. Reefs cannot exist in the vicinity of this point source discharge.

19.9 Summary

Although there are many types of organisms that build reefs, we typically think only of coralgal reefs when the word is mentioned. These spectacular living systems are a very important aspect of the marine environment. They provide protection to the coast and abundant resources for the people who live near them.

Coral reefs are quite delicate systems, in that they must have very high water quality, high temperatures, and clear water for photosynthesis, and many environmental factors can lead to their demise. Some of the byproducts of the increase in coastal development are having negative impacts on coral reefs. Pollution from the atmosphere, human waste disposal, and other natural hazards are all causing problems for these reefs.

Suggested reading

Birkeland, C. (ed.) (1996) *Life and Death of Coral Reefs*. New York: Chapman Hall.

Darwin, C. (1842) *The Structure and Distribution of Coral Reefs*. Reprint. New York: Dover, 1962.

Endean, R. (n.d.) *Australia's Great Barrier Reef*. St Lucia: University of Queensland Press.

Fagerstrom, J. A. (1987) *The Evolution of Reef Communities*. New York: Wiley Interscience.

Hopley, D. (1982) *The Geomorphology of the Great Barrier Reef*. New York: John Wiley & Sons.

Maxwell, W. G. H. (1968) *Atlas of the Great Barrier Reef*. New York: Elsevier.

20 Coastal erosion

20.1 Introduction

The erosion of the coast is a topic that is of concern throughout the world. Estimates of the proportion of coasts that are currently eroding range up to 70%. When combined with the increasing pressure to develop and populate the coast, this becomes a serious problem. For a variety of reasons, there are few areas or environments that are spared from this problem.

In this chapter, we consider the various causes of erosion and the specific environments where erosion tends to be most severe, and then give some case history examples. Some suggestions for mitigation and restoration are also presented, although these are not the focus of this chapter.

Before we go further on this topic, it is important to define our terms. You might think that erosion is pretty straightforward; everyone has seen wind blow sand on a beach or dust from a field, we are familiar with erosion of channel banks on a stream, or waves attacking the bluffs of sediment or rock along a coastal area. In all cases, sediment is taken from one place by some process and moved to another, typically to another environment. Therefore, we should be able to say that erosion is the removal of material from one place and its transport away to another location or another environment. There is a net loss of sediment or rock.

In general, this is a workable definition, but when we are dealing with the coast there are some situations where this definition is not applicable. The most discussed situation deals with storm-generated washover, where during storms large waves and elevated water levels cause sediment to be carried from the surf zone, beach, or even dunes, across to the back or landward side of the island. This is the mechanism whereby barrier islands migrate landward. Is this also erosion, because sediment is removed from one place and carried to another? Most coastal scientists would say it is not. The sediment is not really lost because it stays within the barrier island. Here it would be more appropriate to consider the shoreline as retreating but to avoid using the term "erosion." On the other hand, if sediment is removed from the beach during a storm and carried offshore or transported a significant distance along the shoreline, this would be considered as erosion.

20.2 Causes of erosion

A wide variety of processes and features can contribute to erosion in the coastal zone. The obvious ones are storms and their associated waves, strong currents, wind, and gravity. Commonly these factors act in combination. Important but small in scale is bioerosion, where organisms directly cause erosion. Some other factors in erosion are rising sea level and the presence of various human modifications, such as structures, along the coast. The rise of sea level makes it possible for processes such as waves and currents to continually attack sediments and rocks along the advancing shoreline. Structures of rock, steel, and concrete that are built along the coast can be a major factor in erosion, both directly and indirectly.

20.3 Coastal erosion factors

In this section we discuss the major factors that result in erosion along the coast. We should remember that erosion takes place along interior shorelines as well as along the open marine coast. Large estuaries and lakes may also experience removal of sediment from a shoreline and subsequent transport to other places. This is particularly common along the shorelines of the Great Lakes and also such large estuaries as Chesapeake Bay, the Bay of Fundy, and Pamlico Sound in North Carolina. The discussion here is organized around the individual factors that result in coastal erosion.

20.3.1 Waves

Chapter 6 on waves showed how waves interact with sediment and cause it to move; typically to cause temporary suspension of the sediment. The larger the waves, the more sediment will be moved.

Large waves are the result of strong winds. Such strong winds also cause the water level along the shoreline to rise in the form of a storm surge or storm tide. In addition, the waves typically approach the shoreline at an angle, producing longshore currents. The larger the waves, the faster the longshore currents. These are the conditions along the shoreline during a storm: large waves, elevated water level, and strong longshore currents.

Such highly energetic conditions will cause significant removal of sediment or rock depending on the nature of the coastal material. The materials removed may be transported to two or three different places depending upon the composition and morphology of the coast. A low-lying sandy coast will experience sediment being: (i) washed over the beach in a landward direction; (ii) carried offshore by return surge of the waves; and (iii) transported along the coast by longshore currents. If there are cliffs of sediment or rock, or if there are large dunes, then sediment will not washover the beach.

These are the typical erosional conditions that occur during storms along all coasts. The end product of such storm conditions is typically some type of erosional scarp on or just landward of the beach (Fig. 20.1), with the magnitude of the scarp being proportional to the strength of the storm. The post-storm beach is generally narrower and steeper than it was before the storm. Low-lying coasts may exhibit a smooth beach profile, with large washover fans deposited landward of the beach and, perhaps, the entire barrier island (Fig. 20.2).

Even under conditions of shoreline retreat and washover of sediment, there might be significant loss of sediment offshore or alongshore, resulting in net erosion. This can be confirmed by comparing surveys of the beach and nearshore after the storm with those measured before the storm. A quantitative determination of the loss and gain when the two profiles are compared will show the net change produced by the storm (Fig. 20.3).

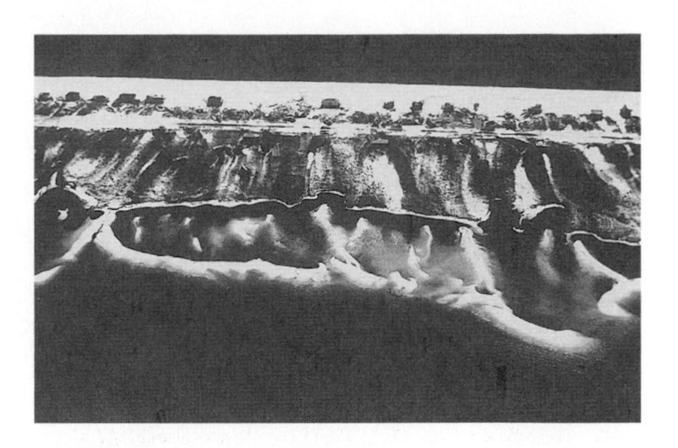

Fig. 20.2 Storm-generated washover fan landward of the beach. The processes that produce these features transport large volumes of sediment to the landward side of the island or even into the water behind the back barrier.

Fig. 20.1 Photograph of a small erosional scarp on a beach after a storm. These features are temporary and generally last only days or weeks before swash processes smooth them out.

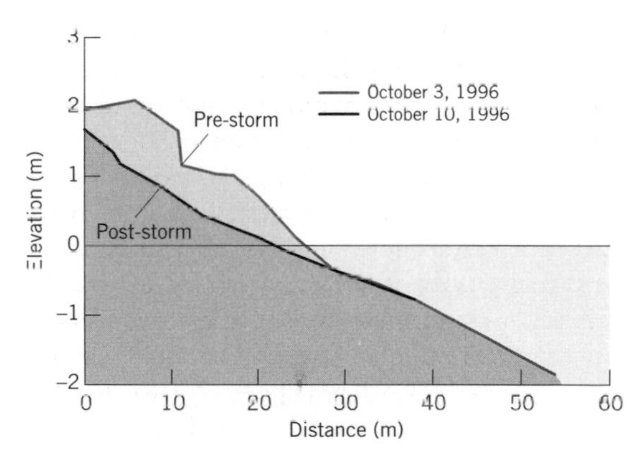

Fig. 20.3 Comparison of pre- and post-storm beach and nearshore profiles showing places where erosion took place as the result of the large waves during Tropical Storm Josephine.

Fig. 20.4 Landward migration of dunes burying the landward forest. This phenomenon is essentially a wind-generated version of a washover. Instead of taking only hours or a day or so, this may take years.

Fig. 20.5 Blowout in a large coastal dune. This feature is caused by a small unvegetated part of a dune being a site of continued scour by the wind. Eventually a large amphitheater is developed: the blowout.

20.3.2 Wind

Removal and transport of sand by the wind may take place during nonstorm as well as storm conditions. Some coastal areas experience wind of sufficient strength to move dry sand most of the time. If sufficient sand is available, these are generally places where coastal dunes are well developed. During storms, the rate of sediment movement is accelerated. The absence or removal of vegetation along the coast can lead to extensive wind erosion.

There are two common types of wind erosion along a coast with well developed dunes. In one case the absence of vegetation causes dune sand to be blown landward and the entire dune or dune system migrates landward, commonly burying forests (Fig. 20.4). This is not erosion in the sense that we defined the term above. In a way, this type of sediment transport by wind is comparable to washover by waves. Dune migration, or blowover as it is sometimes called, takes place over years, whereas washover takes place in just hours.

The other type of wind erosion along a coastal dune complex is associated with local removal of vegetation and subsequent removal of the dune sand. This typically forms a large amphitheater-shaped scar called a blowout (Fig. 20.5).

20.3.3 Currents

Strong currents are also important agents of coastal erosion. In some cases these currents carry sediment that is removed from its resting place by waves. The waves cause the sediment to become temporarily suspended and the currents move it. This situation is common in the nearshore zone during storm conditions. The strong longshore currents may move at a meter per second, a velocity that is capable of transporting large volumes of sand.

Currents may also be the agent of both removal and transport in some coastal environments. This is common where the currents are confined to channels, in tidal inlets, in estuaries, or on tidal flats. As the channel migrates or increases in size, erosion will take place, with sediment removed and transported to another location. One of the most common situations where this phenomenon is displayed is in a migrating tidal inlet. Such an unstable inlet is eroded on one side of the inlet channel, with accumulation on the other side (Fig. 20.6). The sediment removed from the channel wall is carried into or out of the inlet (erosion), while the other side is accumulating sediment from littoral drift caused by longshore currents. If such an unstable inlet is used for navigation or if its migration is threatening property, it is stabilized by the construction of jetties.

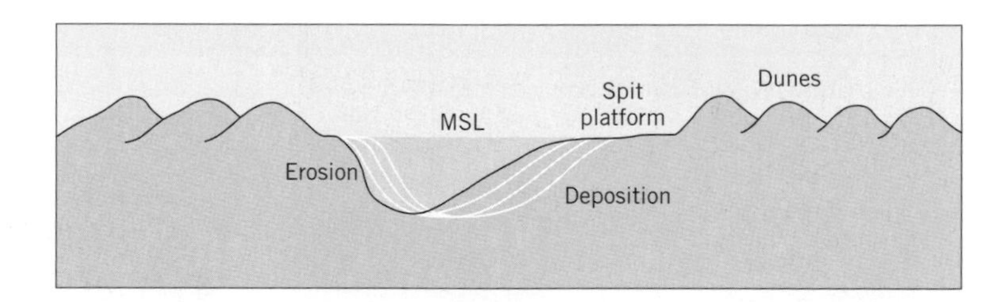

Fig. 20.6 Diagram of an inlet cross section showing how the inlet channel moves as the result of erosion on the downdrift side, while sediment accumulates on the updrift side.

20.3.4 Sea-level rise

Sea level has been rising at an accelerated rate over about the past century. This increase in sea level also results in the landward migration of the shoreline. As the shoreline is displaced landward, there is an increased opportunity for erosion by waves and currents. This condition is the single most important factor in the widespread erosion of our present shoreline throughout the world.

Although the annual rate of global sea-level rise is about 2.5 mm, there are indications that this will increase over the next century. More importantly for this discussion, there are some locations where sea-level rise is much higher – up to 10 mm yr^{-1} in some locations. This rate leads to disastrous results.

We now consider two different coastal environments to demonstrate what a 10 mm yr^{-1} sea-level rise can do. First, consider a typical beach. A common slope for a beach and the adjacent surf zone area is about 1:50. That is, there is a 1 m vertical change over a horizontal distance of 50 m. If we consider a 10 mm yr^{-1} (0.01 m) sea-level rise, it would rise a meter per century. All other things being equal, this would displace the shoreline 50 m landward. Think about the open coastal areas with which you are familiar. How close to the shoreline are the houses, commercial buildings, even the roads? Along most developed coasts, many of these would be gone under the scenario used above.

Another environment where sea-level rise has a major impact is coastal wetlands, especially marshes. As we learned in Chapter 14, coastal marshes accumulate sediment vertically as the result of being supplied by spring tides or floods caused by storms. The vegetation of these marshes is delicately balanced in position relative to sea level. At the present global rate of rise of about 2.5 mm, it is possible for many marshes to keep up with sea-level rise because they are only required to accumulate a like amount of sediment per year. If, on the other hand, the rate of sea level rise was four times greater (10 mm yr^{-1}) there is no way that the marshes could keep up; they would literally drown and erode. This is the situation that is currently taking place along the coast of Louisiana (Fig. 20.7) because of its abnormally high rate of sea-level rise.

Fig. 20.7 Marshes experiencing drowning due to sea-level rise along the coast of Louisiana. As drowning takes place the marsh grass dies and makes these areas vulnerable to erosion.

20.3.5 Human modification of the coast

Because of erosion problems and other types of coastal instability, we have built various structures to "protect" and stabilize the coast. Most prevalent among these structures are sea walls that are built to keep the shoreline from moving landward and to keep property from being eroded.

Unfortunately, these structures also cause erosion. Waves that eventually hit directly on the seawalls

Fig. 20.8 Waves hitting directly on a seawall due to the absence of any beach protection. Eventually the seawall will fatigue and fail from continued pounding by waves. After failure and removal of the seawall, erosion will take place.

(Fig. 20.8) cause scouring at the base of the structure. This scour is a form of erosion because sediment is removed and transported away. In some situations this condition can also lead to the destruction of the sea wall itself.

Another example of structures that cause erosion is jetties at inlets. They stabilize the inlets but they cause erosion to the beaches adjacent to them. The presence of the jetties prevents sediment from moving across the inlet to the beaches downdrift. As a result, there is erosion at this location (Fig. 20.9), and it can be severe.

20.3.6 Bioerosion

Some organisms themselves can cause erosion. Although the erosion caused by an individual organism is very small, the cumulative effect of thousands of individuals can be significant. Some organisms have the ability to physically erode surfaces by boring into them for protection or by scraping across them in search for food. Other organisms have chemical secretions that will react with rock surfaces to cause erosion. Algae, sponges, and bivalves are among the most common bioeroders. The most common place for this activity to take place is where the water level meets the rocky coast. Here notches (Fig. 20.10) are a common product of bioerosion.

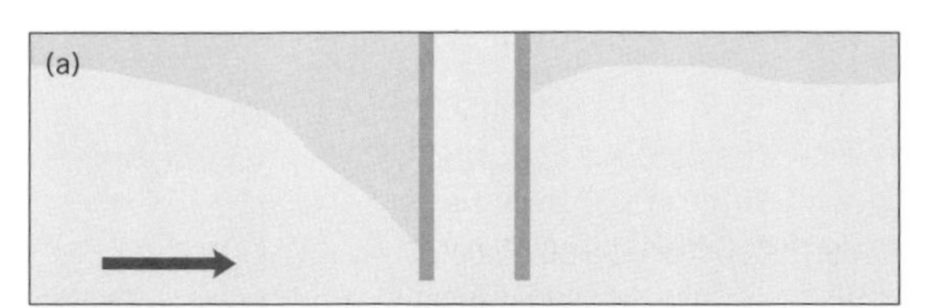

Fig. 20.9 (a) Sketch and (b) photograph showing the downdrift erosion associated with a structured tidal inlet. The jetties prohibit the longshore transport of sediment across the inlet.

Fig. 20.10 Notch formed by bioerosion in limestone cliff at sea level. Although these notches are at sea level they are not the result of physical processes.

20.4 Example scenarios of coastal erosion

This section considers only a few examples of situations where various processes interact to produce coastal erosion. Note that the combination

Box 20.1 Erosion at Sargent Beach, Texas: a natural phenomenon

Sargent Beach is a small community of dominantly single-family residences located southwest of Freeport on the east-central portion of the Texas coast. Most of the development is in the form of small weekend and recreational homes built on stilts to guard against storm surges from hurricanes and other severe events. Most of this development, which was initially several city blocks deep inland from the shoreline, took place in the early 1950s. It lies between the Gulf Intracoastal Waterway (GIWW) and the open Gulf of Mexico.

Historically there were reports of about 2 km of land separating these two features as recently as the 1930s, and an aerial photo from 1943 shows about half that distance. This part of the Texas coast is underlain by muddy deposits from an old river delta. Erosion of this coastal area has been taking place for several decades but has become a major problem since the late 1960s. The once 2 km wide area between the GIWW and the shoreline has been reduced to about 250 m (0.25 km) as five blocks of houses have been lost by the retreating shoreline.

There are at least two contributing factors to this erosion problem. First, the natural situation of muddy, deltaic sediments is such that when waves break up the sediment, unlike sandy material that would remain at or near the shoreline, the muddy sediment is carried away in suspension. Second, the mouth of the Brazos River, which empties several kilometers northeast or updrift of Sargent Beach, was redirected and jettied in the 1920s in order to provide a better commercial harbor for the Freeport area. This combination of conditions meant not only that the Sargent Beach area was susceptible to erosion and removal of most of the eroded sediment, but also that there was no mechanism for longshore transport of sediment to replenish this area because of the jetties at the mouth of the Brazos River. Secondary complications are provided by the increase in the rate of sea-level rise, which enhanced erosion, and by the damming of the Brazos and other rivers, which has reduced the amount of sediment that reaches the coast.

Obviously, the many homeowners that lost their property and those remaining ones that had their homes in serious jeopardy wanted some relief. For many years a variety of efforts were aimed at protecting the remaining portion of Sargent Beach. The cries of owners of small, fairly inexpensive recreational homes do not typically bring much positive response from either state or federal governmental agencies. As erosion caused the shoreline to recede at a rate of 15 m per year, it became apparent that it would not be long before the GIWW would be threatened by this shoreline retreat. If this narrow strip of land were breached by a hurricane, it would intersect with this critical commercial waterway, clog it with sediment, and halt shipping. Pretty soon, the government began to take notice and did some economic analyses of the potential problem. The GIWW is a major commercial shipping thoroughfare, transporting 25% of all petrochemicals produced in the United States, as well as other petroleum and non-petroleum commodities. It was determined that closure of this transport route would result in a loss of $25 million per day. With the projected price of a protective structure for Sargent Beach being tagged at $60 million, the solution became obvious.

There were two choices for solving the problem: cut another channel for the GIWW, which would be through a nature preserve, or construct some type of seawall or similar structure along the open Gulf. Because of time constraints and the difficulties of getting a permit for a new channel through a preserve, the structure option was selected. Eventually a 12 km long protective structure of granite blocks and steel sheet piling was constructed. Subsequent storms have provided a veneer of sand to make its appearance bearable for the present.

One of the most remarkable aspects of this project is the timetable of planning and construction. The typical length of time from the beginning of public outcry to construction is 18–20 years. The Sargent Beach project was initially considered in 1987 and construction was completed in 1998; about half the normal time for such a project.

Although there is considerable confidence in the situation along this coastal area at the present time, and new homes are being constructed, this is still not a permanent solution. We have only bought some time.

(Most of the information from which this discussion was taken came from *Texas Shores*, vol. 52, no. 1, 1999, the publication of the Sea Grant Program of Texas.)

of coastal composition and morphology is an important consideration, along with the various direct and indirect processes that operate on a given coast.

20.4.1 Leading edge coastal cells

A complicated set of circumstances exists along leading edge coasts that leads to removal of much of the coastal sediment that is available. One of the best places to see how this happens is along the California coast of the United States, where this phenomenon was first recognized by Douglas Inman of the Scripps Institute of Oceanography.

Here there are two primary sources of sediment for the open coast: from rivers that drain to the coast and directly from erosion of the bluffs along the coast. The sediment is moved along the coast by longshore currents, typically with a net transport from north to south. The unusual situation presented by leading edge coasts is one of a very narrow continental margin and the presence of the heads of submarine canyons almost at the shoreline. This allows the submarine canyons to capture the sediment being transported in the littoral drift (Fig. 20.11) and carry it to great depths by gravity. As a result the sediment is lost from the coast and cannot return.

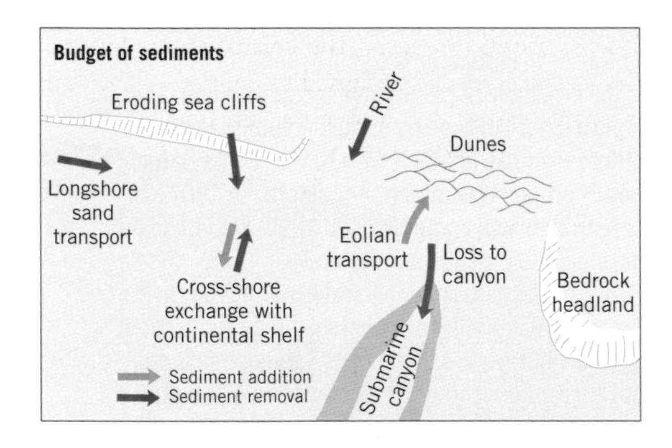

Fig. 20.11 Diagram by P. D. Komar showing the elements and direction of sediment transport along the California coast. This is essentially a sediment budget diagram. A similar pattern would take place on the Atlantic and Gulf coasts but with the absence of the submarine canyon element.

Fig. 20.12 Erosion of high cliffs by wave attack at the base. When waves scour the base of these cliffs the friable or soft strata fail and tumble to the base of the cliff. This situation is very common along the west coast of the United States.

20.4.2 Undercutting high cliffs

Some high-energy coasts have storm waves crashing directly on the base of rocky cliffs. This can lead to extensive erosion, especially if the rocks are not very resistant (Fig. 20.12). A good example of this is present along the southwestern coast of Victoria in Australia. Here near-vertical cliffs about 30–50 m high rise above the shoreline. The strata that form the cliffs are composed of friable sandstone and limestone. As large waves developed by storms in the Southern Ocean crash against the base of these bluffs, the rocks are eroded. The erosion of the base of the bluffs leads to collapse of the entire bluff in sporadic but spectacular events.

20.4.3 Storm erosion of beaches

The most widespread and commonly the most spectacular type of coastal erosion is associated with beaches. There are two general scenarios here: one involves a combination of beach erosion and washover, and the other involves only the erosion of the beach and adjacent foredune.

In places where dunes are small or absent, storms typically cause a combination of erosion of the beach and washover of the beach, whereby a significant amount of sediment is transported landward. The post-storm beach tends to be narrow and steep,

Fig. 20.13 Small beach scarp and washover sediment due to a tropical storm at Egmont Key, Florida. This is a good example of how a storm can effect a low barrier island through both beach erosion and landward sediment transport across the island.

Fig. 20.14 Scarp in the large dunes produced by a hurricane on the east coast of Florida. This contrasts to Fig. 20.13 in that the high elevations along this coast prevent washover but scarping of the dunes is common.

with a small scarp. This scarp was formed by waves after the formation of the washover fan landward of the beach as the storm surge subsided. These conditions are quite common along the Gulf Coast of the Florida peninsula (Fig. 20.13).

Some coasts have well developed dunes that prohibit washover even during severe storms, such as hurricanes. As a consequence, the front of the dune is severely scarped, as well as the beach being eroded (Fig. 20.14). Such post-storm profiles are common along the panhandle and east coasts of Florida and the Texas coast after hurricanes.

20.5 Summary

Coastal erosion is a huge problem, in that most of the world is experiencing this phenomenon. Permanent or near-permanent removal of sediment and rock from the shoreline areas of the globe are the result of many processes; some that are slow and small in scale, and others that are very short and intense. The most important global problem is the increase in sea-level rise over the past century or so. Hurricanes and other severe storms are a major problem in some areas and vary with both time and location. Such storms can cause as much damage in a day as may take place over centuries under so-called normal conditions.

Human activities have also contributed to our erosion problems as the result of the building of various structures along the coastal zone. Most of these structures are designed to control erosion but in many cases they also cause erosion. Dams on rivers may be far removed from the coast, but they can have a significant effect by prohibiting sediment from reaching the coast.

Suggested reading

Bird, E. C. F. (1985) *Coastline Changes*. Chichester: Wiley Interscience.

Pethick, J. (1984) *An Introduction to Coastal Geomorphology*. London: Edward Arnold.

Pilkey, O. H. (ed.) *Living with the . . . Shore*. Durham, NC: Duke University Press. (Several books of various dates about most of the coastal states of the USA.)

Pilkey, O. H. & Neal, W. J. (1984) *Living with the Shore*. Durham, NC: Duke University Press.

21 Human interaction with coastal dynamics

21.1 Introduction

This book has described a number of aspects of a broad variety of coastal environments. These discussions have emphasized primarily the natural state of the environments and have demonstrated their fragile nature. We have also noted that most of the world's population lives along or near the coast. This is probably more true in the United States than in any other developed country. The combination of fragile environments and dense population has produced many problems that are growing in number and becoming more severe with time. Domestic, industrial, and even recreational use of the coast tends to produce extensive unnatural alterations that have long-term deleterious effects.

The majority of the problems associated with human occupation of the coast are directly or indirectly associated with construction of some kind (Fig. 21.1). Other major types of problems concern water and sediment quality, loss of habitat, and changes in hydrodynamics. In some circumstances more than one of these problems are related in a cause and effect fashion. The following discussion concentrates on our interaction with the natural coastal system. An effort is made to consider both the developed countries of North America and Europe and the developing countries. Most of the critical problems rest with the former, however, because of their long-term industrialization.

Fig. 21.1 Example of construction activities along the coast, in this case Sydney, Australia, a large city built on an estuary. Such cities have bridges, seawalls, harbors, and other major construction.

21.2 Construction on the coast

The original coastal settlements came into being in order to take advantage of various attributes of a particular location. Perhaps it was a natural harbor, protection from adversaries, a desirable food supply, or some other benefit to the population that attracted the first residents to a particular coastal location. These coastal settlements utilized natural conditions without any significant modification, even in those areas that were densely populated. Eventually the pressures of increasing populations, larger vessels, industrialization, and other factors resulted in human modification of the coastal environment to better suit our needs. Initially most of these modifications were undertaken without regard for their impact on the natural coastal system. Surprisingly, this attitude continued into the twentieth century and only since the 1970s has our approach to coastal construction experienced major changes in philosophy. It is important that we understand how various types of construction will interfere with coastal dynamics and, as a result, how the problems are generated. We also need to know about alternative methods of coastal construction. This understanding permits us to make proper coastal management decisions in the future, whether it be as a private citizen or as a representative of a governmental unit.

21.2.1 History of coastal construction

Construction along all the environments of the coastal zone has essentially been the domain of the engineering community, specifically coastal engineering, which is commonly treated as a specialization within civil engineering. The practice of this specialty has covered three periods of unequal duration: (i) exploitation and utilization of the coast, which lasted the longest; (ii) the development of protection from coastal hazards, such as hurricanes or erosion, which is next in duration; and (iii) preservation and attempts to achieve harmony between nature and coastal utilization, the stage that we have only recently entered. Most of the developing countries are still in the first stage and, unfortunately, do not show many signs of having learned from the mistakes of North America, Europe, and Japan.

21.2.2 Materials

Attempts to protect or stabilize the coast have utilized nearly any type of material that one could imagine. Some are successful and some are not, and some are legally permitted and others are not. Regardless of material, it is generally the design of the structure(s), the materials used, and the location that determines success or failure. As might be expected, cost is a major factor in the design, scale, and materials used for this type of construction.

Poured concrete, metal sheet piling, and wood (Fig. 21.2) are all utilized in various ways to provide impermeable types of construction. This approach is termed hard construction and is generally for wave protection from erosion or channel stability. Concrete is utilized in much the same fashion in coastal protection as it is in buildings or other large structures. Reinforcing rods and tie-downs are added for both strength and stability. Major factors in this type of construction are the thickness of the structure, the depth to which the footings are placed, and the upper elevation of the structure. Sheet piling can be utilized in a similar fashion but has an advantage of being able to be driven or jetted many meters beneath the sediment surface, thereby preventing undermining by wave action. Wood is the least expensive and can also be driven or jetted into the sediment, but it has a low strength.

Concrete is also fabricated into large elements called dolos or tetrapods (Fig. 21.3), which are composed of four appendages with a common origin. They range in size and are placed in interlocking positions to provide the greatest stability for protection. Probably the largest in the world are used in Japan, where individual tetrapods are over 10 m in diameter and weigh many tons each. **Rip-rap** is a term applied to various sizes of boulders (Fig. 21.4) that are commonly used in different

(a)

(b)

(c)

Fig. 21.2 Construction materials commonly used: (a) poured concrete seawall; (b) sheet piling for a seawall around a marina; and (c) wood being used for a seawall. Cost is typically the controlling factor in which material is used for seawalls.

Fig. 21.3 Large concrete dolos, which are commonly used to build large, wave-resistant coastal structures. These are used along harbor entrances where waves are typically large. The combination of their mass and the very irregular surface they provide dissipates wave energy. (Courtesy of O. H. Pilkey.)

Fig. 21.4 Boulder rip-rap used for coastal protection. This was a common type of protection throughout most of the twentieth century but is less common now. The irregular surface absorbs wave energy and stabilizes the shoreline, although this is not an aesthetically pleasing method.

arrangements along or in combination with other structural material to provide protection. The basic approach is to build structures of blocks of various types of rock. Generally each one is cube shaped but they may have irregular shapes too. The rock may be limestone, granite, or generally whatever is environmentally acceptable, available, and has reasonable strength. Cost is also generally a factor. Typically the design of the structure will call for blocks of

Fig. 21.5 Gabions, wire baskets filled with rocks, used to protect dunes from coastal erosion. Here the main protection is provided by the large mass of these baskets.

specific sizes placed in designated parts of the structure. Properly done, this is a sophisticated type of construction, with each piece of rock placed individually. Some types of coastal construction utilize smaller rock pieces, generally large cobbles or small boulders that are placed in large wire baskets called **gabions** (Fig. 21.5). These rectangular containers are then placed in various configurations and locations so as to protect the coast from wave attack.

All of the above described materials have been used in coastal construction for over a century. More recently the development of synthetic fabrics has provided additional materials for this purpose. One of the most widely used is in the form of large, somewhat sausage-shaped, slightly porous plastic bags (Fig. 21.6) called **longard tubes**. These are filled with sand and placed strategically for protection in somewhat the same way as gabions. Other bags have been filled with a mixture of cement and aggregate and placed along the shore. The water mixes naturally with the dry contents of the bag, which hardens to form a concrete mass to protect the coast.

The least expensive type of protection is also the most aesthetically unpleasing, and typically is ineffective. This is the dumping of scrap debris from construction in areas where erosion is a problem. Broken concrete, asphalt, bricks, and other similar rubble have been used. Old car bodies (Fig. 21.7)

Fig. 21.6 Longard tubes of plastic filled with sand to protect the beach from further erosion and to act as groins. Sand is pumped into the bags, which are porous enough to permit water to escape.

Fig. 21.7 Car bodies dumped along an eroding bluff to retard the rate of coastal erosion. Although it is fairly effective, this ugly and environmentally bad approach to shoreline protection is typically carried out illegally.

Fig. 21.8 Snow fences placed along the back part of the beach to trap sand and help to form dunes. The fence filters wind and allows sand to accumulate in much the same way that it does along highways during winter in the high latitudes.

have also been a popular material for protection in some areas. This casual and environmentally unacceptable approach to coastal protection is illegal in the USA and has all but ceased.

Fencing is also used in coastal stabilization, primarily to prevent or retard the movement of wind-blown sediment in much the same fashion in which fences are used to keep snow from roads in the winter. In fact, the original approach to the problem used the same type of wooden lath fence that is used for snow (Fig. 21.8). Now it is more common for a plastic mesh or even biodegradable material to be used for the fencing. The posts are still metal or preferably wood, which will biodegrade.

Some other synthetic materials that provide a soft or indirect approach to coastal stabilization have also been tried. A rubberized material that is sprayed on the sediment surface has been used for dune stabilization, with the intent that it will hold the sediment until vegetation can become established. It is a similar approach to that used along roadways and other areas where erosion or construction has exposed the soil to potential mass wasting. Another material that has been tried is a synthetic sea grass that is designed to dissipate wave energy and thereby permit sediment to accumulate. Although installed at a number of locations, it has yet to be demonstrated as effective.

Many types of material are still in use and are approved for coastal protection. In nearly all cases, it is not the type of material used but the design and location of the construction that determines its success. Cost is generally the limiting factor in these considerations, and that is certainly applicable to the materials chosen.

21.2.3 Hard coastal protection

Probably the most common types of protection for property along the coast are within the spectrum of structures that are designed to protect and stabilize the coast. This general procedure is typically referred to as hardening the coast. The construction itself may be poured concrete, metal sheet piling, wooden timbers, various sizes of boulders, or anything else that someone believes will stabilize a threatened shoreline. All of these approaches are attempts at keeping waves and currents from moving sediment or eroding rock from its present location at or near the shoreline. Various types of structures have been designed and constructed, including seawalls, groins, jetties, and breakwaters.

Seawalls

The landward movement of the shoreline is a normal and natural process along many coasts. Limited sediment supply, rising sea level, or simply the washing over of a barrier during storms may all cause shoreline retreat. These processes only become a problem when there are obstacles in the path of the moving shoreline. When these obstacles are expensive buildings or roads, they need protection and, in most situations, they get it. In most cases, this protection has been some type of hard structure until about the 1980s.

Anyone who has been to the shore has seen some type of **seawall**. These vertical or sloping structures are generally placed parallel to the shoreline in order to attempt to stop, or at least retard, erosion and the landward displacement of the shoreline. The need for seawalls is the presence of buildings, roads, or other man-made structures that are deemed to require protection from this shoreline movement. Seawalls may be constructed of virtually any type of material, ranging from plastic bags filled with sand to poured concrete armored with large rip-rap boulders. They are one of the most controversial of all coastal structures but are still being built in many areas. The problems caused by seawalls include scour from wave attack and eventual loss of beach, reflection of wave energy that may cause problems elsewhere, and their generally unsightly appearance.

Fig. 21.9 A vertical concrete seawall, a typical approach to shoreline erosion. All incident wave energy is reflected but the wall is stressed and may fail under constant wave attack.

Seawalls are typically limited in their extent. Even if the structure is successful in holding the shoreline position, the adjacent shoreline beyond the seawall will continue to retreat. This dislocation in the coast can also cause problems, especially with the continual longshore supply of beach sediment.

Vertical and impermeable seawalls (Fig. 21.9) cause the greatest problems along a coast because they must withstand the full impact of waves. Permeable structures such as slotted walls or rip-rap walls permit some wave energy to be absorbed through the structure, and sloping walls will dissipate some of the wave energy. The problem is that while these structures are in place, waves scour at their base, eventually resulting in failure of the structure (Fig. 21.10). A good example of this scour is a seawall constructed in front of a high-rise condominium along the central Gulf Coast of Florida. After less than 10 years the water was nearly 3 m deep at the seawall, where there was a beach prior to construction a few years previously. Extreme storm conditions will produce similar results and will also overtop the walls, causing erosion behind or landward of them. Modern poured seawalls have tie-downs (Fig. 21.11) that anchor them at an angle to the shore, but they commonly fail. Seawalls are very expensive, they are typically temporary, they are aesthetically unpleasing, and they seem to cause as many problems as they solve. There are places where seawalls are required, however, because there

Fig. 21.10 A failed seawall after a storm. Sometimes the wall itself fails and is disrupted by waves, but in other cases waves scour under the seawall, which also causes it to fail.

Fig. 21.11 A vertical concrete seawall with tie-downs on the landward side. Excavation by storms has removed the material from the back of the wall, exposing the tie-downs, which are metal rods anchored by concrete.

Box 21.1 Galveston seawall

Many coastal engineers consider the Galveston seawall to be one of the engineering wonders of the world. It is, without a doubt, the most imposing seawall of its kind in the United States. But why is it there? There is nothing else even remotely like it along the entire Gulf of Mexico coast.

In 1900 there was a tremendous hurricane that had landfall near Galveston, Texas, a thriving coastal city of the late nineteenth century. As one might expect, there was no warning system that long ago. The storm generated a surge of about 5 m, with large waves superimposed on top of that. Much of the town was wiped out and 5000 people lost their lives. It was the worst natural disaster in the United States.

Because of the size and vitality of the city, it was decided to make preparations and modifications so that a repeat event would not produce the same results. Many of the large and better constructed buildings, including churches, schools, and municipal buildings among others, were still structurally sound. The decision was made to raise the elevation of the town to insure that a storm tide would not flood the entire city. Existing buildings were jacked up and filled underneath in order to elevate them. The entire community was raised about 2 m.

The other major protection from hurricane attack was the construction of the Galveston seawall. The increased elevation of the city made it necessary to protect this elevated surface from erosion by waves. The wall has a curved design to permit some dissipation of wave energy as run-up on the poured concrete wall takes place. Conditions along this coast eroded the beach from the front of the seawall and a sloping ramp of large concrete slabs and rocks was placed at the foot of the seawall (see Fig. 21.12). Consequently, even modest storms caused waves to impact directly on the seawall. Another aspect of the wall is that it is limited in its extent. As a consequence there is a major "end effect" on the west end, where the unprotected portion of this barrier island has eroded way past the position of the seawall (see Fig. 21.13).

The most recent part of this story is the construction of a beach in front of the seawall. The borrow area was just offshore and near the entrance to the ship channel west of the seawall. A wide and beautiful beach was built in 1995 for both protection from future hurricanes and recreational purposes. The project was very popular with most residents and visitors because now Galveston had a wide and long beach. Unfortunately, the beach has not performed as well as expected. Considerable erosion has already taken place but future renourishment programs are planned. Such projects, when approved and supported by the US Army, Corps of Engineers, are continued for a 50-year lifetime, so there is a long future for the Galveston beach.

Fig. 21.12 Galveston seawall, one of the oldest and most successful coastal protection projects, showing a close-up of the design and construction. This structure has a stone platform at the base and a curved face to absorb some of the wave energy during severe storms.

Fig. 21.13 The end effect of Galveston seawall, showing shoreward movement of the adjacent unprotected coast, where erosion is rapid. This downdrift erosion is a major problem of all seawalls. (Courtesy of University of Texas, Bureau of Economic Geology.)

is expensive property such as hotels in jeopardy. That being the case, the structures need to be designed and located properly.

Large and expensive or historical buildings, even entire towns, may require protection. One of the best examples of a qualified success in this type of construction is the famous Galveston seawall in Texas (Fig. 21.12). This structure was initially built after a severe hurricane in 1900 killed over 5000 people and nearly destroyed the town. The wall was constructed in sections with an upper elevation of 5.2 m above mean low water. The basic construction included wooden pilings, sheet piling and rip-rap, with the bulk of the wall being a curved poured concrete surface. It is impermeable, but rip-rap at the base and the curved face absorb some of the wave energy. In one of the most remarkable engineering and construction projects of its time much of the town was raised up to the surface of the crest of the wall. Thousands of buildings, including large stone churches, were raised an average of over 2 m to accommodate it. Although the Galveston seawall experienced significant damage from a hurricane in 1915 with a storm surge of 3.96 m, there have been no subsequent major problems during numerous hurricanes. The structure supports a six-lane highway and is also the focal point for waterfront activities in the community. It is limited in its length, however, and has caused a large offset in the coast (Fig. 21.13) because of shoreline retreat at the end where there is no stabilization. This is a typical condition where seawalls end and a soft shoreline exists.

Breakwaters

Some structures that are similar in configuration to seawalls are placed beyond the shoreline itself. These are called **breakwaters** because they are designed to break up the wave energy and prevent it from reaching the shoreline and eroding it. These offshore structures may have a variety of configurations and locations depending upon the need. The simplest design is a linear and shore-parallel structure (Fig. 21.14a) that is designed to protect a beach or other coastal feature from wave attack. This type of breakwater is used to try to build up beaches for both recreation and coastal protection. It is fairly widespread in Japan but until recently was uncommon in the United States. In recent years these structures have been used extensively in Louisiana in attempts to protect this rapidly eroding coast.

Some breakwaters are attached to the shore (Fig. 21.14b) and may be designed to form a protective harbor for mooring vessels. This type of construction is fairly common in southern California and

(a)

(b)

Fig. 21.14 Examples of (a) a shore-parallel, offshore breakwater and (b) an attached breakwater. Both are designed to protect the coast, although the parallel type is for beach protection from wave attack and the attached type is typically to protect moorings for boats.

Fig. 21.15 Sediment accumulation on the mainland behind the breakwater. This accumulation is called a salient.

also along the Great Lakes, such as the waterfronts of Chicago, Cleveland, and Toronto, where natural harbors are absent. Because of their position in at least moderately deep water, breakwaters must be large to withstand attack by waves. They generally have vertical walls of poured concrete but some also have large rip-rap or dolos. Unlike seawalls, most breakwaters withstand the rigors of intense storms.

There are problems associated with breakwaters, especially concerning the littoral transport system of sediment along the coast. Virtually all breakwaters interfere with this system and limit or prevent sediment movement along the shoreline. In the case of detached breakwaters that are more or less parallel to the coast the result is the "breakwater effect." By design, the breakwater limits wave energy from reaching the shoreline. Away from the breakwater at either end there is no influence and waves approach the shore without hindrance. They produce longshore currents that carry sediment along the surf zone depending upon the direction of wave approach. Landward of the breakwater this phenomenon is absent or limited. The result is that the littoral transport system carries sediment to the sheltered area landward of the breakwater but cannot carry it beyond. This produces a large bulge of sediment or **salient** in the shoreline behind the breakwater (Fig. 21.15). Accumulation of sediment is, of course, the objective of the structure in many cases, especially if the protected shoreline is a recreational beach. Problems can arise because the trapping of sediment commonly causes erosion downdrift from the breakwater by failing to keep the littoral drift system operating without interruption. Additional problems result from the shoreline prograding all the way to the breakwater, thereby creating a safety hazard for beachgoers. If multiple breakwaters are present, deep holes can develop from wave action between them.

Attached breakwaters produce similar but typically more severe problems. The attachment of a structure to the shore is essentially like building a dam along the shoreline. It interferes with the littoral system and causes large volumes of sediment to accumulate updrift from the "dam" (breakwater)

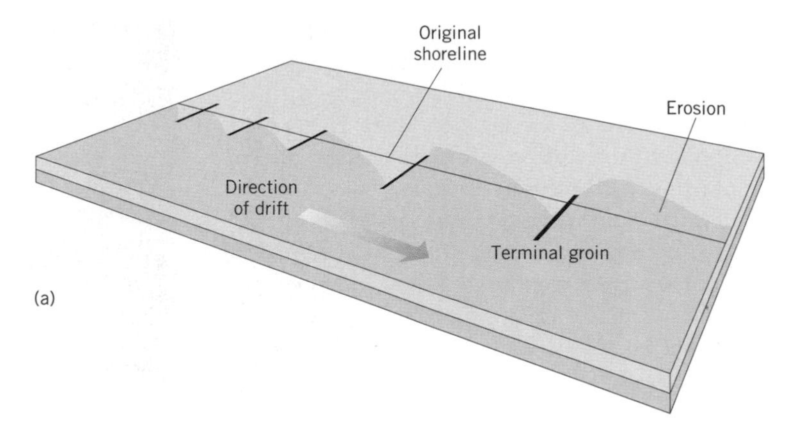

Fig. 21.16 (a) Diagram and (b) photograph of a groin field along a beach. Notice that there is an accumulation on the updrift side and erosion on the downdrift side of each groin.

(Fig. 21.14b). The biggest problem is downdrift of the structure, where sediment is prevented from accumulating, thereby causing erosion. Attached breakwaters are generally associated with some type of harbor and the harbor itself commonly receives much of the sediment that would be transported in the littoral system. This sediment accumulation in the harbor causes navigatonal problems and mooring problems. In the past few decades this problem has been somewhat alleviated by installing sediment bypassing systems where attached breakwaters are located. The bypassing systems may be a permanently installed dredge and pump operation or simply a regular program of dredging. These systems are logistically difficult to keep operational due to both the stress of operating pumps in a saltwater environment and legal challenges from adjacent landowners affected by the sand bypassing activity. They are also very costly.

Groins and jetties

Along with seawalls, **groins** are the most utilized type of construction in attempts to stabilize the coast. These are typically short structures that are attached perpendicular to the shoreline. They commonly extend across at least part of the beach and out into the surf zone. Construction can be of essentially the same variety of materials that are used seawalls. Groins are typically grouped along the coast in what are termed "groin fields" (Fig. 21.16).

The basic idea is that the groins will trap sediment that moves through the littoral transport system and thereby maintain a beach in areas that would otherwise experience erosion. In theory this scheme makes some sense, but in practice it rarely accomplishes the goal. The typical result of groin installation includes a small-scale version of the attached breakwater problem: updrift accumulation and downdrift erosion. In some situations, erosion proceeds and the groins become detached from the beach, where they have no positive effect.

The problem with groins is primarily with the design. At any given location there is a proper design length, elevation, and spacing that will permit sediment to accumulate and eventually bypass the buried groin without causing significant downdrift erosion. The ideal groin field becomes buried and out of sight, except perhaps after a major storm that causes some temporary erosion. In reality, this condition is rarely achieved because of the lack of detailed localized data on wave climate and longshore transport rates.

The best examples of apparently proper design and construction of groins tend to be along the North Sea coasts of the Netherlands and Germany, where coastal protection has been ongoing for many centuries. Here the severe winter storms and the fairly high rate of littoral drift require the groins to be large. Many are hundreds of meters long and stand several meters high. Testimony to their success is the fact that many have become buried, there is little asymmetry to the sediment accumulation on each side (Fig. 21.17), and many have lasted for nearly a century without major maintenance.

Jetties (Fig. 21.18) are much like groins in all respects, except that they are typically larger and are located at tidal inlets. They are built for the purpose of stabilizing one or both sides of the inlet from shifting its position and preventing large volumes of sediment from filling in the inlet. Jetties are constructed in order to stabilize the inlet and permit its continued navigation. In some cases, deepening of the inlet accompanies jetty construction to accommodate the passage of deep draft vessels.

Fig. 21.18 Jettied inlet to prevent movement and infilling of the inlet. These structures are primarily designed to stabilize the inlets and permit navigation.

Like groins, jetties cause problems of interruption of the littoral drift system. Sediment accumulates on the updrift side and is prevented from reaching the downdrift side of the inlet, thus inducing erosion (Fig. 21.19) – the same pattern as in groins. Because

Fig. 21.17 A large groin in the North Sea showing burial and a symmetrical sand accumulation on both sides. These large groins work well, allow for downdrift sediment transport, and last for many decades without the need for major repairs.

Fig. 21.19 Inlet jetties showing extensive accumulation of littoral sand on the updrift side and erosion on the downdrift side where no sand is being supplied because of failure of sediment to bypass the inlet.

jetties tend to be quite long, up to a kilometer or more, the amount of longshore transported sediment accumulating on the updrift side of an inlet may be tremendous. In some cases the accumulation becomes so large that it extends into the inlet through the seaward end of the jetties as sediment tries to bypass the inlet. Intentional sediment bypassing by mechanical means is becoming a part of many jettied inlet maintenance programs. A good example of this problem is illustrated by the inlet south of Ocean City in Maryland, which was cut by a hurricane in 1933. Jetties were constructed shortly thereafter to stabilize the inlet in light of a littoral drift rate of about 140,000 m^3 per year. No bypass system was installed and as a result the shoreline on Assateague Island, the downdrift part of the system, has retreated about a kilometer through washover and blowover of the barrier island (Fig. 21.20).

Another type of inlet problem requiring construction of a jetty to prevent spit growth into the inlet occurred at Clearwater Pass (Fig. 21.21) on the central Gulf coast of Florida, where the barrier island, Sand Key, was being extended northward into the inlet, causing it to narrow. In this situation the encroachment of the barrier spit on the inlet did not produce navigational problems but it did cause problems for an existing drawbridge over the inlet.

Fig. 21.20 Looking north across Assateague Inlet just south of Ocean City, Maryland. The north, developed side has been stabilized and the south side has not. The shoreline has moved hundreds of meters in less than a century. (Photograph by O. H. Pilkey.)

(a)

(b)

Fig. 21.21 Photographs showing Clearwater Pass, Florida, (a) before jetty construction (1974) and (b) a few years after construction (1979). Arrows indicate the same buildings on both photographs.

The amount of water moving through Clearwater Pass during the tidal cycle remained fairly consistent over a few decades up to the early 1970s, meaning that the inlet cross section should remain constant. If extension of the barrier spit is causing the inlet to narrow, then it must also deepen in order to maintain a constant cross-sectional area. Such deepening occurred and caused the pilings that support the bridge to be undermined. A long jetty was constructed to stabilize the inlet width and keep the bridge from collapsing. Unfortunately, although the jetty has stabilized the inlet and the accumulation on the updrift side has provided an excellent park, the bridge stability problem was not resolved and a new bridge was constructed in 1996.

21.2.4 Soft coastal protection

During the past few decades there has been a growing trend away from hard construction and toward so-called soft means of coastal protection. This approach tries to avoid any use of foreign material in the coastal environment that is being protected, and it does not incorporate any of the traditional types of hard construction. The advantages are the aesthetic appearance of the natural materials used and their compatibility with natural coastal processes. In many cases the cost is below hard construction, but longevity is sometimes a problem. We consider a few of the increasing variety of options for this approach to coastal protection using some of the more successful examples.

Beaches

One of the first approaches to soft construction was placement of nourishment sand on eroding beaches (Fig. 21.22). This practice of **beach nourishment** has been used since about the beginning of the twentieth century but has only become widespread and involved sophisticated design since the 1970s. The first nourishment efforts were typically local and without real planning. Generally, an eroding beach was supplied with the closest available sand. It might come from an adjacent dune, from just offshore, or from a nearby shoaling inlet. Generally, little attention was paid to damage to the environment, to the texture of the nourishment material, or to design of the nourished beach area. The common result was environmental damage, rapid removal of the nourishment sand, and a overall waste of money. This led to a general lack of confidence in this approach to beach protection.

More recently there has been an increase in the sophistication of beach nourishment and the environmental damage during the process is being minimized. Engineers design the size and shape of the beach to be compatible with the adjacent bathymetry and wave conditions. Borrow material is selected to be the same size as or larger than ambient beach material and without mud, and the borrow site must be located where removal will cause minimal damage to benthic environments.

(a)

(b)

Fig. 21.22 Construction of a beach nourishment project includes (a) pumping sand onto the beach in a slurry and (b) grading it to the design specifications.

This approach has greatly increased the cost of beach nourishment but has also provided a better product, one that has more longevity and that has a good appearance. There have, however, been several nourishment projects that have not lasted very long. In some cases nearly all of the sand placed on the beach was removed in only a couple of years, while the design called for a lifetime of about 10 years. An example is nourishment of the beaches south of Port Canaveral on the east coast of Florida, where the nourishment material was too fine grained and most was eroded away in less than two years. Another situation of early removal of much nourishment material took place at Myrtle Beach, South Carolina, as the result of Hurricane Hugo in 1989. This was just a matter of bad timing shortening the project lifetime. Though occasional hurricanes are expected along this coast, it is not

Fig. 21.23 Oblique aerial photos of Miami Beach: (a) before nourishment and (b) several years after nourishment. This is one of the largest and most successful beach nourishment projects in the world. (Courtesy of US Army, Corps of Engineers.)

possible to predict when hurricanes will pass a particular coast. Obviously, nourishment projects will experience significant sediment loss from this type of storm.

On the other hand, some projects have been quite successful and cost-effective. Probably the best example is Miami Beach (Fig. 21.23) in Florida, where for several years there was essentially no beach in front of the many luxury hotels along this famous tourist coast. A multiyear project of nourishing over 15 km of beach with millions of cubic meters of sand at a cost of $65 million was completed in 1980. The borrow area was the shallow offshore area where the sediment was extracted by a suction dredge and pumped onshore. Here the sediment was placed carefully by earthmovers and bulldozers to conform to the design criteria. Today the beach is still almost as wide as after completion of the project, with stabilizing vegetation now present on the backbeach area. It is without doubt one of the most successful of all such projects. Other smaller but similar projects on Captiva Island (Fig. 21.24) and Sand Key on the Gulf coast of Florida have performed in a similar fashion. This approach to beach protection has almost totally replaced seawalls and other hard measures of protection, partly because of legislation but also because of their cost-effective nature.

Fig. 21.24 Beach at Captiva Island, Florida, 10 years after nourishment. Although beach nourishment has a finite lifetime and renourishment is commonly necessary after 5–7 years, this project has lasted more than 20 years with only minor renourishment.

A major problem associated with beach nourishment is the characteristics of the source material: its texture, location, and cost. The sediment must be essentially free of mud and organic matter in order to avoid problems of turbidity and fine sediment blanketing benthic communities. It must be coarse grained enough to be stable under the existing wave climate but not dominated by coarse shell, which limits the recreational value of the beach. This has been a problem at Mullet Key near the mouth of

Tampa Bay and at Marco Island in Florida. Proximity to the nourishment site is always a consideration, due to its effect on cost. One of the best sources of borrow material is from the ebb tidal deltas of tidal inlets. These sediment bodies contain large mud-free accumulations of sand and shell. The dynamic nature of their surface is typically unvegetated and contains few types of benthic organisms, thereby making it possible to acquire a dredging permit and minimizing the environmental damage.

Current costs for beach nourishment projects including all aspects of design and construction range from about $3 to $13 per cubic meter. Depending upon the volume of sediment required per linear meter of beach, the cost ranges from about $0.5 to $3.0 million per kilometer of beach.

Dunes

Dunes themselves are a form of natural coastal protection, but they are also commonly protected by soft means. Wind activity causes migration and wave attack produces erosion of the seaward side. Protection and stabilization of dunes (Fig. 21.25) can reduce erosion and prevent their landward movement or destruction. Various active measures to solve these problems have been used at least back to the 1960s. Since that time there has been an evolution in techniques and environmental compatibility that has proven to be successful.

Dunes have always been recognized as a critical factor in protection of the coast. Maintaining them and facilitating their growth is a primary objective of coastal management. This is made difficult because of the extremely fragile nature of dunes, especially their stability. A look at a well worn foot path or a blowout shows how unvegetated dune sand can be mobilized, even in a small space.

Fig. 21.25 Example of dunes that have been stabilized by plantings and burial of brush fences in a checkerboard patterns. This approach has been successful and inexpensive.

Fig. 21.26 Eolian sand being trapped by fencing. Eventually the fences will be buried if the project is successful. Many of these fences have biodegradable plastic between the posts.

The twofold objective of the dune management effort is simple: build dunes and protect them from destruction. Initial efforts at dune building were to use various types of fence to trap wind-blown sand (Fig. 21.26). The type and placement of fence varied, with the simplest configuration being one line of fencing at the back of the beach just in front of the dune. Many other more complicated arrangements, including boxwork fences, have also been used. If successful, the dune grows and is at least partially held in place by the fencing, which in some cases is buried, with a second level of fence being installed to try to increase the dune height. This activity is not without problems even though it does produce important dunes. The anchoring provided by the fences prevents any dune migration during washover and the result may cause accelerated erosion and removal of the sediment on the seaward side of the dune by wave attack.

Vegetation by itself and in concert with fencing has proven to be the best dune stabilizer and is also a means of producing the dunes. The combination of root structures, which hold sediment, and the leaves, grass blades, and other structures of the plants, which baffle winds to trap sediment and hold it in place, provides an effective means for keeping dunes from eroding except in very severe storms. Well planned dune building has become widespread along many coastal areas. One of the most simple but effective techniques is practiced along the Texas coast, where residents take their old Christmas trees to the beach after the holidays to be used as sand trapping agents in front of the existing dunes. This approach is very low cost, it is apparently effective, and the trees are biodegradable.

Another effective and inexpensive approach to dune stabilization is practiced along the European coast of the North Sea, especially in the Netherlands. Here, where coastal protection and control has been practiced for many centuries, the Dutch are using an important natural resource, retired people, to help to protect and build dunes. Extensive dune protection is carried out by placing small fence-like structures of shrub twigs and branches in shallow ditches (Fig. 21.27) that are dug in various patterns at the base of and throughout the foredunes. Retired coastal residents provide the bulk of the labor force and for it they receive a supplement to their pensions – everyone gains! The short fences are quite effective at trapping sand and also at stabilizing existing dunes.

There is also some important prevention and maintenance that need to be incorporated into dune

Fig. 21.27 Use of brush rows along the Netherlands coast to trap wind-blown sand. Retirees are commonly employed to construct these as a supplement to their pensions.

Fig. 21.28 Boardwalk over fragile, vegetated coastal dunes. Such structures are typically required over vegetated portions of the nourished backbeach.

stabilization. It is one thing to build dunes and stabilize them with vegetation. We must also be concerned with maintaining that stability. Included in this scheme is the construction of walkovers (Fig. 21.28) to prevent the destruction of vegetation. Where walkovers are impractical, the use of wooden planking along the footpath protects the surface from wind erosion even in the absence of plants. As the dunes are being built, the paths for people should be oriented at an angle to the dominant wind direction to prevent erosion, not just perpendicular to the beach as is so often the case. After the dunes reach a desired size and are stabilized with vegetation, the walkovers are constructed.

21.2.5 Backbarrier and estuarine construction

The coastal environments that are protected from open marine conditions tend to experience only small waves and tidal currents. These backbarrier and estuarine environments tend to be characterized by low physical energy, except for those such as the Bay of Fundy where tidal currents are very strong. In general, such environments are at or near equilibrium; any changes in their sediment distribution or physical size and shape that take place tend to be modest in scope and at a slow rate. Construction in backbarrier or estuarine areas tends to be for either of two purposes: to provide transportation or to provide buildings and related structures. Both of these can cause tremendous problems for the natural coast, but both can take place in harmony with the environment if done properly.

Causeways

Most coastal metropolitan areas include the need to transport large numbers of people across coastal bays or tidal wetlands for purposes of commuting to work or accessing recreational areas on barrier islands. Whether it be road or railway, the thoroughfare is typically in the form of a causeway over the water surface. Tunnels are very expensive and are only used where the bay is too deep to make causeways cost-effective.

The easiest and least expensive construction approach is by building a roadbed through the placement of fill on the shallow coastal aquatic environment. Bridges for boat traffic and short overpasses for circulation are commonly provided as breaks in the otherwise solid construction (Fig. 21.29). Additional bridges may be built in the event of a local area that is too deep for fill-type construction. This approach to causeway construction is replete with problems. It directly affects bottom communities by burying some organisms and by destroying others through the dredging required to obtain the fill material for the causeway.

Fig. 21.29 Fill-type causeway with a bridge that permits boat traffic and some circulation. This type of causeway is not permitted any more because of its negative influence on tidal flux. Current causeways must be on pilings to allow for full circulation across their paths.

These effects are actually minor compared to the problems the causeway causes to tidal circulation. The causeway itself acts much like a dam and greatly inhibits tidal flushing. This limitation generates two major problems for the coastal bay: (i) the restricted circulation produces problems of water quality; and (ii) there is a reduction in the volume of tidal exchange, which can cause inlets to become unstable.

Most coastal bays are quite productive and are high in nutrients. They also commonly receive considerable chemical pollution from the adjacent populated areas through surface runoff and, in some areas, industrial waste. Restrictions in circulation caused by a fill-type causeway cutting across such a bay produce important deleterious changes in water quality. **Eutrophication** takes place due to an inability to remove decaying organic material and replace oxygen deficient water with well oxygenated water. The inability to disperse pollutants results in a decrease in the habitat's quality, causing death, inability to reproduce, or stunting of various species in the bay.

The above effects of causeway construction are direct and obvious. Changes in tidal flux caused by the partitioning of the bay by the causeway may cause indirect responses to the inlets served by the tidal flow into and out of the bay. A reduction in tidal flow upsets the equilibrium condition that controls inlet cross section and stability. The consequence is commonly a decrease in inlet size, migration of the inlet, and, in extreme cases, inlet closure.

This scenario has taken place at multiple sites on the Gulf coast of Florida, where both inlets and fill causeways that connect the mainland to barrier islands are abundant. Dunedin Pass deteriorated from a width of over 500 m and a maximum depth of 6 m to instability and near closure during a 60-year period. The beginning of the problem followed the completion in 1922 of the Clearwater Causeway south of the inlet, after which the inlet cross section decreased greatly. After the construction of another causeway on the other side of the inlet about 15 km north of the first causeway, there was an increase in the rate of inlet instability and a decrease in cross section (Fig. 21.30). Similar conditions have arisen at other Florida locations and in New Jersey, Texas,

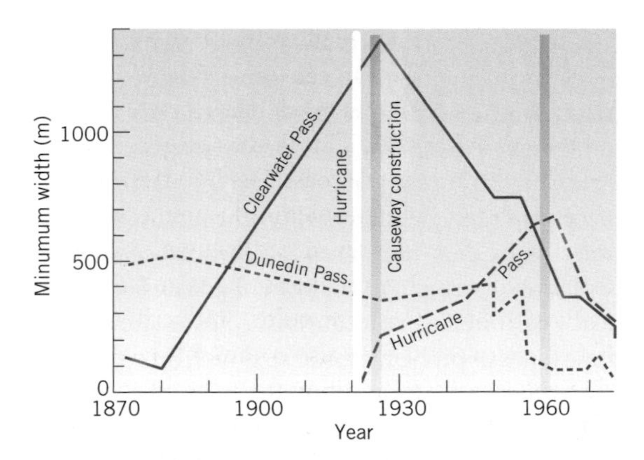

Fig. 21.30 Graph showing the influence of the construction of two fill-type causeways on the Dunedin Pass on the Florida Gulf coast. After a time lag there was a decrease in the inlet cross section after construction of a fill causeway. (Modified from M. A. Lynch-Blosse & R. A. Davis, 1977, Stability of Dunedin and Hurricane Passes, Florida. In *Coastal Sediments '77*. Reston, VA: American Society of Civil Engineers.)

and other northern Gulf coast locations. Most of the potentially affected inlets in these areas have been stabilized by jetties, thus making the response subtle: typically an increase in sediment accumulation in the affected inlet. Placing causeways on pilings, which is now required construction practice, or cutting additional tidal relief channels in fill causeways may reduce or eliminate the negative impacts of these common structures.

21.2.6 Dredging

There are many places around the coast that are dredged (Fig. 21.31) as part of our attempt to modify the environment to better suit our needs. Most dredging is related to the need to deepen or construct channels to permit boat traffic or an increase in the draft of boats. This is nearly always associated with estuaries and tidal inlets, and in some cases with coastal lagoons. Other dredging is associated with harbor and marine construction and maintenance, the mining of shell or sand for construction or nourishment, and increasing the area of upland properties for development. Some of these dredging activities have only modest effects on the coastal environment in which they occur, but

(a)

(b)

Fig. 21.31 Photographs showing (a) a dredge in operation and (b) a close-up of the rotating cutter head that cuts through the sediment and sucks it up to the dredge or onto a barge for transport. These huge dredges cost millions of dollars each and are now operated by computerized global positioning system to maintain their position and take only the appropriate sediment.

others cause some serious problems. We consider some of each category.

There are three primary concerns with channel dredging, especially in coastal bays such as estuaries: (i) the impact on bottom-dwelling organisms; (ii) turbidity and suspended sediment caused by the dredging; and (iii) disposal of the dredged material. Prior to the late 1960s and early 1970s there was little or nothing done to address these problems. Dredging destroys some of the benthic community and changes its environment, as well as severely disturbing an area equivalent to that covered by the channel or the part of the channel that is being dredged. Most coastal bays support a diverse and abundant bottom community that can be disturbed

Fig. 21.32 Aerial photograph showing a plume of sediment turbidity due to dredging. Because dredging tends to be confined to sandy bottoms, this plume of suspended sediment only lasts for hours to a day or so.

or destroyed by the dredging process. Some recent studies have demonstrated the resilience of these organisms, which recover quite rapidly from the initial destruction of the community. In less than a year the environment returns to its original character; in fact some species become more dense than they were prior to dredging, suggesting that the habitat has been improved.

The second major problem with dredging is the turbidity and suspended sediments that are produced during the dredging process in estuaries or harbors that have significant mud content in the bottom sediment (Fig. 21.32). This is a twofold problem. First, turbidity reduces sunlight to photosynthetic organisms in the water column and on the bottom. This is often a temporary problem lasting only as long as the dredging activity: typically weeks or months. The eventual settling of the suspended sediment that produced the turbidity is another matter. This fine clay- and silt-sized sediment will accumulate on the floor of the estuary or other locations near the dredging process. It may bury some organisms and it may clog the filtering system of others.

Another aspect of the dredging process is the disturbance of sediments that may contain material harmful to the environment. In industrial areas it may include various toxic pollutants that have accumulated with the sediment on the floor of the bay. Disturbance by dredging may **oxidize** and mobilize certain pollutants that were stabilized or bound to clays during their undisturbed **reduced** state. More commonly, the problem is that the disturbed sediments will contain abundant organic matter. Releasing this organic matter into the water column places high oxygen demand on the environment in order to oxidize the organics. This process causes depletion of oxygen in the water and can lead to the death of animals, including fish. Fortunately, all of these kinds of problems associated with dredging tend to be short in duration, typically weeks to months.

The third and most problematic negative aspect of dredging is the disposal of the dredged material. It is common for millions of cubic meters of material to be removed during a single project. This material must be placed somewhere and it cannot have a negative impact on the environment. In times prior to environmental awareness and restrictions, the dredge spoil was placed in upland areas or on low wetland areas for land reclamation, thereby causing pollution and destroying wetland environments (Fig. 21.33). Now dredge spoil may be used to create islands within the bay or dumped offshore in deep water. The offshore dumping option can present problems for benthic communities and it is costly because of the transport distances involved. The creation of islands may be a viable alternative. Dikes are constructed and filled with the dredged spoil, which creates another wetland area and

Fig. 21.33 Placing dredge spoil in an environmentally sensitive location such as a salt marsh on the north island of New Zealand. Such practices are not allowed and this construction company was fined for their efforts.

attracts abundant wildlife, especially birds. These impoundment islands can become important new habitats and recreational areas for picnics, fishing, and beach activities. However, drainage of the turbid water from these impoundment sites and wave erosion of containment dikes can reintroduce easily suspended mud to a coastal bay. Such sediment pollution can be a chronic problem, as is the case along the Intracoastal Waterway in Laguna Madre, Texas.

21.2.7 Dredge and fill construction

Waterfront property along coastal bays is scarce and very expensive. For many years it was common practice to dredge from intertidal or shallow bay environments and dispose of the material nearby to create upland areas that could be used for the development of residential (Fig. 21.34) or even industrial property. Many industrial complexes near metropolitan harbors have been constructed on this type of fill material. Included are the metropolitan areas of San Francisco, Long Beach/Los Angeles, Miami, and Boston. Residential areas constructed in this scenario are generally characterized by narrow, filled areas separated by finger canals. These not only destroy considerable natural bay environment but they also create problems of pollution of bays because of poor circulation, accompanied by an increased influx of products from human occupation.

Fig. 21.34 Dredge and fill construction producing finger canals for residential development. Also no longer permitted, this practice reduced tidal prism through inlets and caused severe water quality problems because of the lack of circulation.

An additional problem associated with the **dredge and fill** practice of construction is that it reduces the area of the coastal bay and therefore reduces the tidal flux through inlets serving the bay. Such a condition of reduced tidal prism causes tidal inlets to become unstable and may even lead to their closure. At the least, the inlet instability has required structures to stabilize its migration and maintain a tidal flow. This has happened in a number of locations along the Florida Gulf coast. Fortunately, these practices of dredging and construction are no longer permitted, but there are many coastal areas that retain extensive areas of finger canals.

Within these canals poor water quality commonly develops, in part, because most such construction is in coastal bay settings where tidal range is low, causing circulation to also be low. Virtually all the canals have a single opening, providing no opportunity for through-flowing tidal circulation. The problem is compounded by a high rate of nutrient and pollution influx from lawns, gardens, storm-water runoff, and other products of human occupation. These conditions lead to anoxic conditions and greatly reduce the number of fish, crabs, clams, and other animals that can live in the canal environment. In some cases, phytoplankton blooms flourish in these canals due to the stagnant, nutrient-rich water. Attempts to alleviate the problem have included aeration pumps to help add oxygen to the water and large diameter pipes to connect adjacent canals through the fingers. Although the practice of finger canal construction has essentially been stopped in most states in the United States, it is still practiced in other parts of the world, such as Australia and New Zealand.

21.3 Mining the coast

The term mining generally calls to mind gold mines, open pit coal mines, or other large-scale removal of valuable earth resources. The definition of the

Fig. 21.35 Mining the beach for sand on a Caribbean island. The beach may be the only source of sand for construction purposes, although the removal of sand from the beach is illegal. (Photograph by D. Bush.)

term includes removal of any earth material for consumption while making a profit. Some of the materials and rocks of the coastal zone are currently being mined, a few on a large scale. The products that are or have been taken include sand, gravel, heavy minerals, shell, beach rock, and salt.

Sand is the most important commodity currently being mined on the coast (Fig. 21.35). It is taken for a wide range of uses, especially for beach nourishment and for construction. Initially, nourishment material was taken from any close place where high-quality sand was available, typically dunes and nearby beaches. Due to increasing recognition that dunes protect coastal development from storm damage, coastal dunes are tightly protected from alteration. Thus, virtually all beach nourishment sand now comes from one of three sources: offshore beyond the zone of regular wave action, ebb-tidal deltas off inlets, and upland sources away from the active coastal zone.

Glass and foundry sand for industry is taken from dunes in many parts of the world. These uses require very pure quartz sand, especially for the glass industry. There are still many locations where sand is taken directly from the foredunes but most mining is restricted to the landward inactive dunes. Less and less of this activity is taking place as coastal management regulations are adopted and enforced, although some mining is still being conducted on the active beach. This is generally limited in volume and is still permitted because of old long-term agreements. Most common sites are at fillets of sand near structures or headlands and at the end of prograding spits. These locations minimize the problems to the beach area but the result is removal of a valuable resource from that environment.

Sand is very scarce in some areas, causing great problems for construction. This is particularly the case in some developing island countries that lack a major river system that can provide a sand source. This situation has led to mining of the active beach or any other available and accessible sand source. One bizarre example is on the north coast of Puerto Rico, where sand is scarce and construction is booming. The beaches are having their sand "rustled" in the dark of night by end loaders working in tandem with trucks to remove sand. Authorities are now patrolling the beach to prevent this illegal mining operation.

Heavy minerals are those that have a high density as compared to quartz, calcium carbonate, or other minerals. They commonly accumulate as placer deposits or concentrations in backbeach areas as erosion occurs on the beach and adjacent dunes. Many of these minerals are valuable commodities, such as ilmenite, rutile, zircon, garnet, and magnetite. Concentrations of only 1 or 2% by volume can be economic and typically are found in old backbeach and beach ridge deposits along the coast. These deposits are currently being mined using dredges, with the desirable minerals separated by hydraulic processes, a mining activity that produces temporary damage to the environments involved in the form of long shallow pits. Typically less than 3% of the sand volume is removed, permitting backfilling with no significant loss of sediment. Strict reclamation regulations require the companies to return the mined area to its condition prior to excavation. The most extensive accumulations of heavy minerals are along the west coast of the North Island of New Zealand, where black sand (Fig. 21.36) of nearly 90% magnetite (iron oxide) comprises beaches and dunes. Dredges separate the economic commodity, which is pumped in a slurry onto ships for transport to Japan, where it is processed in the steel industry.

Fig. 21.36 Black iron-rich sand on the west coast of the North Island of New Zealand. This sand is mined and shipped to Japan for use in the steel industry.

Here the large volumes of sand removed create lakes along the coast.

Salt has been an important commodity since the beginning of civilization. It is typically produced in various evaporite areas, in interior desert environments, and also in coastal environments. Conditions necessary for the formation of halite, common salt, include a high rate of evaporation as compared to precipitation, and a source of sodium and chlorine. Both elements are abundant in seawater and many coastal lagoons provide an evaporitic environment.

Economic deposits of halite occur in natural settings and also in evaporite ponds constructed along the coast (Fig. 21.37). Most of these halite-forming areas are in low latitudes; for example, along coastal areas of the Persian Gulf and the southern Mediterranean Sea. Some of the constructed ponds are also located in the mid-latitudes, such as along the southern part of San Francisco Bay in California.

Fig. 21.37 Salt ponds for producing commercial products on the southern coast of Spain. Evaporation under arid conditions produces precipitation of halite.

Some of the rocky coasts provide important materials for quarrying. Uses range from crushed stone for aggregate or road metal to dimension stone that is used in various types of construction. In most cases, the use of coastal sources of rock is uneconomic because of logistics or distances to the area of use. Typically this type of quarrying is done only where no other source is available, such as on small islands. Probably the most extensive quarrying of the coast was carried out in the Caribbean during the early Spanish occupation. Many of these islands and the east coast of Florida have abundant beachrock and eolianite as the only available rock. Many of the original fortifications on these islands were built from this type of limestone. The fort at St Augustine, Florida, the oldest city in the United States, is made from local beachrock.

21.4 Water quality

Throughout time, the human attitude to aquatic environments has been that rivers, lakes, and the ocean are places where waste can be disposed without problems; "out of sight, out of mind" or "dilution is the solution" approaches. Unfortunately, these practices have been quite prevalent in coastal waters and in the ocean as a whole. Problems of water quality caused by these attitudes are not new, they have been around at least since the industrial revolution. With increasing human populations and waste production, coastal waters have been less able to diluting the loads of human and industrial waste. Also, some waste simply accumulates, reaching toxic levels after years of input. Environmental awareness, along with the new ordinances and regulations that have followed, have changed our attitudes toward the water environment. Unfortunately, we are still overcoming some of the attitudes and practices of the past, but progress has been made.

21.4.1 Nutrient loading

Coastal waters tend to be shallow, especially in bays. Most of these coastal waters are naturally productive because of the high amount of nutrients that is present. These nutrients encourage phytosynthesis by diatoms, various algae, and seagrasses. Animals that feed on this plant matter are thus also abundant. Nutrients come from the chemical weathering of rocks and soil and are carried by streams to estuaries, where they support a dense and generally diverse community of organisms. This is a primary reason for estuaries being nurseries for the open marine environment.

As humans began to occupy the coast, especially the margins of coastal bays, the natural system changed and water quality deteriorated. Domestic waste has always been a problem in coastal waters because of nutrient loading and also because of coliform bacteria and other potentially dangerous pollutants. As agricultural development of the land began to utilize fertilizers, the runoff contributed considerable amounts of nutrients (nitrogen and phosphorus) to the coast. The same is true for runoff from residential areas around coastal bays.

Around many highly populated coastal areas the adjacent waters are overloaded with nutrient materials, causing extreme rates of productivity and algal blooms. The shallow bays become choked with vegetation that dies and decays, causing high oxygen demand and thus threatening fish and other animals that take their oxygen from the water. The process keeps getting worse because the decay makes more and more nutrients available as raw materials for photosynthesis. This snowballing effect eventually stifles all life in the estuary except for huge quantities of plant material.

The only solution to this condition is to restrict use of nutrient-bearing compounds and materials, and to better control the runoff discharge into coastal waters. Regulation of use is a difficult problem unless fertilizers are restricted in availability. The better way to solve the problem is by discontinuing the practice of discharging runoff via storm drains into coastal bays. Retention ponds can capture the nutrients, and the retention waters can be put to good use as irrigation material. This practice also helps to recharge the groundwater supply. These approaches are being followed in scattered coastal areas, especially in Florida, but costs are high. Future infrastructure plans will include some portions of this scheme but it will be some time before it is a widespread method of dealing with storm water runoff.

The other primary type of nutrient loading has been a problem for much longer. Sewage discharge, ranging from untreated to tertiary treated, is currently entering coastal waters in varying amounts. In less developed parts of the world, like Southeast Asia, raw sewage enters coastal waters on a regular and large-scale basis. In most developed countries only tertiary treated waters are discharged unless there are problems with the volume of sewage being directed to the treatment plants. As coastal development continues to grow, overloading of these facilities is becoming a regular occurrence.

Industrial pollution is another major source of coastal pollution, although present regulations have stopped blatant dumping. Many industries use toxic chemicals, which until recently were simply dumped into storm sewers or buried without regard to their ultimate destination. Various heavy metals, acids, organic compounds, and other pollutants may be dumped in this fashion (Fig. 21.38). These chemicals may find their way into sediments, be concentrated in certain organisms, or be distributed to other areas by tidal currents. Although the presence of such pollutants in any coastal environment is harmful to the environment, it is a temporary condition that can be corrected, sometimes surprisingly quickly.

The shallow coastal waters of the western end of Lake Erie are about midway between Detroit, Michigan, and Toledo, Ohio, two cities dominated by heavy industry. Tremendous pollution from the auto and glass industries among others resulted in this coast becoming essentially a biological desert during the 1960s. Strict regulations on waste disposal caused the coastal waters of the lake to gradually clean up and various organisms, including game fish returned. After a period of less than ten years, the environment returned to its pre-industrial quality.

Fig. 21.38 Water colored by chemical pollutants as the result of industrial waste eminating from a paper mill along the coast of Spain.

Fig. 21.39 Sedimentary fluorite on the floor of intertidal Tampa Bay from the effluent of a phosphate beneficiation plant. This chemical pollution has been eliminated by prohibiting the discharge of waste from acidization processes into the estuary.

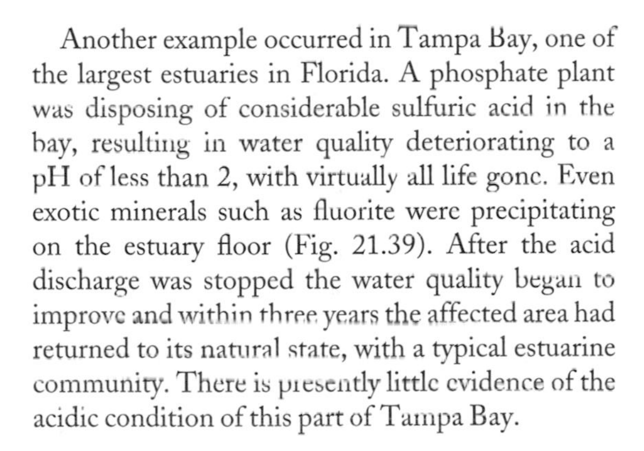

Another example occurred in Tampa Bay, one of the largest estuaries in Florida. A phosphate plant was disposing of considerable sulfuric acid in the bay, resulting in water quality deteriorating to a pH of less than 2, with virtually all life gone. Even exotic minerals such as fluorite were precipitating on the estuary floor (Fig. 21.39). After the acid discharge was stopped the water quality began to improve and within three years the affected area had returned to its natural state, with a typical estuarine community. There is presently little evidence of the acidic condition of this part of Tampa Bay.

21.5 Summary

The coastal zone comprises numerous environments, many of which are fragile. It is been occupied by humans for millennia, generally with little or no major modifications other than harbors and ports. Exceptions like Venice in Italy or the nearly completely engineered coast of the Netherlands can be found but generally people lived in harmony with the coast until the nineteenth century. This changed dramatically in two distinctly different times, for somewhat related reasons. First, the industrial revolution led to a great expansion of population, industrialization, and shipping, all utilizing the coast. Much more recently, essentially after the Second World War, an explosive increase in use of the coast for residences and recreation took place. Some of these activities have been in harmony with the dynamics of the coast but others have not.

Dense populations with numerous large buildings and protective structures have interfered with many natural processes, such as littoral drift and tidal flow, and we have dumped numerous unnatural materials in coastal waters. Fortunately, the coast and its natural systems are quite resilient. Correcting our mistakes can return much of the natural environments to near their original condition. At this stage, we have reached the level of recognizing the problems that have been created; the future must include not only stopping them but also correcting past mistakes.

Suggested reading

Beatley, T., Brower, D. J. & Schwab, A. K. (1994) *An Introduction to Coastal Zone Management*. Covelo, CA: Island Press.

Coastal Engineering Research Center (1984) *Shore Protection Manual*, two volumes. Vicksburg, MS: US Army Corps of Engineers.

Dean, C. (1999) *Against the Tides*. New York: Columbia University Press.

National Research Council (1995) *Beach Nourishment and Protection*. Washington, DC: National Academy Press.

Pilkey, O. H. (ed.) *Living with the . . . Shore*. Durham, NC: Duke University Press. (Several books of various dates about most of the coastal states of the USA.)

Pilkey O. H. and Dixon, K. (1996) *The Corps and the Shore*. Covelo, CA: Island Press.

Glossary

ablation Term given to the collective processes of ice wastage, including melting, sublimation, and ice calving into water.

abrasion Mechanical removal of rock through grinding, scraping, and other physical processes.

accretionary beach A beach where the dominant condition is deposition of sediment or stability; erosion is absent or quite limited.

active margin Type of continental margin coinciding with the edge of a lithospheric plate where two plates are colliding. Because these margins are largely confined to the rim of the Pacific, this type of margin is also termed a Pacific margin.

Afro-trailing edge coasts Coast defining the passive margin of a continent in which the opposite side of the continent is also a passive margin (e.g. east and west coasts of Africa). The lack of a collision zone means that there are no major mountain systems to supply large quantities of sediment to the coast, as compared to Amero-trailing edges.

aggradation Vertical building process due to the deposition of sediment.

aggrading barriers Barriers that build vertically in a regime of rising sea level and generate the same approximate arrangement of environments as they did following their formation and stabilization.

air compression Air trapped by breaking waves between the water and a cliff face becomes extremely highly pressurized, and when compressed into the crevices of rocks leads to enlargement of cracks and ultimately to shattering of the rocks.

alpine glacier Glacier occupying a mountain valley. Also called a valley glacier.

Amero-trailing edge coasts Coast defining the passive margin of a continent in which the opposite side of the continent is a collision margin (e.g. east coast of North and South America). This coast is typically low-lying and depositional, and receives large quantities of sediment.

amphidromic point Nodal point at the center of an amphidromic system around which the tidal wave rotates. At the amphidromic point the tidal range is zero.

amphidromic system Portion of an ocean basin or coastal sea within which a tidal wave rotates.

anticyclones Dome of high-pressure air producing a clockwise-circulating air mass.

aphelion Position of the Earth's orbit when it is farthest from the sun.

apogee Position of the Earth's orbit when it is closest to the sun.

asthenosphere Semi-plastic layer in the upper mantle extending from the base of the lithosphere to a depth of several hundred kilometers. The plastic nature of this layer is due to partial melting.

Atlantic margin Type of continental margin coinciding with the middle of a lithospheric plate. Because of the lack of tectonism, it is also termed a passive margin. Atlantic margins are wide and gently sloping due to long-term sediment deposition and subsidence.

atoll Ring-shaped chain of coral reefs and islands that surround a lagoon in the open ocean. Atolls, commonly formed over sinking volcanoes, have no landmass.

avulsion Change in the course of a river or stream that steepens its gradient and shortens its length.

backshore Landward portion of the beach, extending from the high-water line to the base of the dunes or cliff face. This region of the beach is covered by water only during storms and exceptionally high spring tides. Also called a backbeach.

bar migration The landward movement of a subtidal or intertidal sandbar toward the shoreline. The onshore migration is due to the action of breaking and shoaling waves as well as wave-generated currents. On ebb-tidal deltas flood-tidal currents augment the landward migration of bars.

barrier interior Middle of the barrier extending landward from the foredune ridge through the secondary dune system to the rear of the barrier. This region commonly consists of one or more of these elements: isolated dunes, sand sheets, washovers, blowouts, ridge and swales, ponds, and salt marsh.

barrier reef Coral reef surrounding an island or offshore of a landmass. The reef is commonly cut by surge channels and separated from the land by a lagoon.

barrier rollover The landward migration of a barrier, accomplished primarily through the process of storm overwash.

barrier spit Elongated, wave-built accumulation of sand that builds through longshore sediment transport. Attached to the mainland or a larger sediment accumulation at the updrift end.

barrier stratigraphy The sequence and composition of the layers of sediment comprising the barrier.

barycenter Center of mass of the Earth–Moon system. The barycenter is within the Earth due to the much greater mass of the Earth compared to that of the Moon.

basal slip Type of glacier movement facilitated by water acting as a lubricant that reduces the friction between the base of the ice and the underlying bedrock or sediment surface.

basalt Dark-colored, fined-grained igneous rock comprising the oceanic crust.

bayhead delta Delta formed at the head of an estuary or embayment due to sediment deposition by a river entering the estuary.

beach Deposit of unconsolidated sediment, ranging from boulders to sand, formed by wave and wind processes along the coast. The beach extends from the base of the dunes, cliff face, or change in physiography seaward to the low-tide line.

beach ridges Vegetated former foredune ridges marking previous beach positions. Low areas between ridges are called swales.

beachrock Lithified beach sediment, dominated by carbonate sediment, that forms in low latitudes due to cementation.

bedform Sediment accumulation form with a systematic morphological pattern, which is related to the flow conditions between the fluid and sediment interface.

benches See platforms.

berm The flat upper surfaces of the steplike features that may occupy the backshore. See backshore.

bioturbation Disturbance of the sediment or substrate by organisms through burrowing, boring, or grazing.

brackish Water with intermediate salinity between freshwater and saltwater.

capillary wave The first wave to develop when the wind begins to blow over a water surface. The wavelength is less than 1.7 cm and the restoring force is primarily surface tension.

carbonate minerals Minerals formed by the bonding between the carbonate ion $(CO_3)^{-2}$ and positive ions such as calcium ($Ca\ CO_3$).

centrifugal force Apparent force acting outward on an object moving in a curved pathway, such as the Moon traveling around the Earth.

coast Zone extending from the ocean inland across the region directly influenced by marine processes.

collision coasts Coastline coinciding with the convergence of two lithospheric plates (e.g. the west coast of North and South America).

continental drainage Pattern in which rivers drain a continent.

continental drift Movement of continents across the Earth's surface by plate tectonic processes.

continental glaciers glacial ice of considerable thickness covering a large portion of a continent (see ice sheet).

continental margin The underwater edge of a continent including the continental shelf, continental slope, and continental rise.

continental rise Wedge of sediment deposited at the base of the continental slope mostly by sediment gravity flows.

continental shelf Shallow, seaward-sloping platform that is part of the continent, extending from the shoreline to the continental slope.

continental slope Relatively steeply dipping portion of the continental margin seaward of the continental shelf and extending to the continental rise.

convergent boundary Zone along which two lithospheric plates converge.

coppice mounds Small accumulation of wind-blown sand around a plant.

coral reef Calcareous organic reef that is resistant to wave action. The surface of the reef consists of living coral, calcareous algae, and other attached organisms. The reef framework is composed of living and dead coral, lithified coral debris, and carbonate sand.

co-range lines Lines on a map of an amphidromic system connecting points of equal tidal range.

co-tidal lines Synoptic lines on a map of an amphidromic system showing locations where the tidal wave progresses around the amphidromic point.

core The innermost region of the Earth beginning at the base of the mantle. Thought to be composed primarily of iron and nickel.

Coriolis effect Apparent deflection of moving particles on a rotating Earth. In the northern hemisphere moving particles are deflected to the right, whereas in the southern hemisphere they are deflected to the left. There is no apparent deflection at the equator.

crevasse splay Fan-shaped deposit associated with a delta or river floodplain. Commonly formed when floodwaters breach a levee, allowing sediment to flow from the river into the adjacent lowland.

crust Outermost shell of the Earth, which ranges in thickness from 5 to 7 km under ocean basins and up to 35 km beneath continents.

cuspate foreland A broad triangular projection of the shoreline commonly referred to as a cape (e.g. Cape Hatteras). It may be more than 25 km across and associated with converging longshore transport systems and/or former river deltas.

cuspate spit Triangular accumulation of sand that extends from a barrier or mainland shoreline into a bay or lagoon. Characterized by converging sand transport cells.

cyanobacteria Microscopic blue-green algae.

cyclogenesis Process of cyclone development (low-pressure system). Cyclonic circulation is formed along an advancing frontal system in which the two air masses have a slight component of differential movement.

cyclone Weather system characterized by central low pressure compared to the surrounding airmass. Circulation is counterclockwise.

delta Accumulation of sediment that is deposited at the mouth of a river.

delta front Subaqueous, high-energy portion of the delta beyond the delta plain that is characterized by active sand deposition grading seaward into deeper prodelta mud.

delta plain The landward part of a delta that consists of subaqueous and subaerial environments, including distributary channels, levees, crevasse splays, and interdistributary bays.

desert pavement Thin lag deposit consisting of closely packed pebbles and larger gravel that protects the underlying surface from deflation.

detritus feeders Organisms that feed on organic debris.

dissipative beach Beach with a flat to low gradient profile such that wave energy is expended over a wide area.

distributaries System of secondary channels near the mouth of a river that distribute its sediment and water discharge.

diurnal tides Tides consisting of one high tide and one low tide each tidal day of approximately 24 hours and 25 minutes.

divergent boundary Zone along which two lithospheric plates diverge (e.g. mid-ocean ridges).

drowned river valley The flooded mouth of a river. Caused by valley enlargement during sea-level lowering and subsequent drowning of the valley during the Holocene transgression.

drumlin A teardrop-shaped accumulation of glacial sediment ranging from 15 to 50 m in height and 0.5 to 2 km in length. It represents an equilibrium bed configuration in which the flowing ice molded the underlying water-saturated till into large mounds. Some may be erosional remnants.

dunes see sand dunes.

easterlies Planetary-scale surface winds that blow from the northeast in the Tropics. Also referred to as the "trade winds."

ebb-tidal current Tidal current associated with the falling tide and usually flowing in a seaward direction out of a bay or tidal inlet.

ebb-tidal delta Arcuate to elongate-shaped shoal on the seaward side of a tidal inlet. Formed by ebb-tidal currents and modified by waves and flood-tidal currents.

eccentricity One of the variations in the Earth's orbital characteristics. It is a measure of the degree to which the Earth's orbit around the Sun is circular versus elliptical.

end moraine Ridge of glacial sediment marking the former front of a glacier. It forms during periods of ice sheet stabilization. The glacier continuously transports sediment toward the terminus, where it melts out and is deposited. As this process continues through time, the accumulating sediment forms an end moraine.

eolianites Lithified dune sand. Usually the sand is of Quaternary age and composed of calcium carbonate, as is the cementing agent.

eolian Having to do with the wind.

epifauna Animals that live on the ocean bottom, as contrasted with infauna, which live within the substrate.

equilibrium tide Theoretical tide based on the Earth's surface being completely enveloped by water, with no intervening continents or other landmasses. The oceans are extremely deep and uniform, and the two tidal bulges remain fixed toward and away from the Moon.

erratics Large boulders that are moved by glaciers and transported far from their origin.

estuary Semi-enclosed coastal body of water that has a free connection to the open ocean and within which seawater is measurably diluted by freshwater derived from land drainage.

eustatic sea-level change Worldwide change in sea level resulting from a change in the volume of the oceans or the size of the ocean basins.

evaporite minerals Minerals that precipitate from water during evaporation.

failed third arm Relic feature of the rifting process in which a spreading mantle plume beneath a land mass breaks apart a continent, producing a failed rift valley. It extends from the coast toward the interior of a continent. Failed third arms may evolve into sites of continental drainage.

fecal pellets Aggregates of sedimentary particles resulting from excretion by organisms.

Ferrel cell Atmospheric convection cell existing between 30 and 60° latitude. This cell is formed due to sinking air at 30° latitude and rising air at 60° latitude.

fetch Distance over which the wind blows.

firn Granular ice formed by the recrystalization of melting snow. Composition is intermediate between snow and glacial ice.

fjord A long and narrow U-shaped flooded valley connected to the ocean or sea. Sometimes described as an arm of the sea that extends into a coastal mountain range. Formed by alpine glaciers that deepened existing mountain valleys below sea level.

flaser bedding Type of stratification consisting of ripple-cross strata alternating with discontinuous mud drapes.

floodplain Broad flat land bordering a stream or river. It is covered by sediment deposited during periods of floods.

flood-tidal current Tidal current associated with the rising tide and usually flowing in a landward direction into a bay or tidal inlet.

flood-tidal delta Horseshoe to multilobate shaped sand shoal located landward of a tidal inlet, formed by flood-tidal currents and modified by ebb-tidal currents. Some flood deltas are a product of storm processes.

fluvial delta See delta.

flying spit Type of spit that occurs along straight to slightly irregular shorelines and extends at an acute angle to the beach. They are found along semi-protected shorelines and may contain recurved ridges.

foredunes Vegetated coastal dune system that forms the landward border to the beach. Also called the fore dune ridge or primary dune.

foreshore The intertidal portion of the beach, extending to the landward change in slope. Also called the forebeach or beach face. Includes the swash zone, the part of the foreshore over which the waves uprush and backwash as each one meets the shore.

fringing reef Coral reef that is contiguous with the adjacent shoreline.

funnel-shaped embayment Deeply embayed coastline (100 km wide) associated with macrotides. Commonly a river discharges at the head of the embayment, such as the Bay of Bengal, India, Gulf of Cambay, India, and Cook Inlet, Alaska.

glacial rebound Uplift of land following deglaciation due to the mass of the ice being removed from the land surface; an isostatic response of the lithosphere.

Gondwanaland Southern part of Pangaea comprising South America, Africa, India, Australia, and Antarctica.

gravity waves Waves whose velocity of propagation is controlled primarily by gravity. Most wind-generated waves are gravity waves.

Hadley cell Atmospheric convection cell existing between the equator and 30° latitude. This cell is formed due to rising air in the equatorial region and descending air at 30° latitude.

heavy minerals Minerals with a specific gravity greater than 2.85. They include garnet, ilmenite, magnetite, sphene, and zircon.

hermatypic corals Reef-building corals that have a symbiotic relationship with a blue-green alga, *Zooxanthallae*.

high marsh Portion of a salt marsh above mean high water.

horse latitudes High-pressure region occurring at 30° latitude. Formed by a descending airmass that compresses, producing high pressure, dry air, and variable winds. So named due to early sailing ships being becalmed in these latitudes and having to throw horses and livestock overboard when the ship ran out of water.

hurricane Tropical cyclonic storm with wind velocities exceeding 117 km h^{-1}. The term is used for storms occurring in the North Atlantic.

ice age Cycle of glaciation, which began approximately 2.4 million years ago, when fluctuations in the worldwide climate caused the periodic advance of huge ice sheets in high latitudes, followed by a general retreat of the glacial ice. Coincident with the waxing and waning of ice sheets was the fall and rise of sea level, by as much as 100 m.

ice sheets Glacial ice covering millions of square kilometers and reaching several kilometers in thickness. Because of their great surface area, they are also called continental glaciers.

ice wedging Physical weathering process whereby the expansion of water freezing in the cracks and crevices of rock wedges the rock apart.

igneous rocks Rock formed from the crystalization of magma.

infauna Organisms that live within the sediment bottom (see also epifauna).

intertropical convergence zone (ITCZ) Zone of converging northeast trade winds and southeast trade winds. This zone occurs at about 5° north of the equator.

island arc Group of volcanic islands that have a curvilinear arrangement and are associated with the subduction of an oceanic plate beneath an oceanic plate (e.g. Aleutian Islands).

isobars Line connecting points of equal barometric pressure on a weather map.

isostasy Equilibrium condition whereby portions of the Earth's crust are compensated (floating) by denser material below.

isotopes Atoms with the same number of protons but different numbers of neutrons.

lagoon Shallow coastal body of seawater that is separated from the open ocean by a barrier or coral reef. The term is commonly used to define the shore-parallel body of water behind a barrier island or barrier spit.

land breeze Breeze that blows offshore due to the cooling of the landmass during the evening and night.

landward margin The back side of a barrier, which might abut an intertidal sand or mud flat, a salt marsh, or an open water area associated with a lagoon, bay, or tidal creek.

Laurasia Northern part of Pangaea, comprising most of North America, Eurasia, and Greenland.

leeward reef Leeward side of a coral reef that receives relatively low wave energy.

lenticular bedding Type of stratification consisting of discontinuous lens-shaped sand beds within a mud sequence.

lithosphere Outer shell of the Earth, consisting of ocean crust, continental crust, and the rigid upper portion of the mantle.

lithospheric plates The outer shell of the Earth is broken into eight major and several minor lithospheric plates.

littoral drift Sediment that has been transported in the nearshore zone by the longshore current. It is the material that has been moved, not the process of movement. The term describing the process is called longshore sediment transport.

longshore bar Shore-parallel sand bar in the nearshore zone that is formed by wave processes.

longshore current Current in the nearshore zone that is produced by angular wave approach.

longshore sediment transport Movement of sediment along the coast in the surf and breaker zones by wave suspension and the longshore current.

low marsh Portion of salt marsh extending from mid-tide to just below high water.

magma Hot molten rock capable of flow.

mangal A swamp covered with woody shrubs and trees typically called mangroves.

mantle Portion of the Earth between the crust and core.

mantle plumes Hot columns of mantle rock that rise toward the surface through the mantle. The top of the plume may be mushroom-shaped.

marginal sea coasts Continental coastline landward of an island-arc system.

marine terraces Wave cut platforms that have been elevated by a drop in sea level or tectonic uplift.

marsh Low-lying vegetated wetlands occurring in the upper intertidal to supratidal zone. Salt marshes occur in protected environments, such as behind barriers. In these regions salt grasses and succulents colonize them.

mean sea level Average water level position measured over a period of 19 years, which takes into account natural tidal oscillations.

mean tide Average tidal conditions between spring tide and neap tide.

megaripples Bedform with a wavelength greater than 0.6 m.

mesotidal Tidal range between 2 and 4 m.

metamorphic rock Rock that has been transformed as a result of heat and pressure without melting.

microtidal range Tidal range less than 2 m.

Milankovitch cycles Cyclical changes in the Earth's orbital characteristics. The additive result of these cycles was shown by Milutin Milankovitch to be linked to climatic changes on the Earth.

mixed energy coast Coast in which the morphology has developed through a combination of wave and tidal processes.

mixed estuary An estuary that produces a vertically homogenized water column with a gradient of increasing salinity toward the ocean.

mixed tides Semi-diurnal tide with unequal successive high and low tides.

monsoon Seasonal winds in low latitudes that produce intense rainfall. These zones usually have wet summers and dry winters.

mudcracks Desiccation cracks in mud, usually with a polygonal structure.

natural levee Ridge of sediment that parallels both banks of a river or distributary. It forms during flood events when water and sediment flow beyond the banks of the river and sediment is deposited.

neap tide Smallest tidal range during a lunar cycle. Occurs when the Earth, Moon, and Sun are at right angles to one another (quadratic position) during the first and third quarters.

nearshore bar Sand bars occurring in the nearshore zone (see also longshore bar).

nearshore zone Zone from the shoreline seaward to a point just beyond the breakers.

neo-trailing edge coasts Coastlines along linear seas that have been created by spreading zones. These coasts are geologically young, such as the Red Sea.

northeasters Type of extratropical cyclone that travels northward along the east coast of the United States and Canada, producing strong winds and waves from the northeast.

obliquity The Earth's axis of rotation is inclined 23.5° with respect to a line drawn perpendicular to a plane containing the Earth's orbit. This is commonly referred to as the Earth's tilt.

oceanic reef Coral reefs associated with volcanic islands in ocean basins.

ocean trenches The deepest part of the oceans. Narrow deep troughs that parallel a continent or island arc and are formed by the subduction of an oceanic plate.

overwash Sediment that is transported from the beach across a barrier, and is deposited in an apron-like accumulation along the backside of the barrier. Overwash usually occurs during storms when waves break through the frontal dune ridge and flow toward the marsh or lagoon.

outwash plains Braided stream deposit beyond the margin of a glacier. It is formed from meltwater flowing away from the glacier, depositing mostly sand and fine gravel in a broad plain.

oxbow lake Lake formed when a river cuts through a meander bend. The river eventually abandons the curved-channel region, producing an arcuate lake.

Pangaea Single continent that existed about 200 million years ago prior to the breakup and drifting of the continents to their present position.

Panthalassa Name given to the ocean that surrounded Pangaea.

passive margin Type of continental margin coinciding with the middle of a lithospheric plate, and hence with no tectonic plate interaction and little tectonic activity. Because these margins are found rimming the Atlantic Ocean, this type of margin is also termed an Atlantic margin.

perigee The position of the moon's orbit when it is closest to the Earth.

perihelion The position of the Earth's orbit when it is closest to the sun.

photosynthesis An organic process whereby organisms containing chlorophyll convert water and carbon dioxide into carbohydrates in the presence of sunlight.

plastic flow Movement within a glacier in which the ice behaves in a plastic-like manner and is not fractured.

plate tectonics Theory concerning the dynamics of the Earth's surface. The lithosphere is broken into several large plates that move due to convection within the mantle. The theory accounts for earthquake activity, mid-ocean ridges, deep-sea trenches, mountain building, volcanism, and the distribution of landmasses.

platforms Wave-cut flat or gently sloping surfaces along a coast.

Pleistocene Epoch A time during the Quaternary extending from 1.65 million to 10,000 years ago, when ice sheets advanced and receded through much of the mid and high latitudes.

plucking A glacial excavation process whereby ice wedging loosens pieces of rock from the bedrock and they are incorporated into the ice and transported with the glacier.

plunging breaker Curled-shaped breaking wave that breaks all at once. The classic wave for surfers due to the elliptical-shaped pocket of air beneath the crest.

point bar Arcuate accumulation of sand or sand and gravel forming on the inside of channel meanders.

polar cell Weak circulation cell that occurs between 60 and 90° latitude. It results from upper air masses moving northward and descending at the poles, while at the same time surface air flows south.

polar easterlies Part of the global air circulation pattern where winds blow from the polar high toward the subpolar low.

polar front At about 50–60° latitude the polar easterlies meet the westerlies, establishing the polar front. This convergent zone produces a near permanent boundary separating the cold dense polar air from the warm tropical air mass.

pneumatophores Root system of a mangrove that actively respires. Much of the root system is above water level.

precession Wobble of the Earth's spin axis, like a spinning top slowing down. The Earth's axis of rotation is presently oriented at the North Star, and in approximately 10,000 years (half of its precession orbit) it will be pointing to the star Vegas.

prevailing winds Winds that blow most of the time and are associated with global air flow patterns.

prodelta Seawardmost portion of the delta, which is commonly the thickest unit of the delta. It contains fine sediments.

progradation Seaward accretion of the shoreline caused by the deposition of sediment.

prograding barriers Type of barrier that builds seaward during its evolution due to an abundant sediment supply.

quarrying Hydraulic plucking process whereby pieces of rock are removed from a cliff face or bedrock exposure, having been loosened or separated from the parent rock by biological, chemical, or physical processes.

recurved spit Hook-shaped spit that builds into a bay or along the coast through the addition of curved beach ridges.

reef Shallow offshore ridge or mound that is built by organisms and resistant to wave action.

reflective beach Beach containing a relatively steep beachface on which little of the wave energy is dissipated before it breaks on the beachface.

retrograding barriers Topographically low barrier that migrates onshore primarily through overwash activity during storms. Barrier sands are underlain by marsh, lagoonal, and mainland sediments.

ridge and runnel Inshore bar that migrates onshore and repairs a beach following storm-induced erosion. Commonly they are less than 1 m in height and are fronted by a steep face (slipface) along their landward margin.

rip current Seaward-directed current that flows through the breaker zone, almost perpendicular to the beach, and quickly dissipates. The current is fed by the longshore current and may reach velocities up to several kilometers per hour.

river-dominated delta A delta that develops an elongate, digitate shape due to the overwhelming power of the river depositing sediment, with relatively little of the sediment being reworked by wave action or tidal currents.

roundness Smoothness versus number of sharp corners exhibited by a grain. Its value can be expressed mathematically by the ratio of the average radius of the grain's corners or edges to that of the maximum inscribed circle.

Saffir–Simpson scale The strength of a hurricane is formally classified using the Saffir–Simpson scale and is based on the maximum wind velocity, barometric pressure, storm surge level, and expected damage.

salinity Measure of the dissolved solids in seawater.

sand Grains with a size range between 0.0625 and 2.0 mm.

sand dunes Mounds or ridges of sand. They are formed from sand that is transported and deposited by the wind.

sand shadows Small accumulation of sand that is deposited in the sheltered area behind an obstruction, such as a rock, shell, or piece of debris.

scarp Steep erosional face of a sedimentary deposit, such as the dune scarp that is formed during storm-induced erosion of the foredune.

sea arch A rock arch found offshore of a rocky coast. Formed due to differential erosion of layered rocks or jointed bedrock.

sea breeze Breeze that blows onshore due to the warming of the landmass during the day.

sea level The height of the sea surface above a certain datum point.

sea level highstand For the Quaternary Period, the position of sea level during interglacials, when ice sheets had a minimum size and sea level reached its highest elevation.

sea-level lowstand For the Quaternary Period, the position of sea level during times of maximum glaciation when ice sheets had their greatest extent and sea level reached its lowest elevation.

sea stack Pinnacle of rock rising above the sea floor that sits offshore of a rocky coast. The stack is separated from the adjacent shoreline due to differential erosion.

sea waves Waves formed at the generated site under the influence of local winds.

sediment core Vertical core taken through a sedimentary deposit. Cores are used to study the stratigraphy and depositional history of the accumulation sequence.

sediment gravity flow Down-slope movement of sediment and water under the influence of gravity (e.g. turbidity current).

sedimentary rock Rock that consists of the weathered particles of pre-existing rocks or minerals that have been precipitated from water. The sedimentary particles are transported, deposited, and lithified into rock.

seiche A standing wave in a coastal or semi-enclosed body of water. The wave sloshes back and forth with a period of minutes to hours.

semi-diurnal tidal inequality Tides consisting of two high tides and two low tides each lunar day that reach unequal heights.

semi-diurnal tides Two high and two low tides during a tidal day.

setup A rise in water elevation above still water in the surf zone due to breaking waves piling up water.

shelf reef Reefs that rise from the floor of the continental shelf, as contrasted to oceanic reefs, which are associated with volcanic islands and rise from the ocean floor.

significant wave height The average height of the highest one-third of waves in a given area.

silicate minerals Minerals that have a silicon–oxygen tetrahedron as their basic structure. These minerals include quartz, feldspar, and mica, which are common constituents of beach sand.

sorting Standard deviation of the size distribution. Essentially, how many size classes define the majority of the population. The fewer the size classes the better the sorted sediment, and vice versa.

sphericity The tendency of a particle to be equidimensional along all axes.

spilling breakers A type of breaking wave in which the crest peaks and then water cascades down the front of the wave. This wave continues to break along a wide section of the nearshore zone.

spring tide The tide that is produced when the Earth, Moon, and Sun are aligned, coinciding with new and full moon conditions. Astronomic tides are at a maximum.

standing wave A wave in which the water surface oscillates vertically between fixed points where there is no movement. They commonly occur in a bay or a semi-enclosed coastal body of water.

storm beach A flat featureless beach that is produced by storm erosion.

stratified estuary An estuary in which there exists a distinct gradient between the overlying freshwater and the underlying saltwater.

stratified sediment Layered sediments that are produced by a flowing fluid (e.g. water, air).

stratigraphy The study of the composition and relationship of layered sediments or rocks.

stromatolites Thinly laminated, organically formed structures that usually consist of carbonate particles. They form due to sand particles adhering to blue-green algae.

subduction Tectonic process involving the descension of a lithospheric plate into the mantle.

submarine canyons Large-scale, V-shaped valleys formed by the erosion caused by sedimentary gravity flows. They extend from the edge of the continental shelf to the base of the continental slope.

surf zone The zone landward of the breaker zone where breaking waves create a turbulent flow of water toward the beach.

swell waves Waves that have moved away from the generation zone and have similar periods, lengths, and heights. These are the waves that break along a beach when the wind isn't blowing.

tectonic plates The outer shell of the Earth, which is broken into distinct irregular lithospheric plates that converge, diverge, and slip past one another.

Tethys Sea The ocean that separated Laurasia from Gondwanaland prior to the breakup of Pangaea during the Mesozoic Era.

tidal bore Rapid rise of the tide that propagates up an estuary, commonly in the form of a breaking wave.

tidal bundles Repetitive depositional sequences that are related to lunar cycles, and in which tidal energy varies with changes in tidal range.

tidal channel A channel dominated by reversing tidal flow caused by the rise and fall of the tides.

tidal cycle Periodic rise and fall of the tide.

tidal delta See ebb- and flood-tidal delta.

tidal flat Relatively flat-lying, low-energy intertidal environment that may be incised by tidal channels. It is composed of mud to sand sized sediment.

tidal inlet An opening in the shoreline through which water penetrates the land, thereby providing a connection between the ocean and bays, lagoons, and marsh and tidal creek systems. The main channel of a tidal inlet is maintained by tidal currents.

tidal wave A wave associated with tidal propagation within an amphidromic system, e.g. the counterclockwise rotation of the tidal wave in the north Atlantic Ocean. The term should not be confused with tsunami.

tide-dominated coast Coast where the morphology is primarily a product of tidal processes.

tide-dominated delta A river delta that occurs primarily in funnel-shaped macrotidal embayments. Characteristically, sand bodies are oriented parallel to the strong tidal currents that flow into and out of the embayment.

tide-dominated estuary A funnel-shaped estuary dominated by tidal flow.

till Unsorted, unlayered sediment that is deposited directly by the ice.

time–velocity asymmetry Condition in which maximum ebb- and flood-tidal currents do not occur at midtide and are of unequal velocity.

tombolos Intertidal or supratidal accumulation of sediment connecting the beach to an island. The sediment builds from the beach toward the island due to wave sheltering and convergence of longshore currents.

trailing edge coasts Coast coinciding with the middle of a tectonic plate and facing a spreading zone (ocean ridge). A type of passive margin.

transform boundary A boundary between two lithospheric plates that are shearing past one another.

transform fault A fault separating two lithospheric plates that are sliding past one another. The crest of an oceanic ridge is offset by a transform fault.

tropical storm A tropical cyclone with a maximum wind velocity between 61 and 115 km h^{-1}.

tsunami A long-period, high-velocity wave produced by a disturbance of the sea floor, volcanic eruption, or meteor impact.

turbidity current A type of sedimentary gravity flow in which the sediment is kept in suspension by turbulence. Sediment and water move down-slope under the influence of gravity.

undertow Water from the swash zone returning seaward at the bottom of the water column.

upwelling The process of cold, deep, nutrient-rich water rising toward the surface as a result of coastal currents flowing offshore or alongshore.

valley glaciers See alpine glaciers.

wave base The depth to which sediment is disturbed by surface waves. This depth is usually defined as equal to half the wavelength.

wave diffraction Lateral transfer of energy along the wave crest. Most noticeable where wave crests gradually develop behind wave-sheltered areas.

wave-dominated coast Coast where the morphology is primarily a product of wave processes.

wave-dominated delta A cuspate river delta in which promontories are smoothed by wave action.

wave-dominated estuary An estuary in which wave processes dominate over tide-induced sediment transport. This type of estuary is commonly fronted by a barrier system.

wave hammer The force of a wave breaking directly on a rock cliff.

wave height The vertical difference between the wave crest and the adjacent trough.

wavelength The horizontal distance between successive wave crests.

wave period The time interval between the passage of two successive wave crests.

wave reflection This phenomenon occurs when a wave breaks against a vertical structure and its energy is reflected off the face of the cliff, seawall, or other structure. Waves approaching a vertical structure at an angle will be reflected off the wall at the same angle (Snell's law).

wave refraction Bending of the wave crest due to changes in celerity along the crest, caused by the wave moving into shallow water.

weather front Narrow transition zone (25–250 km wide) between two large air masses that have different densities.

weathering The collection of processes involved in the breakdown of rock at the Earth's surface.

welded barrier A barrier that is attached to the mainland at both ends.

westerlies A belt of prevailing winds in the midlatitudes (30–60°) that flow from west to east.

wind Air in motion relative to the Earth's surface.

windward reef Coral reef on the windward side of an island or coast.

Wisconsinan Ice Age Last of the major glaciations in North America, spanning the period from about 70,000–90,000 up to 10,000 years before present.

wrack lines A debris line found along the upper shore, marking the greatest excursion of the wave up the beachface or berm during the previous spring high tide or storm high tide. The wrack commonly consists of seaweed, eelgrass, marsh grass, or driftwood.

Index

Page numbers in *italics* refer to figures; those in **bold** refer to tables.